INFRARED BAND HANDBOOK

INFRARED BAND HANDBOOK

Edited by

Herman A. Szymanski

Chairman, Chemistry Department
Canisius College

PLENUM PRESS
NEW YORK • 1963

Library of Congress Catalog Card Number 62-15543

227 West 17th St., New York 11, N. Y.

Printed in the United States of America

CONTENTS

INTRODUCTION

Recent rapid advances in the techniques and instrumentation of infrared spectroscopy have made the determination of absorption spectra an increasingly more important and accurate method of analysis. However, the very refinements that have served to enhance the attractiveness of the method have also created the need for a greatly expanded information program. The Infrared Band Handbook is designed to play a many-sided role in this program. Perhaps its primary use will be in the identification of unknown compounds, but it will also be of value as a guide in assigning group frequencies, as a key to most advantageous sample conditions, as a tool in comparing series of related compounds for similar vibrations, and finally as an index to the literature.

The particular arrangement of the data that has been evolved is based on a number of premises:

1. That band position is the primary parameter for identifying unknown compounds from their infrared spectra. The identification of a compound from group frequencies alone is often difficult since many bands are not group frequencies.
2. That modern instrumentation and techniques make possible very accurate reporting of band position, and that, if the sample conditions are stated, this accuracy is reproducible.
3. That much of the literature does not report complete spectra, but concerns itself with the influences of physical and chemical changes on band position. This information is frequently lost to the spectroscopist.

The Handbook therefore presents data arranged by wavenumber, in steps of 1 cm^{-1}, and does not limit itself to complete spectra.

Sources

While the most accurate data are found in the current journals, it seemed desirable to include the widest selection of compounds in this initial volume. For this it was necessary to use data from older sources as well as from the current literature. This is especially true for many common organic compounds whose spectra have not been determined recently, but which form a necessary background, for instance, for the assignment of group frequencies.

The primary source of data for these common compounds was the American Petroleum Institute Project 44 series. This series represents accurate spectra of highly purified compounds, and therefore even early spectra of this group can be considered satisfactory. The Project 44 series, which shall be referred to as the API file, began issuing spectra in 1943. The file has been updated, and much of the early work has been redetermined quite accurately. However, some compounds have not been redetermined, and a few of these older entries have been included in the Handbook for completeness. It has been our experience in comparing these early data with those of later work that they are quite similar, except that the accuracy of the earlier work, especially in the 3600-2500 cm^{-1} region, is much lower. But all incorrect spectra have been replaced with other data by the API workers. The only other source of older data has been several textbooks.

The remaining Handbook data have been taken from the literature published since 1957, although secondary references to earlier work will be found in the reference list. While the API file of course gives only complete spectra, data from the current literature may cover either single bands or the complete spectrum of a compound. In either case, only tabular data were used and in the interest of accuracy no attempt was made to measure band position in printed spectra. It should be noted that the ASTM indexing method utilized only printed spectra, so that in this respect the Handbook forms a complement to the ASTM index.

Accuracy of Band Position

As stated above, data have been entered in the Handbook in 1 cm^{-1} steps. While this arrangement is warranted by the accuracy of much of the information now being reported in the literature, we do not mean to imply this accuracy for all the data in the Handbook. A reasonable estimate of the accuracy of each entry can be made on basis of the date of the reference, the nature of the monochromator used, and the region of the spectrum involved. The following considerations will serve as a guide.

If the reference is of recent origin and only a NaCl prism spectrophotometer was used in the 3600-600 cm^{-1} region, the accuracy of band position from 3600 to 2500 cm^{-1} can be assumed to range from ± 5 cm^{-1} to ± 10 cm^{-1}. From 2500 cm^{-1} to 600 cm^{-1} the accuracy can be assumed to range from ± 5 cm^{-1} to ± 1 cm^{-1}, with the higher accuracy corresponding to the lower wavenumbers. If the reference is to API data prior to 1950, the accuracy should be assumed to be even lower. For example, in the 3600-2500 cm^{-1} region it probably is ± 10 cm^{-1} unless a grating was used, in which case it would be closer to ± 5 cm^{-1}.

In general, where CaF_2, LiF, or grating monochromators are used, the accuracy in the 3600-2500 cm^{-1} region can be assumed to range from ± 5 cm^{-1} to ± 1 cm^{-1}, i.e., to fall within the same limits assumed for the 2500-600 cm^{-1} region. Some literature sources list the accuracy of their data, and this information is entered in the Handbook whenever it appeared pertinent.

Measured with a NaCl monochromator, many hydrocarbons have a series of common bands, particularly in the 3500-2000 cm^{-1} region, where the band position cannot be measured as accurately as in the 2000-600 cm^{-1} region. Obviously, these bands are not particularly useful for identification, and to conserve space in the Handbook, only some typical examples of such bands have been entered. The reader may assume that all compounds of the following types entered in the Handbook have the following bands in common:

Band	Intensity	Path Length or Concentration	Physical State or Solvent
	CH_3 and CH_2 Vibrations		
3370 $\pm$ 10 cm^{-1}	MWSh	0.169	liquid
3215 $\pm$ 10 cm^{-1}	SSh	0.169	liquid
3180 $\pm$ 10 cm^{-1}	SSh	0.169	liquid
2970 $\pm$ 10 cm^{-1}	S	0.0088	liquid
2920 $\pm$ 10 cm^{-1}	S	0.0088	liquid
2880 $\pm$ 10 cm^{-1}	S	0.0088	liquid

Band	Intensity	Path Length or Concentration	Physical State or Solvent
2850 ± 10 cm^{-1}	S	0.0088	liquid
2660 ± 10 cm^{-1}	MSSh	0.169	liquid
2415 ± 10 cm^{-1}	M	0.728	liquid
2320 ± 10 cm^{-1}	M	0.728	liquid
2295 ± 10 cm^{-1}	S	0.728	liquid
	OH of Alcohols		
3350 ± 10 cm^{-1}	S	5%	CCl_4

Intensity Data

It is important for the reader to understand how intensity data were entered into the Handbook. Of all the information in the Handbook, intensity is the one most difficult to give in a generally meaningful way.

It is not intended for intensities of one compound to be compared to those of another, but only for intensities of all bands of a single compound to be compared to each other. Of course, in some instances different compounds can be compared to each other, but the reader should do this only after he has assured himself that the sample conditions were sufficiently similar to make the comparison meaningful.

The intensity data were prepared in the following manner. Where the source listed relative intensity for each band of a compound and only a single sample condition was used, the relative intensity data and physical conditions under which the sample was determined are listed. In general, no cell path length or concentrations are given unless such data are pertinent to band position. Where several path lengths or concentrations were used to complete the spectrum but the author has converted intensities to a relative scale, the relative intensity scale is used and concentrations and/or path lengths are not stated. API data presented a special problem since in many instances the sample had been determined using several path lengths and/or concentrations but no relative intensity data were given. For these spectra the intensity reported is that corresponding to the single concentration and/or path length which, in a given region of the spectrum, brings the greatest number of bands to an intensity from 5% to 95%. To indicate which bands are comparable in this situation, either the concentration or the path length is entered after the physical state code as an identifying mark. Since in this case either the path length or concentration is listed, but not both, the reader must consult the original reference for complete sample data.

Buffalo, N.Y. H.A. Szymanski

EXPLANATION

The information listed to the left of the STRUCTURAL FORMULA of each compound has the following significance:

The first line gives the WAVE NUMBER in cm^{-1} and a code designation indicating the INTENSITY of the band (see Intensity Code below). For gases the band center or branch is identified whenever possible.

The second entry gives either a code letter indicating the PHYSICAL STATE in which the spectrum was measured (see Physical State Code below) or the solvent used for the sample (followed by the concentration, in brackets, and cell thickness, where pertinent).

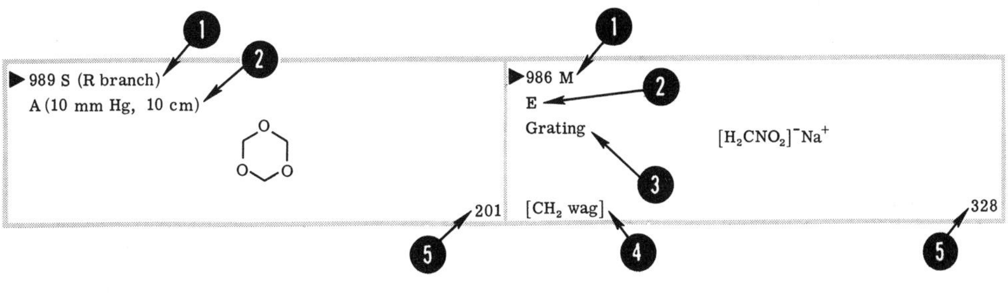

The next line is reserved for SPECIAL INFORMATION, such as the dispersive element used, the material of the prism, etc. In the absence of an entry the reader may assume that an NaCl prism was used.

The bracketed entry on the last line indicates the STRUCTURAL GROUP to which the vibration was assigned in the original reference, which may belong to either the original or the isomerized form of the compound, and the mode of vibration, where pertinent.

The number in the lower right-hand corner is the REFERENCE NUMBER and pertains to the list of source material on pages 429-434.

Intensity Code

(Intensity is described relative to all other bands in a given spectrum, except where it is necessary to use different path lengths and/or concentrations to record all bands. In such cases the path length and/or concentration is stated on the second line.)

S = Strong, **M** = Medium, **W** = Weak, **V** = Very, **B** = Broad, **Sh** = Shoulder, **Sp** = Sharp

Physical State Code

(The temperature is given after the code letter if it is other than 25°C.)

- **A** = Vapor or gaseous state (followed by the pressure and path length, where pertinent)
- **B** = Sample run as a liquid (followed by the cell thickness, in millimeters, where pertinent)
- **C** = Solid, powder
- **D** = Solid, film
- **E** = Nujol oil
- **F** = Fluorocarbon oil
- **G** = KBr disk
- **H** = Solution – solvent not specified
- **I** = Solid state – method not specified
- **K** = Polyethylene bagging
- **M** = Hexachlorobutadiene mull
- **N** = Sample above room temperature and run as a liquid
- **O** = Single crystal
- **P** = KCl disk
- **Q** = KI disk

EXPLICATION

Les renseignements situés à gauche de la FORMULE DE CONSTITUTION de chaque composé ont la signification suivante:

La première ligne donne l'INDICE DE L'ONDE en cm^{-1} et une désignation-code indiquant l'INTENSITE de la bande (voir Code-Intensité ci-dessous). Pour les gaz, le milieu de la bande ou l'embranchement est identifié quand c'est possible.

La seconde inscription donne soit une lettre-code indiquant l'ETAT PHYSIQUE dans lequel le spectre fut mesuré (voir Code-Etat Physique ci-dessous) ou le solvant utilisé comme échantillon, entre parenthèses, et la cellule épaisse, là où c'est à propos.

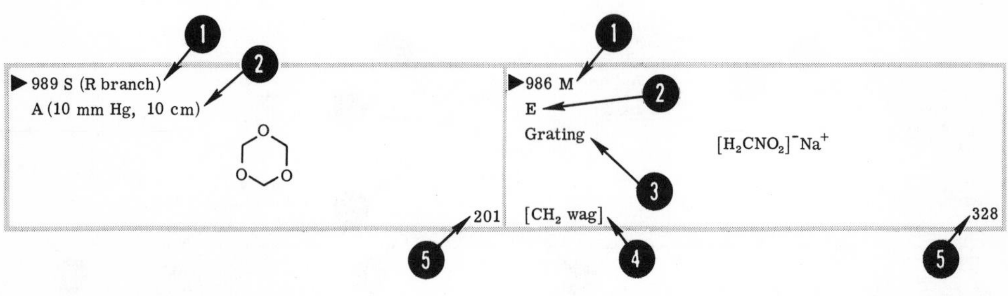

La ligne suivante est réservée pour des RENSEIGNEMENTS SPECIAUX, tels que l'élément du composé utilisé, la nature du prisme, etc. En l'absence d'une inscription, le lecteur peut assumer qu'on a employé un prisme NaCl.

L'inscription entre crochets de la dernière ligne indique le GROUPE DE CONSTITUTION auquel la vibration fut assignée dans la référence originale, laquelle peut appartenir ou à l'originale ou à la forme isomèrisée du composé, et du genre du vibration, là où c'est à propos.

Le nombre qui se trouve à l'extrémité droite, en bas, est le NUMERO DE REFERENCE et se rattache à la liste de matériel source aux pages 429-434.

Code-Intensité

(L'intensité est décrite relative à toutes les autres bandes d'un spectre donné, sauf là où il faut s'en servir des differentes longueurs de traces et/ou des concentrations differentes pour enregistrer toutes les bandes. Dans de tels cas, la longueur de trace et/ou la concentration est donnée dans la seconde ligne.)

S = fort, **M** = moyen, **W** = faible, **V** = très, **B** = large, **Sh** = epaulement, **Sp** = aigu

Code-Etat Physique

(Si la température est ature que 25°C, elle est donnée après la lettre-code.)

A = Vapeur ou état gazeux (suivi de la pression et la longueur du trace, là où c'est à propos)
B = Echantillon éprouvé en tant que liquide (suivi de l'épaisseur de la cellule, en millimètres, là où c'est à propos)
C = Solid, en poudre
D = Solide, pellicule
E = Huile de Nujol
F = Huile de fluorocarbone
G = Disque KBr
H = Solution—solvant non specifié
I = Etat solide—méthode non specifié
K = Ensachement polyéthylène
M = Mélange hexachlorobutadiene
N = Echantillon au-dessus de la température normale d'intérieur et utilisé en tant que liquide
O = Cristal simple
P = Disque KCl
Q = Disque KI

ПОЯСНЕНИЯ

Слева СТРУКТУРНОЙ ФОРМУЛЫ каждого соединения приводятся данные, имеющие следующее значение:

1 **На первой строке** указываются НОМЕР ВОЛНЫ в $см^{-1}$ и код, обозначающий ИНТЕНСИВНОСТЬ полосы. (См. ниже код интенсивности.) Для газов там, где это представляется возможным, обозначается центр или ветвь полосы.

На второй строке приводится буква кода, указывающая на ФИЗИЧЕСКОЕ СОСТОЯНИЕ, в котором измерялся спектр (см. ниже код физического состояния), или растворитель, применявшийся для пробы. Затем, в скобках, указывается концентрация, а там, где это существенно, толщина ячейки.

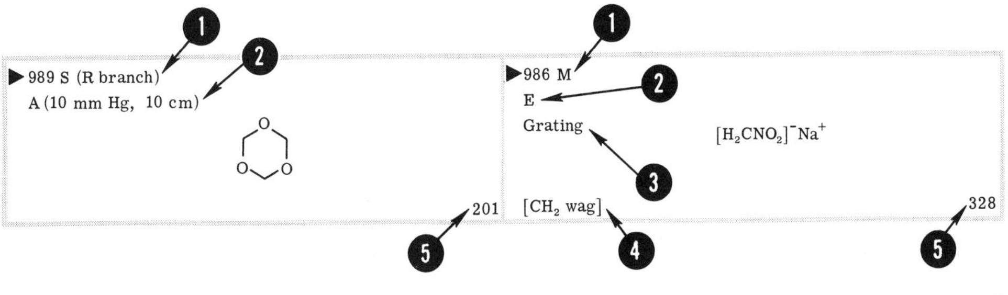

На следующей строке приводятся ОСОБЫЕ ДАННЫЕ, как-то: применявшийся элемент дисперсии, материал призмы и т. д. Если третья строка отсутствует, следует считать, что применялась призма NaCl.

На последней строке, в скобках, указывается СТРУКТУРНАЯ ГРУППА, к которой вибрация была отнесена в первоначальной ссылке. Эта группа может относиться к первоначальной или к изомеризованной форме соединения, а также там, где это существенно, к типу вибрации.

Номер, напечатанный справа внизу, является НОМЕРОМ ССЫЛКИ. Он относится к списку источников, упоминающихся на стр. 429-434.

Код интенсивности

(Интенсивность указывается в отношении всех прочих полос данного спектра, за исключением тех случаев, когда следует применять иные длины пробега и/или концентраций для записи всех полос. В таких случаях длина пробега и/или концентрации приводятся на второй строке.)

S = сильная, **M** = средняя, **W** = слабая, **V** = очень, **B** = широкая, **Sh** = плечо, **Sp** = острая

Код физического состояния

(Температура следует за буквой кода в том случае, если она не равняется 25°.)

- **A** = Паро- или газообразное состояние (с указанием давления и длины пробега там, где это существенно)
- **B** = Проба в жидком состоянии (с указанием толщины ячейки в миллиметрах там, где это существенно)
- **C** = Твердое тело, порошок
- **D** = Твердое тело, пленка
- **E** = Масло Нюжоль
- **F** = Фтороуглеводородное масло
- **G** = Диск KBr
- **H** = Раствор—без указания растворителя
- **I** = Твердое состояние—без указания метода
- **K** = Полиэтиленовая упаковка
- **M** = Шестихлоробутадиеновая смесь
- **N** = Температура пробы выше комнатной; проба в жидком состоянии
- **O** = Монокристалл
- **P** = Диск KCl
- **Q** = Диск KI

ERKLÄRUNG

Die links von der STRUKTURFORMEL einer jeden Verbindung stehenden Angaben haben die folgende Bedeutung:

Auf der ersten Zeile steht die WELLENZAHL in cm^{-1} sowie ein oder zwei Kennbuchstaben, die die INTENSITÄT des Bandes angeben (siehe Intensitäts-Kennbuchstaben unten). Nach Möglichkeit ist für Gase die Bandmitte oder jeweilige Bandseite mitgeteilt.

Die zweite Eintragung enthält entweder einen Kennbuchstaben, der den PHYSIKALISCHEN ZUSTAND bezeichnet, in dem das Spektrum ermittelt wurde (siehe Zustands-Kennbuchstaben unten), oder das für die Probe benutzte Lösungsmittel (sowie nachstehend die Konzentration in eckigen Klammern und die Zellendicke, falls zutreffend).

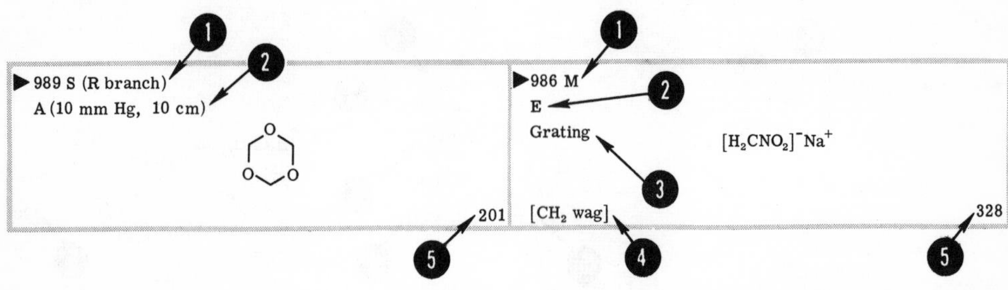

Die nächste Zeile ist für BESONDERE ANGABEN vorbehalten, wie z. B. das benutzte Dispersionselement, das Prismenmaterial usw. Sofern keine Angaben gemacht sind, kann der Leser annehmen, dass ein NaCl-Prisma benutzt wurde.

Die Eintragung in eckigen Klammern auf der letzten Zeile gibt die STRUKTURGRUPPE an, der die Schwingung im Originalbericht zugeteilt worden war—diese kann entweder der ursprünglichen oder der isomerischen Form der Verbindung angehören —, sowie die Schwingungsart, falls zutreffend.

Die Zahl in der rechten unteren Ecke stellt die BEZUGSZAHL dar und bezieht sich auf die Quellenliste auf S. 429-434.

Intensitäts-Kennbuchstaben

(Die Einstufung der Intensität bezieht sich auf alle anderen Bänder eines gegebenen Spektrums, ausser in Fällen, wo verschiedene Bahnlängen und/oder Konzentrationen erforderlich waren, um alle Bänder zu erfassen. In diesen Fällen sind Bahnlänge und/oder Konzentration auf der zweiten Zeile angegeben.)

S = Stark, **M** = Mittel, **W** = Schwach, **V** = Sehr, **B** = Breit, **Sh** = Schulter, **Sp** = Scharf

Zustands-Kennbuchstaben

(Die Temperatur ist nach dem Kennbuchstaben angeführt, falls sie nicht 25°C betrug.)

A = Dampf- oder Gaszustand (dahinter Druck und Bahnlänge, falls zutreffend)
B = Prüfung erfolgte im flüssigen Zustand (dahinter die Zellendicke, falls zutreffend)
C = Festzustand, pulverförmig
D = Festzustand, Filmform
E = Nujol-Öl
F = Fluorkohlenstoff-Öl
G = KBr-Scheibe
H = Lösung—Lösungsmittel nicht angegeben
I = Festzustand—Methode nicht angegeben
K = Polyäthylen-Beutel
M = Hexachlorbutadien-Mull
N = Prüfung oberhalb Zimmertemperatur im flüssigen Zustand
O = Einkristall
P = KCl-Scheibe
Q = KI-Scheibe

▶3610 S
C_6H_{12}, C_2Cl_4 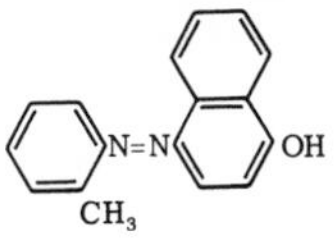
[OH] 35

▶3600 S
CCl_4 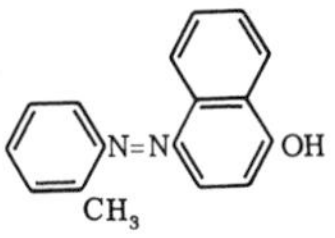
[OH] 35

▶3580 S
CH_2Cl_2 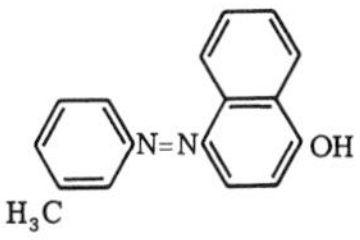
[OH] 35

▶3580 S
CH_2Cl_2 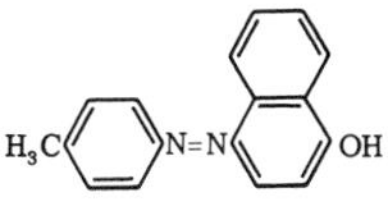
[OH] 35

▶3580 S
CH_2Cl_2 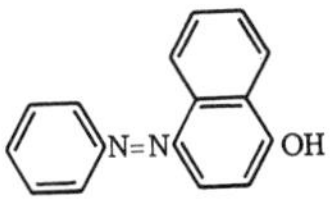
[OH] 35

▶3580 M
CH_2Cl_2 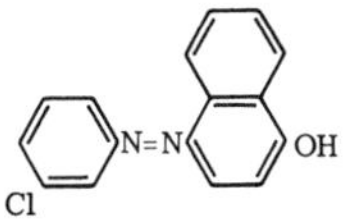
[OH] 35

▶3575 S
CH_2Cl_2 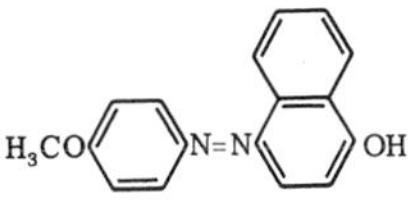
[OH] 35

▶3575 MS
$CHCl_3$ 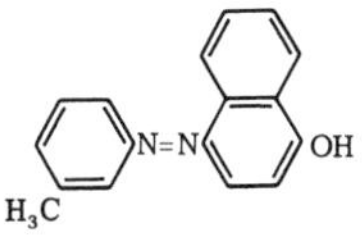
[OH] 35

▶3570 S
$CHCl_3$ 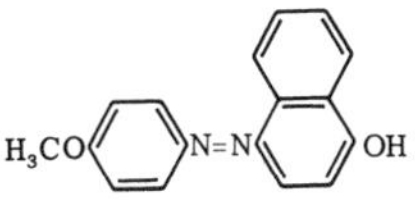
[OH] 35

▶3570 MS
CH_2Cl_2, $CHCl_3$ 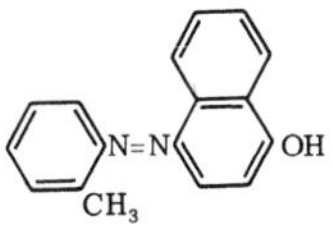
[OH] 35

▶3570 MS
$CHCl_3$ 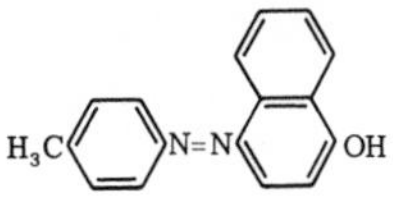
[OH] 35

▶3570 MS
$CHCl_3$ 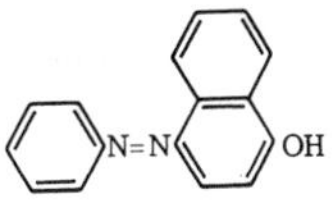
[OH] 35

▶3570 VW
$CHCl_3$ 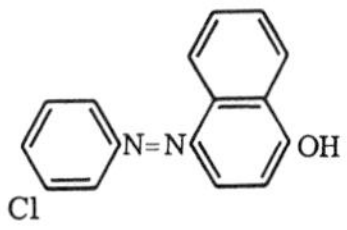
[OH] 35

▶3560 M
CH_2Cl_2, $CHCl_3$ 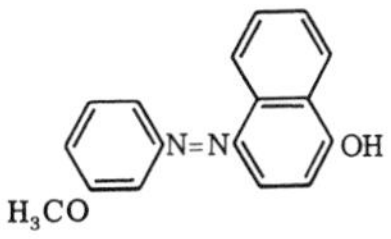
[OH] 35

▶3540 MW
CCl_4 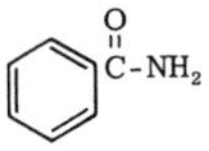
29

▶3538
$CHCl_3$ 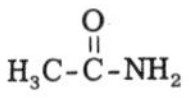
474

▶3525 MW
$CHCl_3$ 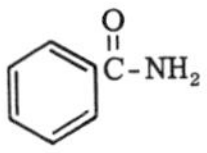
29

▶3520 MW
CCl_4 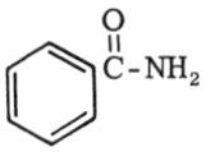
29

▶3520 MW
CCl_4
CaF_2 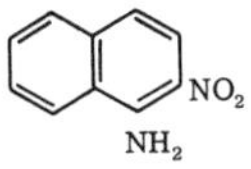
[NH_2] 19

▶3519
$CHCl_3$
CaF_2 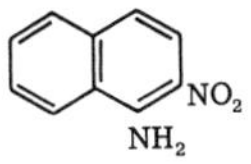
[NH_2] 19

▶3517
CCl_4
CaF_2
NH_2 O_2N NO_2
$[NH_2]$ 19

▶3516 M
CCl_4
H_3C-O-C(=O)- OH =O O H_3C
[OH] 475

▶3516 Sh
$CHCl_3$
CaF_2
NO_2 NH_2
$[NH_2]$ 19

▶3511 M
CCl_4, $CHCl_3$
CaF_2
Cl NH_2 NO_2
$[NH_2]$ 19

▶3511
$CHCl_3$
CaF_2
NH_2 O_2N NO_2
$[NH_2]$ 19

▶3510
$CHCl_3$, CCl_4
N NH_2
327

▶3510
$CHCl_3$
NH_2 N
327

▶3507 M
CCl_4
H_3C CH_3 OH O =O CH_3
[OH] 475

▶3504
F_3C-C(=O)-OH
477

▶3504 MW
$(C_2H_5)_2O$
CaF_2
Cl NH_2 NO_2
$[NH_2]$ 19

▶3504 Sh
$(C_2H_5)_2O$
CaF_2
NO_2 NH_2 NO_2
$[NH_2]$ 19

▶3501 MW
C_6H_6
CaF_2
Cl NH_2 NO_2
$[NH_2]$ 19

▶3501 MW
C_6H_6
CaF_2
NO_2 NH_2
$[NH_2]$ 19

▶3500
C_6H_6
CaF_2
NH_2 O_2N NO_2
$[NH_2]$ 19

▶3500
H_5C_2-C(=O)-NH_2
465

▶3497 MW
$C_6H_5NO_2$
CaF_2
Cl NH_2 NO_2
$[NH_2]$ 19

▶3495
C_5H_5N
CaF_2
Cl NH_2 NO_2
$[NH_2]$ 19

▶3490
$C_6H_5NO_2$
CaF_2
NO_2 NH_2
$[NH_2]$ 19

▶3488 W
C_5H_5N
CaF_2
NO_2 NH_2 NO_2
$[NH_2]$ 19

▶3487 MW
CH_3CN
CaF_2
Cl NH_2 NO_2
$[NH_2]$ 19

▶3487 Sh
$CHCl_3$
CaF_2
Cl NH_2 NO_2
$[NH_2]$ 19

▶3484 M
$C_5H_5NO_2$
CaF_2
[NH_2] 19

▶3481
479

▶3480 W
CCl_4 [0.3 g/ml]
2

▶3480
[C=O overtone] 478

▶3480
$CHCl_3$, CCl_4
CaF_2
[NH_2] 470

▶3479 MW
$(C_2H_5)_2O$
CaF_2
[NH_2] 19

▶3478 S
$CHCl_3$
CaF_2
[NH_2] 19

▶3478 M
CH_3CN
CaF_2
[NH_2] 19

▶3477 M
$(C_2H_5)_2O$
CaF_2
[NH_2] 19

▶3476 S
C_2Cl_4
CaF_2
[NH_2] 19

▶3475 S
CCl_4
CaF_2
[NH_2] 19

▶3475 M
C_6H_6
CaF_2
[NH_2] 19

▶3474 M
C_5H_5N
CaF_2
[NH_2] 19

▶3474 MW
CH_3CN
CaF_2
[NH_2] 19

▶3473 M
$C_6H_5NO_2$
CaF_2
[NH_2] 19

▶3470
CH_3NH_2
480

▶3464 MW
CH_3CN
CaF_2
[NH_2] 19

▶3460 W
CCl_4 [0.007 g/ml]
[NH] 2

▶3457 MS
C_2Cl_4, $CHCl_3$
CaF_2
[NH_2] 19

▶3457 M
CH_3CN
CaF_2
[NH_2] 19

▶3456 S
C_6H_6
CaF_2
NO_2 NH_2 O_2N NO_2
[NH_2] 19

▶3455 MW
C_5H_5N
CaF_2
Cl NH_2 NO_2
[NH_2] 19

▶3454
$CHCl_3$
CaF_2
NH_2
[NH_2] 470

▶3450 M
E
N NH_2
327

▶3450 Sh
C_5H_5N
CaF_2
CaF_2
NO_2 NH_2 O_2N NO_2
[NH_2] 19

▶3450
OH
483, see also 36, 44, 482

▶3443 M
$(C_2H_5)_2O$
CaF_2
NO_2 NH_2
[NH_2] 19

▶3437
P
CaF_2
NO_2 NH_2 O_2N NO_2
[NH_2] 19

▶3435 M
C_6H_5N
CaF_2
NH_2 O_2N NO_2
[NH_2] 19

▶3435 M
E
NH_2 N
327

▶3420 M
$CHCl_3$
NH_2 N
327

▶3420
CCl_4
N NH_2
327

▶3420 MW
CCl_4
O C-NH_2
29

▶3420
$CHCl_3$
$H_3C-\overset{O}{\overset{\|}{C}}-NH_2$
474

▶3410 M
$CHCl_3$
O C-NH_2
29

▶3401 MS
CCl_4 [5%]
$CH_3\underset{OH}{CH}(CH_2)_2CH_3$
318

▶3400 S
$CH{=}N{-}NH{-}\overset{S}{\overset{\|}{C}}{-}NH_2$
380

▶3400 S
H_3C CH_3 $CH{=}N{-}NH{-}\overset{S}{\overset{\|}{C}}{-}NH_2$
380

▶3400 S
$CH{=}N{-}NH{-}\overset{S-CH_3}{\overset{|}{C}}{-}NH_2$
380

▶3400
CH_2-NH_2
383

▶3400
$CHCl_3$
NH_2 N
327

▶3395 MW
C_5H_5N
O_2N–C_6H_3(NO_2)–NH_2, NO_2
19

▶3383 MW
C_5H_5N
CaF_2
Cl, NH_2, NO_2
[NH_2] 19

▶3394
CCl_4
CaF_2
NH_2
[NH_2] 470

▶3382 M
CH_3CN
CaF_2
H
N-CH
O_2N, NO_2
19

▶3393
CCl_4
CaF_2
NH_2, O_2N, NO_2
[NH_2] 19

▶3381
C_6H_6
CaF_2
NH_2, O_2N, NO_2
[NH_2] 19

▶3392 M
$(C_2H_5)_2O$, CCl_4, $CHCl_3$
CaF_2
Cl, NH_2, NO_2
[NH_2] 19

▶3381 M
$C_6H_5NO_2$
CaF_2
Cl, NH_2, NO_2
[NH_2] 19

▶3391 MS
CCl_4 [5%]
$CH_3CH_2C(CH_3)_2$
OH
194

▶3380 MWSh
B (0.169)
$(CH_3)_2CHCHCH(CH_3)_2$
CH_2
CH_3
301, see also 472

▶3391
$CHCl_3$
CaF_2
NH_2, O_2N, NO_2
[NH_2] 19

▶3380 MW
CCl_4
N=N, OH, CH_3
[NH] 35

▶3388 M
C_6H_6
CaF_2
Cl, NH_2, NO_2
[NH_2] 19

▶3380 MSh
A
O
H-C≡C-C-H
18

▶3385 VVS
B (0.055)
CaF_2
H_3C, CH_3, N, H
344

▶3380 VS
$CHCl_3$
CaF_2
H
N-CH_3
O_2N, NO_2
[NH] 19

▶3384 S
C_6H_6
CaF_2
H
N-CH_3
O_2N, NO_2
[NH] 19

▶3380 M
CCl_4, C_6H_6, $CHCl_3$
CaF_2
NO_2 H O
N-C-CH_3
O_2N, NO_2
[NH] 19

▶3383 S
CCl_4
CaF_2
H
N-CH_3
O_2N, NO_2
[NH] 19

▶3380 M
C_6H_{12}
N=N, OH, OCH_3
[NH] 35

▶3379 VVS
B (0.055)
CaF_2
H_3C, CH_2CH_3, CH_3, N, H
343

▶3378 M
$C_6H_5NO_2$
CaF_2
H, N-CH_3, O_2N, NO_2
[NH]
19

▶3376
$CHCl_3$
CaF_2
NH_2
[NH_2]
470

▶3373 MW
CH_3CN
Cl, NH_2, NO_2
19

▶3372 MS
G
H O S H, N-C-C-N, H C_6H_{11}
[NH_2]
31

▶3370 M
C_6H_{12}, CCl_4, C_2Cl_4
N=N, OH, Cl
[NH]
35

▶3370 M
C_2Cl_4, CCl_4
N=N, OH, OCH_3
[NH]
35

▶3368 S
$C_6H_5NO_2$
CaF_2
O_2N, NH_2, NO_2
[NH_2]
19

▶3367 VS
V
$(CH_3)_2CHOH$
460

▶3367 VS
CCl_4 [5%]
$CH_3CH_2CHCH_2CH_3$ (OH)
196

▶3367 S
CCl_4 [5%]
$(CH_3)_2CHCH_2CH_2OH$
195

▶3367 MS
CCl_4 [25%]
$CH_3CH_2CH(CH_3)CH_2OH$
321

▶3365 M
$(CH_3)_2CO$
CaF_2
H, N-CH_3, O_2N, NO_2
[NH]
19

▶3364 M
$CHCl_3$
CaF_2
NO_2, NH_2, NO_2
[NH_2]
19

▶3363 MW
$CHCl_3$
CaF_2
NO_2, NH_2
[NH_2]
19

▶3362 M
C_6H_6, $C_6H_5NO_2$, $(C_2H_5)_2O$
CaF_2
NO_2, NH_2, NO_2
19

▶3362 M
CCl_4, C_6H_6
CaF_2
NO_2, NH_2
[NH_2]
19

▶3361 M
C_2Cl_4
CaF_2
NO_2, NH_2, NO_2
[NH_2]
19

▶3360 W
CH_2Cl_2, $CHCl_3$
N=N, OH
[NH]
35

▶3360 W
CH_2Cl_2, $CHCl_3$
N=N, OH, CH_3
[NH]
35

▶3360 M
$C_6H_5NO_2$
CaF_2

NO_2 NH_2

$[NH_2]$ 19

▶3360

CH_3NH_2

480

▶3360 MW
C_5H_5N
CaF_2

H N-CH_3 O_2N NO_2

[NH] 19

▶3359 MW
$C_6H_5NO_2$
CaF_2

NO_2 H O N-C-CH_3 O_2N NO_2

[NH] 19

▶3358 M
C_5H_5N
CaF_2

NO_2 NH_2 NO_2

$[NH_2]$ 19

▶3358 M
CCl_4
CaF_2

NO_2 NH_2 NO_2

$[NH_2]$ 19

▶3356 VS
CCl_4 [5%]

$CH_3CH_2CH_2OH$

314

▶3350 W
CCl_4 [0.007 g/ml]

O C-NH CH_3

[NH] 2

▶3350 M
CH_2Cl_2

N=N OH Cl

[NH] 35

▶3350 M
CH_2Cl_2

N=N OH Cl

[NH] 35

▶3350 M
CH_2Cl_2, $CHCl_3$

H_3C N=N OH

[NH] 35

▶3350 M
CH_2Cl_2, $CHCl_3$

N=N OH H_3CO

[NH] 35

▶3350 MS
CH_2Cl_2

N=N OH OCH_3

[NH] 35

▶3349 MW
CH_3CN
CaF_2

NO_2 H O N-C-CH_3 O_2N NO_2

[NH] 19

▶3348 MS
CH_3CN
CaF_2

NH_2 O_2N NO_2

$[NH_2]$ 19

▶3348 M
CH_3CN
CaF_2

NO_2 NH_2

$[NH_2]$ 19

▶3344 MS
$CHCl_3$, C_2Cl_4, C_6H_6
CaF_2

NO_2 NH_2 O_2N NO_2

$[NH_2]$ 19

▶3343 MW
$(C_2H_5)_2O$
CaF_2

Cl NH_2 NO_2

$[NH_2]$ 19

▶3342 Sh
$(C_2H_5)_2O$
CaF_2

NO_2 NH_2 NO_2

$[NH_2]$ 19

▶3340 MSh
B (0.169)

$(CH_3)_2CHCH_2CH_2CH(CH_3)_2$

137, see also 339

▶3340 M E, G, $CHCl_3$ [NH] 35	▶3330 MS C_2Cl_4 [NH] 35
▶3340 VS B (0.169) $H_2C=C(CH_3)CH_2CH_2CH_3$ 206	▶3330 M $CHCl_3$ [NH] 35
▶3340 M $CHCl_3$ [NH] 35	▶3330 M E 327
▶3340 W CH_2Cl_2, $CHCl_3$ [NH] 35	▶3329 M P, CH_3CN, C_5H_5N CaF_2 19
▶3340 W CH_2Cl_2, $CHCl_3$ [NH] 35	▶3325 S G [NH] 31
▶3339 M CH_3CN CaF_2 [NH_2] 19	▶3325 A NH_2-NH_2 485
▶3336 M $(C_2H_5)_2O$ CaF_2 [NH_2] 19	▶3321 M P CaF_2 [NH_2] 19
▶3335 VS A $HC\equiv C-C(=O)-H$ [≡CH] 18	▶3320 MS E, G, CH_2Cl_2 [NH] 35
▶3334 M $CHCl_3$, C_6H_6 CaF_2 [NH] 19	▶3320 M E, G [NH] 35
▶3334 E, $CHCl_3$ [NH_2] 327 ▶3334 CCl_4 327	▶3320 E, $CHCl_3$ 327

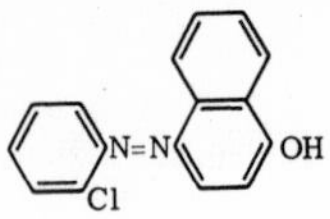

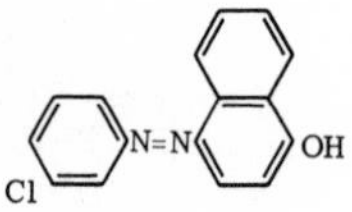

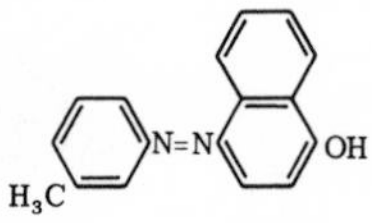

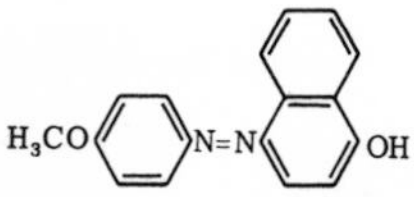

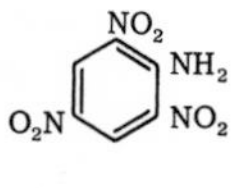

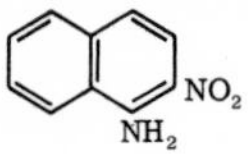

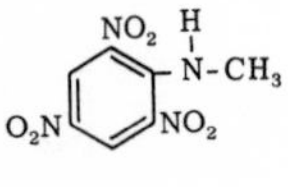

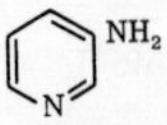

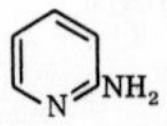

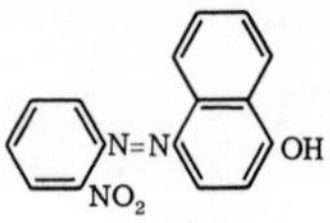

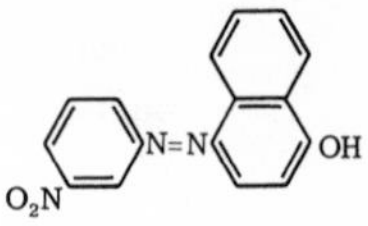

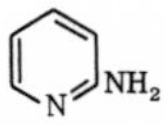

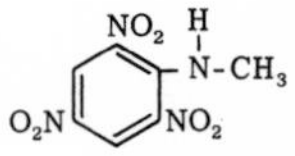

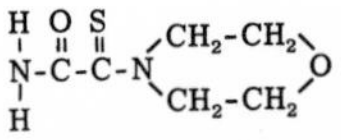

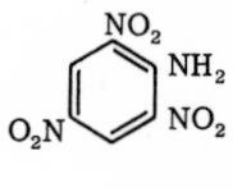

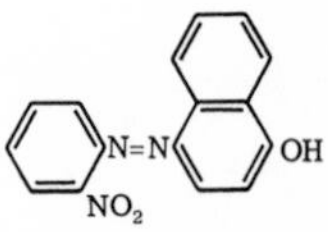

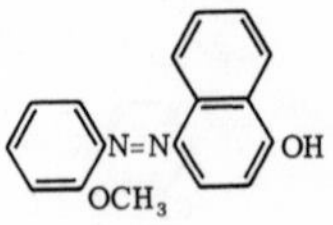

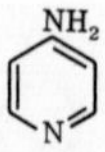

▶3314 M
C_5H_5N
CaF_2

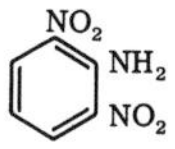

$[NH_2]$ 19

▶3314

NH_2-NH_2

485

▶3311

HC ≡ N

486

▶3311 MW
C_5H_5N
CaF_2

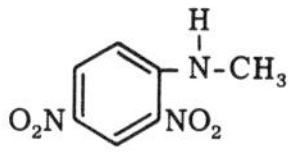

[NH] 19

▶3305 MW
$(CH_3)_2CO$
CaF_2

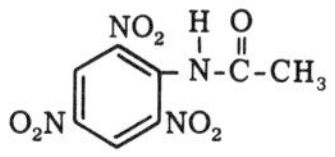

[NH] 19

▶3304 MW
$CHCl_3$
CaF_2

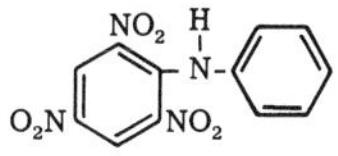

[NH] 19

▶3303 MW
C_6H_6
CaF_2

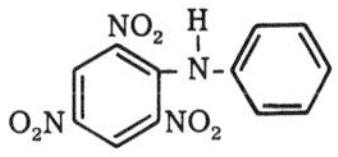

[NH] 19

▶3300 MS
G

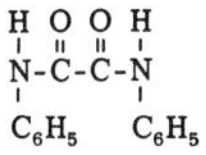

[NH] 31

▶3300 M
E, G

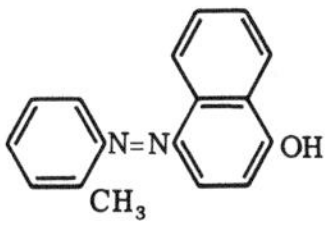

[NH] 35

▶3300 M
$CHCl_3$

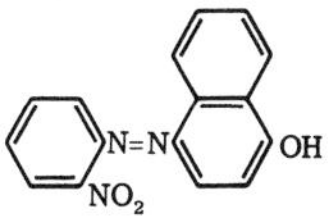

[NH] 35

▶3300 MW
CH_3CN
CaF_2

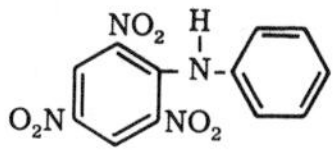

[NH] 19

▶3300
C_5H_5N

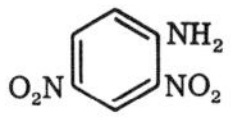

19

▶3297.0 M
B (0.2)

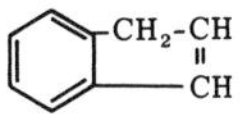

9

▶3295 M
P
CaF_2

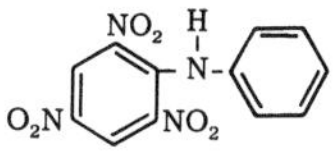

[NH] 19

▶3292 MW
C_5H_5N
CaF_2

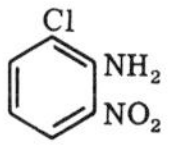

$[NH_2]$ 19

▶3290 S
G

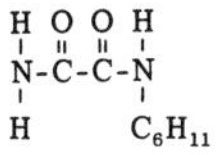

[NH] 31

▶3288 W
C_5H_5N
CaF_2

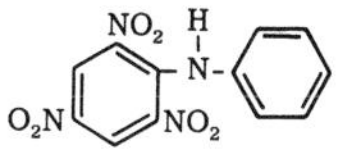

[NH] 19

▶3285 MW
C_5H_5N
CaF_2

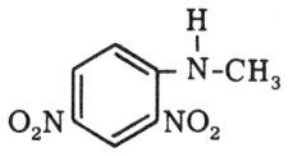

[NH] 19

▶3281 Sh
C_5H_5N
CaF_2

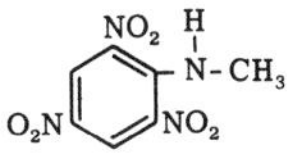

[NH] 19

▶3280 S
G

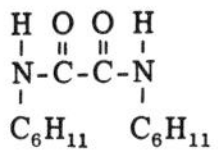

[NH] 31

▶ 3280 S G [NH] 31	▶ 3250 M E, G [NH] 35
▶ 3280 M E, G 35	▶ 3250 M E, G [NH] 35
▶ 3280 M E, G 35	▶ 3244 M G [NH] 31
▶ 3280 M E, G [NH] 35	▶ 3240 M E, G [NH] 35
▶ 3279 S B 379	▶ 3236 MS G [NH] 31
▶ 3270 M E, G [NH] 35	▶ 3236 W C_5H_5N CaF_2 [NH] 19
▶ 3267 MS G [NH] 31	▶ 3228 S G [NH] 31
▶ 3260 MS G [NH] 31	▶ 3220 M E, G [NH] 35
▶ 3260 M E, G [NH] 35	▶ 3220 M P CaF_2 [NH] 19
▶ 3250 M E, G [NH] 35	▶ 3210 $CHCl_3$ 327 ; ▶ 3205 M E 327

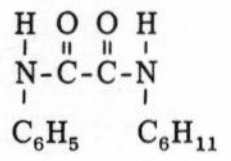

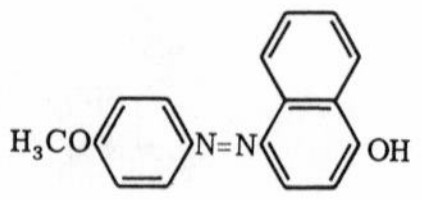

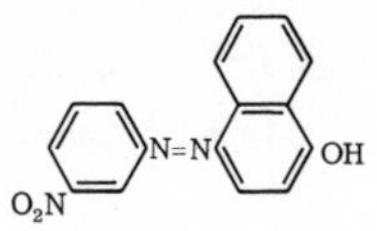

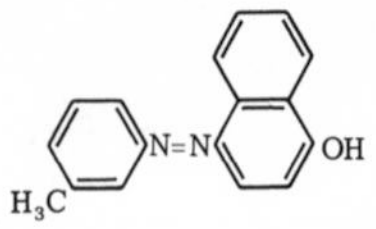

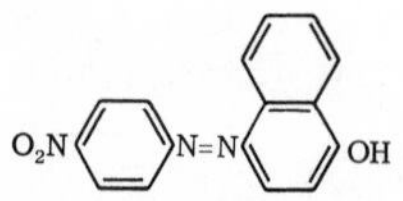

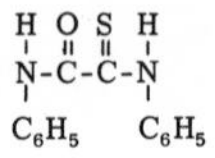

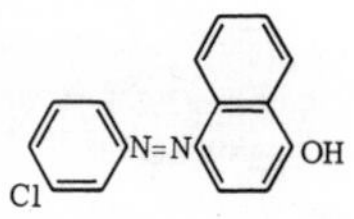

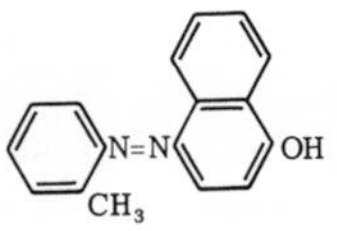

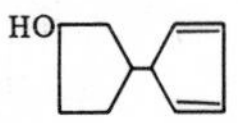

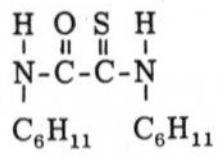

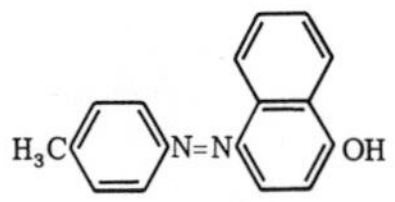

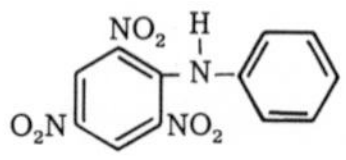

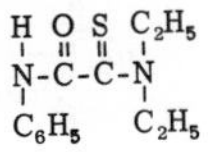

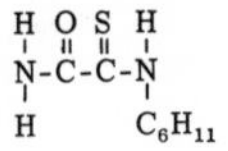

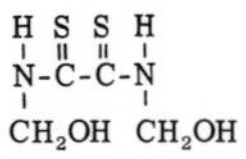

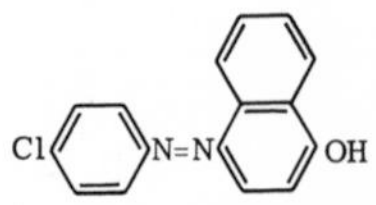

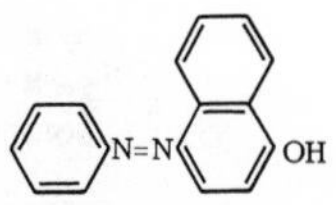

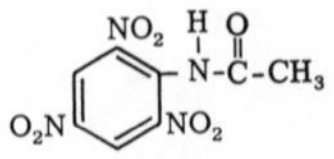

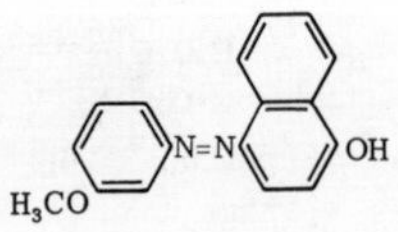

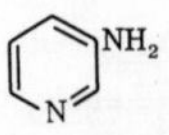

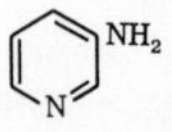

▶3200 S

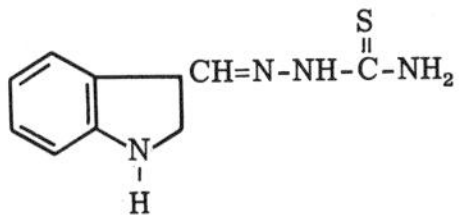

$CH{=}N{-}NH{-}\overset{S}{\overset{\|}{C}}{-}NH_2$ (on indoline ring, N–H)

380

▶3200 S

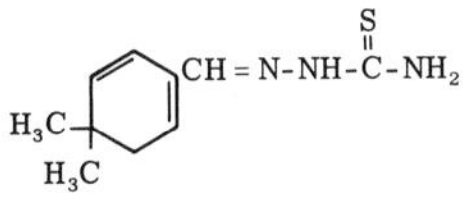

H_3C, H_3C (ring) $CH = N{-}NH{-}\overset{S}{\overset{\|}{C}}{-}NH_2$

380

▶3200 S

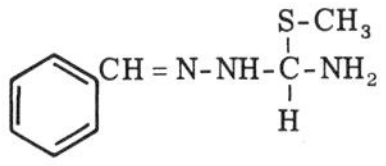

$S{-}CH_3$

(phenyl) $CH = N{-}NH{-}C{-}NH_2$

H

380

▶3200 M

B (0.20)

H_3C (cyclopentane) CH_3

(trans)

308

▶3200 MS

G

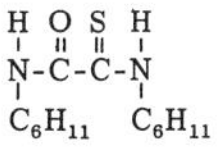

H O S H

N-C-C-N

C_6H_{11} C_6H_{11}

[NH]

31

▶3200 MS

G

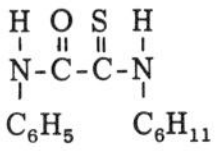

H O S H

N-C-C-N

C_6H_5 C_6H_{11}

[NH]

31

▶3200

$H_5C_2{-}\overset{O}{\overset{\|}{C}}{-}NH_2$

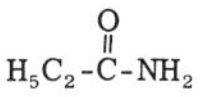

465

▶3195 WSh

B (0.065)

$H_2C{=}CHCH_2C(CH_3)_3$

273

▶3193 MSh

B (0.2398)

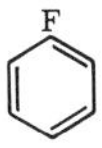

F (benzene ring)

226

▶3191 W

A (20.7 mm Hg, 40 cm)

Grating

CH_3

$CH_3CH_2C{-}CH_2CH_2CH_3$

CH_3

112

▶3185 SSh

B (0.007)

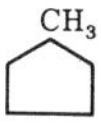

CH_3 (cyclopentane)

171

▶3185 MSh

B (0.15)

$(CH_3)_2CHCH_2NO_2$

326

▶3185 M

E

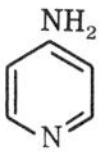

NH_2 (pyridine, N)

327

▶3180 Sh

B (0.169)

$(CH_3)_2CHCH_2CH_2CH(CH_3)_2$

137

▶3180 VSSh

B (0.169)

$(CH_3)_2CHCH(CH_2CH_3)_2$

136

▶3180 S

B (0.169)

CH_3

$(CH_3)_3CCH_2CHCH_2CH_3$

285

▶3180 SSh

B (0.169)

$(CH_3)_3CCH(CH_2CH_3)_2$

300, see also 78

▶3180 VSSh

B (0.169)

$(CH_3)_2CHCHCH(CH_3)_2$

CH_2

CH_3

301, see also 472

▶3180 VSSh

B (0.169)

CH_3

$CH_3CH_2C - CHCH_2CH_3$

CH_3 CH_3

299

▶3180 SSh

B (0.169)

$(CH_3)_2CHCH_2CHCH_2CH_3$

CH_3

138, see also 334

▶3180 VS
B (0.169)
$(CH_3)_2CHCH(CH_3)_2$
144, see also 210

▶3180 MS
G
H O S / N-C-C-N (morpholine: CH_2-CH_2, O, CH_2-CH_2) / H
[NH] 31

▶3180 SSh
B (0.169)
$(CH_3)_2CHCH(CH_3)CH_2CH_2CH_3$
139, see also 338

▶3180 MSh
B (0.169)
$CH_3CH_2C(CH_3)_2$ - $CH_2CH(CH_3)CH_2CH_3$
302

▶3180 M
CCl_4
2-aminopyridine (NH_2, N)
327

▶3179 MSh
B (0.2398)
$C_6H_5CF_3$
219

▶3176 MSh
B (0.2398)
tetrafluorobenzene (F, F, F, F)
222

▶3175 M
E
3-aminopyridine (NH_2, N)
327

▶3175
$CHCl_3$
4-aminopyridine (NH_2, N)
327

▶3170 MSh
B (0.169)
$CH_3CH_2CH(CH_3)$-$CH(CH_3)CH_2CH_3$
83

▶3170 M
E, G
Cl-C_6H_4-N=N-$C_{10}H_6$-OH
[NH] 35

▶3170 WSh
B (0.136)
$CH_2(CH_2)_2CH_3$ / $CH_3(CH_2)_2H_2C$-C_6H_3-$CH_2(CH_2)_2CH_3$
279

▶3167 MB
A (17.7 mm Hg, 40 cm)
Grating
$CH_3CH_2CH(CH_3)CH(CH_3)CH_2CH_3$
111

▶3165 SSh
B (0.238)
H_3CH_2C CH_2CH_3 / C=C / H H
(cis)
275

▶3165 S
G
H S S H / N-C-C-N / CH_2OH CH_2OH
[NH] 31

▶3150 S
G
H S S H / N-C-C-N / HO_2CH_2C CH_2CO_2H
[NH] 31

▶3140 MS
G
H S S H / N-C-C-N / C_6H_{11} C_6H_{11}
[NH] 31

▶3140 M
E, G
C_6H_5-N=N-$C_{10}H_6$-OH
[NH] 35

▶3140 M
E, G
H_3C-C_6H_4-N=N-$C_{10}H_6$-OH
35

▶3140 M
E, G
H_3C-C_6H_4-N=N-$C_{10}H_6$-OH
[NH] 35

▶3130 M
E, G

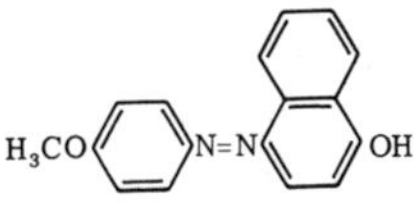
H₃CO
N=N
OH

[NH] 35

▶3130 M
E, G

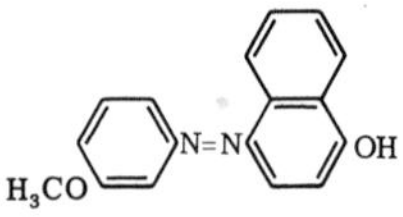
N=N
OH
H₃CO

[NH] 35

▶3115 SSp
B (0.065)

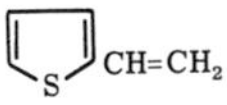
S
CH=CH₂

212

▶3115 MSSp
B (0.028)

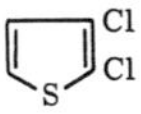
Cl
Cl
S

162

▶3110 S
B (0.2)

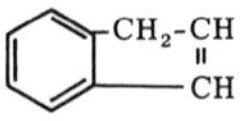
CH₂-CH
CH

9

▶3100 VS
B (0.0576)

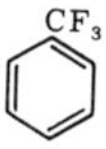
CF₃

219

▶3100 W
E

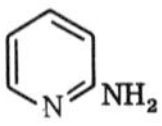
N
NH₂

327

▶3100

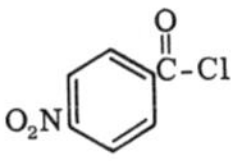
O
C-Cl
O₂N

[CH] 385

▶3098 S
B (0.2398)

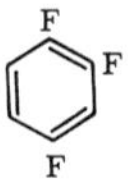
F
F
F

224

▶3096 VS
B (0.10)

S

199

▶3096 M
E

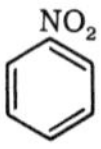
NO₂

[CH] 20

▶3095 SSp
B (0.064)

$H_2C{=}CHCHCH_2CH_3$
CH_3

167

▶3095 S
CS_2

$(C_5H_5)_2Os$

[symm. CH stretch] 450

▶3095 M
$CHCl_3$

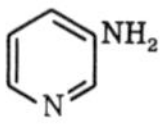
NH₂
N

327

▶3095 M
CCl_4

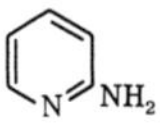
N
NH₂

327

▶3095 VS
CCl_4 [31%]

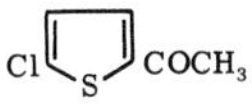
Cl
S
COCH₃

372

▶3093 SSp
CS_2 [0.169]

$CH_2{=}CHCH{=}CHCH{=}CHCH{=}CH_2$

202

▶3090 S
A (4.2 mm Hg, 40 cm)

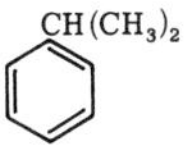
CH(CH₃)₂

66, see also 121

▶3090 M
E

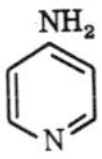
NH₂
N

327

▶3090 MW
A (3.2 mm Hg, 40 cm)

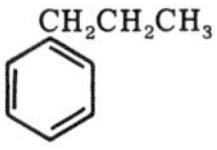
CH₂CH₂CH₃

67

▶3090

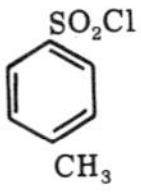
SO₂Cl
CH₃

389

▶3090
E or N

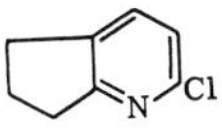
N
Cl

41

▶3090
E or N
O
C-O-C_2H_5
N 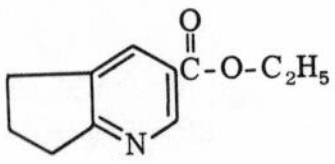
41

▶3088 VS
B (0.2398) 
F
F
225

▶3088 VS
B (0.2398) 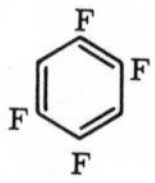
F
F
F
F
222

▶3087 VS
B (0.2398) 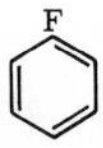
F
226

▶3087 SSh
B (0.055)
CaF_2 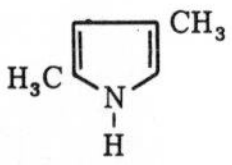
CH_3
H_3C
N
H
344

▶3085 MSp
B (0.0153) 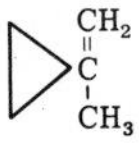
CH_2
C
CH_3
288

▶3085 SSh
B (0.064)
$H_2C{=}C(CH_2CH_3)_2$
163

▶3085 VS
B (0.065)
$H_2C{=}CHCH_2C(CH_3)_3$
273

▶3080 W
CCl_4 [0.007 g/ml] 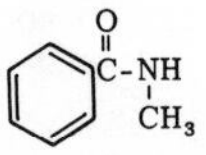
O
C-NH
CH_3
2

▶3085 M
$CHCl_3$ 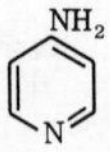
NH_2
N
327

▶3085 MWSh
B (0.008)
$H_2C = CH(CH_2)_6CH_3$
271

▶3085 WSh
B (0.008)
$H_2C = CH(CH_2)_7CH_3$
270

▶3085 VS
B (0.0576) 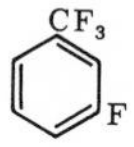
CF_3
F
218

▶3085
E or N
$(CH_2)_3$
N 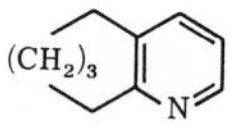
41

▶3083 S
CS_2
$(C_5H_5)_2$ Fe
[symm. CH stretch]
450

▶3080 MW
CCl_4 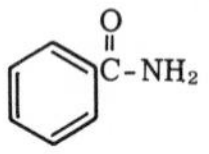
O
C-NH_2
29

▶3080
E or N 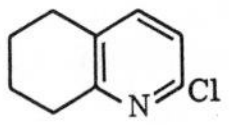
N
Cl
41

▶3080
E or N 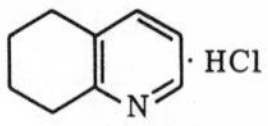
· HCl
N
41

▶3080 M
G 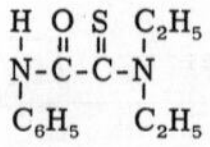
H O S C_2H_5
N-C-C-N
C_6H_5 C_2H_5
[NH]
31

▶3078 S
CS_2
$(C_5H_5)_2Ru$
[symm. CH stretch]
450

▶ 3077 MSSp
B (0.0088)

$H_2C{=}C(CH_3)CH_2CH_2CH_3$

206

▶ 3077 VS
B (0.2398)

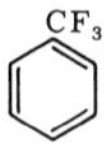

CF_3

219

▶ 3076 M
A (4.2 mm Hg, 40 cm)
Grating

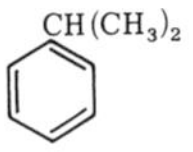

$CH(CH_3)_2$

121, see also 66

▶ 3075 M
A (3.2 mm Hg, 40 cm)
Grating

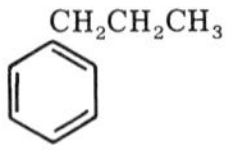

$CH_2CH_2CH_3$

122

▶ 3075 WSh
B (0.0088)

$H_2C{=}CH(CH_2)_8CH_3$

352

▶ 3075

CH_3-NH-Cl

487

▶ 3074 S
B (0.063)

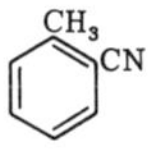

CH_3
CN

312

▶ 3070 M
E

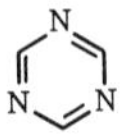

N
N N

14

▶ 3070
E or N

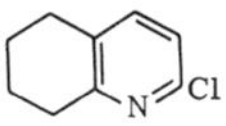

N Cl

41

▶ 3069 VS
B (0.2398)

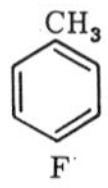

CH_3
F

220

▶ 3069 VS
B (0.2398)

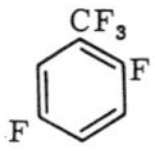

CF_3
F
F

217

▶ 3069 VS
B (0.0576)

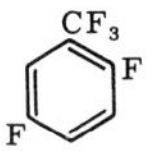

CF_3
F
F

217

▶ 3068.5 VS
B (0.03)

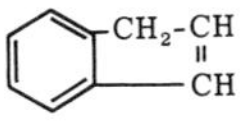

CH_2-CH
=
CH

9

▶ 3068 M
E

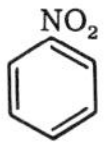

NO_2

[CH] 20

▶ 3067 VS
B (0.2398)

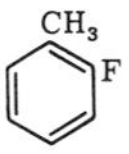

CH_3
F

221

▶ 3067 VS
B (0.2398)

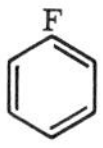

F

226

▶ 3067 VS
B

$(CH_3)_2CHCH_2CH(CH_3)CH_2CH_3$

334, see also 138

▶ 3067 M
B

$C_6H_5(CH_3)SiCl_2$

86

▶ 3067 VS
B (0.08)

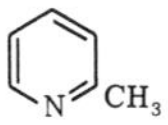

N CH_3

198

▶ 3064 VS
B (0.2398)

F
F

225

▶3060

[CH of C=C] 542

▶3050 M
A
$C_6H_5CH_2CH_3$
70

▶3060
E or N 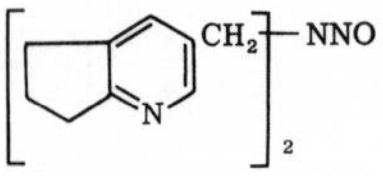
$[C_5H_6\text{-pyridyl-}CH_2]_2NNO$
41

▶3050 M
G
$HO_2CH_2C\text{-}NH\text{-}C(S)\text{-}C(S)\text{-}NH\text{-}CH_2CO_2H$
[NH] 31

▶3060 VS
A (24.1 mm Hg, 40 cm) 
C_6H_6
309

▶3050 W
CCl_4 [0.3 g/liter]
$C_6H_5C(O)NHCH_3$
2

▶3058 S
B (0°C)
$(CH_3)_2CHCH(CH_3)CH(CH_3)_2$
284, see also 135

▶3050
E or N
$(CH_2)_3$-pyridine-COOH
41

▶3058
CH_3Br
363

▶3050
$C_6H_5CH(C(=O)C_6H_5)CH_2C(=O)H$
489

▶3056
CH_3Br
363

▶3049 S
B (0.003)
CH_3NO_2
311

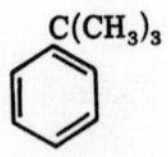

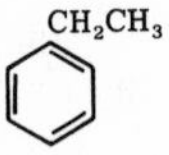

▶3055 S
N 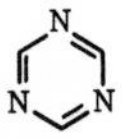
1,3,5-triazine
14

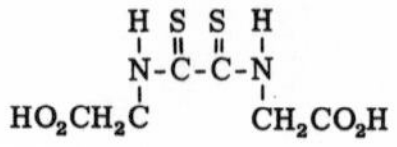

▶3049
CH_2I_2
491

▶3053 VS
B (0.2398) 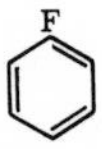
C_6H_5F
226

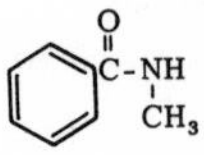

▶3049
CH_2Cl_2
376

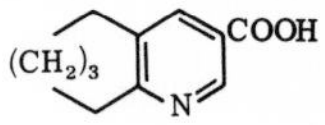

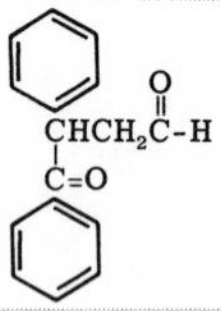

▶3052 S
A (6.9 mm Hg, 40 cm)
Grating 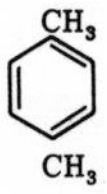
$CH_3C_6H_4CH_3$
123, see also 68

▶3049 S
A (8.3 mm Hg, 40 cm) 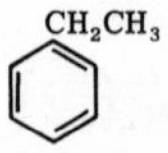
$C_6H_5CH_2CH_3$
70

▶3050 VS
B (0.033)
$C_6H_5C(CH_3)_3$
133

▶3045 VS
B (0.2398) 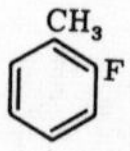
$CH_3C_6H_4F$
221

▶3043 VS
B (0.2398) 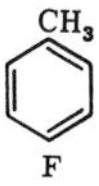
CH_3-C6H4-F
220

▶3033
A
$CHCl_3$
672, see also 408

▶3040 SSp
B (0.09)
CH_2=CHCH=CHCH=CH_2
203

▶3032 VS
CS_2 (0.169)
CH_2=CHCH=CHCH=CHCH=CH_2
202

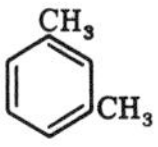

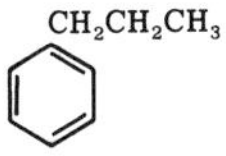

▶3040 S
A (6.3 mm Hg, 40 cm)
CH_3, CH_3
69

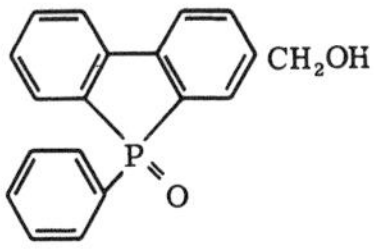

▶3032 M
A (20.7 mm Hg, 40 cm)
Grating 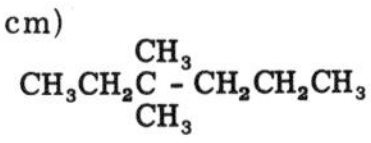
$CH_3CH_2C(CH_3)_2$-$CH_2CH_2CH_3$
112

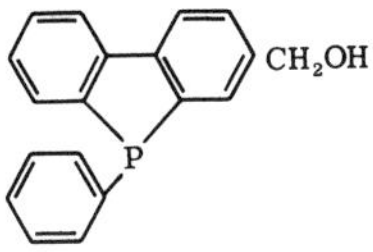

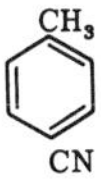

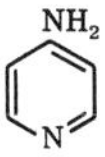

▶3040 M
A (3.2 mm Hg, 40 cm)
$CH_2CH_2CH_3$
67, see also 122

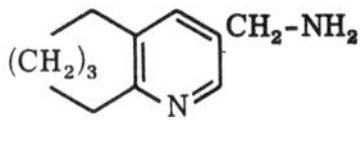

▶3030 VSSh
B (0.06) 
128

▶3030 SSp
A 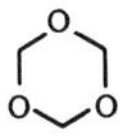
O, O, O
201

▶3040 M
E
CH_2OH, P, O
34

▶3030 S
A (7.6 mm Hg, 40 cm) 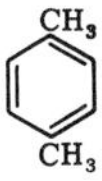
CH_3, CH_3
68, see also 123

▶3040 M
E
CH_2OH, P
34

▶3030
E or N 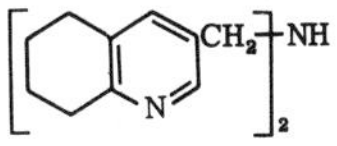
[CH_2]-NH, N, 2
41

▶3040 VS
B (0.063)
CH_3, CN
[CH] 214

▶3040 S
E
NH_2, N
327

▶3030 SSh
CS_2 [20%] 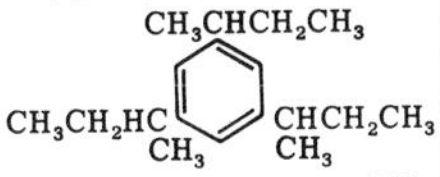
$CH_3CHCH_2CH_3$, CH_3CH_2HC, CH_3, $CHCH_2CH_3$, CH_3
280

▶3030 SSh
CCl_4 [20%] 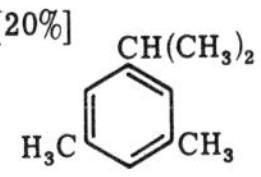
$CH(CH_3)_2$, H_3C, CH_3
277

▶3040
E or N
CH_2-NH_2, $(CH_2)_3$, N
41

▶3030 VS
B (0.2398) 
F, F
225

▶3030 VS
B (0.08) 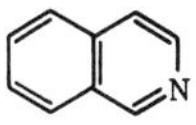
N
197

▶3038 S
A (4.2 mm Hg, 40 cm)
Grating
$CH(CH_3)_2$
121, see also 66

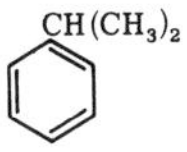

▶3025.6 S
B (0.03) 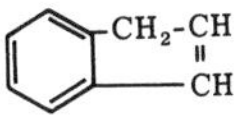
CH_2-CH=CH
9

▶3037 M
A (3.2 mm Hg, 40 cm)
Grating
$CH_2CH_2CH_3$
122, see also 67

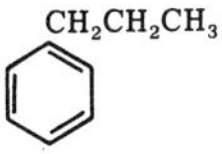

▶3025 S
A (6.9 mm Hg, 40 cm)
Grating 
CH_3, CH_3
123, see also 68

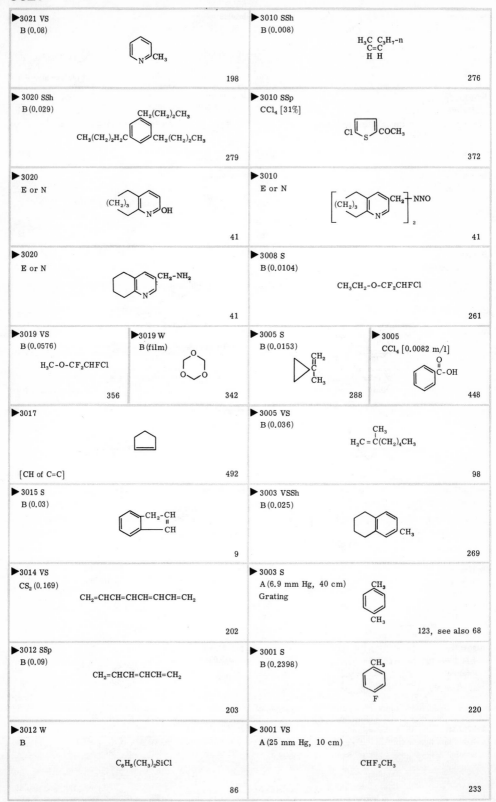

▶3021 VS B (0.08) N CH$_3$ 198	▶3010 SSh B (0.008) H_3C C_3H_7-n C=C H H 276
▶3020 SSh B (0.029) $CH_2(CH_2)_2CH_3$ $CH_3(CH_2)_2H_2C$ $CH_2(CH_2)_2CH_3$ 279	▶3010 SSp CCl_4 [31%] Cl S $COCH_3$ 372
▶3020 E or N $(CH_2)_3$ N OH 41	▶3010 E or N [$(CH_2)_3$ N CH_2]$_2$ NNO 41
▶3020 E or N N CH_2-NH_2 41	▶3008 S B (0.0104) CH_3CH_2-O-CF_2CHFCl 261
▶3019 VS B (0.0576) H_3C-O-CF_2CHFCl 356	▶3005 S B (0.0153) CH_2 C CH_3 288
▶3019 W B (film) O O O 342	▶3005 CCl_4 [0.0082 m/l] O C-OH 448
▶3017 [CH of C=C] 492	▶3005 VS B (0.036) CH_3 H_2C = C$(CH_2)_4CH_3$ 98
▶3015 S B (0.03) CH_2-CH CH 9	▶3003 VSSh B (0.025) CH_3 269
▶3014 VS CS_2 (0.169) CH_2=CHCH=CHCH=CHCH=CH_2 202	▶3003 S A (6.9 mm Hg, 40 cm) Grating CH_3 CH_3 123, see also 68
▶3012 SSp B (0.09) CH_2=CHCH=CHCH=CH_2 203	▶3001 S B (0.2398) CH_3 F 220
▶3012 W B $C_6H_5(CH_3)_2SiCl$ 86	▶3001 VS A (25 mm Hg, 10 cm) CHF_2CH_3 233

▶3000
C_6H_6 [0.0172 m/l] 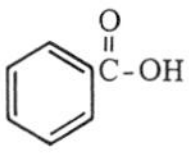
(benzoic acid: C_6H_5-C(=O)-OH)
448

▶2994 VS
B
(2,5,5-trimethylmorpholine: H_3C, O, N-H, H_3C CH_3)
[CH] 26

▶3000
E or N 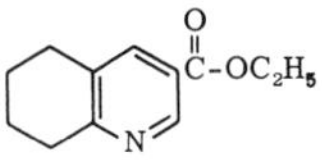
(ring structure with N; -C(=O)-OC_2H_5)
41

▶2994 VS
B (0.003)
$(CH_3)_2CHNO_2$
366

▶3000
CCl_4 [0.0128 m/l] 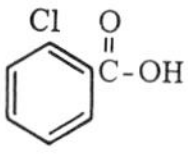
(Cl; C(=O)-OH on benzene ring)
448

▶2990
C_2H_5-C(=O)-O-C(=O)-C_2H_5
436

▶3000
C_6H_6 [0.0044 m/l] 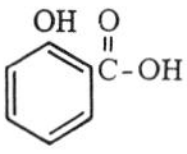
(OH; C(=O)-OH on benzene ring)
448

▶2990 VS
A (100 mm Hg, 15 cm)
CH_3CH_2Cl
88

▶2996 M
CCl_4 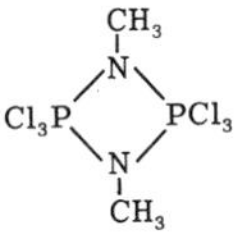
(Cl_3P, PCl_3, N-CH_3, N-CH_3 ring)
348

▶2990 VS
A (8.3 mm Hg, 40 cm)
(CH_2CH_3 on benzene ring)
70

▶2995 VS
B (0.065) 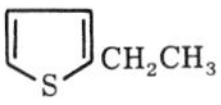
(thiophene ring, S; CH_2CH_3)
213

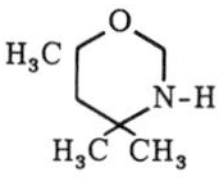

▶2985 VS
CCl_4 [5%]
CH_3CH_2OH
315

▶2995 S
A (141 mm Hg, 15 cm)
CH_3-C≡CH
87

▶2985 S
B
(HO on cyclopentane ring linked to cyclopentene ring)
379

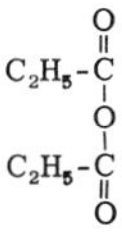

▶2995 VS
B (0.036) 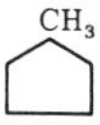
(CH_3 on cyclopentane ring)
104

▶2982 VS
B (0.0104)
n-H_7C_3-O-CF_2CHFCl
263

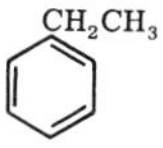

▶2995 VS
B
CH_3CH_2CH=$CH(CH_2)_3CH_3$
99

▶2982 S
B (0.2398)
(CF_3, F on benzene ring)
218

▶2995 VS
CS_2 [20%] 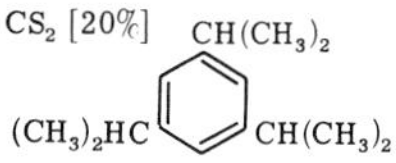
($CH(CH_3)_2$, $(CH_3)_2HC$, $CH(CH_3)_2$ on benzene ring)
278

▶2995 VS
B (0.025) 
(cyclooctatetraene ring)
200

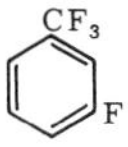

▶2982
CH_3F
493

▶2981 W
B (0.2398)
F, F, F, F (tetrafluorobenzene ring)
222

▶2980 VS
A (4.2 mm Hg, 40 cm)
$CH(CH_3)_2$ (benzene ring)
66, see also 121

▶2980 VS
A (2.5 mm Hg, 40 cm)
$(CH_3)_2CHCH(CH_3)CH_2CH_2CH_3$
72, see also 115

▶2980 S
A (0.5 mm Hg, 40 cm)
$(CH_3)_2CHC(CH_3)_2-CH_2CH_3$
90

▶2980 S
B
H_5C_2, O_2N, O, $N-C_2H_5$ (ring)
[CH]
26

▶2980
O_2N (benzene ring) $CH=CH-C(=O)-O-C_2H_5$
[CH]
576

▶2979 S
A (25 mm Hg, 10 cm)
CHF_2CH_3
233

▶2979 S
P
$[(CH_3)_2P-BCl_2]_3$
30

▶2979 S
B
H_3C, O_2N, O, $N-C_3H_7$-i (ring)
[CH]
26

▶2977
$C_2H_5-C(=O)-C_2H_5$
494

▶2976 S
B (0.003)
CH_3NO_2
311

▶2976 VS
B (0.0576)
$H_3C-O-CF_2CHFCl$
356

▶2976 VS
B (0.003)
$C_3H_7NO_2$
392

▶2976 VS
B (0.003)
$CH_3CH_2CH(NO_2)CH_3$
324

▶2976 VS
C (0.003)
$(CH_3)_3CNO_2$
325

▶2975 VS
B (0.036)
$H_2C=CH(CH_2)_3CH(CH_3)_2$
97

▶2975 VS
B (0.036)
$CH_3CH=CH(CH_2)_4CH_3$
(cis)
100, see also 160

▶2975 VS
B (0.015)
$H_2C=CHC(CH_3)_3$
129

▶2975 VS
B (0.025)
$CH_3CH_2SCH_3$
209

▶2975 VS
B
$(CH_3)_2C=C(CH_3)_2$
65

▶2975 VS
B

$CH_3CH=CHCH(CH_3)_2$

130

▶2973 VS
A (4 mm Hg, 40 cm)
Grating

CH_3
$CH_3CH_2CHCHCH_2CH_3$
CH_3

111

▶2973 VS
B (0.0104)

n-H_9C_4-O-CF_2CHFCl

264

▶2973 S
A (18.8 mm Hg, 40 cm)
Grating

CH_3
$(CH_3)_3CCHCH_2CH_3$

110, see also 71

▶2973 M
A

$(CH_3)_2SO$

[CH stretch]

13

▶2972

CH_3-NH-Cl

487

▶2972

CH_3Br

363

▶2971 VS
B (0.008)

H CH_2CH_3
C=C
H_3C CH_3
(trans)

428, see also 165

▶2971 S
A (3.2 mm Hg, 40 cm)
Grating

$CH_2CH_2CH_3$ (on benzene ring)

122, see also 67

▶2971 MSSp
CS_2 (0.169)

$H_2C=CHCH=CHCH=CHCH=CH_2$

202

▶2970 VS
A (4.2 mm Hg, 40 cm)
Grating

$CH(CH_3)_2$ (on benzene ring)

121

▶2970 VS
CS_2 [20%]

$CH(CH_3)_2$
H_3C (benzene ring) CH_3

277

▶2970 VS
A (10 mm Hg, 58 cm)

CH_3 (on cyclopentane ring)

105

▶2970 VS
A (15.9 mm Hg, 40 cm)
Grating

$CH_3CH_2CH(CH_2)_3CH_3$
CH_3

119, see also 106

▶2970 VS (band center)
A (10 mm Hg, 58 cm)

$CH_3CH=CH(CH_2)_4CH_3$
(cis and trans)

101

▶2970 VS
B (0.0088)

CH_3
$(CH_3)_3CC-CH(CH_3)_2$
CH_3

306

▶2970 VS
B (0.0088)

$(CH_3)_2CHCHCH(CH_3)_2$
CH_2
CH_3

301, see also 472

▶2970 VS
B (0.0088)

CH_3
$(CH_3)_3CCHC(CH_3)_3$

307

▶2970 VS
B (0.0088)

CH_3
$CH_3CH_2C-CHCH_2CH_3$
CH_3 CH_3

299

▶2970 S
A (10 mm Hg, 40 cm)

CH_3
$(CH_3)_3CCHCH_2CH_3$

71, see also 110

▶2970 S
CS_2 [20%]
$CH_3CHCH_2CH_3$ / CH_3CH_2HC (benzene ring) $CHCH_2CH_3$ / CH_3 CH_3
280

▶2970
$CH_3CH_2CH_2C(=O)\text{-}O\text{-}CH_3$
495

▶2969 VS
A (3 mm Hg, 40 cm)
Grating
$CH_3(CH_2)_2CH(CH_3)(CH_2)_2CH_3$
118

▶2969 VS
A (3 mm Hg, 40 cm)
Grating
$(CH_3CH_2)_2CHCH_2CH_2CH_3$
117

▶2969 VSSp
B (0.0088)
$CH_3CH=CHCH_2CH_3$
(cis and trans)
208 (cis), 207 (trans)

▶2968 VS
A (2.5 mm Hg, 40 cm)
Grating
$(CH_3)_2CHCH(CH_3)CH_2CH_2CH_3$
115, see also 72

▶2967 VS
A (2 mm Hg, 40 cm)
Grating
$(CH_3)_2CH(CH_2)_4CH_3$
120, see also 294

▶2967 VS
A (2 mm Hg, 40 cm)
$CH_3(CH_2)_6CH_3$
91

▶2967 VS
CCl_4 [5%]
$CH_3CH_2CH_2OH$
314

▶2967 VVS
B (0.0088)
$H_2C=C(CH_3)CH_2CH_2CH_3$
206

▶2967
CH_2I_2
491

▶2966 VS
A (1 mm Hg, 40 cm)
Grating
$(CH_3)_2CHCH_2CH_2CH(CH_3)_2$
113

▶2966 VS
A (2.5 mm Hg, 40 cm)
Grating
$(CH_3)_2CHCH_2CH(CH_3)CH_2CH_3$
114, see also 429

▶2966 VS
B (0.105)
Grating
$CH_3CH_2C(CH_3)_2\text{-}CH_2CH_3$
148, see also 149, 336

▶2966 VS
CCl_4
Grating
$(CH_3CH_2)_3CH$
82, see also 143

▶2966 VS
B (0.008)
$(CH_3)_2CHCH_2CH(CH_3)_2$
151, see also 141, 150

▶2965 VS
A (1.5 mm Hg, 40 cm)
Grating
$(CH_3)_3C(CH_2)_3CH_3$
116

▶2965 VS
A (12 mm Hg, 15 cm)
$H_2C=CHCH(CH_3)_2$
131

▶2965 VS
B (0.008)
H_3C CH_3 / $C=C$ / H CH_2CH_3
(cis)
274

▶2965 VS
B
$H_2C=CH(CH_2)_3CH_3$
496

▶2965 VVS
B (0.0088)

$CH_3(CH_2)_2CH=CH(CH_2)_2CH_3$
(trans)

205

▶2965 S
P

$[(CH_3)_2\text{-}P\cdot BH_2]_3$

30

▶2965

CH_3F

493

▶2964 VS
B (0.2398)

CH_3
F

221

▶2963 S
A (25 mm Hg, 10 cm)

CHF_2CH_3

233

▶2963 S
B

O
H_5C_2
$N\text{-}C_3H_7\text{-}i$
O_2N

[CH] 26

▶2962 VS
A (100 mm Hg, 10 cm)

$CH_3\text{-}CCl_3$

251

▶2962 VS
CCl_4 (0.104)
Grating

$(CH_3)_3CCH(CH_3)_2$

146, see also 147

▶2962 VVS
B (0.005)

$(CH_3)_2CHCH(CH_3)_2$

210, see also 144

▶2962 S
B

O
H_5C_2
$N\text{-}CH_3$
O_2N

[CH] 26

▶2961 VS
B (0.010)

CH_3
$(CH_3)_2CHCHCH_2CH_3$

153, see also 142, 152

▶2960 VS
B (0.0088)

CH_3
$(CH_3)_2CHCHCH_2CH_2CH_3$

139, see also 338

▶2960 VS
B (0.0088)

CH_3 CH_3
$CH_3CH_2CH\text{-}CHCH_2CH_3$

83

▶2960 VS
B (0.0088)

$H_2C=CCH_2C(CH_3)_3$
CH_3

96

▶2960 VS
B (0.0088)

$(CH_3)_2CHCH_2CH_2CH(CH_3)_2$

137, see also 339

▶2960 VS
B (0.0088)

$(CH_3)_3CCH_2CHCH(CH_3)_2$
CH_3

303

▶2960
$CHCl_3$

NH_2
N

327

▶2960 VS
B (0.008)

$H_2C=CHCH_2C(CH_3)_3$

273

▶2960 VS
B (0.0088)

$(CH_3CH_2)_2CHCH_2CH_2CH_3$

140

▶2961 VS
B (0.008)

$(CH_3)_2CHCH_2CH(CH_3)_2$

141, see also 150, 151

▶ 2960 VS
B (0.0088)

$(CH_3)_2CHCH(CH_2CH_3)_2$

136

▶ 2960 VS
B (0.0088)

$(CH_3)_2CHCHCH(CH_3)_2$
CH_3

135, see also 284

▶ 2960 VS
B (0.0088)

CH_3 CH_3
$CH_3CH_2C-CH_2CHCH_2CH_3$
CH_3

302

▶ 2960 VS
B (0.0088)

CH_3
$(CH_3)_3CC-CH_2CH_2CH_3$
CH_3

304

▶ 2960 VS
B (0.008)

H_3CH_2C CH_2CH_3
$C=C$
H H
(cis)

275

▶ 2960 VS
B (0.0088)

$(CH_3)_2CHCH(CH_3)_2$

144, see also 210

▶ 2960 VS
B (0.0088)

$(CH_3CH_2)_3CH$

143, see also 156

▶ 2960 VS
B (0.0088)
Grating

CH_3
$(CH_3)_2CHCHCH_2CH_3$

142, 152, see also 153

▶ 2960 VS
B (0.0088)

$(CH_3)_3CCH_2CH_3$

145

▶ 2959 VS
CCl_4 [5%]

$(CH_3)_2CHCH_2CH_2OH$

195

▶ 2959 VS (band center)
A (2.4 mm Hg, 43 cm)

$(CH_3)_3CH$

373

▶ 2959 VS
CCl_4 [5%]

$CH_3CHCH_2CH_3$
OH

319

▶ 2959 VS
CCl_4 [5%]

$CH_3(CH_2)_4OH$

393

▶ 2959 VS
CCl_4 [5%]

$CH_3(CH_2)_7OH$

317

▶ 2959 VS
B (0.051)

$CH(CH_3)_2$
$CH(CH_3)_2$

296

▶ 2959 W
B

$(C_6H_5)_2(CH_3)SiCl$

86

▶ 2958 S
B (0.106)
Grating

$(CH_3)_2CHCH_2CH(CH_3)_2$

150, see also 141, 151

▶ 2958 M
B (0.2398)

CF_3
F
F

217

▶ 2958 SSp
CCl_4 [25%]

CH_3
$H_2C=CCO_2CH_3$

193

▶ 2957 VS
B (0.008)
Grating

$CH_3CH_2CH(CH_2)_2CH_3$
CH_3

157, see also 81

▶2957 VS
B (0.006)
Grating

$(CH_3)_2CH(CH_2)_3CH_3$

158, see also 80

▶2957 VS
B (0.0085)
Grating

$CH_3(CH_2)_5CH_3$

84

▶2957 VS
CCl_4 (0.0017)

CH_3, CH_3 (cyclopentane) (cis)

170

▶2957 S
B (0.2398)

CF_3 (benzene)

219

▶2956 VS
CCl_4 (0.0017)

CH_2CH_3 (cyclopentane)

171

▶2955 VS
B (0.028)
CaF_2

$CH_3(CH_2)_5CH(C_3H_7)(CH_2)_5CH_3$

281

▶2955 W
B (0.2398)

F (benzene)

226

▶2955 VS
B (0.005)

$(CH_3)_3CCH_2CH_2CH_3$

154, see also 155

▶2955 S
B

$[(C_2H_5)_2P\cdot BH_2]_3$

30

▶2954 S
B (0.008)

$CH_3CH_2CH(CH_3)(CH_2)_2CH_3$

81, see also 157

▶2954 VS
B

H_5C_2, O_2N, O, $N-C_3H_7-n$ (ring)

[CH]

26

▶2953 M
B (0.2398)

CF_3, F (benzene)

218

▶2953 S
B

H_5C_2, O_2N, O, $N-C_5H_{11}-n$ (ring)

[CH]

26

▶2953 VS
B (0.007)

$CH_3CH_2C(CH_3)_2-CH_2CH_3$

149, see also 148, 336

▶2953 VS
A (100 mm Hg, 69 cm)

CH_3-CCl_3

251

▶2952 VS
B (0.006)

$(CH_3)_2CH(CH_2)_3CH_3$

80, see also 158

▶2952 VS
B (0.011)

$(CH_3)_3CCH(CH_3)_2$

147, see also 146

▶2952 VS
CCl_4 (0.0017)

CH_3 (cyclopentane)

171

▶2952 VS
CCl_4 (0.0017)

CH_3, CH_3 (cyclopentane) (trans)

170

▶2952

$CH_3CH_2CH_2-C(=O)-O-CH_3$

495

▶2950 VS
B (0.033)
$(CH_3)_3CCH(CH_2CH_3)_2$
78, see also 300

▶2950 VS
A (100 mm Hg, 15 cm)
$H_2C=C(CH_3)CH_2CH_3$
92

▶2950 VS
B (0.0088)
$(CH_3)_3C(CH_2)_2CH(CH_3)_2$
365, see also 292

▶2950 VS
B (0.033)
Grating
$(CH_3)_3CCH(CH_3)CH(CH_3)_2$
293

▶2950 VS
B (0.015)
$H_2C=CHCH_2C(CH_3)_3$
102, see also 273

▶2950 VS (band center)
A (2 mm Hg, 58 cm)
$H_2C=CH(CH_2)_5CH_3$
289

▶2950 VS
B (0.033)
$C(CH_3)_3$
133

▶2950 VS
B (0.003)
CH_3
$CH(CH_3)_2$
132

▶2950 VS
B
H_3C CH_3
(trans)
308

▶2950 VS
CCl_4 [5%]
CH_3OH
316

▶2950 VS
B (0.0088)
$(CH_3)_2CHCH_2CH(CH_3)CH_2CH_3$
138, see also 334

▶2950 VS
B (0.0088)
$(CH_3)_3CCH_2CH_2C(CH_3)_3$
305

▶2950 S
A (3.2 mm Hg, 40 cm)
$CH_2CH_2CH_3$
67, see also 122

▶2950 VS
B (0.003)
$(CH_3)_2CHCH_2NO_2$
326

▶2950 VSSh
B (0.025)
200

▶2950 S
P
$[(C_2H_5)_2P\cdot BI_2]_3$
30

▶2949 VS
B (0.008)
$(CH_3CH_2)_3CH$
156, see also 143

▶2948 VS
B
H_3C O N-H H_3C CH_3
[CH]
26

▶2947 VS
B (0.008)
$(CH_3)_3CCH_2CH_2CH_3$
155, see also 154

▶2945 VSSh
B (0.0104)
n-C_4H_9-O-CF_2CHFCl
264

▶2943.8 S
B (0.03)
CH_2-CH = CH
9

▶2943 M
B
H_5C_2, O_2N — morpholine ring — N-C_2H_5
[CH] 26

▶2942 S
B
H_5C_2, O_2N — morpholine ring — N-C_4H_9-n
[CH] 26

▶2941 VS
B (0.051)
$CH(CH_3)_2$ / $CH(CH_3)_2$ (benzene ring)
298

▶2941 VS
B (0.025)
CH_3CH_2-S-CH_2CH_3
172

▶2941 VS
B
H_3C, O_2N — morpholine ring — N-C_5H_{11}-n
[CH] 26

▶2941 S
A (4.2 mm Hg, 40 cm)
Grating
$CH(CH_3)_2$ (benzene ring)
121, see also 66

▶2941 S
(pyridine ring, N) C_4H_9-i
455

▶2941 M
CCl_4
CH_3, N, Cl_3P, PCl_3, N, CH_3
384

▶2940 VS
B (0.033)
$(CH_3)_3CCH_2CH(CH_3)CH_2CH_3$
79

▶2940 VS
B (0.0153)
H_3C, CH_3 (cyclohexane ring), CH_2CH_3
215

▶2940
$CHCl_3$
NH_2 (pyridine ring, N)
327

▶2940 S
A (706 mm Hg, 40 cm)
CH_3 / CH_3 (benzene ring)
68

▶2940 S
A (6.3 mm Hg, 40 cm)
Grating
CH_3, CH_3 (benzene ring)
69

▶2940 S
B (0.102)
H_3C, CH_3, CH_3 (cyclopentane ring)
124

▶2940 VS
B (0.063)
CH_3 / CN (benzene ring)
214

▶2940 S
B (0.063)
CH_3, CN (benzene ring)
312

▶2940 S
P
$[(C_2H_5)_2P \cdot BCl_2]_3$
30

▶2940 S
P
$[(C_2H_5)_2P \cdot BBr_2]_3$
30

▶2940 S
B (0.114)
CH_3, CH_3, CH_3 (cyclopentene ring)
126

▶2940 MW
CCl_4 [0.04 g/ml]
(benzene ring) C(=O)-N$(CH_3)_2$
3

▶2940 W
CCl_4 [0.007 g/ml]
(benzene ring) C(=O)-NH-CH_3
[CH] 2

▶2939 VS
B (0.008)

H CH_2CH_3
C=C
H_3C CH_3
(trans)

428, see also 165

▶2939 S
A (3.2 mm Hg, 40 cm)
Grating

$CH_2CH_2CH_3$ (on benzene ring)

122, see also 67

▶2939 S
A (2.5 mm Hg, 40 cm)
Grating

CH_3
$(CH_3)_2CHCHCH_2CH_2CH_3$

115

▶2939 VS
B (0.0088)

$CH_3CH = CHCH_2CH_3$
(cis and trans)

208 (cis), 207 (trans)

▶2938 S
B

H_3C, O_2N (on ring with O), N-C_4H_9-n

[CH] 26

▶2938 VVS
B (0.0088)

CH_3
$H_2C=CCH_2CH_2CH_3$

206

▶2937 S
A (3 mm Hg, 40 cm)
Grating

CH_3
$CH_3(CH_2)_2CH(CH_2)_2CH_3$

118

▶2937 S
A (4 mm Hg, 40 cm)
Grating

CH_3
$CH_3CH_2CHCHCH_2CH_3$
CH_3

111

▶2937 S
B

H_3C, O_2N (on ring with O), N- C_3H_7-i

[CH] 26

▶2936 VS
A (6.9 mm Hg, 40 cm)
Grating

CH_3 (on benzene ring), CH_3

123, see also 68

▶2936 S
B

H_3C, O_2N (on ring with O), N- C_3H_7-n

[CH] 26

▶2936 M
B

H_3C, O_2N (on ring with O), N-CH_3

[CH] 26

▶2936

O
‖
C_2H_5-C-C_2H_5

494

▶2935 VS
B (0.0088)

$H_2C = CH(CH_2)_7CH_3$

270

▶2935 VSSh
B (0.008)

H_3C CH_3
C=C
H CH_2CH_3
(cis)

274

▶2935 VS
B (0.06)

128

▶2935 MVB
B (0.106)
Grating

$(CH_3CH_2)_3CH$

82

▶2935 VS
B (0.064)

$H_2C=CHCHCH_2CH_3$
CH_3

167

▶2934 VS
A (2 mm Hg, 40 cm)
Grating

$(CH_3)_2CH(CH_2)_4CH_3$

120, see also 294

▶2934 VS
A (15.9 mm Hg, 40 cm)
Grating

$CH_3CH_2CH(CH_2)_3CH_3$
CH_3

119, see also 106

▶2934 VS
A (3 mm Hg, 40 cm)
Grating
$(CH_3CH_2)_2CHCH_2CH_2CH_3$
117

▶2934 VS
B (0.025)
$CH_3CH_2SCH_3$
209

▶2934 VS
B (0.2398)
CH_3
F
221

▶2933 VS
B (0.025)
CH_3
269

▶2933 VSB
A (2 mm Hg, 40 cm)
$CH_3(CH_2)_6CH_3$
91

▶2933 VS
B (0.0088)
$(CH_3)_2C=CHC(CH_3)_3$
204

▶2933 VVS
B (0.0088)
$CH_3(CH_2)_2CH=CH(CH_2)_2CH_3$
(trans)
205

▶2933 VS
CCl_4 [5%]
$CH_3(CH_2)_5OH$
320

▶2933 VS
CCl_4 [5%]
$CH_3(CH_2)_4OH$
393

▶2933 VS
CCl_4 [5%]
$CH_3CH(OH)(CH_2)_2CH_3$
318

▶2933 VS
CCl_4 [5%]
$CH_3CH_2CH(OH)CH_2CH_3$
196

▶2933 VS
CCl_4 [5%]
CH_3CH_2OH
315

▶2933 VS
CCl_4 [5%]
$CH_3CH_2CH_2OH$
314

▶2933 VS
B
$(CH_3)_2CHOH$
460

▶2933 VS
B (0.003)
$(CH_3)_2CHNO_2$
366

▶2933 VS
C (0.003)
$(CH_3)_3CNO_2$
325

▶2933 S
N
C_3H_7-n
455

▶2933 VS
B (0.003)
$CH_3CH_2CH(NO_2)CH_3$
324

▶2932 VS
B
$(C_2H_5\text{-}NH\text{-}B\text{-}N\text{-}C_2H_5)_3$
[CH]
32

▶2932 VSSp
CCl_4 [25%]
$H_2C=C(CH_3)CO_2CH_3$

▶2931 VS
B (0.2398) 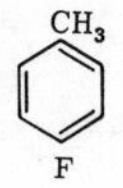
m-fluorotoluene structure (CH_3, F)
220

▶2931 VS
B (0.010)
Grating
$(CH_3)_2CHCH(CH_3)CH_2CH_3$
152, see also 142, 153

▶2930 SSh
CS_2 [20%] 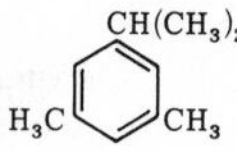
$CH(CH_3)_2$, H_3C, CH_3 (ring)
277

▶2930 VS
A (2 mm Hg, 40 cm)
$CH_3CH_2CH(CH_3)(CH_2)_3CH_3$
106, see also 119

▶2930 VS
A (2 mm Hg, 40 cm)
$(CH_3)_2CH(CH_2)_4CH_3$
294, see also 120

▶2930 M
B 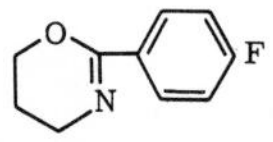
(O, N ring)–C_6H_4–F
[CH] 26

▶2929 VS
B (0.008)
$(CH_3CH_2)_3CH$
156, see also 143

▶2929 M
B 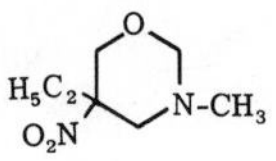
H_5C_2, O_2N, N-CH_3 (O ring)
[CH] 26

▶2928 S
B 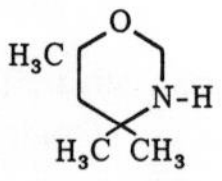
H_3C, N-H, H_3C CH_3 (O ring)
[CH] 26

▶2928 M
B 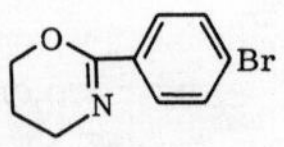
(O, N ring)–C_6H_4–Br
[CH] 26

▶2928 M
B 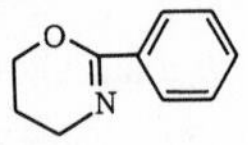
(O, N ring)–C_6H_5
[CH] 26

▶2927 VS
B (0.0085)
Grating
n-$CH_3(CH_2)_5CH_3$
84

▶2927 S
A (1 mm Hg, 40 cm)
Grating
$(CH_3)_2CHCH_2CH_2CH(CH_3)_2$
113

▶2927 S
B 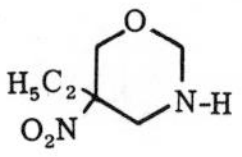
H_5C_2, O_2N, N-H (O ring)
26

▶2927 MWSp
B (0.0104)
CH_3CH_2-O-CF_2CHFCl
261

▶2927 S
B 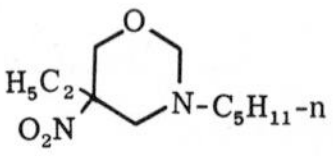
H_5C_2, O_2N, N-C_5H_{11}-n (O ring)
[CH] 26

▶2927 S
D 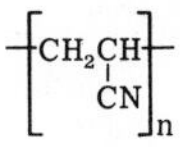
$[CH_2CH(CN)]_n$
[tert. CH] 23

▶2926 VS
B (0.008)
Grating
$CH_3CH_2CH(CH_3)(CH_2)_2CH_3$
157, see also 81

▶2926 S
B 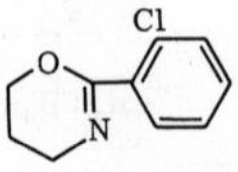
(O, N ring)–C_6H_4–Cl
[CH] 26

▶2925 VS
B (0.028)
CaF_2
$C_{26}H_{54}$
(5, 14-di-n-butyloctadecane)
283

▶2925 VS
B (0.008)

$H_2C = CH(CH_2)_6CH_3$

271

▶2925 VS
B (0.028)
CaF_2

$CH_3(CH_2)_5CH(C_3H_7)(CH_2)_5CH_3$

281

▶2925 VS
B
CaF_2

CH_3
H_3C
N
H

344

▶2925 S
CS_2 [19 g/l]

HO CH_3
$CH_2CH{=}CHCH_3$
O
(trans)

430

▶2925 S
B (0.2398)

CF_3
F

218

▶2925 MS
G

H S S H
N-C-C-N
HO_2CH_2C CH_2CO_2H

[HCH] 31

▶2924 VS
CCl_4 [5%]

$(CH_3)_2CHCH_2CH_2OH$

195

▶2924 VS
CCl_4 [5%]

$CH_3CHCH_2CH_3$
OH

319

▶2924 VS
CCl_4 [5%]

CH_3
$CH_3CH_2CHCH_2OH$

321

▶2924 VS
CCl_4 [5%]

$CH_3(CH_2)_7OH$

317

▶2924 M
B

O
H_3C $N{-}C_2H_5$
O_2N

[CH] 26

▶2924 VS
B (0.0088)

$CH_3CH = CHCH_2CH_3$
(trans)

207

▶2922 MB
CCl_4
Grating

$(CH_3CH_2)_3CH$

82, see also 143, 156

▶2922 S
B

$[(C_2H_5)_2P \cdot BH_2]_3$

[CH] 30

▶2922 VS
B (0.0178)

103

▶2921 S
B

O
N
Cl

[CH] 26

▶2920 VS
A (2 mm Hg, 40 cm)

$CH_3(CH_2)_6CH_3$

89, see also 91

▶2920 VS
A

CH_3
$CH_3CH_2CHCH_2CH_3$

95, 332

▶2920 MW
M

$[H_2CNO_2]^-Na^+$

[CH_2 asymm. stretch] 328

▶2920 S
B

O
H_5C_2 $N{-}C_3H_7{-}n$
O_2N

[CH] 26

▶2920 S B 5-methyl-5-nitro-3-(n-hexyl)tetrahydro-1,3-oxazine: H_3C, O_2N, $N\text{-}C_6H_{13}\text{-}n$ [CH] 26	▶2917 MSp B (0.0104) $n\text{-}H_7C_3\text{-}O\text{-}CF_2CHFCl$ 263
▶2920 MS G $C_6H_{11}NH\text{-}C(=O)\text{-}C(=S)\text{-}NHC_6H_{11}$ [HCH] 31	▶2917 CCl_4 [0.00658 m/l] OH, C(=O)-OH (salicylic acid) 448
▶2920 MS G $H_2N\text{-}C(=O)\text{-}C(=S)\text{-}NHC_6H_{11}$ [HCH] 31	▶2915 VS A (12 mm Hg, 15 cm) $H_2C=CHCH(CH_3)_2$ 131
▶2920 MS G $C_6H_5NH\text{-}C(=O)\text{-}C(=S)\text{-}NHC_6H_{11}$ [HCH] 31	▶2915 VS B (0.0088) $(CH_3)_3CCH(CH_3)C(CH_3)_3$ 307
▶2920 MS CS_2 [20%] benzene with $CH_3CHCH_2CH_3$, $CH_3CH_2HC(CH_3)$, $CHCH_2CH_3(CH_3)$ 280	▶2915 VS B (0.028) CaF_2 $(C_2H_5)_2CH(CH_2)_{20}CH_3$ 282
▶2919 S B 5-ethyl-5-nitro-3-(n-butyl)tetrahydro-1,3-oxazine: H_5C_2, O_2N, $N\text{-}C_4H_9\text{-}n$ [CH] 26	▶2915 VS B $CH_3CH=CH(CH_2)_2CH=CH(CH_2)_2CO\text{-}NH(CH_3)$ 497
▶2918 M B (0.2398) benzene: CF_3, F, F 217 | ▶2918 VS B (0.2398) benzene: F, F, F, F 222	▶2915 S A (2.5 mm Hg, 40 cm) Grating $(CH_3)_2CHCH_2CH(CH_3)CH_2CH_3$ 114, see also 429
▶2918 VS CCl_4 [0.00658 m/l] OH, C(=O)-OH (salicylic acid) 448	▶2915 MW B (0.2398) fluorobenzene: F 226
▶2918 S P $[(CH_3)_2P\cdot BCl_2]_3$ [CH] 30	▶2914 S B (0.2398) benzene: F, F, F 224
▶2918 MS G $C_6H_5NH\text{-}C(=O)\text{-}C(=O)\text{-}NHC_6H_{11}$ [HCH] 31	▶2914 MB CCl_4 (0.104) Grating $(CH_3)_3CCH(CH_3)_2$ 146, see also 147

▶2914 MS
G
H O O H / N-C-C-N (H, C_6H_{11})
[HCH] 31

▶2912 VS
B
H_5C_2, O_2N, N-C_6H_{13}-n (oxazine ring)
[CH] 26

▶2912 MS
G
H O O H / N-C-C-N (C_6H_{11}, C_6H_{11})
[HCH] 31

▶2911 S
P
$[(C_2H_5)_2P \cdot BCl_2]_3$
[CH] 30

▶2910
CCl_4 [0.0128 m/l]
Cl, C(=O)-OH (benzene ring)
448

▶2910 S
A (2 mm Hg, 40 cm)
Grating
$CH_3CH_2C(CH_3)_2-CH_2CH_2CH_3$
112

▶2910 S
P
$[(CH_3)_2P \cdot BH_2]_3$
[CH] 30

▶2910 S
P
$[(C_2H_5)_2P \cdot BI_2]_3$
[CH] 30

▶2909 M
B (0.005)
Grating
$(CH_3)_3CCH_2CH_2CH_3$
154, see also 155

▶2908 VS
B (0.055)
CaF_2
H_3C, CH_2CH_3, CH_3, N-H (pyrrole ring)
343

▶2908 MS
G
H S S H / N-C-C-N (C_6H_{11}, C_6H_{11})
[HCH] 31

▶2908 M
A
$(CH_3)_2SO$
[CH symm. stretch] 13

▶2907 S
CCl_4 [25%]
$H_2C=C(CH_3)CO_2CH_3$
193

▶2907 S
B
H_3C, O_2N, N-C_5H_{11}-n (oxazine ring)
[CH] 26

▶2907 S
B
HO (cyclopentyl)-(cyclopentenyl)
379

▶2907 S
C_3H_7-n, N (pyridine ring)
431

▶2907 M
M
$[(CH_3)_2CNO_2]^- Na^+$
[CH_3 asymm. stretch] 328

▶2905 VS
B
$(CH_3)_2C=C(CH_3)_2$
65

▶2905 M
B (0.106)
Grating
$(CH_3)_2CHCH_2CH(CH_3)_2$
150, see also 141, 151

▶2905 VSSh
B (0.008)
$H_2C=CHCH_2C(CH_3)_3$
273

▶2905 S
P
$[(C_2H_5)_2P \cdot BBr_2]_3$
[CH]
30

▶2905 SB
E
$(C_2H_5)_2NH_2^+(C_2H_5)_2NN_2O_2^-$
498

▶2904 M
A (4.2 mm Hg, 40 cm)
Grating
$CH(CH_3)_2$
121, see also 66

▶2902 M
A (775 mm Hg, 10 cm)
CHF_2CH_3
233

▶2902
O
$C_2H_5-C-C_2H_5$
494

▶2900 VS
B (0.064)
H_3C CH_3
C=C
H CH_2CH_3
(cis)
164

▶2900 VVS
B (0.064)
$CH_3CH=CH(CH_2)_4CH_3$
(cis)
160

▶2900 S
N $C(CH_3)_3$
455

▶2900 M
B (0.2398)
CF_3
F
F
217

▶2900 VVS
B (0.064)
H_3CH_2C H
C=C
H CH_2CH_3
(trans)
168

▶2900 VVS
B (0.064)
H_3C H
C=C
H $CH_2CH_2CH_3$
(trans)
169

▶2900 VVS
B (0.064)
$C_{17}H_{34}$
(1-heptadecene)
159

▶2900 VVS
B (0.064)
H CH_2CH_3
C=C
H_3C CH_3
(trans)
165, see also 428

▶2900 VVS
B (0.064)
$H_2C=C(CH_2CH_3)_2$
163

▶2900 VVS
B (0.064)
$(CH_3)_2C=CHCH_2CH_3$
166

▶2899 W
B
$C_6H_5(CH_3)SiCl_2$
86

▶2899 VS
CCl_4 [5%]
CH_3CH_2OH
315

▶2899 VS
B (0.003)
$(CH_3)_2CHNO_2$
366

▶2899 VS
B (0.003)
$CH_3(CH_2)_2CH_2NO_2$
310

▶2898 VS
B (0.2398)
F
F
225

▶ 2893 S
B

H_3C O N-H H_3C CH_3

[CH] 26

▶ 2892 MW
B (0.2398)

F

226

▶ 2890 S
B (0.0104)

n-H_7C_3-O-CF_2CHFCl

263

▶ 2890 VS
B (0.0088)

CH_3
CH_3CH_2C - $CHCH_2CH_3$
CH_3CH_3

299

▶ 2890 Sh
A (0.5 mm Hg, 40 cm)

CH_3
$(CH_3)_2CHC$ - CH_2CH_3
CH_3

90

▶ 2890 M
CS_2 [20%]

$CH(CH_3)_2$
$(CH_3)_2HC$ $CH(CH_3)_2$

278

▶ 2890 VS
B (O°C)

CH_3
$(CH_3)_2CHC$ - CH_2CH_3
CH_3

341

▶ 2890 S

$(C_6H_5)_3CH$

[CH stretch] 499

▶ 2890 VS
A (8.2 mm Hg, 40 cm)

CH_2CH_3

70

▶ 2890 M
A (7.6 mm Hg, 40 cm)

CH_3
CH_3

68, see also 123

▶ 2890 MSh
B (0.063)

CH_3 CN

312

▶ 2890 VS
B (0.0088)

$(CH_3)_2CHCHCH(CH_3)_2$
CH_2
CH_3

301

▶ 2887 S
A (2 mm Hg, 40 cm)
Grating

CH_3
CH_3CH_2C - $CH_2CH_2CH_3$
CH_3

112

▶ 2887 MW
A (775 mm Hg, 10 cm)

CHF_2CH_3

233

▶ 2887 VS
B (0.03)

CH_2-CH
‖
CH

9

▶ 2885 S
A (1 mm Hg, 40 cm)
Grating

CH_3
$(CH_3)_3CCHCH_2CH_3$

110, see also 71

▶ 2884 VS
B (0.008)
Grating

$(CH_3CH_2)_3CH$

82, see also 143, 156

▶ 2884 S
A (2.5 mm Hg, 40 cm)
Grating

CH_3
$(CH_3)_2CHCHCH_2CH_2CH_3$

115, see also 72

▶ 2884 VS
A (3 mm Hg, 40 cm)
Grating

$(CH_3CH_2)_2CHCH_2CH_2CH_3$

117

▶ 2884 M
A (4.2 mm Hg, 40 cm)
Grating

$CH(CH_3)_2$

121, see also 66

▶2884 W
B (0.2398)
CF_3
219

▶2884 M
B
O_2N O N
[CH]
26

▶2884 VS
A (2.5 mm Hg, 40 cm)
Grating
$(CH_3)_2CHCH_2CHCH_2CH_3$ (CH_3)
114, see also 429

▶2883 S
A (3 mm Hg, 40 cm)
Grating
$CH_3(CH_2)_2CH(CH_2)_2CH_3$ (CH_3)
118

▶2882 VS
A
$CH_3CH_2CH(CH_2)_3CH_3$ (CH_3)
119

▶2882 VS
B
$(CH_3)_2CHOH$
460

▶2882 VS
B (0.2398)
F F F
224

▶2882 VSSp
B (0.0104)
$n\text{-}H_9C_4\text{-}O\text{-}CF_2CHFCl$
264

▶2882 VS
B (0.0088)
$(CH_3)_2CHCH_2NO_2$
326

▶2882 MSpSh
B (0.0576)
$CH_3CH_2\text{-}O\text{-}CF_2CHFCl$
261

▶2882 VS
B (0.008)
$(CH_3)_2CH(CH_2)_2CH(CH_3)_2$
339, see also 137

▶2882 VS
B (0.003)
$CH_3CH_2CHCH_3$ (NO_2)
324

▶2882 MSB
B (0.105)
Grating
$CH_3CH_2C(CH_3)_2\text{-}CH_2CH_3$
148, see also 149, 336

▶2882 VS
B (0.003)
$C_3H_7NO_2$
392

▶2882 M
A (1 mm Hg, 40 cm)
$(CH_3)_2CHCH_2CH_2CH(CH_3)_2$
113

▶2882 MSh
CCl_4 [5%]
$CH_3CH_2C(CH_3)_2$ (OH)
194

▶2881 M
A (3.2 mm Hg, 40 cm)
Grating
$CH_2CH_2CH_3$
122, see also 67

▶2881 S
A (2 mm Hg, 40 cm)
Grating
$(CH_3)_2CH(CH_2)_4CH_3$
120, see also 294

▶2880 VSSh
B (0.008)
H CH_2CH_3
C=C
H_3C CH_3
(trans)
428, see also 165

▶2880 VS
B (0.0088)
$(CH_3)_2CHCHCH_2CH_2CH_3$ (CH_3)
139, see also 338

▶2880 VS
B (0.008)
$(CH_3CH_2)_2CHCH_2CH_2CH_3$
140

▶2880 VSSh
B (0.0088)
$(CH_3)_3CCH(CH_3)C(CH_3)_3$
307

▶2880 VS
B (0.0088)
$CH_3CH_2CH(CH_3)-CH(CH_3)CH_2CH_3$
83

▶2880 M
A (3.2 mm Hg, 40 cm)
$CH_2CH_2CH_3$ (on benzene ring)
67, see also 122

▶2880 VSSh
B (0.0088)
$(CH_3)_3CCH(CH_2CH_3)_2$
300, see also 78

▶2880 M
B
H_5C_2, O_2N (morpholine ring), N-C_2H_5
[CH]
26

▶2880 VS
B (0.0088)
$(CH_3)_2CHCH(CH_3)CH_2CH_3$
142, see also 152, 153

▶2879 VSSp
B (0.0088)
$CH_3CH=CHCH_2CH_3$
(cis)
208

▶2880 VS
B (0.008)
$(CH_3)_2CHCH(CH_3)_2$
144, see also 210

▶2879 M
A (1.5 mm Hg, 40 cm)
Grating
$(CH_3)_3C(CH_2)_3CH_3$
116

▶2880 VS
B (0.0088)
$(CH_3)_2CHCH(CH_2CH_3)_2$
136

▶2878 VS
B (0.008)
$CH_3CH_2CH(CH_3)(CH_2)_2CH_3$
81, see also 157

▶2880 VS
B (0.0088)
$(CH_3)_2CHCH(CH_3)CH(CH_3)_2$
135, see also 284

▶2878 S
B
H_3C, O_2N (morpholine ring), N-C_3H_7-i
[CH]
26

▶2880 SSh
A (2 mm Hg, 40 cm)
$CH_3CH_2CH(CH_3)(CH_2)_3CH_3$
106, see also 119

▶2878 M
B (0.2398)
F (on benzene ring)
226

▶2880 M
A (6.9 mm Hg, 40 cm)
CH_3, CH_3 (on benzene ring)
123, see also 68

▶2878 VS
B (0.0088)
$H_2C=C(CH_3)CH_2CH_2CH_3$
206

▶2880 S
B (0.0088)
$(CH_3)_2CHCH_2CH_2CH(CH_3)_2$
137, see also 339

▶2878 M
B
(dihydrooxazine ring: O, N) – (benzene ring) – F
[CH]
26

▶ 2878 M
B

O N

[CH] 26

▶ 2877 VS
B
Grating

$(CH_3)_2CHCH(CH_3)_2$

210, see also 144

▶ 2877 SSh
B (0.056)

$CH_2{=}CHCH{=}CHCH{=}CH_2$

203

▶ 2877 VS
B (0.0088)

$CH_3(CH_2)_2CH = CH(CH_2)_2CH_3$
(trans)

205

▶ 2877 MW
A (775 mm Hg, 10 cm)

CHF_2CH_3

233

▶ 2876 VS
B (0.010)
Grating

CH_3
$(CH_3)_2CHCHCH_2CH_3$

152, see also 142, 153

▶ 2875
CCl_4 [0.0082 m/l]

O
C-OH

448

▶ 2875 VS
B

$H_3CH_2C\ \ CH_2CH_3$
C=C
H H
(cis)

275

▶ 2875 VS
B (0.008)
Grating

$CH_3CH_2CH(CH_2)_2CH_3$
CH_3

157, see also 81

▶ 2875 VS
B (0.003)

$H_3C\ \ CH_3$
C=C
$H\ \ CH_2CH_3$
(cis)

274

▶ 2875 VSSh
B (0.008)

$H_3C\ \ C_3H_7$-n
C=C
H H
(cis)

276

▶ 2875 VS
B (0.008)

$H_2C = CH(CH_2)_7CH_3$

270

▶ 2875 S
B (0.0576)

$H_3C\text{-}O\text{-}CF_2CHFCl$

356

▶ 2874 S

$CH(C_2H_5)_2$
N

431

▶ 2874 VS
CCl_4 [5%]

$CH_3CH(CH_2)_2CH_3$
OH

318

▶ 2874 VS
CCl_4 [5%]

$CH_3CH_2CH_2OH$

314

▶ 2874 VS
CCl_4 [5%]

$CH_3CH_2CHCH_2CH_3$
OH

196

▶ 2874 VSSh
B (0.010)

CH_3
$(CH_3)_2CHCHCH_2CH_3$

153, see also 142, 152

▶ 2874 S
B (0.006)
Grating

$(CH_3)_2CH(CH_2)_3CH_3$

80, 158

▶ 2873 M
B

O
H_5C_2 N-CH_3
O_2N

[CH] 26

▶2873 M
B
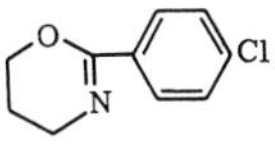
[CH]
26

▶2872 S
B
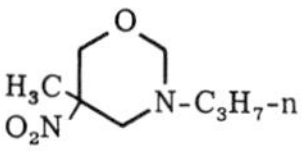
[CH]
26

▶2872 M
B
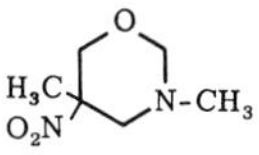
[CH]
26

▶2871 MS
CCl_4 (0.104)
Grating
$(CH_3)_3CCH(CH_3)_2$
146, see also 147

▶2870 VS
CCl_4 [0.00658 m/l]
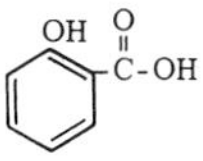
448

▶2870 SSh
CS_2 [20%]
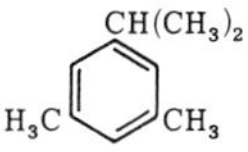
277

▶2870 S
B
$[(C_2H_5)_2P \cdot BH_2]_3$
[CH]
30

▶2870 VVS
B (0.055)
CaF_2
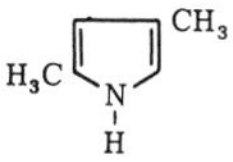
344

▶2870 M
CS_2 [20%]
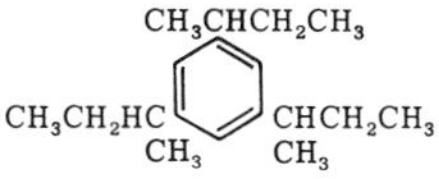
280

▶2869 MW
B (0.2398)

222

▶2869 VS
B (film)
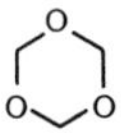
342

▶2869 S
A
HC≡C-C(=O)-H
[O=C-H]
18

▶2868 VS
CCl_4 (0.0017)
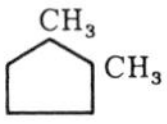
(cis)
170

▶2868 S
A (2 mm Hg, 40 cm)
Grating
$CH_3(CH_2)_6CH_3$
91, see also 89

▶2867 M
B (0.2398)
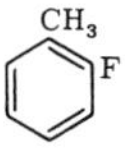
221

▶2866 VS
CCl_4 (0.0017)
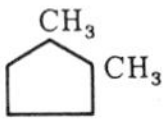
(trans)
170

▶2866 VS
CCl_4 (0.0017)
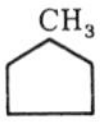
171

▶2866 VS
CCl_4 (0.0017)
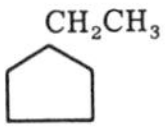
171

▶2865 VS
CCl_4 [5%]
$CH_3CH(OH)CH_2CH_3$
319

▶2865 VS
CCl_4 [5%]
$CH_3(CH_2)_4OH$
393

▶2865 VS
B (0.028)
$CH_3(CH_2)_5CH(C_3H_7)(CH_2)_5CH_3$
281

▶2865 VS
B
$(CH_3)_2C=C(CH_3)_2$
65

▶2865 VS
B (0.025)
CH_3
269

▶2865 VS
B
$CH_3CH=CH(CH_2)_2CH=CH(CH_2)_2CO\text{-}NH\text{-}CH_2CH(CH_3)_2$
497

▶2865 VS
CCl_4 [5%]
$(CH_3)_2CHCH_2CH_2OH$
195

▶2865 S
N, $C_4H_9\text{-i}$
455

▶2863 MB
CCl_4
Grating
$(CH_3CH_2)_3CH$
82

▶2862 VS
B (0.008)
$(CH_3CH_2)_3CH$
156, see also 82, 143

▶2860 M
B
O, H_3C, O_2N, $N\text{-}C_2H_5$
[CH] 26

▶2860 VS
B (0.029)
$CH_2(CH_2)_2CH_3$, $CH_3(CH_2)_2H_2C$, $CH_2(CH_2)_2CH_3$
279

▶2859 S
B
Grating
$CH_3CH_2C(CH_3)_2\text{-}CH_2CH_3$
148, see also 149, 336

▶2859 S
P
$[(C_2H_5)_2P\cdot BCl_2]_3$
30

▶2857 VS
CCl_4 [5%]
$CH_3(CH_2)_4OH$
393

▶2857 M
CCl_4 [5%]
$CH_3(CH_2)_7OH$
317

▶2857 VS
CCl_4 [5%]
$CH_3(CH_2)_5OH$
320

▶2857 S
B
$(C_2H_5\text{-}NH\text{-}B\text{-}N\text{-}C_2H_5)_3$
[CH] 32

▶2857 SSh
C (0.003)
$(CH_3)_3CNO_2$
325

▶2857 S
$C_3H_7\text{-n}$, N
431

▶2857 S
$C_4H_9\text{-i}$, N
431

▶2857 S
N, $C_4H_9\text{-i}$
455

▶2857 S $CH(C_2H_5)_2$ (4-pyridyl) 431	▶2850 S A $CH_3-NHCH_2CH_2NH-CH_3$ 501
▶2857 S (2-pyridyl) $CH(C_2H_5)_2$ 455	▶2849 VSSp A (1,3,5-trioxane) 201
▶2857 S (2-pyridyl) $CH(C_4H_9\text{-}i)_2$ 455	▶2847 MW $[H_2CNO_2]^-Na^+$ [CH_2 symm. stretch] 328
▶2855 W M $[(CH_3)_2CNO_2]^-Na^+$ [CH_3 symm. stretch] 328	▶2846 S B H_5C_2, O_2N … $N-C_6H_{13}\text{-}n$ [CH] 26
▶2855 S $[(C_2H_5)_2P\cdot BI_2]_3$ [CH] 30	▶2846 MS G $H_2N-C(=O)-C(=S)-NH-C_6H_{11}$ [HCH] 31
▶2855 S B (0.008) $H_2C=CH(CH_2)_6CH_3$ 271	▶2846 M G $H_2N-C(=O)-C(=O)-NH-C_6H_{11}$ [HCH] 31
▶2852 S P $[(C_2H_5)_2P\cdot BBr_2]_3$ [CH] 30	▶2846 M B (2-(4-bromophenyl)-dihydro-oxazine) Br [CH] 26
▶2851 M B O_2N (2-(2-nitrophenyl)-dihydro-oxazine) [CH] 26	▶2845 S B H_3C, O_2N … $N-C_6H_{13}\text{-}n$ [CH] 26
▶2850 VS B (0.028) CaF_2 $C_{26}H_{54}$ (5, 14-di-n-butyloctadecane) 283	▶2845 M G $C_6H_5-NH-C(=O)-C(=O)-NH-C_6H_{11}$ [HCH] 31
▶2850 VS B (0.028) CaF_2 $(C_2H_5)_2CH(CH_2)_{20}CH_3$ 282	▶2844 M B H_3C, O_2N … $N-C_3H_7\text{-}i$ [CH] 26

▶2843 VS
B (0.0088)

$CH_3(CH_2)_2CH=CH(CH_2)_2CH_3$
(trans)

205

▶2843 M
B

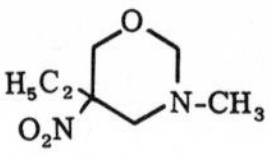
O
H5C2
N-CH3
O2N

[CH] 26

▶2842 M
G

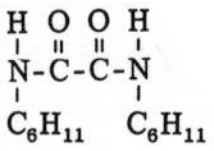
H O O H
N-C-C-N
C6H11 C6H11

[HCH] 31

▶2842

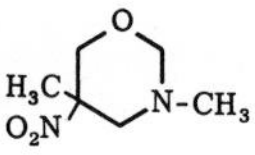
O
H3C
N-CH3
O2N

[CH] 26

▶2841 S

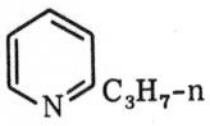
N
C3H7-n

455

▶2840 M
G

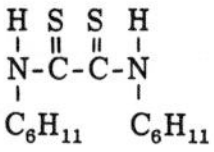
H S S H
N-C-C-N
C6H11 C6H11

[HCH] 31

▶2836 MSSh
B (0.056)

$CH_2=CHCH=CHCH=CH_2$

203

▶2835 S

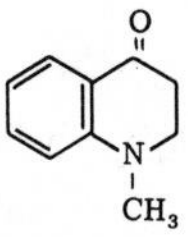
O
N
CH3

502

▶2835 M
CCl_4 [31%]

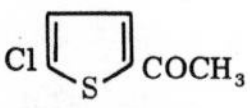
Cl
S
COCH3

372

▶2835 Sh
B

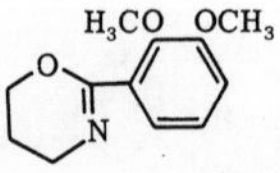
H3CO OCH3
O
N

[CH] 26

▶2833 VS
CCl_4 [5%]

CH_3OH

316

▶2833 S

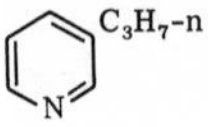
C3H7-n
N

431

▶2832 S
B (0.20)

H3C
CH3
(trans)

308

▶2828 S
B (0.2398)

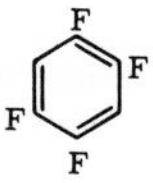
F
F
F
F

222

▶2826 M
B

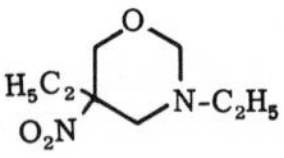
O
H5C2
N-C2H5
O2N

[CH] 26

▶2824 S
B

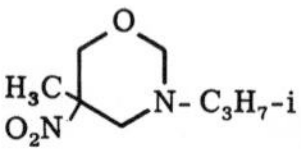
O
H3C
N- C3H7-i
O2N

[CH] 26

▶2822 MS
B (0.2398)

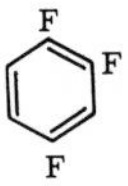
F
F
F

224

▶2822 S
A (775 mm Hg, 10 cm)

CHF_2CH_3

233

▶2821 M
B

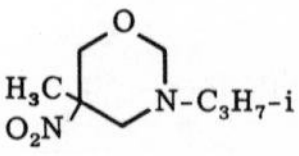
O
H3C
N-C3H7-i
O2N

[CH] 26

▶2820 S
A

CH_3NH_2

480

▶2819 M
B

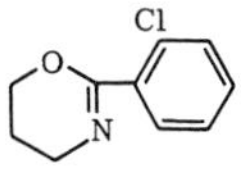

[CH] 26

▶2817 S
B

379

▶2815 S
L

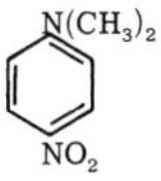

21, see also 626

▶2813 W
B (0.2398)

225

▶2812 S
B

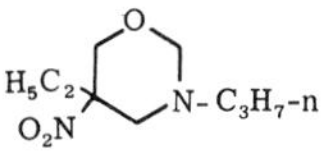

[CH] 26

▶2810 S
A

$(CH_3)_3N$

[CH symm. stretch] 505

▶2810 M
B

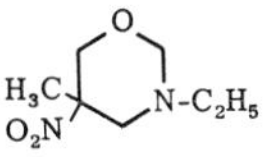

[CH] 26

▶2807 S
A (775 mm Hg, 10 cm)

CHF_2CH_3

233

▶2806 MSh
B (0.2398)

220

▶2804 S
B (0.2398)

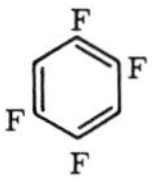

222

▶2802 S
A

$(CH_3\text{-}SiH_2)_2N\text{-}CH_3$

504

▶2800 S
A

$(CH_3)_2N\text{-}BCl_2$

503

▶2796 MWVB
B (0.011)

$(CH_3)_3CCH(CH_3)_2$

146, see also 147

▶2793 S
B (0.003)

CH_3NO_2

311

▶2790 VSSp
B (film)

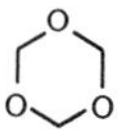

201

▶2787 S
B

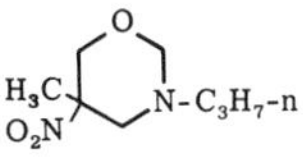

[CH] 26

▶2786 S
B (0.2398)

225

▶2785 M
B

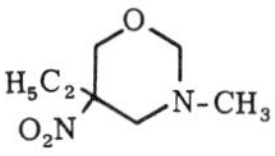

[CH] 26

▶2781 MW
B (0.2398)

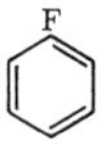

226

▶2780 M
A

$HC\equiv C\text{-}C(=O)\text{-}H$

18

▶2780
C_6H_6 [0.0172 m/l]
C-OH, O
448

▶2775
C_6H_6 [0.0128 m/l]
Cl, O, C-OH
448

▶2774 M
B
O, H_5C_2, N-C_2H_5, O_2N
[CH] 26

▶2771 MW
B (0.2398)
F, F, F, F
222

▶2771.0 S
B (0.2)
CH_2-CH = CH
9

▶2769 M
B
O, H_3C, N-C_6H_{13}-n, O_2N
[CH] 26

▶2766 M
A (775 mm Hg, 10 cm)
F_2, F_2, F_2, F_2
243

▶2765 M
B
O, H_3C, N-C_4H_9-n, O_2N
[CH] 26

▶2761 S
B
O, H_5C_2, N-C_3H_7-n, O_2N
[CH] 26

▶2757 S
A (775 mm Hg, 10 cm)
CHF_2CH_3
233

▶2755 S
B (0.15)
$(CH_3)_2CHNO_2$
366

▶2750 M
C
KH_2PO_4
28

▶2750 WSh
A (2 mm Hg, 40 cm)
$CH_3CH_2CH(CH_2)_3CH_3$ / CH_3
106, see also 119

▶2750 W
A (2 mm Hg, 40 cm)
$(CH_3)_2CH(CH_2)_4CH_3$
294, see also 120

▶2747 MSh
B (0.003)
$CH_3(CH_2)_2CH_2NO_2$
310

▶2745 MW
B (0.065)
$H_2C{=}CHCH_2C(CH_3)_3$
273

▶2745 SSh
B (0.169)
$(CH_3)_3CCH_2CH_3$
145

▶2744 S
A (775 mm Hg, 10 cm)
CHF_2CH_3
233

▶2744 MW
B (0.104)
$CH_3CH_2CH(CH_2)_2CH_3$ / CH_3
81, see also 157

▶2741 SSh
B (0.101)
CH_3 / $(CH_3)_2CHCH$-CH_2CH_3
153, see also 142, 152

▶2740 MSh
B (0.238)

$H_3CH_2C\ CH_2CH_3$
C=C
H H
(cis)

275

▶2736 MW
B (0.20)

CH_2CH_3 (cyclopentane)

171

▶2740 VSSh
B (0.169)

$(CH_3)_3CCH_2CH_2C(CH_3)_3$

305

▶2732 S
A (200 mm Hg, 10 cm)

$CF_3CF_2CF_3$

235

▶2740 VSSh
B (0.169)

CH_3
$CH_3CH_2C - CHCH_2CH_3$
$CH_3\ CH_3$

299

▶2732 MSh
B (0.10)

CH_3 (tetralin ring)

269

▶2740 MSSh
B (0.066)

$H_2C{=}CH(CH_2)_8CH_3$

352

▶2731 S
B (0.105)

$(CH_3)_2CHCH_2CH(CH_3)_2$

151, see also 141, 150

▶2740 MSh
B (0.157)

$(CH_3)_3CCH(CH_2CH_3)_2$

78, see also 300

▶2730 VSSh
B (0.169)

CH_3
$(CH_3)_2CHCHCH_2CH_3$

142, see also 152, 153

▶2740 M
B (0.006)

$(CH_3)_2CH(CH_2)_3CH_3$

80, see also 158

▶2730 SSh
B (0.169)

$(CH_3CH_2)_3CH$

143, see also 82, 156

▶2740 VS
B (0.114)

CH_3
H_3C (cyclopentene ring)
CH_3

127

▶2730 SSh
B (0.008)

$CH_3CH = CHCH_2CH_3$
(trans)

207

▶2738 MW
B (0.101)

$CH_3(CH_2)_5CH_3$

85

▶2730 VSSh
B (0.169)

CH_3
$(CH_3)_2CHCHCH_2CH_2CH_3$

139, see also 338

▶2738 SSh
B (0.40)

$CH_3CH_2SCH_3$

209

▶2730 VSSh
B (0.169)

$(CH_3)_2CHCH(CH_2CH_3)_2$

136

▶2736 S
B (0.20)

CH_3
CH_3 (cyclopentane ring)
(cis)

170

▶2730 VSSh
B (0.169)

$(CH_3)_2CHCHCH(CH_3)_2$
CH_3

135, see also 284

▶2730 VSSh
B (0.169)

$(CH_3)_2CHC(CH_3)_2-CH_2CH_2CH_3$

134

▶2730 VSSh
B (0.169)

$(CH_3)_2CHCH(CH_2CH_3)CH(CH_3)_2$

301, see also 472

▶2730 VSSh
B (0.169)

$(CH_3)_3CC(CH_3)_2-CH_2CH_2CH_3$

304

▶2730 M
B (0.068)

$H_3C(H)C=C(CH_3)(CH_2CH_3)$

(cis)

274

▶2730 W
B (0.028)
CaF_2

$C_{26}H_{54}$
(5, 14-di-n-butyloctadecane)

283

▶2730 VSSh
B (0.169)

$(CH_3CH_2)_2CHCH_2CH_2CH_3$

140

▶2730 S
B

$CH_3(CH_2)_5CH(C_3H_7)(CH_2)_5CH_3$

281

▶2730 S
B (0.169)

$CH_3CH_2CH(CH_3)-CH(CH_3)CH_2CH_3$

83

▶2730 MW
B (0.136)

$CH(CH_3)_2$, $(CH_3)_2HC$, $CH(CH_3)_2$ (on benzene ring)

278

▶2730 M
B (0.20)

CH_3 (on cyclopentane ring)

171

▶2730 SSh
B (0.169)

$(CH_3)_3CCH(CH_2CH_3)_2$

300, see also 78

▶2730 SSh
B (0.169)

$CH_3CH_2C(CH_3)_2-CH_2CH(CH_3)CH_2CH_3$

302

▶2730 SSh
B (0.243)

$H_2C=CH(CH_2)_6CH_3$

271

▶2730 MSh
B (0.238)

$H_2C=C(CH_3)C(CH_3)_3$

272

▶2730 VSSh
B (0.169)

$(CH_3)_2C=CHC(CH_3)_3$

204

▶2730 MSpSh
B (0.064)

$(CH_3)_2C=CHCH_2CH_3$

166

▶2730 MW
B (0.136)

$CH_2(CH_2)_2CH_3$, $CH_3(CH_2)_2H_2C$, $CH_2(CH_2)_2CH_3$ (on benzene ring)

279

▶2729 VSSp
B (0.20)

CH_3, CH_3 (on cyclopentane ring)
(trans)

170

▶2727 MSh
B (0.112)

$(CH_3)_3CCH_2CH_2CH_3$

155, see also 154

▶2725 VSSh
B (0.169)

$(CH_3)_2CHCH(CH_3)_2$

144, see also 210

▶ 2725 MSh
B (0.169)

CH_3
$(CH_3)_3CCH_2CHCH_2CH_3$

285

▶ 2725 VSSh
B (0.169)

$(CH_3)_3CCH_2CHCH(CH_3)_2$
CH_3

303

▶ 2725 SSh
B (0.169)

CH_3
$(CH_3)_3CC-CH(CH_3)_2$
CH_3

306

▶ 2725 VSSh
B (0.40)

CH_3CH_2-S-CH_2CH_3

172

▶ 2725 S
B (0.169)

$(CH_3)_2CHCH_2CH(CH_3)_2$

141, see also 150, 151

▶ 2725 SSh
B (0.151)

H_3C CH_3
CH_2CH_3

215

▶ 2725 SSh
B (0.169)

$(CH_3)_2CHCH_2CHCH_2CH_3$
CH_3

138, see also 334

▶ 2725 SSh
B (0.169)

CH_3
$(CH_3)_3CCHC(CH_3)_3$

307

▶ 2725 MSh
B (0.157)

CH_3
$(CH_3)_3CCH_2CHCH_2CH_3$

79

▶ 2725 SSh
B (0.169)

CH_3
$CH_3CH_2C-CH_2CH_2CH_3$
CH_3

360

▶ 2725 MSSh
B (0.169)

$CH_3CH=CHCH_2CH_3$
(cis)

208

▶ 2725 VW
B (0.2398)

CF_3
F

218

▶ 2725 SSh
B (0.169)

$(CH_3)_2CHCH_2CH_2CH(CH_3)_2$

137, see also 339

▶ 2720 MSSh
B (0.011)

$(CH_3)_3CCH(CH_3)_2$

147, see also 146

▶ 2720 S
B (0.157)

$(CH_3)_3CCHCH(CH_3)_2$
CH_3

293

▶ 2718 S
A (775 mm Hg, 10 cm)

F_2 F_2
F_2 F_2

243

▶ 2715 SSh
B (0.169)

$(CH_3)_3CCH_2CH_2C(CH_3)_3$

305

▶ 2715 MSSh
B (0.169)

$(CH_3)_3CCH_2CH_3$

145

▶ 2713 W
B (0.955)

F

226

▶ 2711 S
A (775 mm Hg, 10 cm)

CHF_2CH_3

233

▶2710 SSh
B
$(CH_3)_3C(CH_2)_3CH_3$
337

▶2703 MSh
B (0°C)
$(CH_3)_2CHCH(CH_3)CH_2CH_2CH_3$
284, see also 135

▶2702 W
$C_6H_5SiCl_3$
86

▶2700 SSh
B (0.20) 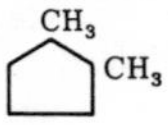
CH_3 CH_3
(trans)
170

▶2700 M
C
SrH_2GeO_4
690

▶2699 VW
B (0.097)
$(CH_3)_2CH(CH_2)_3CH_3$
80, see also 158

▶2698 MSh
B (0.20) 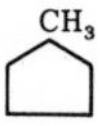
CH_3
171

▶2695 MW
C (0.003)
$(CH_3)_3CNO_2$
325

▶2695 S
B (0.15)
$(CH_3)_2CHNO_2$
366

▶2695 MWSh
B (0.0178) 
103

▶2692 VS
A (775 mm Hg, 10 cm)
CHF_2CH_3
233

▶2690 W
B (0.2398) 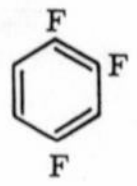
F F F
224

▶2688 W
$(C_6H_5)_2SiCl_2$
86

▶2687 M
B (0.2398) 
F F
225

▶2686 VW
B (0.2398) 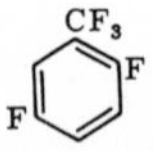
CF_3 F F
217

▶2685 SSp
B (0.20) 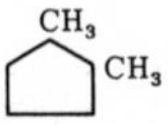
CH_3 CH_3
(cis)
170

▶2680 VSSh
B (0.243)
$H_2C{=}CH(CH_2)_6CH_3$
271

▶2680 VSSh
B (0.169)
$(CH_3)_2CHCH(CH_3)CH_2CH_2CH_3$
139, see also 338

▶2680 VW
B (0.028)
CaF_2
$(C_2H_5)_2CH(CH_2)_{20}CH_3$
282

▶2675 VSSh
B (0.169)
$(CH_3)_2CHCH(CH_3)_2$
144, see also 210

▶2675 S
B (0.157)
$(CH_3)_3CCHCH(CH_3)_2$ with CH_3 on the middle CH
293

▶2673 MSh
B
$(CH_3)_2CHCHCH_2CH_2CH_3$ with CH_3 on the second CH
338, see also 139

▶2672 MW
B (0.101)
$CH_3(CH_2)_5CH_3$
85

▶2670
C_6H_6 [0.0128 m/l]
2-chlorobenzoic acid (Cl, C(=O)-OH)
448

▶2667 M
B (0.10)
tetralin with CH_3
269

▶2665
C_6H_6 [0.0172 m/l]
benzoic acid (C(=O)-OH)
448

▶2665 MSh
B (0.169)
$(CH_3)_3CCH_2CH_3$
145

▶2665 MSSh
B (0.169)
$CH_3CH_2C(CH_3)_2-CH_2CH(CH_3)CH_2CH_3$
302

▶2660 SBSh
B (0.169)
$CH_3CH_2C(CH_3)_2-CH(CH_3)CH_2CH_3$
299

▶2658 M
B (0.2398)
trifluorobenzene (F, F, F)
224

▶2658
CCl_4 [0.0082 m/l]
benzoic acid (C(=O)-OH)
448

▶2655 MS
B (0.169)
$(CH_3)_2CHCH_2CH(CH_3)_2$
141, see also 150, 151

▶2655 MSh
B (0.169)
$(CH_3)_3CCH_2CH_2C(CH_3)_3$
305

▶2655
CCl_4 [0.0128 m/l]
2-chlorobenzoic acid (Cl, C(=O)-OH)
448

▶2652 MS
B (0.105)
$(CH_3)_2CHCH_2CH(CH_3)_2$
151, see also 141, 150

▶2650 WSh
B (0.157)
$(CH_3)_3CCH(CH_2CH_3)_2$
78

▶2649 VS
B (0.2398)
tetrafluorobenzene (F, F, F, F)
222

▶2649 M
B (0.2398)
fluorobenzene (F)
226

▶2647 MS
B (0.20)
ethylcyclopentane (CH_2CH_3)
171

▶2646 M
B (0.2398)
p-difluorobenzene (F, F)
225

▶ 2645 MSSh
B (0.169)

$(CH_3)_3CC(CH_3)_2-CH(CH_3)_2$

306

▶ 2645 MW
B (0.2398)

CF_3, F, F (benzene ring)

217

▶ 2645 SSh
B (0.169)

$(CH_3)_2CHCH_2CH(CH_3)CH_2CH_3$

138

▶ 2645 MSSh
B (0.157)

$(CH_3)_3C(CH_2)_2CH(CH_3)_2$

292, see also 365

▶ 2644 S
B (0.2398)

CF_3 (benzene ring)

219

▶ 2640 SSh
B (0.169)

$(CH_3)_3CC(CH_3)_2-CH_2CH_2CH_3$

304

▶ 2640 SSh
B (0.169)

$(CH_3)_2CHCH_2CH(CH_3)CH_2CH_3$

138

▶ 2637

$(C_5H_4COOH)Ru(C_5H_4COOH)$

450

▶ 2636 MSSh
A (200 mm Hg, 10 cm)

$CF_3CF_2CF_2CF_2CF_3$

237

▶ 2635
C_6H_6 [0.0044 m/l]

OH, C(=O)-OH (benzene ring)

448

▶ 2631 S
A (200 mm Hg, 10 cm)

$CF_3CF_2CF_3$

235

▶ 2630 M
B (0.006)

$(CH_3)_2CH(CH_2)_3CH_3$

80, see also 158

▶ 2630
C_6H_6 [0.0172 m/l]

C(=O)-OH (benzene ring)

448

▶ 2628 VSSp
B (0.20)

CH_3, CH_3 (cyclopentane ring)

(trans)

170

▶ 2627 MB
B (0.20)

CH_3 (cyclopentane ring)

171

▶ 2625 VSSh
B (0.169)

$(CH_3)_2CHCH(CH_2CH_3)CH(CH_3)_2$

301, see also 472

▶ 2625 S
A (775 mm Hg, 10 cm)

F_2, F_2, F_2, F_2 (cyclobutane ring)

243

▶ 2625 S

$(C_5H_4COOH)Fe(C_5H_4COOH)$

450

▶ 2624 S

$(C_5H_5)Fe(C_5H_4COOH)$

450

▶ 2622 SB
B (0.20)

CH_3, CH_3 (cyclopentane ring)

(cis)

170

▶2621 M
B(0.2398)

CF$_3$ and F substituted benzene ring (structure)

218

▶2621 MB
B(0.20)

CH_2CH_3 on cyclopentane ring (structure)

171

▶2620 SBSh
B(0.169)

$CH_3CH_2C(CH_3)_2-CH(CH_3)CH_2CH_3$

299

▶2620 VS
B(0.169)

$(CH_3)_2CHC(CH_3)_2-CH_2CH_2CH_3$

134

▶2620 VSSh
B(0.169)

$(CH_3)_2CHCH(CH_2CH_3)_2$

136

▶2620 S
B(0.169)

$CH_3CH_2CH(CH_3)-CH(CH_3)CH_2CH_3$

83

▶2620 MS
B(0.169)

$(CH_3)_2CHCH_2CH(CH_3)_2$

141, see also 150, 151

▶2620 M
B(0.2398)

H_3C-O-CF_2CHFCl

356

▶2620 S
B(0.728)

$CH_3CH=CHCH_2CH_3$
(trans)

207

▶2620 S
B(0.157)

$(CH_3)_3CCH(CH_3)CH(CH_3)_2$

293

▶2619 MS
B(0.011)

$(CH_3)_3CCH(CH_3)_2$

147

▶2615 M
B(0.2398)

Benzene ring with three F substituents (structure)

224

▶2615 MS
B(0.2398)

F on benzene ring (structure)

226

▶2614 MW
B(0.2398)

CH$_3$ and F substituted benzene ring (structure)

220

▶2612 SSp
B(0.2398)

Benzene ring with CF$_3$ and two F substituents (structure)

217

▶2612 M
B(0.2398)

CF$_3$ on benzene ring (structure)

219

▶2611 S

$(C_5H_5)Ru(C_5H_4COOH)$

450

▶2611 M
B(0°C)

$(CH_3)_2CHCH(CH_3)CH(CH_3)_2$

284, see also 135

▶2610 MSh
B(0.169)

$(CH_3)_3CCH_2CH(CH_3)CH(CH_3)_2$

303

▶2610 VSSh
B(0.169)

$(CH_3)_2CHCH(CH_3)CH_2CH_3$

142, see also 152, 153

▶2610 SSh
B (0.2398)

$F_9C_4\text{-}O\text{-}C_4F_9$

266

▶2610 SSh
B (0.169)

$(CH_3)_3CCH_2CHCH(CH_3)_2$
CH_3

303

▶2610 MW
B (0.2398)

$CH_3CH_2\text{-}O\text{-}CF_2CHFCl$

261

▶2610 MSh
B (0.169)

CH_3
$(CH_3)_3CC\text{ - }CH(CH_3)_2$
CH_3

306

▶2609 SSh
B (0.101)

CH_3
$(CH_3)_2CHCHCH_2CH_3$

153, see also 142, 152

▶2608 S
A (200 mm Hg, 10 cm)

$CF_3CF_2CF_3$

235

▶2606 MSh
B (0.20)

H_3C (cyclopentane ring) CH_3
(trans)

308

▶2605 MSSh
B (0.169)

$(CH_3)_3CCH(CH_2CH_3)_2$

300, see also 78

▶2605 SSh
B (0.169)

$(CH_3CH_2)_3CH$

143, see also 82, 156

▶2605 VSSh
B (0.2398)

$CF_3(CF_2)_5CF_3$

241

▶2605 MSSh
B (0.169)

CH_3
$(CH_3)_3CCHC(CH_3)_3$

307

▶2605 VSSh
B (0.169)

CH_3
$(CH_3)_2CHCHCH_2CH_2CH_3$

139

▶2604 MSh
B

CH_3
$(CH_3)_2CHC\text{ - }CH_2CH_3$
CH_3

341

▶2598.6 S
B (0.2)

(benzene ring)$\text{-}CH_2\text{-}CH$ = CH

9

▶2597 MSh
B

CH_3
$(CH_3)_2CHCHCH_2CH_2CH_3$

338, see also 139

▶2596 M
A (200 mm Hg, 10 cm)

F_2 F_2
F_2 F_2
F_2

245

▶2595 VSSh
B (0.169)

$(CH_3)_2CHCH(CH_3)_2$

144, see also 210

▶2595 SSh
B (0.169)

$(CH_3CH_2)_2CHCH_2CH_2CH_3$

140

▶2594 S
A (200 mm Hg, 10 cm)

$CF_3CF_2CF_2CF_2CF_3$

237

▶2593 MSh
B (0.2398)

CF_3
(benzene ring)

219

▶2593 M
B (0.20)
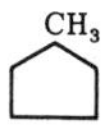
CH3
171

▶2592 S
B (0.2398)
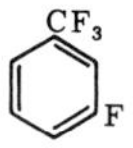
CF3
F
218

▶2590
CCl_4 [0.0082 m/l]
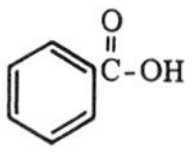
O
C-OH
448

▶2590
CCl_4 [0.00658 m/l]
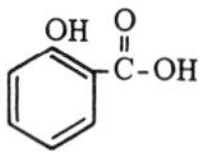
OH
O
C-OH
448

▶2585 MWSh
B (0.169)
$(CH_3)_3CCH_2CH_3$
145

▶2585 VSB
B (0.2398)
$(C_4F_9)_3N$
346

▶2584 WVB
B (0.003)
NO_2
$CH_3CH_2CHCH_3$
324

▶2581 S
B (0.2398)
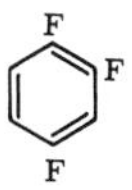
F
F
F
224

▶2579 MS
B (0.2398)
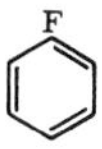
F
226

▶2577 S
A (200 mm Hg, 10 cm)
$CF_3CF_2CF_2CF_2CF_3$
237

▶2571 S
A (200 mm Hg, 10 cm)
$CF_3CF_2CF_3$
235

▶2570
C_6H_6 [0.0044 m/l]
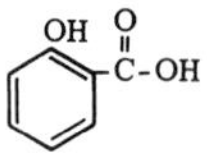
OH
O
C-OH
448

▶2565 MB
B (0.728)
$(CH_3)_3CCH_2CH_2C(CH_3)_3$
305

▶2558 M
B (0.15)
$(CH_3)_2CHCH_2NO_2$
326

▶2557 S
A (200 mm Hg, 10 cm)
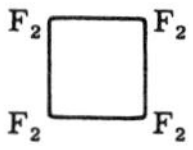
F2
F2
F2
F2
243

▶2555
CCl_4 [0.0128 m/l]
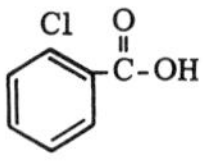
Cl
O
C-OH
448

▶2554 SB
B (0.2398)
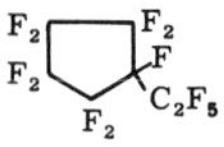
F2
F2
F2
F2
F
C2F5
247

▶2551 VS
B (0.2398)
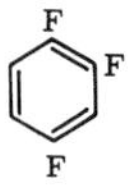
F
F
F
224

▶2550 MSh
B (0.728)
CH_3
$(CH_3)_3CC - CH_2CH_2CH_3$
CH_3
304

▶2550 MWSh
B (0.169)
$(CH_3)_3CCH_2CH_3$
145

▶2548 M
B (0.2398)

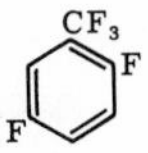

217

▶2546 VW
B (0.2398)

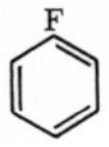

226

▶2544 S

$(C_5H_4COOH)Fe(C_5H_4COOH)$

450

▶2543 MS
B (0.2398)

218

▶2540
CCl_4 [0.0082 m/l], C_6H_6 [0.0172 m/l]

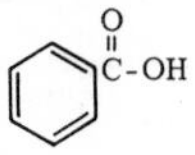

448

▶2539 W
B (0.055)

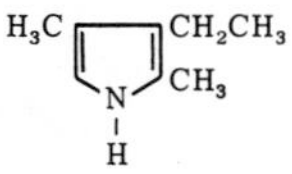

343

▶2538 S

$(C_5H_5)Ru(C_5H_4COOH)$

450

▶2537 VS
B (0.2398)

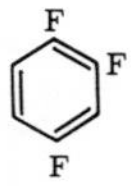

224

▶2534 VS
B (0.2398)

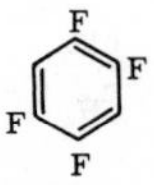

222

▶2532 M
B (0.15)

$(CH_3)_2CHNO_2$

366

▶2530 MSh
B (0.2398)

$H_3C\text{-}O\text{-}CF_2CHFCl$

356

▶2530
C_6H_6 [0.0128 m/l]

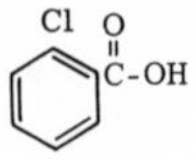

448

▶2530
CCl_4 [0.00658 m/l]

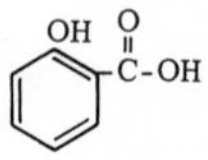

448

▶2529 VS
B (0.210)

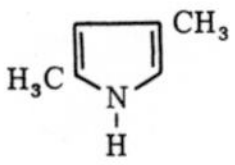

344

▶2529 MSh
B (0.2398)

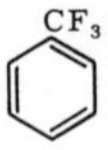

219

▶2527 MW
B (0.2398)

225

▶2525 MS
B (0.2398)

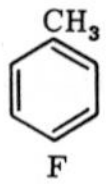

220

▶2522 MS
A (200 mm Hg, 10 cm)

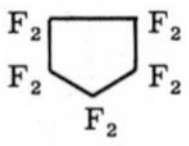

245

▶2521 M
B (0.2398)

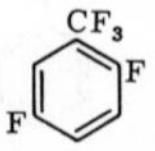

217

▶2520 MWSh
B (0.003)

$(CH_3)_3CNO_2$

325

▶2518 VS A (200 mm Hg, 10 cm) $CF_3CF_2CF_3$ 235	▶2501 MS B (0.2398) CF_3-benzene 219
▶2515 MB B (0.169) $(CH_3)_2CHCH(CH_3)CH_2CH_3$ 142, see also 152, 153	▶2500 VSB B (0.2398) $(C_4F_9)_3N$ 346
▶2513 VW B (0.2398) F-benzene 226	▶2500 MW B (0.169) $(CH_3CH_2)_3CH$ 143, see also 82, 156
▶2511 W B (0.2398) CH_3, F-benzene 221	▶2500 MWB B (0.169) $(CH_3)_3CCH(CH_2CH_3)_2$ 300, see also 78
▶2510 M B (0.2398) CH_3, F-benzene 220	▶2500 MWB B (0.728) $(CH_3)_2CHCH(CH_2CH_3)_2$ 136
▶2508 M A (200 mm Hg, 10 cm) F_2, F_2, F_2, F_2 cyclobutane 243	▶2500 M B (0.15) $(CH_3)_2CHNO_2$ 366
▶2506 MW B (0.2398) F, F, F-benzene 224	▶2500 MW B (0.728) $(CH_3)_3CCH(CH_3)C(CH_3)_3$ 307
▶2505 MSh B (0.169) $(CH_3)_2CHCH_2CH_2CH(CH_3)_2$ 137, see also 339	▶2497 MW B (0.2398) F, F, F-benzene 224
▶2505 MB B (0.728) $(CH_3)_3CCH_2CH_2C(CH_3)_3$ 305	▶2495 MB B (0.2398) n-H_7C_3-O-CF_2CHFCl 263
▶2503 VW B (0.2398) F-benzene 226	▶2494 S A (775 mm Hg, 10 cm) CHF_2CH_3 233

▶2492 W B (0.055) H_3C, CH_2CH_3, CH_3, N, H 343	▶2475 M B (0.003) CH_3NO_2 311
▶2490 VS B (0.2398) $CF_3(CF_2)_5CF_3$ 241	▶2470 M B (0.728) $(CH_3)_3CCH_2CH_2C(CH_3)_3$ 305
▶2490 S B (0.2398) F, CF_3, CF_3, F, F_2, F_2, F_2, F_2 231	▶2470 W B (0.055) H_3C, CH_2CH_3, CH_3, N, H 343
▶2490 MS B (0.2398) F, F, F, F 222	▶2470 MW B (0.169) $(CH_3CH_2)_3CH$ 143, see also 82, 156
▶2485 SB B (0.2398) F_2, F_2, F_2, F, C_2F_5, F_2 247	▶2465 MW B (0.728) $(CH_3)_3CCHC(CH_3)_3$ with CH_3 307
▶2484 W B (0.2398) F 226	▶2465 MS B (0.2398) $H_3C\text{-}O\text{-}CF_2CHFCl$ 356
▶2481 MW A (200 mm Hg, 10 cm) F_2, F_2, F_2, F_2, F_2 245	▶2458 VS A (200 mm Hg, 10 cm) $CF_2Cl\text{-}CF_2Cl$ 257
▶2476 MB B (0.20) CH_2CH_3 171	▶2455 MS A (100 mm Hg, 69 cm) $CH_3\text{-}CCl_3$ 251
▶2475 VSVB B (0.2398) $F_9C_4\text{-}O\text{-}C_4F_9$ 266	▶2452 VSB B (0.2398) F_2, F_2, F_2, F, C_2F_5, F_2 247
▶2475 VS A (200 mm Hg, 10 cm) $CF_3CF_2CF_2CF_2CF_3$ 237	▶2451 M B (0.15) $(CH_3)_2CHNO_2$ 366

▶2450 VSB
B (0.2398)
$(C_4F_9)_3N$
346

▶2450 S
A (385 mm Hg, 10 cm)
CF_3CCl_3
253

▶2434 M
A (200 mm Hg, 10 cm)
F_2 F_2 F_2 F_2 F_2
245

▶2447 VS
B (0.2398) 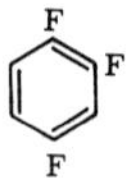
F F F
224

▶2425 MB
B (0.2398)
$n\text{-}H_9C_4\text{-}O\text{-}CF_2CHFCl$
264

▶2445 SSp
B (0.15)
$(CH_3)_2CHCH_2NO_2$
326

▶2445 M
B (0.003)
$C_3H_7NO_2$
392

▶2424 VS
B (0.2398)
F F
225

▶2445 S
B (0.2398)
$CH_3CH_2\text{-}O\text{-}CF_2CHFCl$
261

▶2445 VS
B (0.2398)
$CF_3(CF_2)_5CF_3$
241

▶2422 M
B (0.2398)
CF_3 F
218

▶2443 S
B (0.2398) 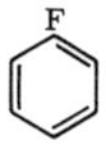
F
226

▶2421 M
A (200 mm Hg, 10 cm)
F_2 F_2 F_2 F_2 F_2
245

▶2442 VS
B (0.2398) 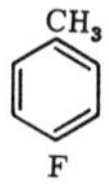
CH_3 F
220

▶2421 MW
B (0.003)
$CH_3CH_2CH(NO_2)CH_3$
324

▶2442 VSSh
A (200 mm Hg, 10 cm)
$CF_3CF_2CF_2CF_2CF_3$
237

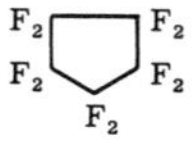

▶2416 S
A (200 mm Hg, 10 cm)
$CF_3CF_2CF_3$
235

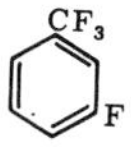

▶2438 W
B (0.20) 
H_3C CH_3
(trans)
308

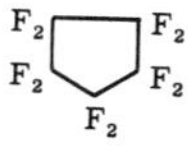

▶2415 VSB
B (0.2398)
CF_3 F F
217

▶2437 MS
A (775 mm Hg, 10 cm) 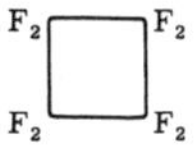
F_2 F_2 F_2 F_2
243

▶2415 M
B (0.40)
$CH_3CH_2\text{-}S\text{-}CH_2CH_3$
172

▶2435 S
B (0.728)
$(CH_3)_2CHCH(CH_3)_2$
144, see also 210

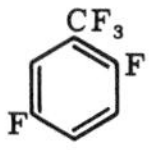

▶2410 VS
B (0.728)
$(CH_3)_2CHCH(CH_2CH_3)_2$
136

▶2410 SB
B (0.728)

$(CH_3)_2CHCH_2CHCH_2CH_3$
CH_3

138, see also 334

▶2407 VW
B (0.2398)

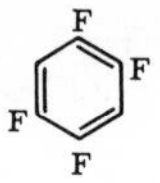
F
F
F
F

222

▶2407 M
A (200 mm Hg, 10 cm)

$CF_2Cl\text{-}CF_2Cl$

257

▶2405 S
B (0.728)

$(CH_3)_3CCH(CH_2CH_3)_2$

300

▶2405 S
B (0.728)

CH_3
$(CH_3)_2CHCHCH_2CH_3$

142, see also 152, 153

▶2404 VS
B (0.2398)

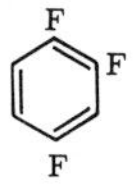
F
F
F

224

▶2400 VS
B (0.2398)

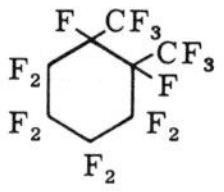
F
CF3
CF3
F
F2
F2
F2
F2

231

▶2400 S
B (0.728)

$CH_3CH{=}CHCH_2CH_3$
(cis and trans)

208 (cis), 207 (trans)

▶2400
E

$LiNO_3$

45

▶2400
E

$NaNO_3$

45

▶2400
E

$Cu(NO_3)_2 \cdot 5H_2O$

45

▶2398 S
B (0.15)

$(CH_3)_2CHNO_2$

366

▶2395 W
B (0.2398)

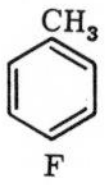
CH3
F

220

▶2395 MW
B (0.2398)

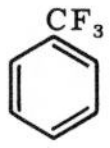
CF3

219

▶2392 S
A (775 mm Hg, 10 cm)

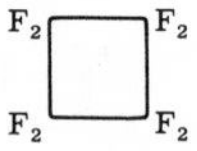
F2
F2
F2
F2

243

▶2390 VS
B (0.2398)

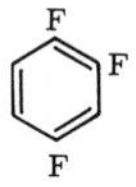
F
F
F

224

▶2390 SB
B (0.728)

$CH_3 \quad CH_3$
$CH_3CH_2C\text{-}CH_2CHCH_2CH_3$
CH_3

302

▶2390 S
B (0.728)

CH_3
$(CH_3)_3CCHC(CH_3)_3$

307

▶2390
E

$Hg(NO_3)_2$

45

▶2390
P

$Zn(NO_3)_2$

45

▶2390 VS
B (0.2398)

F_9C_4-O-C_4F_9

266

▶2389 MW
B (0.2398)

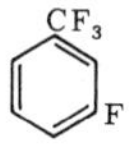

CF_3
F

218

▶2387 MW
C (0.003)

$(CH_3)_3CNO_2$

325

▶2385 S
B (0.728)

CH_3
$(CH_3)_3CC$-$CH_2CH_2CH_3$
CH_3

304

▶2385 S
B (0.728)

$(CH_3)_2CHCH_2CH_2CH(CH_3)_2$

137, see also 339

▶2385 S
B (0.728)

CH_3
$(CH_3)_3CC$ - $CH(CH_3)_2$
CH_3

306

▶2383 MS
A (200 mm Hg, 10 cm)

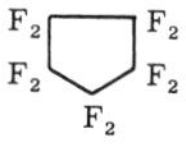

F_2 F_2
F_2 F_2
F_2

245

▶2381 M
B (0.2398)

CF_3

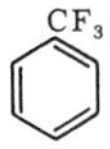

219

▶2380 S
B (0.728)

$(CH_3)_2CHCH(CH_3)_2$

144, see also 210

▶2380 S
B (0.2398)

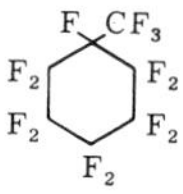

F CF_3
F_2 F_2
F_2 F_2
F_2

232

▶2380 S
B

$[(C_2H_5)_2P \cdot BH_2]_3$

30

▶2380
E

$Ba(NO_3)_2$

45

▶2380 S
B (0.2398)

CF_2CL-$CFCl_2$

255

▶2380 VS
B (0.2398)

$CF_3(CF_2)_5CF_3$

241

▶2375 MW
B (0.2398)

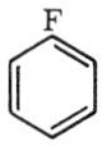

F

226

▶2371 VSSp
A (200 mm Hg, 10 cm)

$CF_3CF_2CF_2CF_2CF_3$

237

▶2370 S
P

$[(CH_3)_2P \cdot BH_2]_3$

30

▶2370
E

KNO_3

45

▶2370
E

$CsNO_3$

45

▶2370
P

$Cu(NO_3)_2 \cdot 5H_2O$

45

▶2369 MW
B (0.2398)

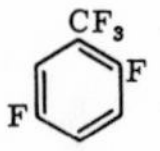

217

▶2367 M
B (0.2398)

218

▶2366 MS
A (200 mm Hg, 10 cm)

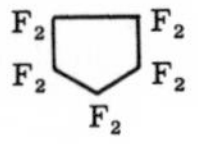

245

▶2365 S
A (385 mm Hg, 10 cm)

CF_3-CF_3

253

▶2363 VS
A (775 mm Hg, 10 cm)

CF_2Cl-CF_2Cl

257

▶2360
E

$RbNO_3$

45

▶2360
P

$Hg_2(NO_3)_2$

45

▶2360
P

$Al(NO_3)_3$

45

▶2358 S
A (200 mm Hg, 10 cm)

$CF_3CF_2CF_3$

235

▶2356 VS
B (0.2398)

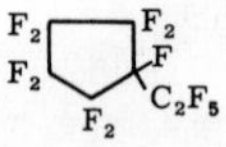

247

▶2356 VS
B (0.2398)

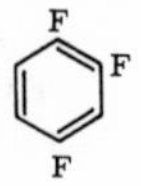

224

▶2354 MW
B (0.2398)

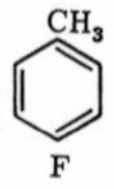

220

▶2353 VS
B (0.2398)

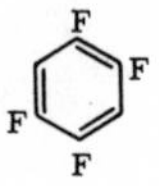

222

▶2351 MS
B (0.2398)

218

▶2351 VW
B (0.2398)

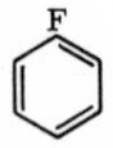

226

▶2351 W
B (0.2398)

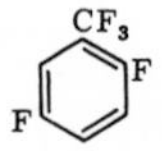

217

▶2350
E

$Be(NO_3)_2$

45

▶2350
E

$Mg(NO_3)_2$

45

▶2350
P

$RbNO_3$

45

▶2350
P

$AgNO_3$

45

▶2349 S B (0.2398) [structure: benzene ring with CF_3] 219	▶2332 S P $[(CH_3)_2P \cdot BH_2]_3$ [BH] 30
▶2347 VS A (775 mm Hg, 10 cm) [structure: cyclobutane ring with F_2 at each corner] 243	▶2331 S A (200 mm Hg, 10 cm) $CF_3CF_2CF_2CF_2CF_3$ 237
▶2345 VS B (0.2398) [structure: benzene ring with F, F, F] 224	▶2330 S B (0.2398) [structure: benzene ring with CF_3 and F] 218
▶2345 S B $[(C_2H_5)_2P \cdot BH_2]_3$ [BH] 30	▶2330 P $Cr(NO_3)_3$ 45
▶2345 C_6H_6 [0.0172 m/l] [structure: benzene ring with C(=O)-OH] 448	▶2328 MW B (0.2398) [structure: benzene ring with CF_3] 219
▶2340 VSB B (0.2398) $(C_4F_9)_3N$ 346	▶2326 W B (0°C) $(CH_3)_2CHCHCH(CH_3)_2$ with CH_3 on central CH 284, see also 135
▶2340 P $La(NO_3)_3 \cdot 3NH_4NO_3$ 45	▶2325 VS B (0.728) $(CH_3)_2CHCHCH(CH_3)_2$ with CH_3 on central CH 135, see also 284
▶2340 VS B (0.2398) [structure: benzene ring with F, F, F, F] 222	▶2325 M B (0.300) CaF_2 $(C_2H_5)_2CH(CH_2)_{20}CH_3$ 282
▶2335 VSSp B (0.169) $CH_2{=}CHCH{=}CHCH{=}CH_2$ 203	▶2320 S B (0.2398) $CF_2Cl\text{-}CFCl_2$ 255
▶2335 MSh B (0.728) $(CH_3)_2CHC(CH_3)_2\text{-}CH_2CH_2CH_3$ (two CH_3 on C) 134	▶2320 VS B (0.2398) [structure: cyclohexane ring with F, CF_3; F_2; F_2; F_2; F_2; F, CF_3] 230

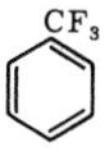

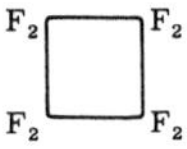

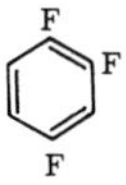

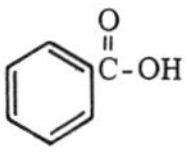

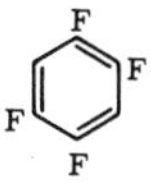

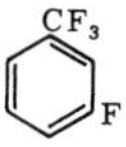

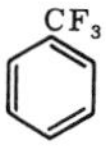

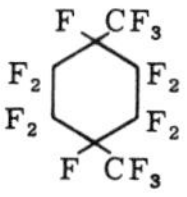

▶2320
E

$Ce(NO_3)_3$

45

▶2320
P

$Ce(NH_4)_2(NO_3)_6$

45

▶2320
P

$Bi(NO_3)_3$

45

▶2320
P

$Ni(NO_3)_2$

45

▶2319 VS
A (775 mm Hg, 10 cm)

$CF_2Cl\text{-}CF_2Cl$

257

▶2316 S
B (0.2398)

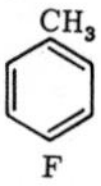

CH_3
F

220

▶2315 VSB
B (0.2398)

F_2 F_2
F_2 F
C_2F_5
F_2

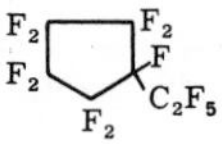

247

▶2314 MS
A (200 mm Hg, 10 cm)

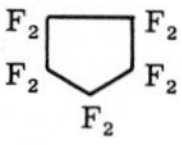

F_2 F_2
F_2 F_2
F_2

245

▶2313 S
B (0.2398)

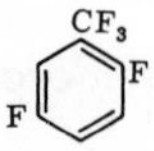

CF_3
F
F

217

▶2313 VSSp
B (0.2398)

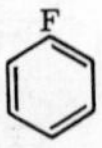

F

226

▶2310 S
B (0.2398)

F CF_3
F_2 F_2
F_2 F_2
F_2

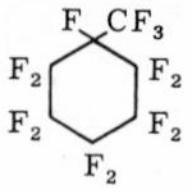

232

▶2310 S
B (0.728)

$CH_3CH{=}CHCH_2CH_3$
(cis)

208

▶2304.8 S
B (0.2)

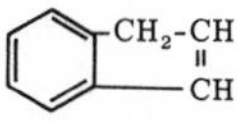

$CH_2\text{-}CH$
$\|$
CH

9

▶2302 S
B (0.2398)

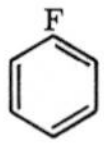

F

226

▶2300
C_6H_6 [0.017 m/l]

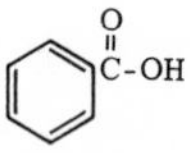

O
$\|$
C-OH

448

▶2300 VSSp
B (0.2398)

F
F
F

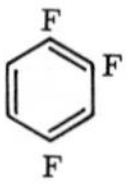

224

▶2300 MW
A (200 mm Hg, 10 cm)

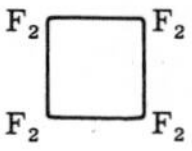

F_2 F_2
F_2 F_2

243

▶2299 MS
B (0.15)

$(CH_3)_2CHCH_2NO_2$

326

▶2296 VSSp
B (0.2398)

F
F

225

▶2295 M
B (0.728)

CH_3
$(CH_3)_3CCH_2CHCH_2CH_3$

285

▶ 2295 S
B (0.728)

$(CH_3)_3CCH(CH_2CH_3)_2$

300, see also 78

▶ 2295 S
B (0.728)

$(CH_3)_3CCH_2CHCH(CH_3)_2$
CH_3

303

▶ 2295 S
B (0.728)

$(CH_3)_3CCH_2CH_2C(CH_3)_3$

305

▶ 2295 S
B (0.728)

$(CH_3)_3CCH_2CH_3$

145

▶ 2295 S
B (0.728)

CH_3
$(CH_3)_3CCHC(CH_3)_3$

307

▶ 2295 S
B (0.728)

CH_3
$(CH_3)_3CC - CH(CH_3)_2$
CH_3

306

▶ 2295 MS
B (0.728)

CH_3
$CH_3CH_2C - CHCH_2CH_3$
CH_3 CH_3

299

▶ 2295 MB
B (0.728)

$(CH_3)_2CHCHCH(CH_3)_2$
CH_2
CH_3

301, see also 472

▶ 2295 M
B (0.728)

CH_3 CH_3
$CH_3CH_2C\text{-}CH_2CHCH_2CH_3$
CH_3

302

▶ 2295 M
B (0.728)

CH_3
$(CH_3)_2CHC\text{-}CH_2CH_2CH_3$
CH_3

134

▶ 2295 M
B (0.728)

CH_3
$(CH_3)_3CC\text{-}CH_2CH_2CH_3$
CH_3

304

▶ 2290 VS
B (0.728)

$(CH_3)_2CHCH(CH_2CH_3)_2$

136

▶ 2290 MBSh
B (0.728)

CH_3
$(CH_3)_2CHCHCH_2CH_3$

142, see also 152, 153

▶ 2288 MW
B (0.003)

CH_3NO_2

311

▶ 2285 MS
B (0.2398)

CF_3
F F

217

▶ 2285 MS
B (0.728)

$(CH_3CH_2)_3CH$

143, see also 82, 156

▶ 2285 M
N

N
N N

14

▶ 2280 MS
A (200 mm Hg, 40 cm)

$CF_3CF_2CF_2CF_2CF_3$

237

▶ 2280 W
B (0.2398)

$CH_3CH_2\text{-}O\text{-}CF_2CHFCl$

261

▶ 2280
P

$Th(NO_3)_4$

45

▶2279 VS
A (775 mm Hg, 10 cm)

CHF_2CH_3

233

▶2279 VS
B (0.2398)

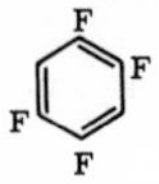

222

▶2278 MW
B (0.10)

199

▶2275
LiF

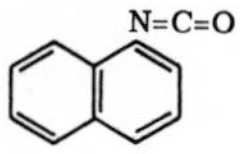

507

▶2274 VS
LiF

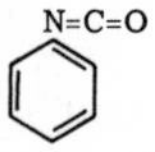

507

▶2273 W
A (55.4 mm Hg, 43 cm)

$(CH_3)_3CH$

373

▶2272 VS
LiF

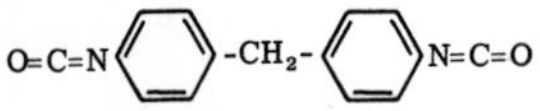

507

▶2271 S
B (0.2398)

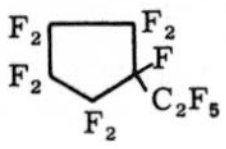

247

▶2270
LiF

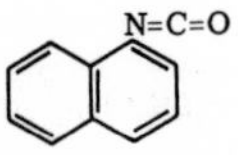

507

▶2270 M
E

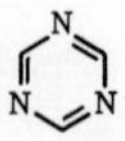

14

▶2270
LiF

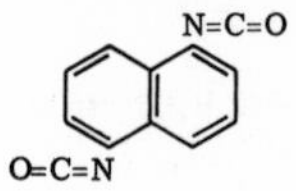

507

▶2268 M
B (0.15)

$(CH_3)_2CHNO_2$

366

▶2267 VS
LiF

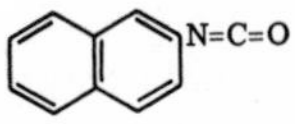

507

▶2265
LiF

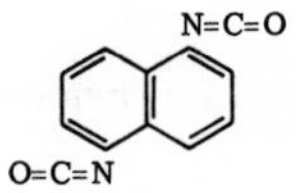

507

▶2265 VS
LiF

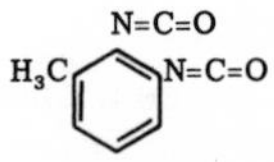

507

▶2265 VS
LiF

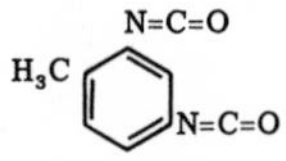

507

▶2265 VS
LiF

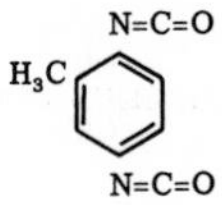

507

▶2264 VS
A (775 mm Hg, 10 cm)

CHF_2CH_3

233

▶2263
LiF

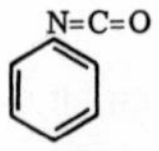

507

▶2263 MS
B (0.2398)

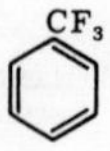

219

▶2263 S
A (775 mm Hg, 10 cm)

$CF_2Cl\text{-}CF_2Cl$

257

▶2263 M
A (200 mm Hg, 10 cm)

$CF_3CF_2CF_3$

235

▶2261 W
B (0.2398)

F

226

▶2261 MW
A (200 mm Hg, 10 cm)

$CF_3CF_2CF_2CF_2CF_3$

237

▶2257 W
B

$C_6H_5(CH_3)SiCl_2$

86

▶2257 VW
B (0.2398)

F
F

225

▶2256 MS
B (0.2398)

CH_3
F

220

▶2255 MS
B (0.728)

CH_3
$(CH_3)_3CC - CH(CH_3)_2$
CH_3

306

▶2255 M
B (0.2398)

CF_3
F

218

▶2255 W
B (0.2398)

F CF_3
F_2 F_2
F_2 F_2
F CF_3

230

▶2254 MS
B (0.2398)

F
F
F
F

222

▶2252 MB
E

$Na_2H_2P_2O_6$

[OH stretch]

28

▶2250 S
B (0.728)

CH_3
$(CH_3)_3CCHC(CH_3)_3$

307

▶2250 MB
B (0.728)

$(CH_3)_2CHCHCH(CH_3)_2$
CH_2
CH_3

301, see also 472

▶2248 MS
B (0.2398)

CF_3

219

▶2247 M
A (775 mm Hg, 10 cm)

F_2 F_2
F_2 F_2

243

▶2246 MW
B (0.2398)

F
F
F

224

▶2240 W
B (0.2398)

F CF_3
F_2 F_2
F_2 F_2
F CF_3

230

▶2240 M
B (0.728)

$(CH_3)_2CHCH_2CHCH_2CH_3$
CH_3

138

▶2240 MSSh
B (0.728)

$(CH_3)_2CHCH(CH_2CH_3)_2$

136

▶2240 M B (0.728) $(CH_3)_2CHCH(CH_3)_2$ 144, see also 210	▶2235-2215 S R-C=C-C≡N [C≡N stretch] 680
▶2240 M B (0.728) $(CH_3)_3CCH_2CHCH(CH_3)_2$ with CH_3 below 303	▶2234 S A (775 mm Hg, 10 cm) CF_2Cl-CF_2Cl 257
▶2240 P $Gd(NO_3)_3$ 45	▶2233 M A (775 mm Hg, 10 cm) CHF_2CH_3 233
▶2239 W A (682 mm Hg, 10 cm) F_2 F_2 F_2 F_2 F_2 (ring) 245	▶2232 M B (0.2398) CF_3, F, F (ring) 217
▶2237 M A (200 mm Hg, 10 cm) $CF_3CF_2CF_2CF_2CF_3$ 237	▶2230 VS B (0.025) CH_3, CN (ring) 214
▶2235 S B (0.728) $CH_3CH{=}CHCH_2CH_3$ (cis) 208	▶2230 MW B (0.728) $(CH_3)_3CCH(CH_2CH_3)_2$ 300, see also 78
▶2235 M B (0.2398) F, F, F (ring) 224	▶2230 P $Pr(NO_3)_3$ 45
▶2235 M B (0.728) $(CH_3)_3CCH_2CH_3$ 145	▶2228 VS B (0.2398) F, F (ring) 225
▶2235 MSSh B (0.2398) $H_3C-O-CF_2CHFCl$ 356	▶2227 M B (0.40) $CH_3CH_2-S-CH_2CH_3$ 172
▶2235 MW B (0.728) $CH_3CH_2C(CH_3)_2-CH_2CH(CH_3)CH_2CH_3$ 302	▶2227 M $CHCl_3$ $N{=}C(CN)_2$, $N(CH_3)_2$ (ring) [CN] 562

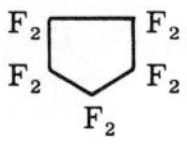

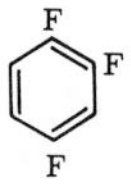

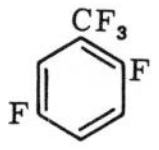

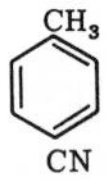

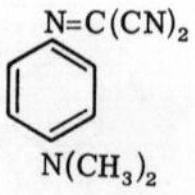

▶2226 VS
B (0.2398)

F, F, F, F (tetrafluorobenzene ring)

222

▶2225 VSSp
B (0.025)

CH_3, CN (benzene ring)

312

▶2224 M
$CHCl_3$

$N{=}C(CN)_2$ (benzene ring) $N{-}C_2H_5$, CH_3

[CN] 562

▶2223 M
$CHCl_3$

$N{=}C(CN)_2$, CH_3 (benzene ring) $N(C_2H_5)_2$

[CN] 562

▶2223 M
$CHCl_3$

$N{=}C(CN)_2$ (benzene ring) $N(C_2H_5)_2$

[CN] 562

▶2222 MS
B (0.2398)

F (benzene ring)

226

▶2221 M
$CHCl_3$

$N{=}C(CN)_2$, CH_3 (benzene ring) $N(CH_3)_2$

[CN] 562

▶2220 M
E

$H_5C_2{-}N(H)$ (benzene ring) $N{=}C(CN){-}C(=O){-}C_2H_5$

[CN] 562

▶2220 M
E

$(H_3C)_2N$ (benzene ring) $N{=}C(CN){-}C(=O){-}CH_3$

[CN] 562

▶2220
E

$Pr(NO_3)_3$

45

▶2218 S
$CHCl_3$

$CH_3N{=}C{-}Fe(CO)_4$

[N=C stretch] 509

▶2217 S

$C_6H_5HSiCl_2$

86

▶2216 MW
A (682 mm Hg, 10 cm)

F_2, F_2, F_2, F_2, F_2 (cyclopentane ring)

245

▶2216 M
CCl_4

$H_3C{-}N(H)$ (benzene ring) $N{=}C(CN){-}C(=O){-}C_2H_5$

[CN] 562

▶2216 M
E

$N{=}C(CN)_2$ (benzene ring) $N(CH_3)_2$

[CN] 562

▶2216 M
E, CCl_4
CCl_4

$H_3C{-}N(H)$ (benzene ring) $N{=}C(CN){-}C(=O){-}CH_3$

[CN] 562

▶2215 MS
B (0.728)

$(CH_3CH_2)_3CH$

143, see also 82, 156

▶2215 VS
B (0.955)

$CF_2Cl{-}CFCl_2$

255

▶2214 M
CCl_4

$H_5C_2{-}N(H)$ (benzene ring) $N{=}C(CN){-}C(=O){-}CH_3$

[CN] 562

▶2214 M
E

$N{=}C(CN)_2$, CH_3 (benzene ring) $N(C_2H_5)_2$

[CN] 562

▶2214 M E $N{=}C(CN)_2$, CH_3, $N(CH_3)_2$ (benzene ring) [CN] 562	▶2210 M B (0.2398) F CF_3, F_2, F_2, F_2, F_2, F CF_3 (cyclohexane ring) 230
▶2213 M CCl_4 $(H_3C)_2N$–ring–$N{=}C(CN)$–$C(=O)$–C_2H_5 [CN] 562	▶2210 MB B (0.2398) $CF_3(CF_2)_5CF_3$ 241
▶2213 M E $N{=}C(CN)_2$, $N(C_2H_5)_2$ (benzene ring) [CN] 562	▶2210 MSVB B (0.728) $(CH_3)_2CHCH_2CHCH_2CH_3$ with CH_3 138, see also 334
▶2212 M E $N{=}C(CN)_2$, N–C_2H_5, CH_3 (benzene ring) [CN] 562	▶2210 M E H_3C–N(H)–ring–$N{=}C(CN)$–$C(=O)$–CH_3 [CN] 562
▶2212 M CCl_4 $(H_3C)_2N$–ring–$N{=}C(CN)$–$C(=O)$–CH_3 [CN] 562	▶2210 M E H_5C_2–N(H)–ring–$N{=}C(CN)$–$C(=O)$–CH_3 [CN] 562
▶2212 Sh $CHCl_3$ $N{=}C(CN)_2$, $N(CH_3)_2$ (benzene ring) [CN] 562	▶2210 Sh $CHCl_3$ $N{=}C(CN)_2$, CH_3, $N(CH_3)_2$ (benzene ring) [CN] 562
▶2211 M A (200 mm Hg, 10 cm) $CF_3CF_2CF_3$ 235	▶2210 Sh $CHCl_3$ $N{=}C(CN)_2$, CH_3, $N(C_2H_5)_2$ (benzene ring) [CN] 562
▶2211 M CCl_4 H_5C_2–N(H)–ring–$N{=}C(CN)$–$C(=O)$–C_2H_5 [CN] 562	▶2209 Sh $CHCl_3$ $N{=}C(CN)_2$, $N(C_2H_5)_2$ (benzene ring) [CN] 562
▶2211 Sh $CHCl_3$ $N{=}C(CN)_2$, N–C_2H_5, CH_3 (benzene ring) [CN] 562	▶2208 S E $K_2S_2O_6$ 28
▶2210 VS B CH_3NCS [NCS stretch] 682	▶2207 M CCl_4 $(H_5C_2)_2N$–ring(CH_3)–$N{=}C(CN)$–$C(=O)$–C_2H_5 [CN] 562

▶ 2205 M
A (200 mm Hg, 10 cm)

$CF_3CF_2CF_2CF_2CF_3$

237

▶ 2204 VS
A (775 mm Hg, 10 cm)

CF_2Cl-CF_2Cl

257

▶ 2203 MW
B (0.2398)

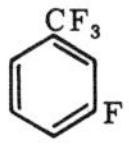

CF_3, F

218

▶ 2203 S
E

$Na_2S_2O_6$

28

▶ 2202 M
E

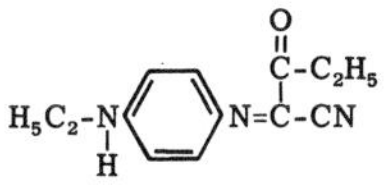

O
C-C_2H_5
H_5C_2-N(H) N=C-CN

[CN] 562

▶ 2202 Sh
E

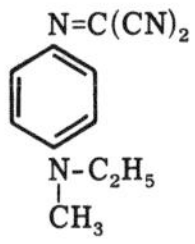

N=C$(CN)_2$
N-C_2H_5
CH_3

[CN] 562

▶ 2200 M
E

O
C-C_2H_5
$(H_5C_2)_2N$ N=C-CN
CH_3

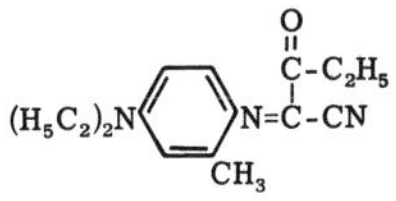

[CN] 562

▶ 2200 MWVB
B (0.2398)

$(C_4F_9)_3N$

346

▶ 2200 M
E

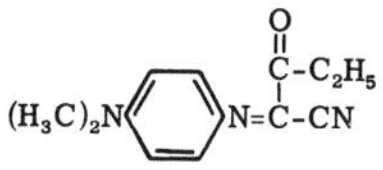

O
C-C_2H_5
$(H_3C)_2N$ N=C-CN

[CN] 562

▶ 2198 M
E

$Na_2S_2O_6 \cdot 2H_2O$

28

▶ 2197
C_6H_6 [0.0172 m/l]

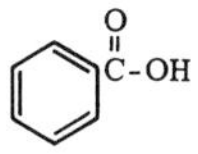

O
C-OH

448

▶ 2196 M
B (0.2398)

F, F, F

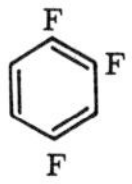

224

▶ 2195 MW
B (0.2398)

F_9C_4-O-C_4F_9

266

▶ 2193 MS
B (0.2398)

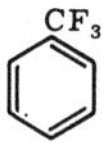

CF_3

219

▶ 2190 MWB
B (0.728)

$(CH_3)_2CHCHCH(CH_3)_2$
CH_3

135, see also 284

▶ 2187 MW
B (0.2398)

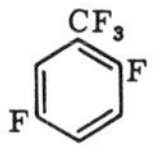

CF_3, F, F

217

▶ 2186 S
$CHCl_3$

$(CH_3)_3CN$=C-Fe$(CO)_4$

[N=C stretch] 509

▶ 2185 M
B (0.728)

CH_3
$(CH_3)_2CHC$ - $CH_2CH_2CH_3$
CH_3

134

▶ 2182 W
B (0.2398)

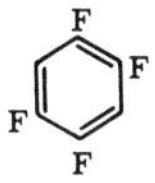

F, F, F, F

222

▶ 2180-2145

RNC

[C≡N stretch] 681

▶2180 VS
A (775 mm Hg, 10 cm) 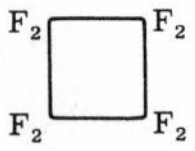
Perfluorocyclobutane (F_2 ring: square with F_2 at each corner)
243

▶2180 MB
B (0.2398)
$CF_3(CF_2)_5CF_3$
241

▶2175 MW
B (0.2398) 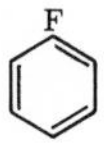
Fluorobenzene (F on benzene ring)
226

▶2174 S
$CHCl_3$ 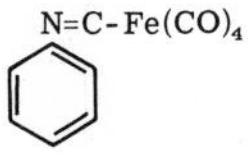
$N{=}C{-}Fe(CO)_4$ on benzene ring
[N=C stretch]
509

▶2173 MW
B (0.2398) 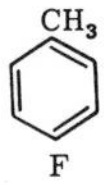
CH_3 and F on benzene ring (para)
220

▶2170 MS
B (0.2398)
$H_3C{-}O{-}CF_2CHFCl$
356

▶2170 M
B (0.728)
$(CH_3)_3CCH_2CHCH(CH_3)_2$ with CH_3 on CH
303

▶2169 W
B
$C_6H_5(CH_3)SiCl_2$
86

▶2165 SSp
B (0.2398) 
CF_3, F, F on benzene ring
217

▶2165 M
B (0.10) 
Thiophene (S ring)
199

▶2165 W
B
$CH_3(CH_2)_5CH(C_3H_7)(CH_2)_5CH_3$
281

▶2164 MW
C (0.003)
$(CH_3)_3CNO_2$
325

▶2160 MS
A (141 mm Hg, 15 cm)
$CH_3{-}C{\equiv}CH$
87

▶2160 MW
B (0.728)
$CH_3CH_2CH(CH_3){-}CH(CH_3)CH_2CH_3$
83

▶2160 SSh
B (0.955)
$CF_2Cl{-}CFCl_2$
255

▶2158 W
B (0.2398) 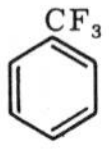
CF_3 on benzene ring
219

▶2157 MB
B (0.2398) 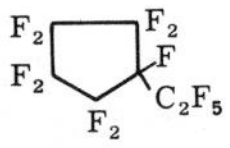
Perfluoroethylcyclopentane (ring with F_2, F_2, F_2, F_2; F and C_2F_5 on one carbon)
247

▶2155 W
B
$C_{26}H_{54}$
(5, 14-di-n-butyloctadecane)
283

▶2155 M
B (0.728)
$(CH_3CH_2)_3CH$
143, see also 82, 156

▶2151.5 VS
P
$K_2Zn(CN)_4$
468

▶2151 MS B (0.2398) p-difluorobenzene (F, F) 225	▶2137 W B (0.2398) benzotrifluoride (CF_3) 219
▶2150 M E, G $Na_4P_2O_6$ 28	▶2135 S $CHCl_3$ $(CH_3)_3GeN{=}C\text{-}Fe(CO)_4$ [N=C stretch] 509
▶2150 MW B (0.2398) F, CF_3, F_2, F_2, F_2, F_2, F_2 (perfluoromethylcyclohexane) 232	▶2135 S $(C_6H_5)_3SiH$ [Si-H stretch] 499
▶2146 VS P $K_2Hg(CN)_4$ 468	▶2132 VW B (0.2398) 1,2,4-trifluorobenzene (F, F, F) 224
▶2145 VS P $K_2Cd(CN)_4$ 468	▶2132 MW B (0.50) $CH_3(CH_2)_2CH_2NO_2$ 310
▶2142 S $CHCl_3$ $(CH_3)_3SnN{=}C\text{-}Fe(CO)_4$ [N=C stretch] 509	▶2129 VW B (0.2398) fluorobenzene (F) 226
▶2141 MSh A (775 mm Hg, 10 cm) F_2, F_2, F_2, F_2 (perfluorocyclobutane) 243	▶2125 VS A $HC{\equiv}C\text{-}C({=}O)\text{-}H$ [C≡C] 18
▶2141 M B (0.50) $C_3H_7NO_2$ 392	▶2122 MW A (775 mm Hg, 10 cm) $CF_2Cl\text{-}CF_2Cl$ 257
▶2140 M B (0.2398) $(CF_3)_2CFCF_2CF_3$ 240	▶2120 SSh B (0.955) $CF_2Cl\text{-}CFCl_2$ 255
▶2140 M B (0.728) $(CH_3)_3CCH_2CHCH(CH_3)_2$ with CH_3 branch 303	▶2119 W B (0.15) $(CH_3)_2CHCH_2NO_2$ 326

▶2119 VSSp
A (600 mm Hg, 10 cm)
$CF_3CF_2CF_2CF_2CF_3$
237

▶2115 S
B (0.2398)
$F_9C_4\text{-}O\text{-}C_4F_9$
266

▶2115 M
B (0.728)
$(CH_3)_3CCH_2CHCH(CH_3)_2$
CH_3
303

▶2114 M
B (0.10) 
S
199

▶2114 M
E
$Na_4P_2O_6$
28

▶2110 MW
B (0.728)
$(CH_3)_3CCH(CH_2CH_3)_2$
300

▶2110 W
B (0.728)
CH_3
$(CH_3)_3CCH_2CHCH_2CH_3$
285

▶2109 MW
B (0.2398) 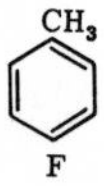
CH_3
F
220

▶2105 W
B (0.2398) 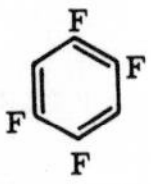
F F F F
222

▶2105 S
B (0.728)
$(CH_3)_3CCH_2CH_2C(CH_3)_3$
305

▶2105 M
G
$Na_4P_2O_6$
28

▶2103 S
B (0.2398) 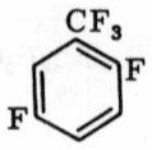
CF_3 F F
217

▶2103 S
B (0.2398) 
CF_3 F
218

▶2102 MS
B (0.2398) 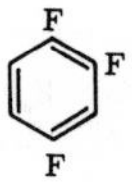
F F F
224

▶2095 MW
B (0.2398)
$CF_3(CF_2)_5CF_3$
241

▶2093 VS
B (0.2398) 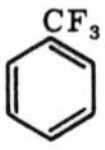
CF_3
219

▶2092 VSSp
A (600 mm Hg, 10 cm)
$CF_3CF_2CF_2CF_2CF_3$
237

▶2090 M
B (0.2) 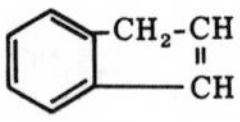
$CH_2\text{-}CH$
CH
9

▶2088 M
E
$Na_2S_2O_6$
28

▶2086 S
B (0.2398) 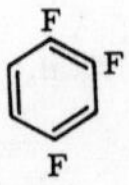
F F F
224

▶2083 M
E

$K_2S_2O_6$

26

▶2083 S
E

$Na_2S_2O_6 \cdot 2H_2O$

28

▶2080
CCl_4 [0.0082 m/1], C_6H_6 [0.0172 m/1]

$C_6H_5C(=O)OH$ (O, C-OH)

448

▶2079 MWB
B (0.15)

$(CH_3)_2CHNO_2$

366

▶2076 S
B (0.2398)

CF_3, F

218

▶2070 WB
B (0.003)

NO_2
$CH_3CH_2CHCH_3$

324

▶2069 S
B (0.2398)

F, F, F, F

222

▶2066 MS
B (0.2398)

CF_3, F, F

217

▶2065 M
B (0.728)

$(CH_3CH_2)_3CH$

143, see also 82, 156

▶2062 M
B (0.50)

$C_3H_7NO_2$

392

▶2061 M
B (0.2398)

F, F, F

224

▶2060 MS
B (0.728)

CH_3
$(CH_3)_3CC\text{-}CH_2CH_2CH_3$
CH_3

304

▶2058 M
B (0.2398)

CF_3, F

218

▶2055 M
B (0.728)

CH_3
$(CH_3)_2CHC - CH_2CH_2CH_3$
CH_3

134

▶2054 S
B (0.2398)

CH_3, F

220

▶2053 W
B

$C_6H_5(CH_3)SiCl_2$

86

▶2052 VW
B (0.2398)

CF_3

219

▶2050 M
B (0.2398)

F CF_3, F_2, F_2, F, CF_3, F, F_3C, F_2

228

▶2050
P

$La(NO_3)_3 \cdot 3NH_4NO_3$

45

▶2049 S
B (0.2)

$CH_2\text{-}CH$
CH

9

▶2049 S
A (600 mm Hg, 10 cm)

$CF_3CF_2CF_2CF_2CF_3$

237

▶2049 M
B (0.50)

$CH_3(CH_2)_2CH_2NO_2$

310

▶2048 VS
B (0.2398)

F
F

225

▶2046 VW
B (0.2398)

F
F
F
F

222

▶2045 VWVB
B (0.243)

$H_2C{=}CH(CH_2)_6CH_3$

271

▶2045 W

$C_6H_5(CH_3)_2SiCl$

86

▶2043 M
B

$VOBr_3$

12

▶2042 M
B (0.2398)

F_2 F_2
F_2 F
C_2F_5
F_2

247

▶2040 VS
B (0.2398)

CF_3
F
F

217

▶2040 S

$(C_6H_5)_3GeH$

[Ge-H]

499

▶2040 SSh
B (0.728)

$CH_3CH{=}CHCH_2CH_3$
(cis)

208

▶2039 VS
A (200 mm Hg, 10 cm)

$CF_3CF_2CF_3$

235

▶2037 MB
B (0.003)

CH_3NO_2

311

▶2037 M
A (775 mm Hg, 10 cm)

F_2 F_2
F_2 F_2

243

▶2037 W
C (0.003)

$(CH_3)_3CNO_2$

325

▶2035 M
B (0.2398)

F CF_3
F_2 F_2
F_2 F_2
F CF_3

230

▶2035 VW
B (0.2398)

F CF_3
CF_3
F_2 F
F_2 F_2
F_2

231

▶2035 W
B

$CH_3(CH_2)_5CH(C_3H_7)(CH_2)_5CH_3$

281

▶2034 MSh
B (0.2398)

CH_3
F

220

▶2031 VS
B (0.2398)

CF_3
F
F

217

▶ 2030 MW
B (0.300)
CaF_2

$C_{26}H_{54}$
(5, 14-di-n-butyloctadecane)

283

▶ 2030 W
B (0.2398)

CH_3CH_2-O-CF_2CHFCl

261

▶ 2028 VS
B (0.2398)

F
F
F

224

▶ 2025 VWB
B (0.243)

H_2C=$CH(CH_2)_7CH_3$

270

▶ 2025 M
B (0.2398)

CF_3
F

218

▶ 2025 MS
B (0.728)

CH_3
$(CH_3)_3CC$-$CH_2CH_2CH_3$
CH_3

304

▶ 2021 S
B (0.2398)

F

226

▶ 2021 MSh
B (0.2398)

F
F

225

▶ 2020 M
B (0.955)

CF_2Cl-$CFCl_2$

255

▶ 2020 MS
B (0.2398)

F
F
F
F

222

▶ 2017 VSSp
A (600 mm Hg, 10 cm)

$CF_3CF_2CF_2CF_2CF_3$

237

▶ 2016 VW
B (0.2398)

CF_3

219

▶ 2015 W
B

$C_6H_5(CH_3)SiCl_2$

86

▶ 2010 S
E

$Na_4P_2O_6$

28

▶ 2010 SB
B (0.2398)

$CF_3(CF_2)_5CF_3$

241

▶ 2010 VS
B (0.955)

F CF_3
F_2 F_2
F_2 F_2
F_2

232

▶ 2008 VW
B (0.2398)

CF_3
F
F

217

▶ 2005 W
B (0.2398)

F CF_3
CF_3
F_2 F
F_2 F_2
F_2

231

▶ 2004 VS
A (775 mm Hg, 10 cm)

CHF_2CH_3

233

▶ 2000 VS
B (0.728)

CH_3CH=$CHCH_2CH_3$
(cis)

208

▶2000 S
G
$Na_4P_2O_6$
28

▶1998 MW
B (0.2398) 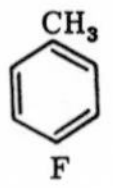
CH_3 / F
220

▶1996 W
B (0.25)
$CH_3CH_2\overset{NO_2}{C}HCH_3$
324

▶1996 MW
B (0.15)
$(CH_3)_2CHNO_2$
366

▶1990 M
B (0.728)
$(CH_3)_3CCH_2CH_2C(CH_3)_3$
305

▶1990 W
B (0.13)
$H_2C{=}CHC(CH_3)_3$
129

▶1982 S
B (0.2398) 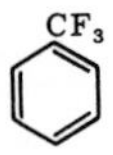
CF_3
219

▶1980 MSpSh
B (0.2398) 
CF_3 / F
218

▶1980 W
$C_6H_5SiCl_3$
86

▶1980 S
B (0.05) 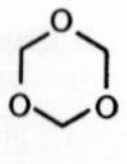
O / O / O
342

▶1980 MW
B (0.50)
$CH_3(CH_2)_2CH_2NO_2$
310

▶1980 M
E 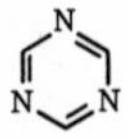
N / N / N
14

▶1979 S
B (0.2398) 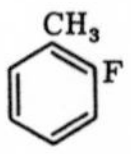
CH_3 / F
221

▶1978 VS
B (0.2398) 
F / F
225

▶1976
CCl_4 [0.0128 m/l] 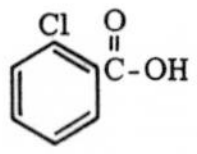
Cl / O / C-OH
448

▶1976 S
A (600 mm Hg, 10 cm)
$CF_3CF_2CF_2CF_2CF_3$
237

▶1976 W
B
$(C_6H_5)_2(CH_3)SiCl$
86

▶1972 W
CCl_4
$(C_6H_5)_3SiCl$
86

▶1971 M
A (775 mm Hg, 10 cm) 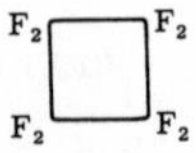
F_2 F_2 / F_2 F_2
243

▶1970
CCl_4 [0.0082 m/l] 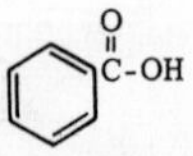
O / C-OH
448

▶1970 MS A (24.1 mm Hg, 40 cm) [benzene] 309	▶1960 M B (0.728) $(CH_3)_3CCH_2CH_2C(CH_3)_3$ 305
▶1969 S B (0.2398) [1,2,3,5-tetrafluorobenzene: F, F, F, F] 222	▶1958 W B $(C_6H_5)_2(CH_3)SiCl$ 86
▶1963 VS B (0.2398) CF_3-benzene 219	▶1957 M N [1,3,5-triazine] 14
▶1963 S A (600 mm Hg, 10 cm) $CF_3CF_2CF_2CF_2CF_3$ 237	▶1955 W B (0.2398) [cyclohexane ring: F, CF_3; CF_3, F; F_2; F_2; F_2; F_2] 231
▶1962 VS B (0.2398) F-benzene 226	▶1955 W B (0.2398) [cyclohexane ring: F, CF_3; F_2; F_2; F_2; F_2; F_2] 232
▶1962 M B (0.2398) [cyclopentane ring: F_2, F_2, F_2, F_2, F, C_2F_5] 247	▶1953 MW B (0.50) $(CH_3)_2CHCH_2NO_2$ 326
▶1961 VW B (0.2398) [benzene ring: F, F, F] 224	▶1952 SSp B (0.2398) [benzene ring: CF_3, F] 218
▶1961 MW C (0.15) $(CH_3)_3CNO_2$ 325	▶1950 S B (0.2398) [benzene ring: CH_3, F] 220
▶1961 W $C_6H_5SiCl_3$ 86	▶1950 SB B (0.2398) F_9C_4-O-C_4F_9 266
▶1960 W B (0.169) $CH_3CH{=}CHCH_2CH_3$ (trans) 207	▶1949 MS A (200 mm Hg, 10 cm) $CF_3CF_2CF_3$ 235

▶ 1949 M B (0.50) $C_3H_7NO_2$ 392	▶ 1939 VS B (0.2398) (fluorobenzene structure) 226
▶ 1948 S n-C_6H_{14} [RuHI{C_2H_4[P(C_2H_5)$_2$]$_2$}$_2$] (trans) [Ru-H] 1	▶ 1938 S n-C_6H_{14} [RuHCl{C_2H_4[P(C_2H_5)$_2$]$_2$}$_2$] (trans) [Ru-H] 1
▶ 1947 CCl_4 [0.0128 m/l] (2-chlorobenzoic acid: Cl, C(=O)-OH) 448	▶ 1934 MS A (200 mm Hg, 10 cm) $CF_3CF_2CF_3$ 235
▶ 1945 S n-C_6H_{14} [RuHBr{C_2H_4[P(C_2H_5)$_2$]$_2$}$_2$] (trans) [Ru-H] 1	▶ 1929 H COF_2 [C=O] 513
▶ 1945 MS B (0.2398) (F, F, F trifluorobenzene) 224 ; ▶ 1945 S B (0.2398) (CH_3, F) 221	▶ 1929 M C (CH_3, CH_3 dimethylnaphthalene) 185
▶ 1945 M B (0.2) (cyclopentene) 627	▶ 1927 M C H_3C (naphthalene) CH_3 179
▶ 1943.2 S B (0.2) (indene: CH_2-CH = CH) 9	▶ 1927 M C (CH_3 CH_3 dimethylnaphthalene) 182
▶ 1941 MS A (200 mm Hg, 10 cm) $CF_3CF_2CF_3$ 235	▶ 1927 CCl_4 [0.0082 m/l] (benzoic acid: C(=O)-OH) 448
▶ 1940 VS A (775 mm Hg, 10 cm) (F_2 F_2 F_2 F_2 cyclobutane) 243	▶ 1926 M B (0.2398) (F, F, F trifluorobenzene) 224
▶ 1940 SB B (0.2398) $(C_4F_9)_3N$ 346	▶ 1925 M B (0.728) $CH_3CH_2C(CH_3)_2$ - $CH(CH_3)CH_2CH_3$ 299

▶1925 S
B (0.10) 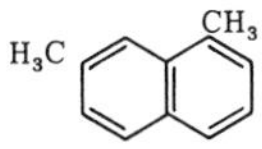
H_3C CH_3
183

▶1923 S
B (0.10) 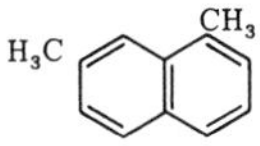
H_3C CH_3
183

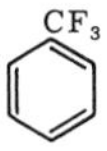

▶1922 S
B (0.10) 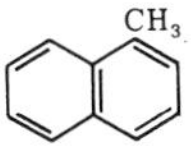
CH_3
192

▶1922 M
C 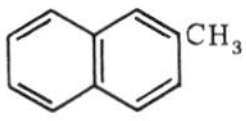
CH_3
191

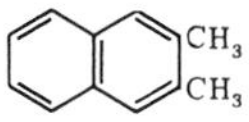

▶1920 M
A (600 mm Hg, 10 cm)
$CF_3CF_2CF_2CF_2CF_3$
237

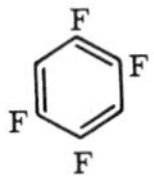

▶1919 MS
B (0.10) 
200

▶1919 M
B (0.063)
CH_3
CN
214

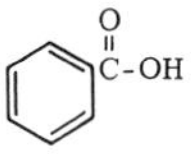

▶1916 M
B (0.151)
CH_2
C
CH_3
288

▶1916 M
B (0.10)
CH_2CH_3
190

▶1916 S
B (0.10)
CH_3
H_3C
184

▶1915.0 S
B (0.2)
CH_2-CH
CH
9

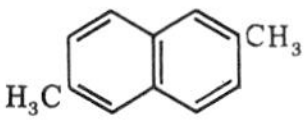

▶1914 VS
B (0.2398)
CF_3
219

▶1914 S
C
CH_3
CH_3
181

▶1912 VS
B (0.2398)
F
F
F
F
222

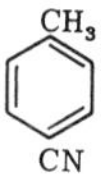

▶1912
CCl_4 [0.0082 m/l]
O
C-OH
448

▶1910 M
C
CH_3
H_3C
180

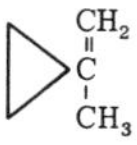

▶1908 W
B
$C_6H_5HSiCl_2$
86

▶1905 MS
B (0.10) 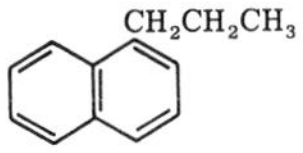
$CH_2CH_2CH_3$
178

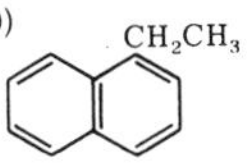

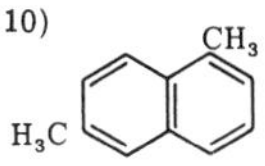

▶1905 W
B
$(C_6H_5)_3SiCl$
86

▶1904 VS
B (0.2398) 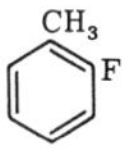
CH_3
F
221

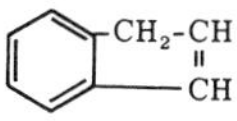

▶1902
C_6H_6 [0.0172 m/l] 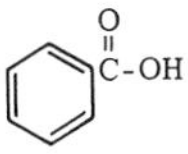
O
C-OH
448

▶ 1901
H
$CF_3-C(=O)-F$
513

▶ 1896 VS
B (0.2398) 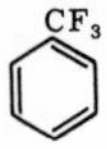
CF_3
219

▶ 1895 S
A (775 mm Hg, 10 cm)
CF_2Cl-CF_2Cl
257

▶ 1895 MS
A (775 mm Hg, 10 cm)
CHF_2CH_3
233

▶ 1893 M
B (0.0576) 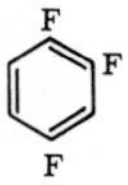
F
F
F
224

▶ 1893 MS
B (0.064) 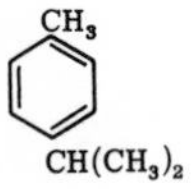
CH_3
$CH(CH_3)_2$
132

▶ 1890 M
B (0.728)
$(CH_3)_3CCH_2CH_2C(CH_3)_3$
305

▶ 1890 M
A (775 mm Hg, 10 cm)
CHF_2CH_3
333

▶ 1890 W
CCl_4
$(C_6H_5)_3SiCl$
86

▶ 1885 M
B (0.2398)
$(CF_3)_2CFCF_2CF_3$
240

▶ 1884.5 S
B (0.2) 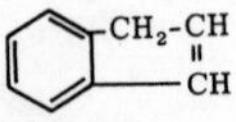
CH_2-CH
CH
9

▶ 1883 M
C 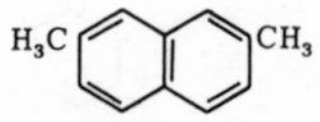
H_3C CH_3
179

▶ 1883 W
B
$C_6H_5(CH_3)_2SiCl$
86

▶ 1882 VVS
B (0.2398) 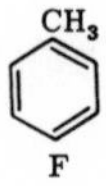
CH_3
F
220

▶ 1881 M
A (775 mm Hg, 10 cm)
CHF_2CH_3
233

▶ 1880 VS
B (0.2398) 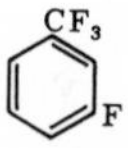
CF_3
F
218

▶ 1880 M
B (0.10) 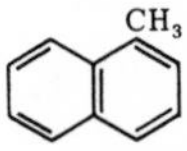
CH_3
192

▶ 1875 M
B (0.50)
$(CH_3)_2CHCH_2NO_2$
326

▶ 1875 W
B 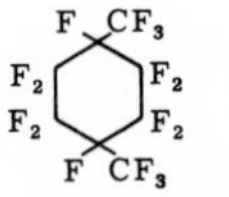
F CF_3
F_2 F_2
F_2 F_2
F CF_3
230

▶ 1875 MSSh
B (0.169)
$CH_2=CHCH=CHCH=CH_2$
203

▶ 1874 W
B (0.2398) 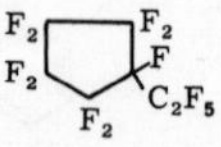
F_2 F_2
F_2 F
C_2F_5
F_2
247

▶1873 VS
A (682 mm Hg, 10 cm) 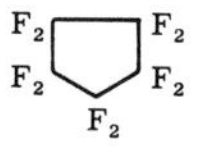
F_2 F_2 F_2 F_2 F_2
245

▶1872
$CHCl_3$ 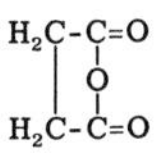
$H_2C-C=O$
O
$H_2C-C=O$
[C=O]
556

▶1872 MS
A (200 mm Hg, 10 cm)
$CF_3CF_2CF_3$
235

▶1871 M
B 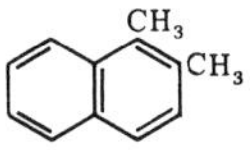
CH_3 CH_3
188

▶1871
A 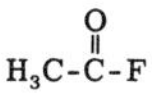
O
H_3C-C-F
[C=O]
513

▶1868 VS
B (0.2398) 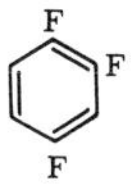
F F F
224

▶1866 MW
A (775 mm Hg, 10 cm) 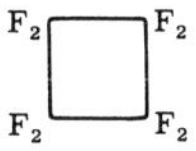
F_2 F_2 F_2 F_2
243

▶1866 MS
B (0.2398) 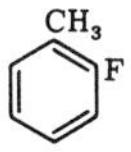
CH_3 F
221

▶1860 VS
B (0.169)
$CH_3(CH_2)_2CH=CH(CH_2)_2CH_3$
(trans)
205

▶1860 VS
B (0.2398) 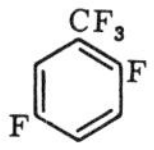
CF_3 F F
217

▶1860 VS
B (0.05) 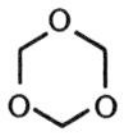
O O O
342

▶1860 M
B (0.2398) 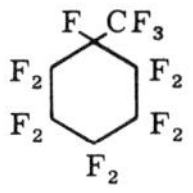
F CF_3 F_2 F_2 F_2 F_2 F_2
232

▶1857 VS
B (0.2398) 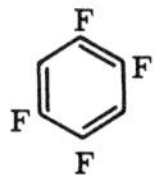
F F F F
222

▶1856.7 S
B (0.2) 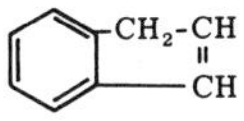
CH_2-CH
CH
9

▶1856 S
A (600 mm Hg, 10 cm)
$CF_3CF_2CF_2CF_2CF_3$
237

▶1856 M
B (0.10) 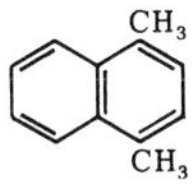
CH_3 CH_3
186

▶1856 S
B (0.955)
$CF_2Cl-CFCl_2$
255

▶1856 M
A (775 mm Hg, 10 cm)
CF_2Cl-CF_2Cl
257

▶1850 W
B (0.2398) 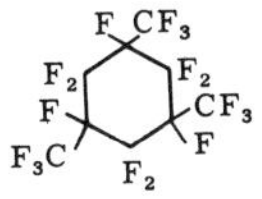
F CF_3 F_2 F_2 F CF_3 F F_3C F_2
228

▶1850 M
B (0.728) 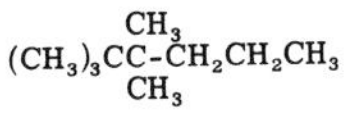
CH_3
$(CH_3)_3CC-CH_2CH_2CH_3$
CH_3
304

▶1850 M
B (0.2398)

$(C_4F_9)_3N$

346

▶1848 M
B (0.10)

CH_3 CH_3

188

▶1846 M
A (775 mm Hg, 10 cm)

$CF_2Cl\text{-}CF_2Cl$

257

▶1845 S
C

CH_3 CH_3

181

▶1845 SB
B (0.2398)

$CF_3(CF_2)_5CF_3$

241

▶1845 MS
B (0.10)

200

▶1845 M
B (0.2398)

F CF_3 F_2 F_2 F_2 F_2 F CF_3

230

▶1842 M
B (0.238)

H_3C CH_3
C=C
H CH_2CH_3
(cis)

274, see also 164

▶1840 M
B (0.169)

$CH_3CH{=}CHCH_2CH_3$
(trans)

207

▶1840 S
B (0.13)

$H_2C{=}CHCH_2C(CH_3)_3$

102

▶1840
H

$H_3C\text{-}C(=O)\text{-}F$

513

▶1838 M
B (0.10)

CH_3

192

▶1838 M
B (0.2398)

CF_3

219

▶1838 W
B

$C_6H_5SiCl_3$

86

▶1838 SSh
B (0.2398)

CF_3

219

▶1836 S
A (600 mm Hg, 10 cm)

$CF_3CF_2CF_2CF_2CF_3$

237

▶1835 MS
B (0.127)

$H_2C{=}CH(CH_2)_3CH(CH_3)_2$

97

▶1835 MB
C

SrH_2GeO_4

690

▶1833

$Fe_3(CO)_{12}$

511

▶1832 W
B (0.003)

CH_3NO_2

311

▶1832 M
B (0.50)

$CH_3(CH_2)_2CH_2NO_2$

310

▶1832 M
B (0.127)

$H_2C{=}CH(CH_2)_5CH_3$

290

▶1832

$Cl_3C{-}O{-}C({=}O){-}O{-}CCl_3$

[C=O] 513

▶1830 VS
B (0.2398)

Fc1ccc(F)c(F)c1F

222

▶1830 W
B

FC1(C(F)(F)F)C(F)(F)C(F)(C(F)(F)F)C(F)(F)C(F)(C(F)(F)F)C1(F)F

228

▶1828 VS
B (0.238)

$H_2C{=}CHCH_2C(CH_3)_3$

273

▶1828 M
C

H_3C-naphthalene-CH_3

179

▶1828 M
B (0.064)

$H_2C{=}CHCH(CH_3)CH_2CH_3$

167

▶1828 MW
A (36 mm Hg, 15 cm)

$H_2C{=}CHCH(CH_3)_2$

131

▶1828
A

$Cl{-}C({=}O){-}Cl$

[C=O] 513

▶1826 S
B (0.2398)

Fc1ccc(F)c(F)c1

224

▶1825.5 S
B (0.2)

C1Cc2ccccc2C=C1 ($CH_2{-}CH{=}CH$)

9

▶1825 VS
A (775 mm Hg, 10 cm)

F_2 F_2 / F_2 F_2 (octafluorocyclobutane)

243

▶1825 M
B (0.728)

$(CH_3)_3CCH_2CH_2C(CH_3)_3$

305

▶1825 MS
B (0.13)

$H_2C{=}CHC(CH_3)_3$

129

▶1825

$H_{11}C_5{-}C{=}O$
O
$H_{11}C_5{-}C{=}O$

[C=O] 514

▶1825 M
B (0.066)

$H_2C{=}CH(CH_2)_7CH_3$

270

▶1824

$CH_3{-}C{=}O$
O
$CH_3{-}C{=}O$

[C=O] 514

▶1821 S
B (0.2398)

Cc1ccccc1F

221

▶1821 S
B (0.243)

$H_2C{=}CH(CH_2)_6CH_3$

271

▶1821 M
B (0.066)
$CH_3(CH_2)_8CH=CH_2$
352

▶1821 W
B
$C_6H_5HSiCl_2$
86

▶1820 S
B (0.955)
$CF_2Cl-CFCl_2$
255

▶1820 S
A (24.1 mm Hg, 40 cm) 
309

▶1820
A
$F_3C-\overset{\overset{O}{\|}}{C}-OH$
522

▶1820
$Cl_3C-\overset{\overset{O}{\|}}{C}-Cl$
[C=O]
553

▶1819 VS
B (0.2398) 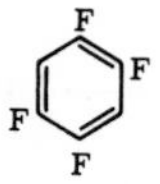
F F F F
222

▶1819 S
A (200 mm Hg, 10 cm)
$CF_3CF_2CF_3$
235

▶1819 MS
A (775 mm Hg, 10 cm)
CHF_2CH_3
233

▶1818 M
B (0.50)
$(CH_3)_2CHCH_2NO_2$
326

▶1818 M
B (0.10) 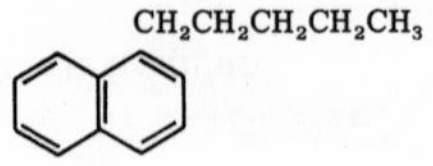
$CH_2CH_2CH_2CH_2CH_3$
174

▶1815
CCl_4 [0.0082 m/1]
$\overset{\overset{O}{\|}}{C}-OH$
448

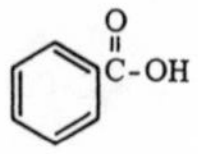

▶1815
$CHCl_3$ 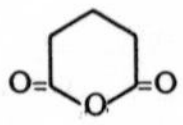
O= O =O
[C=O]
557

▶1815 M
B (0.10) 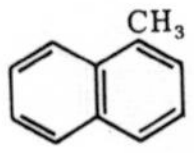
CH_3
192

▶1815
CCl_4
NOCl
[N=O]
555

▶1815 VS
B (0.2398) 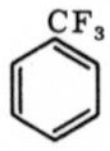
CF_3
219

▶1813 S
B (0.2398) 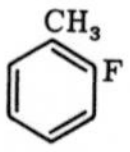
CH_3
F
221

▶1813 S 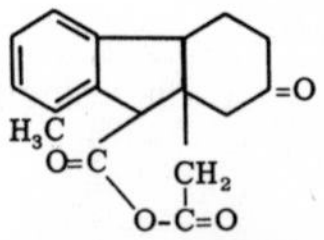
=O
H_3C
O=C
CH_2
O-C=O
[C=O]
556

▶1812 M
B (0.2398) 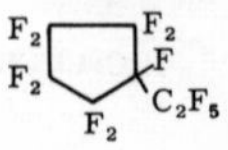
F_2 F_2 F
F_2 C_2F_5
F_2
247

▶1812 S
C 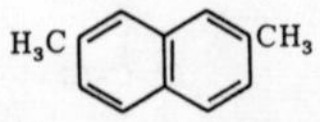
H_3C CH_3
179

▶1812 M
A (775 mm Hg, 10 cm)
CHF_2CH_3
233

▶1812
$H_3C-\overset{O}{\overset{\|}{C}}-Br$
[C=O]
513

▶1810 MS
B (0.2398)
F CF₃ / F₂ F₂ / F₂ F₂ / F CF₃ (ring)
230

▶1810 VS
B (0.056)
$CH_2=CHCH=CHCH=CH_2$
203

▶1810 M
B (0.728)
$(CH_3)_3CCH_2CH_3$
145

▶1810
CCl_4
$F_3C-\overset{O}{\overset{\|}{C}}-OH$
[C=O]
522

▶1808 S
B (0.2398)
CF_3, F (benzene ring)
218

▶1808 M
B
CH_3, CH_3 (naphthalene ring)
186

▶1808 M
A (775 mm Hg, 10 cm)
CHF_2CH_3
233

▶1808
C=O / O / $H_3C-C=O$ (benzene ring)
[C=O]
514

▶1807 MS
B (0.10)
$CH_2CH_2CH_3$ (naphthalene ring)
178

▶1807 M
C
CH_3 (naphthalene ring)
191

▶1807 VS
B
$ClH_2C-\overset{O}{\overset{\|}{C}}-Cl$
[C=O]
554

▶1805 S
CS_2 (0.728)
$CH_2=CHCH=CHCH=CHCH=CH_2$
202

▶1805 M
B (0.10)
S (ring)
199

▶1805 MSh
B (0.05)
O, O, O (ring)
342

▶1805
B
H_2C-O / C=O / H_2C-O
[C=O]
514

▶1804 VS
A (682 mm Hg, 10 cm)
F_2 F_2 / F_2 F_2 / F_2 (ring)
245

▶1802
$H_3C-\overset{O}{\overset{\|}{C}}-Cl$
[C=O]
513

▶1801 M
B (0.10)
CH_2CH_3 (naphthalene ring)
190

▶1800 MW
B (0.2398) 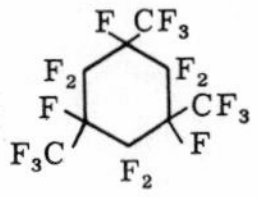
F CF$_3$ / F$_2$ F$_2$ / F CF$_3$ / F$_3$C F / F$_2$
228

▶1800 M
B (0.728)
$(CH_3)_3CCH_2CH_2C(CH_3)_3$
305

▶1800
A
NOCl
[N=O]
555

▶1800
NOBr
[N=O]
555

▶1799
$ClH_2C-C(=O)-Br$
557

▶1798
A
$CF_3CF=CF_2$
[C=C]
558

▶1796 M
C 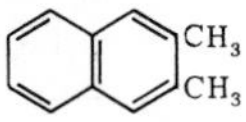
CH_3 / CH_3
181

▶1795 S
A (775 mm Hg, 10 cm) 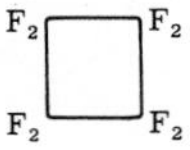
F$_2$ F$_2$ / F$_2$ F$_2$
243

▶1795 S
B (0.0576) 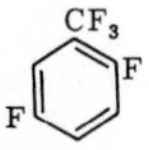
CF$_3$ / F / F
217

▶1790 MSSp
B (0.064)
$H_2C=C(CH_2CH_3)_2$
163

▶1790 MS
B (0.2398) 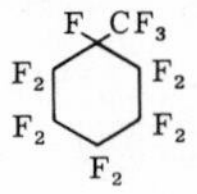
F CF$_3$ / F$_2$ F$_2$ / F$_2$ F$_2$ / F$_2$
232

▶1790
E
$LiNO_3$
45

▶1790
E
$Zn(NO_3)_2$
45

▶1790
$H_{15}C_7-C(=O)-Cl$
[C=O]
513

▶1790 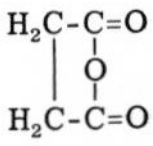
$H_2C-C=O$ / O / $H_2C-C=O$
556

▶1788 M
B (0.2398) 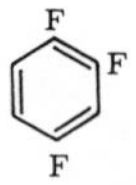
F / F / F
224

▶1787 VS
CCl_4 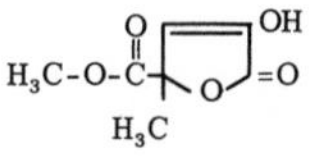
$H_3C-O-C(=O)-$ OH / O / =O / H_3C
[C=O]
475

▶1787 S
C_6H_6 [0.0044 m/l]
OH / $C(=O)-OH$
448

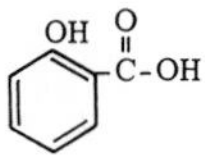

▶1785 SSh
B (0.169)
$CH_3CH=CHCH_2CH_3$
(trans)
207

▶1785
E
$Pr(NO_3)_3$
45

▶1785
E
$Cu(NO_3)_2 \cdot 5H_2O$
45

▶1785
E
$Fe(NO_3)_3$
45

▶1785
E
$Ni(NO_3)_2$
45

▶1785
A
$H_3C-C(=O)-OH$
521

▶1785
E
$NaNO_3$
45

▶1785
P
$Sr(NO_3)_2$
45

▶1785 VS
B (0.0563)
$H_2C=C(CH_3)CH_2CH_2CH_3$
206

▶1784 VS
B (0.2398) 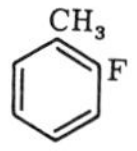
CH_3, F (benzene ring)
221

▶1783 M
B (0.136) 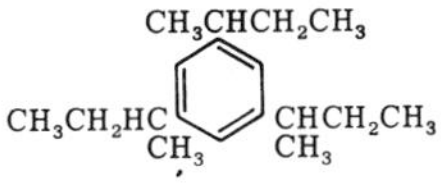
$CH_3CHCH_2CH_3$; $CH_3CH_2HC(CH_3)$; $CHCH_2CH_3$ CH_3 (benzene ring)
280

▶1782 MS
B (0.136) 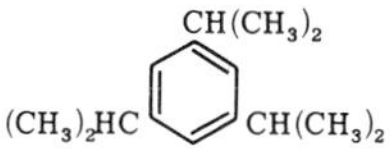
$CH(CH_3)_2$; $(CH_3)_2HC$; $CH(CH_3)_2$ (benzene ring)
278

▶1780 WSh
B
$ClH_2C-C(=O)-Cl$
554

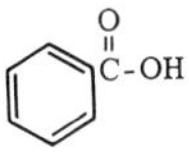

▶1780 S
CCl_4 [0.0082 m/l]
$C(=O)-OH$ (benzene ring)
448, see also 38

▶1780 M
N 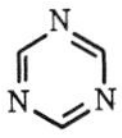
N, N, N (ring)
14

▶1780
E
$Sr(NO_3)_2$
45

▶1780
E
$Hg(NO_3)_2$
45

▶1780
E
$Co(NO_3)_2$
45

▶1780
E
$Ba(NO_3)_2$
45

▶1780
E
$Y(NO_3)_3$
45

▶1780
K
$La(NO_3)_3$
45

▶1780
$HFC=CF_2$
[C=C]
564

▶1780
A
$F_3C-C(=O)-CH_3$
517

▶1779 MS
B (0.10) 
200

▶1778 VS
B (0.2398) 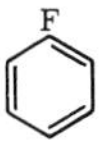
F (benzene ring)
226

▶1776 S
C 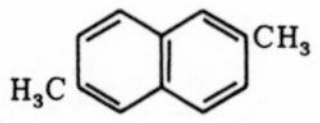
H_3C–naphthalene–CH_3
180

▶1776
$Cl_3C-C(=O)-OCH_3$
[C=O] 560

▶1776 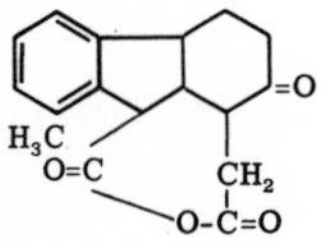
[C=O]
H_3C, O=C, CH_2, O-C=O, =O
556

▶1775 M
B (0.728)
$(CH_3CH_2)_3CH$
143

▶1775
E
$NaNO_3$
45

▶1775
E
$Bi(NO_3)_3$
45

▶1775 M
E 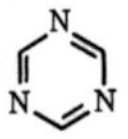
N, N, N
14

▶1775
H 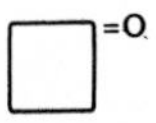
=O
[C=O] 559

▶1775 S
E 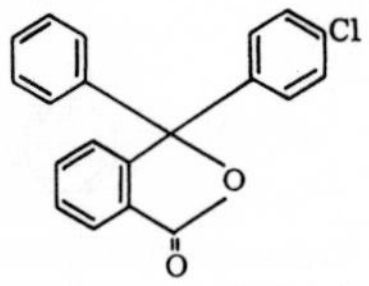
Cl, O, O
39

▶1775
$Cl_2HC-C(=O)-OCH_3$
[C=O] 560

▶1774 S
A (200 mm Hg, 10 cm)
$CF_3CF_2CF_3$
235

▶1773 W
B
$C_6H_5HSiCl_2$
86

▶1773 M
E 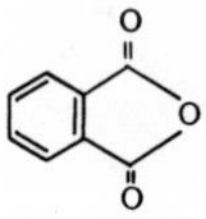
O, O, O
39

▶1773
H 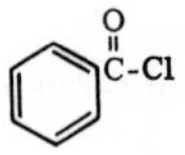
$C(=O)-Cl$
[C=O] 513, see also 38

▶1772
B 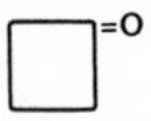
=O
[C=O] 565, see also 559

▶1771 W
B (0.2398) 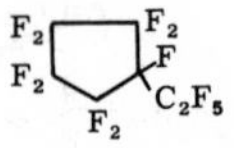
F_2, F_2, F_2, F_2, F, C_2F_5
247

▶1770 S
B (0.136) 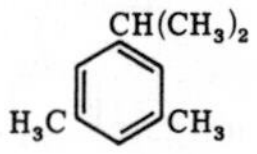
$CH(CH_3)_2$, H_3C, CH_3
277

▶1770 S
E 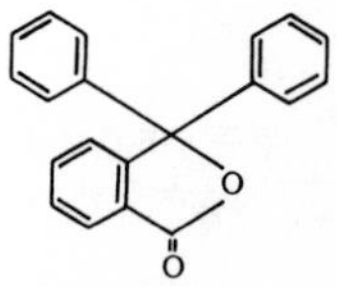
O, O
39

▶1770 MS
B (0.10) 
S
199

▶1770 W
B
$C_6H_5(CH_3)SiCl_2$
86

▶ 1770 W
CCl_4
$(C_6H_5)_3SiCl$
86

▶ 1770
E
$RbNO_3$
45

▶ 1770
E
$Cr(NH_3)_5(NO_3)_3$
45

▶ 1770
P, E
$Ce(NO_3)_3$
45

▶ 1770
P
$Ba(NO_3)_2$
45

▶ 1770
K
$Ca(NO_3)_2$
45

▶ 1770
CCl_4
$Cl_3C\text{-}C(=O)\text{-}OCH_3$
[C=O]
561

▶ 1769 S
C 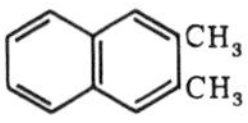
CH_3
CH_3
181

▶ 1768 VS
B (0.2398) 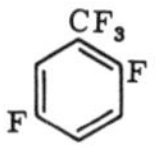
CF_3
F
F
217

▶ 1767 MB
B (0.136) 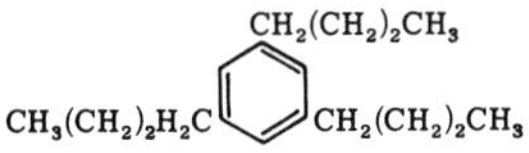
$CH_2(CH_2)_2CH_3$
$CH_3(CH_2)_2H_2C$
$CH_2(CH_2)_2CH_3$
279

▶ 1767 VS
B (0.2398) 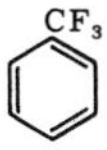
CF_3
217

▶ 1766 S 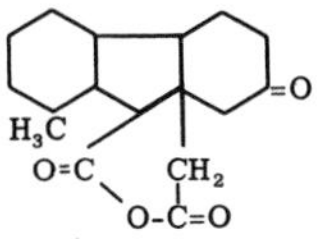
H_3C
O=C
CH_2
O-C=O
=O
556

▶ 1766 S
B (0.2398) 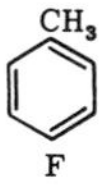
CH_3
F
220

▶ 1766 S
B (0.955)
$CF_2Cl\text{-}CFCl_2$
255

▶ 1766
$CHCl_3$ 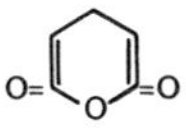
O=
O
=O
557

▶ 1765
P
$La(NO_3)_3$
45

▶ 1765
E
$AgNO_3$
45

▶ 1764 S
B (0.136) 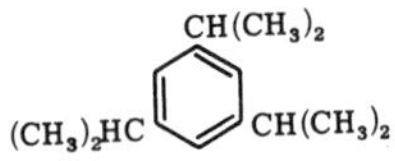
$CH(CH_3)_2$
$(CH_3)_2HC$
$CH(CH_3)_2$
278

▶ 1764 M
B (0.136) 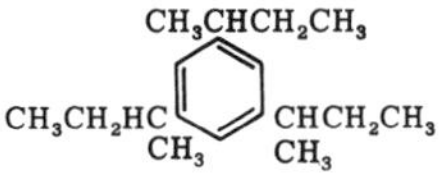
$CH_3CHCH_2CH_3$
CH_3CH_2HC
CH_3
$CHCH_2CH_3$
CH_3
280

▶ 1764 S
E 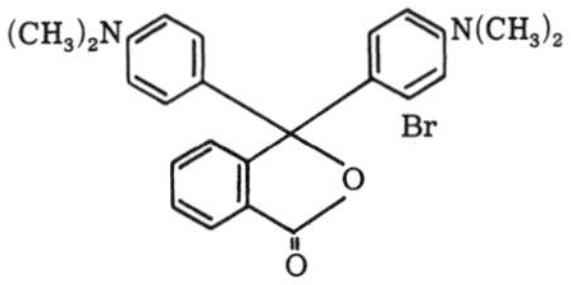
$(CH_3)_2N$
$N(CH_3)_2$
Br
O
O
39

▶1762
H

Cl_3-C-C(=O)-H

[C=O] 519

▶1761 VS
B (0.2398)

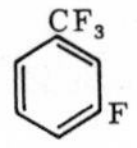

CF_3, F

218

▶1761 S
B (0.151)

CH_2=C(CH_3)– (cyclopropyl)

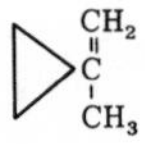

288

▶1760 S
B (0.169)

$CH_3CH=CHCH_2CH_3$
(trans)

207

▶1760 S
CCl_4 [0.005 m/1]

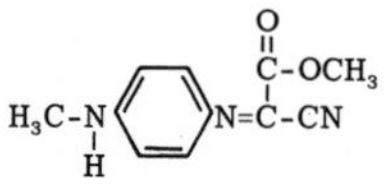

H_3C-N(H)-C_6H_4-N=C(CN)-C(=O)-OCH_3

[C=O] 562

▶1760 S
CCl_4 [0.005 m/1]

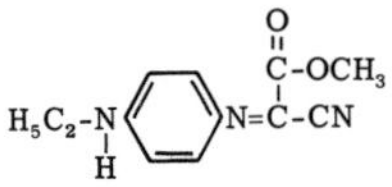

H_5C_2-N(H)-C_6H_4-N=C(CN)-C(=O)-OCH_3

[C=O] 562

▶1760 Sh
CCl_4
Prism-Grating, ±1 cm^{-1}

C(=O)-OH, OCH_3

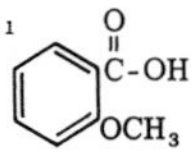

[C=O] 25

▶1760
E

KNO_3

45

▶1760
E

$CsNO_3$

45

▶1760
P

$RbNO_3$

45

▶1760
P

$AgNO_3$

45

▶1760
P

$Cd(NO_3)_2$

45

▶1760
P

$Zn(NO_3)_2$

45

▶1760
P

$Hg_2(NO_3)_2$

45

▶1760
P

$La(NO_3)_3 \cdot 3NH_4NO_3$

45

▶1760

$H_{11}C_5$-C=O
O
$H_{11}C_5$-C=O

[C=O] 514

▶1760

CH_3CHBr
C=O
CH_2Br

[C=O] 566

▶1759 M
B (0.10)

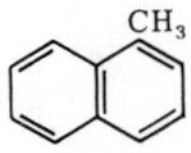

CH_3

192

▶1757 S
E

$(CH_3)_2N$, CH_3, $N(CH_3)_2$, CH_3, O, O

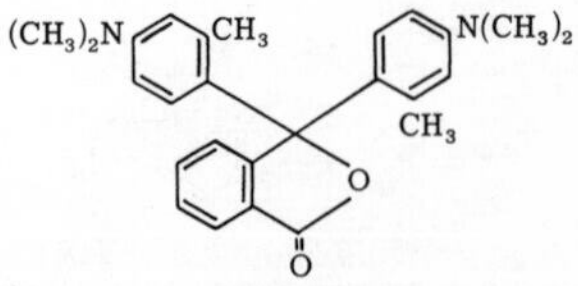

39

▶1757 S
CCl_4
Prism-Grating, ±1 cm^{-1}

C(=O)-OH, Br

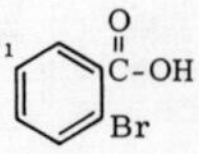

[C=O] 25

▶1757 S
CCl_4 [0.005 m/l]
$(H_3C)_2N$-C₆H₄-N=C(CN)-C(=O)-OCH_3
[C=O] 562

▶1756 S
CCl_4 [0.005 m/l]
H_3C-NH-C₆H₄-N=C(CN)-C(=O)-OC_2H_5
[C=O] 562

▶1756 S
CCl_4 [0.005 m/l]
H_5C_2-NH-C₆H₄-N=C(CN)-C(=O)-OC_2H_5
[C=O] 562

▶1756 S
CCl_4
Prism-Grating, ±1 cm^{-1}
Cl-C₆H₄-C(=O)-OH
[C=O] 25

▶1755 S
CCl_4 [0.0128 m/l]
Cl-C₆H₄-C(=O)-OH
[C=O] 448

▶1755 S
E
$(CH_3)_2N$ … $N(CH_3)_2$
39

▶1755 S
CCl_4
Prism-Grating, ±1 cm^{-1}
F-C₆H₄-C(=O)-OH
[C=O] 25

▶1755
P
$Al(NO_3)_2$
45

▶1754 S
CCl_4 [0.005 m/l]
$(H_3C)_2N$-C₆H₄-N=C(CN)-C(=O)-OC_2H_5
[C=O] 562

▶1754 S
B
Cl_3C-C(=O)-O
379

▶1753 S
CCl_4
Prism-Grating, ±1 cm^{-1}
I-C₆H₄-C(=O)-OH
[C=O] 25

▶1752 S
CH_3, OH, =O, C=O, OCH_3
[C=O] 556

▶1752 S
CCl_4
Prism-Grating, ±1 cm^{-1}
O_2N-C₆H₄-C(=O)-OH
[C=O] 25

▶1752 S
CCl_4
Prism-Grating, ±1 cm^{-1}
NO_2-C₆H₄-C(=O)-OH
[C=O] 25

▶1752
P
$Hg(NO_3)_2$
45

▶1752
A
H_3C-C(=O)-H
[C=O] 519

▶1751 S
B (0.10)
200

▶1751 S
CH_3, OH, =O, C=O, OH
[C=O] 556

▶1751 S
CCl_4
Prism-Grating, ±1 cm^{-1}
OCH_3-C₆H₄-C(=O)-OH
[C=O] 25

▶1751
NC-CH_2-C(=O)-O-C_2H_5
[C=O] 526

▶1750 S
CCl_4
CH_3, OH, =O, C=O, OCH_3
[C=O] 556

▶ 1750 S
CH_2Cl_2 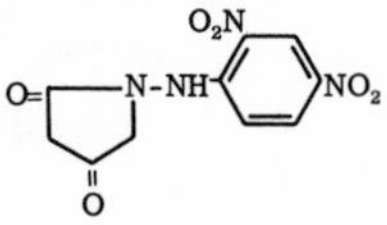
577

▶ 1750 S
CCl_4 [0.005 m/l] 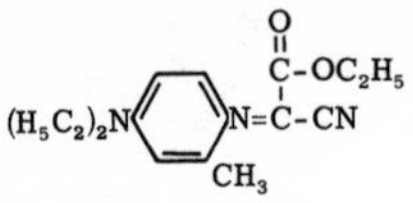
[C=O]
562

▶ 1750 S
E 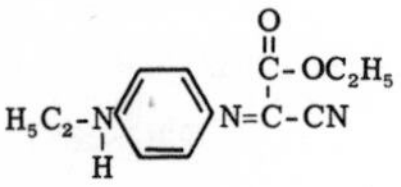
[C=O]
562

▶ 1750 M
C 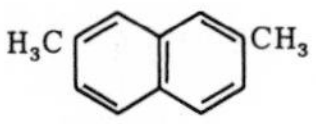
179

▶ 1750 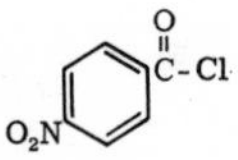
[C=O]
385

▶ 1750 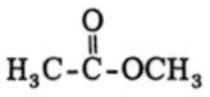
$H_3C-C(=O)-OCH_3$
[C=O]
560

▶ 1750 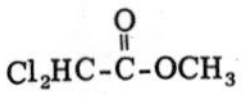
$Cl_2HC-C(=O)-OCH_3$
[C=O]
560

▶ 1749 S
$CHCl_3$ 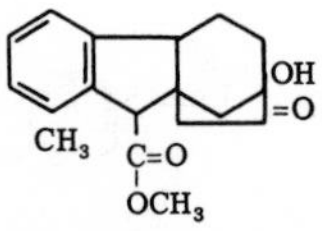
[C=O]
556

▶ 1749 S
E 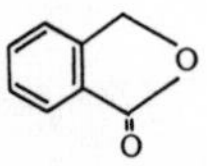
39

▶ 1749 S
n-C_6H_{14} [0.0076 m/l]
Prism-Grating, ±1 cm^{-1} 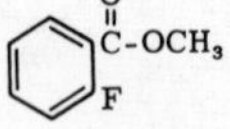
[C=O]
25

▶ 1749 M
B (0.10) 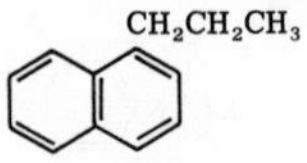
178

▶ 1748 S
CCl_4
Prism-Grating, ±1 cm^{-1} 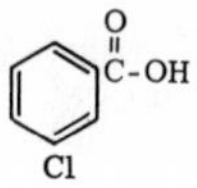
[C=O]
25

▶ 1748 S
CCl_4
Prism-Grating, ±1 cm^{-1} 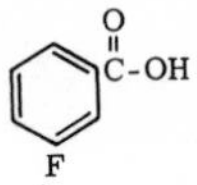
[C=O]
25

▶ 1748 S
CCl_4
Prism-Grating, ±1 cm^{-1} 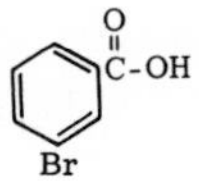
[C=O]
25

▶ 1748 S
CCl_4 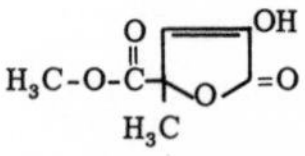
[C=O of ester group]
475

▶ 1748 S
G 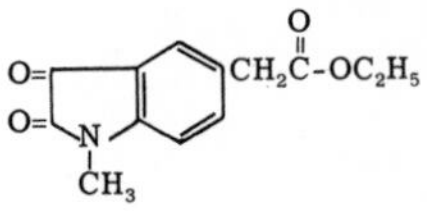
[C=O]
673

▶ 1748 Sh
CCl_4
Prism-Grating, ±1 cm^{-1} 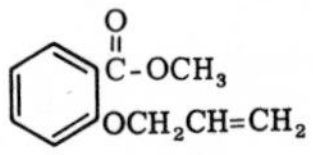
[C=O]
25

▶ 1748 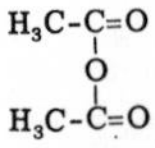
[C=O]
514

▶ 1747 S
CCl_4
Prism-Grating, ±1 cm^{-1} 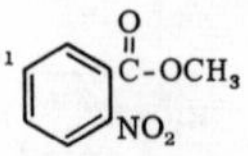
[C=O]
25

▶ 1747 S
CCl_4
Prism-Grating, ±1 cm^{-1} 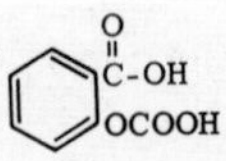
[C=O]
25

▶1747
B
$C_2H_5-\overset{O}{\overset{\|}{C}}-\overset{O}{\overset{\|}{C}}-O-C_2H_5$
[C=O] 674

▶1745 S
G
$CH_2\overset{O}{\overset{\|}{C}}-OC_2H_5$ (isatin derivative)
[C=O] 673

▶1746 S
CCl_4
Prism-Grating, ±1 cm^{-1} 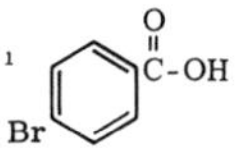
$\overset{O}{\overset{\|}{C}}-OH$, Br
[C=O] 25

▶1745
C=O, O, $H_3C-C=O$
[C=O] 514

▶1746 S
B (0.2398) 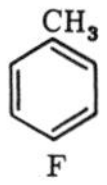
CH_3, F
220

▶1745 S
G
$\overset{O}{\overset{\|}{C}}-N(C_3H_7)_2$, CH_3
[C=O] 673

▶1746
$H_9C_4-O-\overset{O}{\overset{\|}{C}}-\overset{O}{\overset{\|}{C}}-O-C_4H_9$
[C=O] 579

▶1744 S
CCl_4
Prism-Grating, ±1 cm^{-1}
$\overset{O}{\overset{\|}{C}}-OH$
[C=O] 25

▶1745 S
B (0.169)
$CH_3(CH_2)_2CH=CH(CH_2)_2CH_3$
205

▶1744 S
CCl_4
Prism-Grating, ±1 cm^{-1}
$\overset{O}{\overset{\|}{C}}-OCH_3$, Br
[C=O] 25

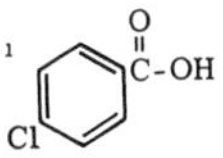

▶1745 VS
B (0.0563)
$CH_3(CH_2)_2CH=CH(CH_2)_2CH_3$
(trans)
205

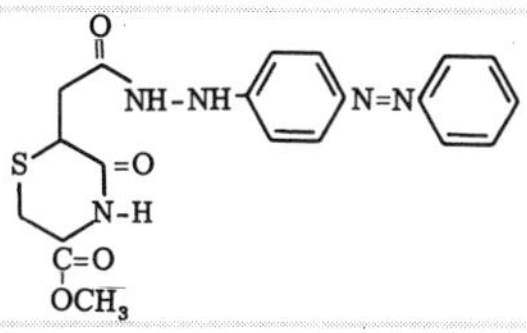

▶1744 S
H
O
[C=O] 674

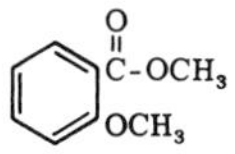

▶1745 S
CCl_4
Prism-Grating, ±1 cm^{-1}
$\overset{O}{\overset{\|}{C}}-OH$, F
[C=O] 25

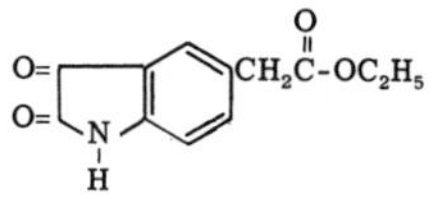

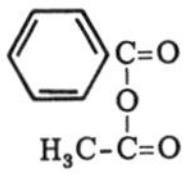

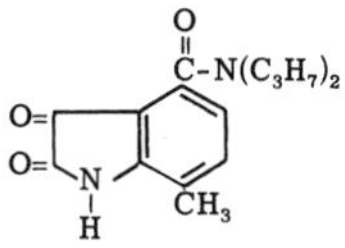

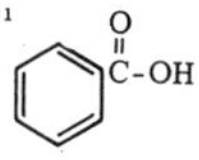

▶1744 S
CCl_4
Prism-Grating, ±1 cm^{-1} 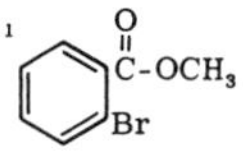
$\overset{O}{\overset{\|}{C}}-OCH_3$, Cl
[C=O] 25

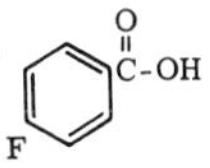

▶1745 S
CCl_4
Prism-Grating, ±1 cm^{-1} 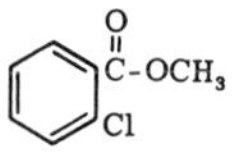
$\overset{O}{\overset{\|}{C}}-OH$, Cl
[C=O] 25

▶1744
B
$C_2H_5-\overset{O}{\overset{\|}{C}}-\overset{O}{\overset{\|}{C}}-OH$
[C=O] 674

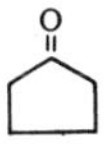

▶1745 S
E
O, NH-NH, N=N, S, =O, N-H, C=O, OCH_3
570

▶1743 S
B (0.2398) 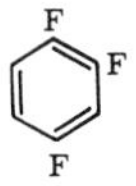
F, F, F
224

▶1745 Sh
CCl_4
Prism-Grating, ±1 cm^{-1}
$\overset{O}{\overset{\|}{C}}-OCH_3$, OCH_3
[C=O] 25

▶1743 M
B (0.2398) 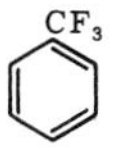
CF_3
219

▶1742 S
CCl_4
Prism–Grating, ±1 cm^{-1}
[C=O] 25

▶1742
H
Br Cl
[C=O] 674

▶1742 S
HO Br Br OH Br Br O O
39

▶1742
A
$H_3C-C(=O)-CH_3$
[C=O] 516

▶1741 S
E
$H_3C-NH-C_6H_4-N=C(CN)-C(=O)-OC_2H_5$
[C=O] 562

▶1741 S
CCl_4
Prism–Grating, ±1 cm^{-1}
C-OH OCH_3
[C=O] 25

▶1741 S
CCl_4
Prism–Grating, ±1 cm^{-1}
$C-OCH_3$ F
[C=O] 25

▶1741
H
[C=O] 674

▶1740 S
CCl_4
Prism–Grating, ±1 cm^{-1}
$C-OCH_3$ I
[C=O] 25

▶1740 S
CCl_4
Prism–Grating, ±1 cm^{-1}
C-OH H_3C
[C=O] 25

▶1740 S
G
C-OH CH_3 N H
[C=O] 673

▶1740 S
E
$(H_3C)_2N-C_6H_4-N=C(CN)-C(=O)-OC_2H_5$
[C=O] 562

▶1740 S
HO OH O O
39

▶1740 S
CCl_4
Prism–Grating, ±1 cm^{-1}
C-OH $N(CH_3)_2$
[C=O] 25

▶1740 M
B (0.2398)
$CH_3CH_2-O-CF_2CHFCl$
261

▶1740
$CHCH_2C(=O)-H$ C=O
[C=O of C(=O)-H] 582

▶1740
CH_2Cl_2
NH-NH N=N S =O N-H C=O OCH_3
580

▶1740
$H_2C-C(=O)-OC_2H_5$
H_2C
H_2C
$H_2C-C(=O)-OC_2H_5$
[C=O] 571

▶1740
A
$H-C(=O)-NH_2$
[C=O] 529

▶1739 S
$CHCl_3$
=O CH_3 C=O C=O OCH_3 OCH_3
[C=O] 556

▶1739 S
CCl_4
Prism–Grating, ±1 cm^{-1}
$C_6H_4(F)$–C(=O)–OH
[C=O] 25

▶1739 M
C (0.15)
$(CH_3)_3CNO_2$
325

▶1738 S
CCl_4
Prism–Grating, ±1 cm^{-1}
$C_6H_4(Br)$–C(=O)–OH
[C=O] 25

▶1738 S
CCl_4
Prism–Grating, ±1 cm^{-1}
$C_6H_4(Cl)$–C(=O)–OH
[C=O] 25

▶1738 S
CCl_4
Prism–Grating, ±1 cm^{-1}
$C_6H_4(NO_2)$–C(=O)–OCH_3
[C=O] 25

▶1738 S
G
O=, O=, N–H (isatin ring) –$CH_2C(=O)$–OC_2H_5
[C=O] 673

▶1737 S
CCl_4
Prism–Grating, ±1 cm^{-1}
H_3CO–C_6H_4–C(=O)–OH
[C=O] 25

▶1737 S
CCl_4 [0.0082 m/l]
C_6H_5–C(=O)–OH
[C=O] 448

▶1737 S
CCl_4
Prism–Grating, ±1 cm^{-1}
O_2N–C_6H_4–C(=O)–OCH_3
[C=O] 25

▶1737
H
Br Br, =O (cyclohexanone)
[C=O] 674

▶1737 M
B (0.0576)
F–C_6H_4–F
225

▶1736 S
CCl_4
Prism–Grating, ±1 cm^{-1}
$C_6H_4(OCH_3)$–C(=O)–OCH_3
[C=O] 25

▶1736 S
(dioxane)
CH_3, C=O, OCH_3, OH, =O
[C=O] 556

▶1736 S
B (0.2398)
CH_3–C_6H_4–F
220

▶1736
H
$CH_2ClCOOH$
[C=O] 674

▶1736 S
A (775 mm Hg, 10 cm)
CF_2Cl-CF_2Cl
257

▶1736 S
B (0.50)
$(CH_3)_2CHCH_2NO_2$
326

▶1736
CF_3-CH_2-O-N=O
[N=O, trans] 584

▶1736
H
C_6H_5–C(=O)–Cl
[C=O] 513, see also 38

▶1736
BrH_2C-C(=O)-OC_2H_5
[C=O] 583

▶1735 S CCl_4 Prism-Grating, ±1 cm^{-1} 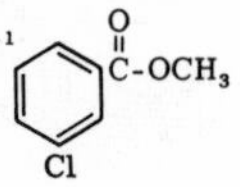 [C=O] 25	▶1733 S CCl_4 Prism-Grating, ±1 cm^{-1} 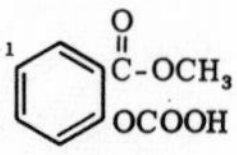 [C=O] 25
▶1735 S 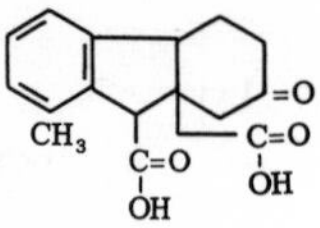 [C=O] 556	▶1733 S CCl_4 [0.005 m/l] 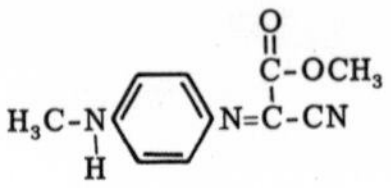 [C=O] 562
▶1735 S G 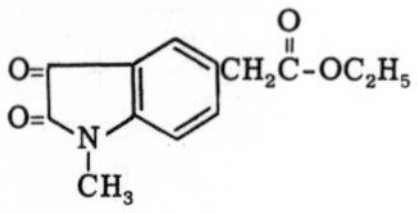 [C=O] 673	▶1733 S CCl_4 Prism-Grating, ±1 cm^{-1} 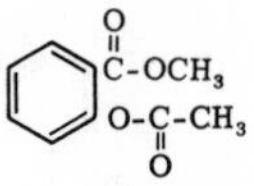 [C=O] 25
▶1735 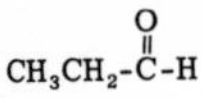 CH_3CH_2-C(=O)-H [C=O] 520 ▶1735 A 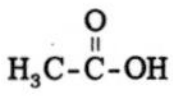 H_3C-C(=O)-OH [C=O] 521	▶1733 S CCl_4 [0.005 m/l] ±2 cm^{-1} 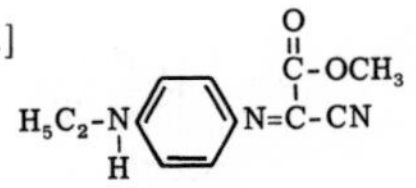 [C=O] 562
▶1734 S CCl_4 Prism-Grating, ±1 cm^{-1} 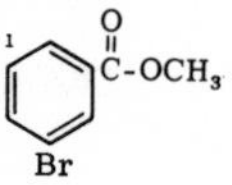 [C=O] 25	▶1733 S CCl_4 Prism-Grating, ±1 cm^{-1} 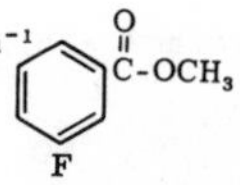 [C=O] 25
▶1734 S 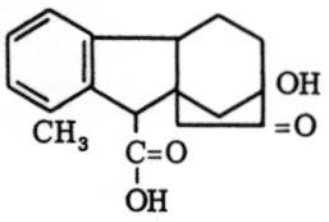 [C=O] 556	▶1733 H 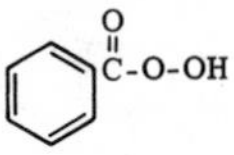 [C=O] 674
▶1734 S 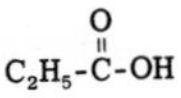 C_2H_5-C(=O)-OH [C=O of monomer] 674	▶1733 MW A (682 mm Hg, 10 cm) 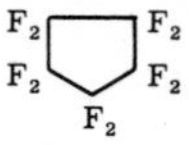 245
▶1734 S CCl_4 Prism-Grating, ±1 cm^{-1} 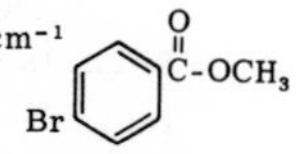 [C=O] 25	▶1733 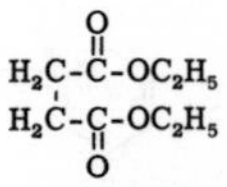 H_2C-C(=O)-OC_2H_5 H_2C-C(=O)-OC_2H_5 [C=O] 571
▶1734 C 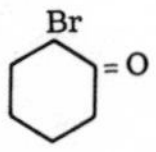 [C=O] 674	▶1733 H 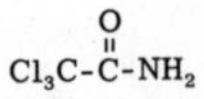 Cl_3C-C(=O)-NH_2 [C=O] 674
▶1734 M B 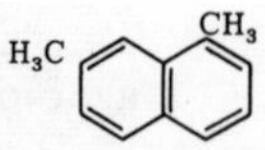 183	▶1732 S B (0.10) 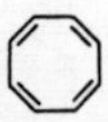 200

▶1732 S
CCl_4
Prism-Grating, ±1 cm^{-1} 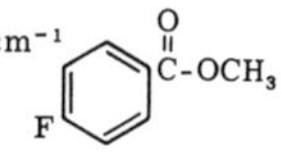
[C=O] 25

▶1732 MS
A (775 mm Hg, 10 cm)
CF_2Cl-CF_2Cl
257

▶1732
$CHCl_3$ 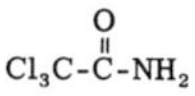
$Cl_3C-C(=O)-NH_2$
531

▶1731 S
E 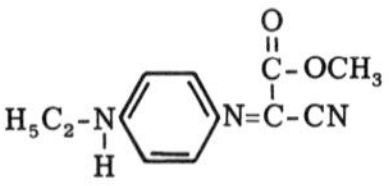
[C=O] 562

▶1731 S
CCl_4
Prism-Grating, ±1 cm^{-1} 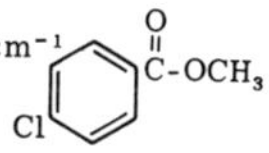
[C=O] 25

▶1731 S
CCl_4 [0.005 m/l] 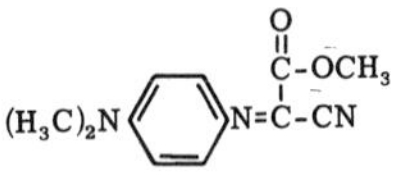
$(H_3C)_2N$
[C=O] 562

▶1731 S
A (600 mm Hg, 10 cm)
$CF_3CF_2CF_2CF_2CF_3$
237

▶1731 S
C 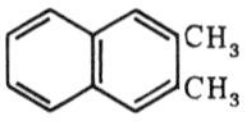
181

▶1731
H
$CH_2BrCOOH$
[C=O] 674

▶1730 VVS
CCl_4 [25%] 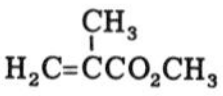
$H_2C=C(CH_3)CO_2CH_3$
193

▶1730 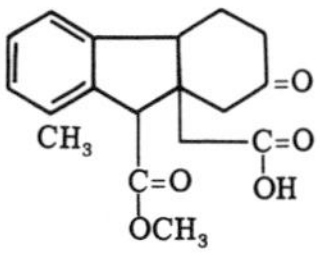
[C=O] 556

▶1730 S
CCl_4
Prism-Grating, ±1 cm^{-1} 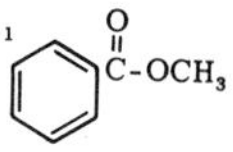
[C=O] 25, see also 38

▶1730 S
E 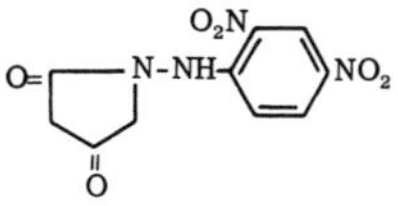
577

▶1730 MW
B (0.064) 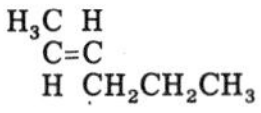
(trans)
169

▶1730 M
B (0.728)
$(CH_3)_3CCH_2CH_3$
145

▶1730
$H_2C=CF_2$
[C=C] 546

▶1730
$Cl_2C=CF_2$
[C=C] 546

▶1730 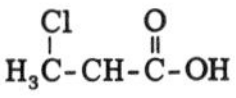
$H_3C-CH(Cl)-C(=O)-OH$
523

▶1728
H 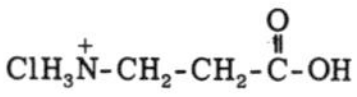
$ClH_3\overset{+}{N}-CH_2-CH_2-C(=O)-OH$
[C=O] 674

▶1728 S
CCl_4
Prism-Grating, ±1 cm^{-1} 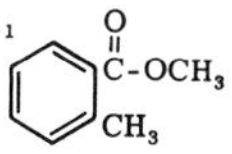
[C=O] 25

▶1728 S
G
O= O= N H; $C-NH_2$ (C=O); CH_3
[C=O] 673

▶1728 S
CCl_4
Prism-Grating, ±1 cm^{-1}
$C-OCH_3$ (C=O); OCH_3
[C=O] 25

▶1727 S
B (0.2398)
F; F; F
224

▶1727 S
CCl_4
Prism-Grating, ±1 cm^{-1}
$C-OCH_3$ (C=O); $N(CH_3)_2$
[C=O] 25

▶1727 S
CCl_4 [0.005 m/1]
$H_3C-N(H)$; $C-OC_2H_5$ (C=O); $N=C-CN$
[C=O] 562

▶1727 S
CCl_4
Prism-Grating, ±1 cm^{-1}
$C-OCH_3$ (C=O); Cl
[C=O] 25

▶1727 SSh
B (0.127)
$H_2C=C(CH_3)(CH_2)_4CH_3$
98

▶1727 M
B (0.2398)
CF_3
219

▶1726 S
E, CCl_4
$H_5C_2-N(H)$; $C-OC_2H_5$ (C=O); $N=C-CN$
[C=O]

▶1726
⌬
H_3C; O=C; CH_2; O-C=O; =O
[C=O] 556

▶1726 S
CCl_4
Prism-Grating, ±1 cm^{-1}
$C-OCH_3$ (C=O); F
[C=O] 25

▶ 1726
H
O; =O
[C=O] 674

▶1726 M
C
CH_3; CH_3
185

▶1725 W
B
F CF_3; F_2; F_2; F; CF_3; F_3C; F; F_2
228

▶1725 S
C_6H_6 [0.0172 m/1]
C-OH (C=O)
448, see also 38

▶1724 S
CCl_4 [0.005 m/1]
$(H_3C)_2N$; $C-OC_2H_5$ (C=O); $N=C-CN$
[C=O] 562

▶1725 S
HO; OH; O; O
39

▶1725 M
B (0.2398)
$H_3C-O-CF_2CHFCl$
356

▶1723 S
CCl_4
Prism-Grating, ±1 cm^{-1}
$C-OCH_3$ (C=O); H_3CO
[C=O] 25

▶1723 S
G
O= O= N H; $CH_2C-OC_2H_5$ (C=O)
[C=O] 673

▶ 1723

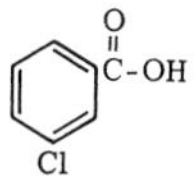

O, C-OH, Cl

585

▶ 1722 S

CCl_4 [0.005 m/l]

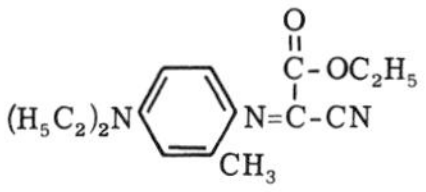

O, C-OC_2H_5, $(H_5C_2)_2N$, N=C-CN, CH_3

[C=O] 562

▶ 1721 W

CH_3

$H_2C=CCO_2CH_3$

[C=O] 527

▶ 1720-1715

A

O

$H_3C-C-NHC_2H_5$

[C=O] 529

▶ 1720 S

B (0.2398)

F, F

225

▶ 1720 VW

B (0.2398)

F, CF_3, CF_3, F, F_2, F_2, F_2, F_2

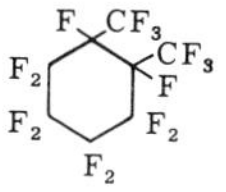

231

▶ 1720 M

B (0.056)

$CH_2=CHCH=CHCH=CH_2$

203

▶ 1720 MS

B (0.955)

$CF_2Cl-CFCl_2$

255

▶ 1720

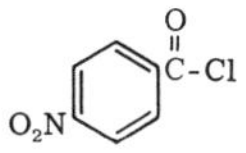

O, C-Cl, O_2N

[C=O] 385

▶ 1720

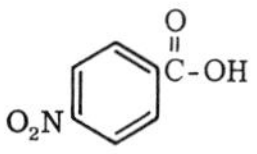

O, C-OH, O_2N

586

▶ 1719 S

E

O, O, O

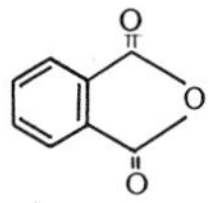

39

▶ 1718 VS

CCl_4

Prism-Grating, ±1 cm^{-1}

O, C-OCH_3, $OCH_2CH=CH_2$

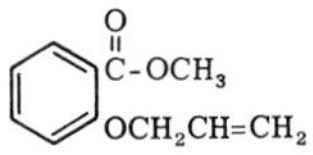

[C=O] 25

▶ 1718 VS

CCl_4

Prism-Grating, ±1 cm^{-1}

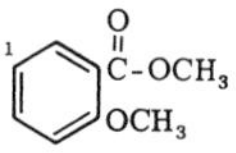

O, C-OCH_3, OCH_3

[C=O] 25

▶ 1718

H

O H

$CHF_2-C-N-C_2H_5$

[C=O] 674

▶ 1718 M

B (0.10)

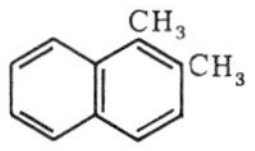

CH_3, CH_3

188

▶ 1718 W

CH_3

$H_2C=CCO_2CH_3$

[C=O] 527

▶ 1718

B

O

$H_3C-C-CH_3$

516

▶ 1717 S

C_6H_6 [0.0128 m/l]

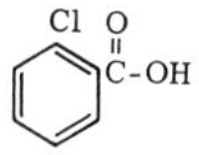

Cl, O, C-OH

448

▶ 1717 M

B (0.10)

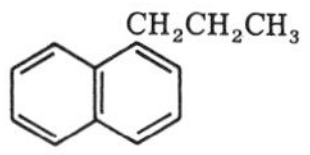

$CH_2CH_2CH_3$

178

▶ 1717

B

O

$H_3C-C-OH$

[C=O] 521

▶1716 S
B (0.2398) 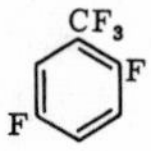
217

▶1712 S
$CHCl_3$ 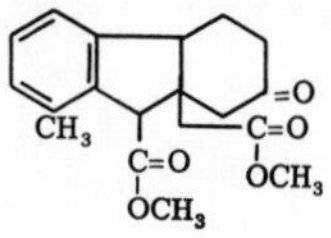
[C=O]
556

▶1715 VS
G 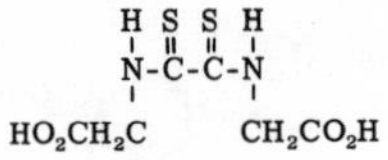
[C=O]
31

▶1712 SSp
B (0.028) 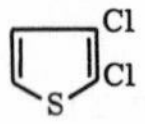
162

▶1715 Sh 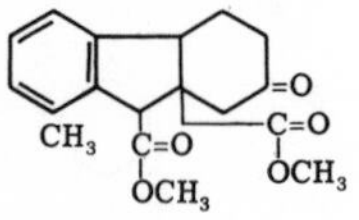
[C=O]
556

▶1711 S
CCl_4
Prism-Grating, ±1 cm^{-1} 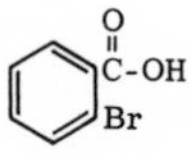
[C=O]
25

▶1715 S
CCl_4
Prism-Grating, ±1 cm^{-1} 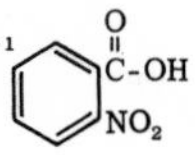
[C=O]
25

▶1711 M
B (0.10) 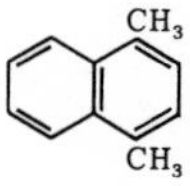
186

▶1715 S 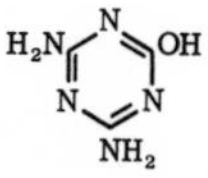
587

▶1710 S
B (0.169)
$CH_3CH=CHCH_2CH_3$ (trans)
207

▶1715 M
B (0.2398) 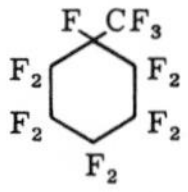
232

▶1710 S
E 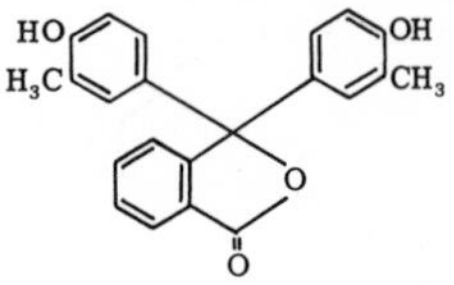
39

▶1714 S
B (0.2398) 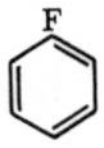
226

▶1710 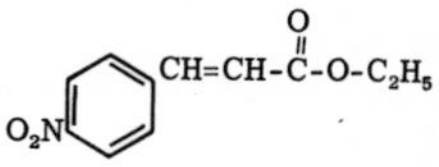
[C=O]
576

▶1713.5 S
B (0.2)
Reference for wavelength
indene + cyclohexanone
9

▶1709 VS
B (0.2398) 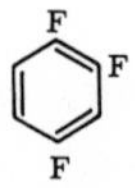
224

▶1713 S
E 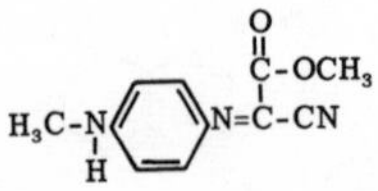
[C=O]
562

▶1709 W
B (0.2398) 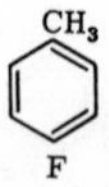
220

▶1712 MS
B (0.10) 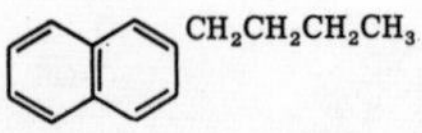
175

▶1709 S
CCl_4
Prism-Grating, ±1 cm^{-1} 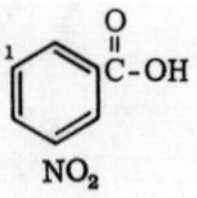
[C=O]
25

▶1709 M
C 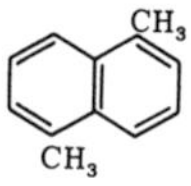
CH_3 CH_3
185

▶1709
$CHCl_3$
O
‖
$H-C-NH_2$
[C=O] 529

▶1708 S
E 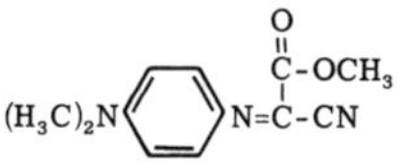
O
‖
$C-OCH_3$
$(H_3C)_2N$ $N=C-CN$
[C=O] 562

▶1708 S
CCl_4
Prism–Grating, ±1 cm^{-1} 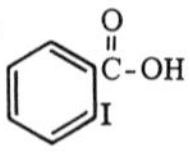
O
‖
C-OH
I
[C=O] 25

▶1708 M
A (200 mm Hg, 10 cm)
$CF_3CF_2CF_3$
235

▶1708 M
C 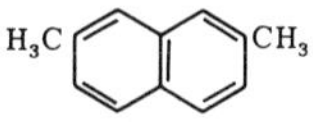
H_3C CH_3
179

▶1707 S
CCl_4
Prism–Grating, ±1 cm^{-1} 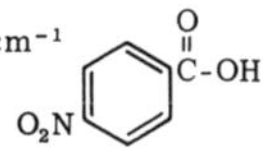
O
‖
C-OH
O_2N
[C=O] 25

▶1707 S
G 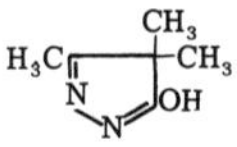
CH_3
H_3C CH_3
N OH
N
[C=C] 6

▶1707 S
CCl_4
Prism–Grating, ±1 cm^{-1} 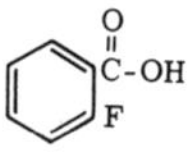
O
‖
C-OH
F
[C=O] 25

▶1707 M
B (0.10) 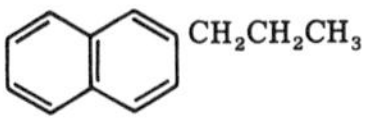
$CH_2CH_2CH_3$
177

▶1707 M
A (50 mm Hg, 10 cm) 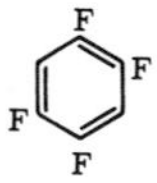
F
F
F
F
223

▶1706 S
CCl_4
Prism–Grating, ±1 cm^{-1} 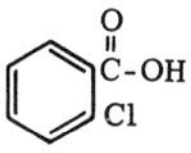
O
‖
C-OH
Cl
[C=O] 25

▶1705 S
B (0.0576)
$(C_4F_9)_3N$
346

▶1705 S
G 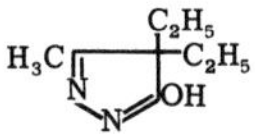
C_2H_5
H_3C C_2H_5
N OH
N
[C=C] 6

▶1705 S
B (0.728)
$(CH_3)_3CCH(CH_2CH_3)_2$
300

▶1705 MSB
B (0.728)
$(CH_3)_3CCH_2CH_3$
145

▶1705
O O
‖ ‖
HO-C-CH=CH-C-OH
(cis)
[C=O] 524

▶1705 W 
N $CH(C_4H_9\text{-}i)_2$
431

▶1704 VS
B (0.2398) 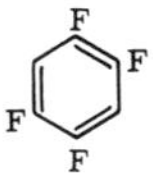
F
F
F
F
222

▶1704 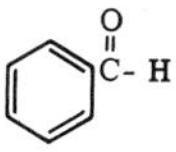
O
‖
C- H
[C=O] 519

▶1703 S CCl_4 Prism-Grating, ±1 cm^{-1} [o-allyloxybenzoic acid: C(=O)-OH, $OCH_2CH=CH_2$] [C=O] 25	▶1700 M CCl_4 [0.0128 m/l] [o-chlorobenzoic acid: Cl, C(=O)-OH] 448
▶1703 S CCl_4 Prism-Grating, ±1 cm^{-1} [C(=O)-OH, F] [C=O] 25	▶1700 M B (0.10) [naphthalene, $CH_2CH_2CH_2CH_2CH_3$] 174
▶1703 S CCl_4 Prism-Grating, ±1 cm^{-1} [C(=O)-OH, Cl] [C=O] 25	▶1700 [O=, =O, H_3C, CH_3] [C=O] 518
▶1702 S CCl_4 Prism-Grating, ±1 cm^{-1} [C(=O)-OH, OCOOH] [C=O] 25	▶1699 S CCl_4 Prism-Grating, ±1 cm^{-1} [C(=O)-OH, F] [C=O] 25
▶1702 S CCl_4 Prism-Grating, ±1 cm^{-1} [C(=O)-OH, Br] [C=O] 25	▶1698 VS B (capillary) [HO, CH_3, $CH_2CH=CHCH_3$, O] (trans) 430
▶1702 W CCl_4 Prism-Grating, ±1 cm^{-1} [C(=O)-OH, OCH_3] [C=O] 25	▶1698 VS B (0.2398) [CF_3, F] 218
▶1702 MW CCl_4 [H_3C, CH_3, OH, O, =O, CH_3] [C=C] 475	▶1698 S CCl_4 Prism-Grating, ±1 cm^{-1} [C(=O)-OH, OCH_3] [C=O] 25
▶1701 MS A (50 mm Hg, 10 cm) [F, F, F, F] 223	▶1698 S CCl_4 Prism-Grating, ±1 cm^{-1} [C(=O)-OH, CH_3] [C=O] 25
▶1700 S E, F [H-N, CF_3, N, HO-C(=O)-CH_2] [C=O] 37	▶1697 S B (0.0576) [CF_3] 219
▶1700 SB B (0.728) $(CH_3)_3CCH_2CH_2C(CH_3)_3$ 305	▶1697 S CCl_4 Prism-Grating, ±1 cm^{-1} [C(=O)-OH, H_3C] [C=O] 25

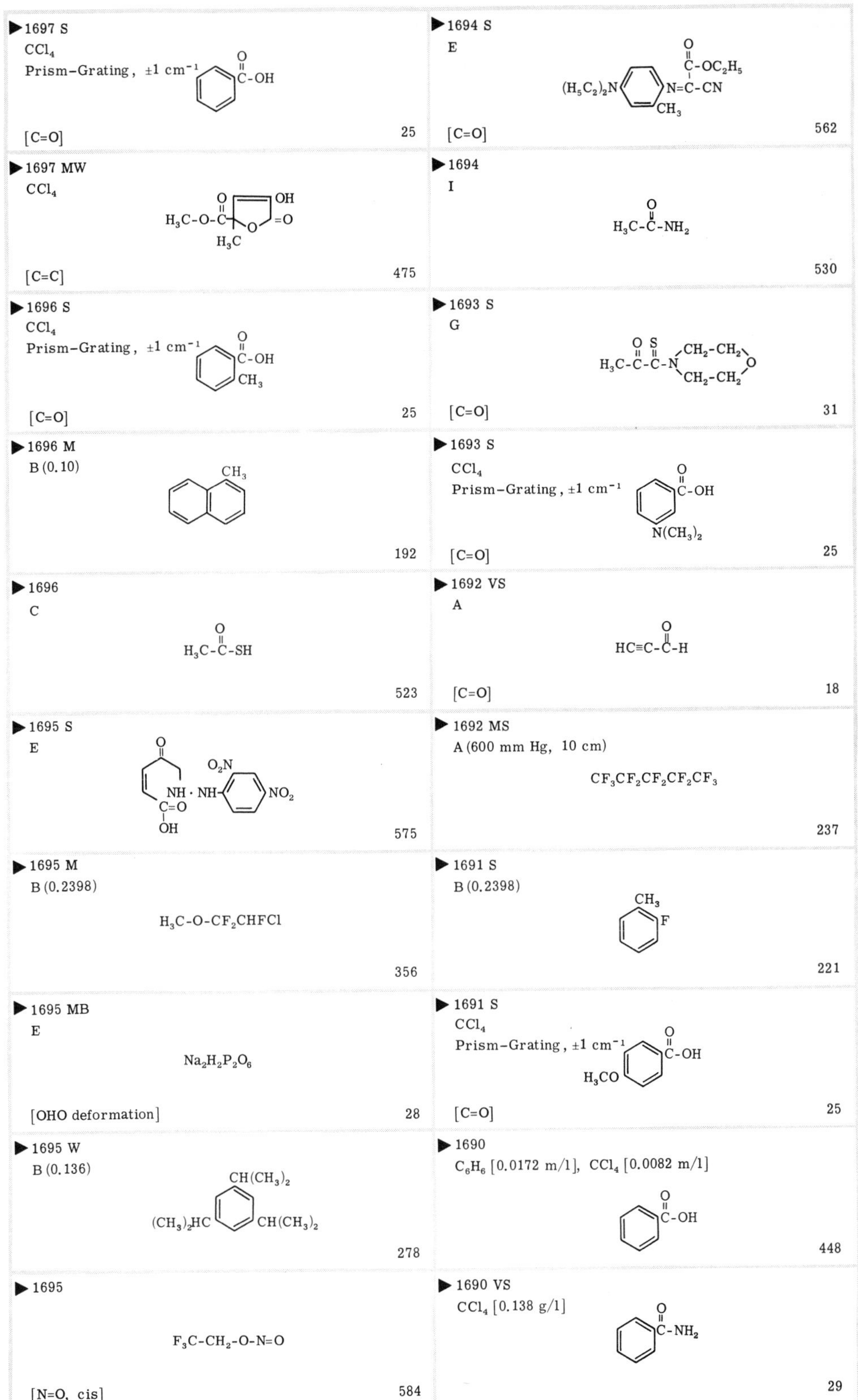

▶ 1697 S
CCl_4
Prism–Grating, ±1 cm^{-1}
[C=O] 25

▶ 1697 MW
CCl_4
[C=C] 475

▶ 1696 S
CCl_4
Prism–Grating, ±1 cm^{-1}
[C=O] 25

▶ 1696 M
B (0.10)
192

▶ 1696
C
$H_3C-C(=O)-SH$
523

▶ 1695 S
E
575

▶ 1695 M
B (0.2398)
$H_3C-O-CF_2CHFCl$
356

▶ 1695 MB
E
$Na_2H_2P_2O_6$
[OHO deformation] 28

▶ 1695 W
B (0.136)
278

▶ 1695
$F_3C-CH_2-O-N=O$
[N=O, cis] 584

▶ 1694 S
E
[C=O] 562

▶ 1694
I
$H_3C-C(=O)-NH_2$
530

▶ 1693 S
G
[C=O] 31

▶ 1693 S
CCl_4
Prism–Grating, ±1 cm^{-1}
[C=O] 25

▶ 1692 VS
A
$HC\equiv C-C(=O)-H$
[C=O] 18

▶ 1692 MS
A (600 mm Hg, 10 cm)
$CF_3CF_2CF_2CF_2CF_3$
237

▶ 1691 S
B (0.2398)
221

▶ 1691 S
CCl_4
Prism–Grating, ±1 cm^{-1}
[C=O] 25

▶ 1690
C_6H_6 [0.0172 m/l], CCl_4 [0.0082 m/l]
448

▶ 1690 VS
CCl_4 [0.138 g/l]
29

▶1690 S
B (0.728)
$(CH_3)_3CC(CH_3)_2CH_2CH_2CH_3$
304

▶1690 S
E, F 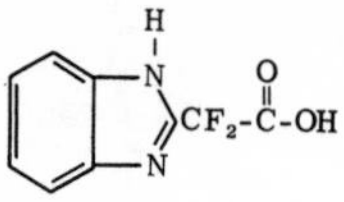
(2-(carboxydifluoromethyl)benzimidazole: CF_2-C(=O)-OH)
[C=O]
37

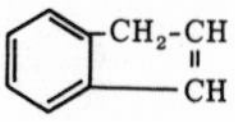

▶1690 M
CCl_4 [0.00658 m/l]
(OH, C(=O)-OH on benzene)
448

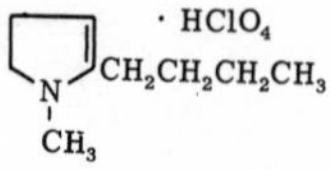

▶1690 SVB
B (0.728)
$(CH_3)_2CHCH_2CH_2CH(CH_3)_2$
137

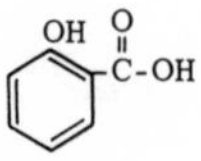

▶1690 M
C_6H_6 [0.0128 m/l]
(Cl, C(=O)-OH on benzene)
448

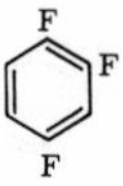

▶1690
(H_2N, C(=O)-OH on benzene)
573

▶1689 VS
CCl_4 [sat.]
(C(=O)-NH_2 on benzene)
29

▶1687 S
E
H_3C-N(H)-C₆H₄-N=C(CN)-C(=O)-OCH_3
[C=O]
562

▶1687 M
C_6H_6 [0.0044 m/l]
(OH, C(=O)-OH on benzene)
448

▶1686 S
E, F
(HO-C(=O)-H_2C)₂ benzimidazole · H_2O
[C=O]
37

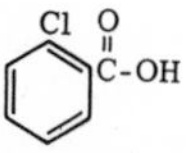

▶1686 S
B (0.2)
CH_2-CH=CH
9

▶1686 S
E
(N-CH_3 pyrroline, $CH_2CH_2CH_2CH_3$) · $HClO_4$
574

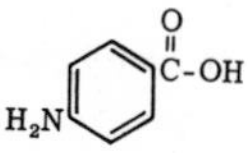

▶1685 W
B (0.2398)
(F, F, F on benzene)
224

▶1685 SVB
B (0.728)
$(CH_3)_3CC(CH_3)_2CH(CH_3)_2$
306

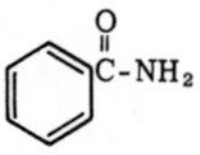

▶1685
H
$CH_3CH=CH-C(=O)-H$
[C=O]
519

▶1684 S
n-C_6H_{14}
CaF_2 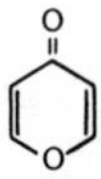
O, O
[C=O]
7

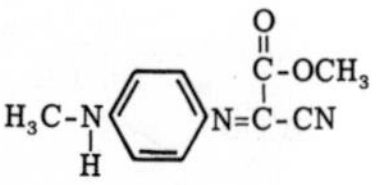

▶1684
CCl_4 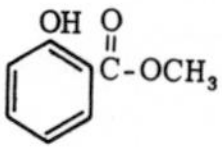
(OH, C(=O)-OCH_3 on benzene)
[-C(=O)-OCH_3]
579

▶1682 S
n-C_6H_{14}
CaF_2 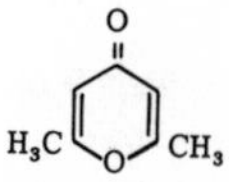
(2,6-dimethyl-4-pyrone: H_3C, CH_3)
[Ring]
7

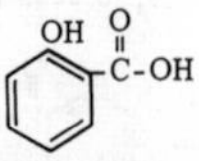

▶1682 M
C 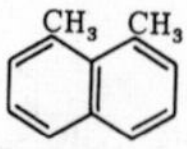
(naphthalene with CH_3, CH_3)
182

▶1682 M
C 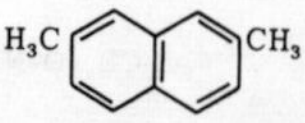
(H_3C-naphthalene-CH_3)
179

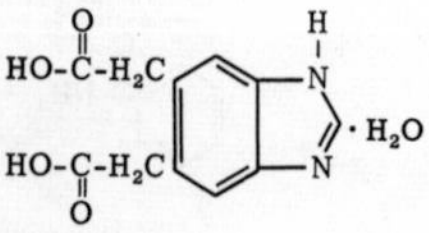

▶1680 M
A (775 mm Hg, 10 cm)

CF_2Cl-CF_2Cl

257

▶1680 S
CH_2Cl_2

O, NH-NH, N=N, S, =O, N-H, C=O, OCH_3

580

▶1680 S
$Cl_2C=CCl_2$
CaF_2

O

[C=O] 7

▶1680

$HO-\overset{O}{\overset{\|}{C}}-CH=CH-\overset{O}{\overset{\|}{C}}-OH$
(trans)

[C=O] 524

▶1679 VS
A (775 mm Hg, 10 cm)

CF_2Cl-CF_2Cl

257

▶1679 S
$Cl_2C=CCl_2$
CaF_2

O, H_3C, CH_3

[Ring] 7

▶1679
$CHCl_3$

$CH_3(CH_2)_n-\overset{O}{\overset{\|}{C}}-NH_2$
(n = 1-10)

[C=O] 530

▶1678 VS
CCl_4 [20 g/l]

$CH_3CH=CH(CH_2)_2CH=CH(CH_2)_2CO-NH-CH_2CH(CH_3)_2$

357

▶1678 S
B (0.2398)

CH_3, F

221

▶1678 SSh
C_6H_6
CaF_2

O

[C=O] 7

▶1678

$=CH_2$

[C=C] 572

▶1677 S
G

$H_2N-\overset{O}{\overset{\|}{C}}-\overset{S}{\overset{\|}{C}}-NH-C_6H_{11}$

[C=O] 31

▶1676 S
E, F

H, N, N, $CF_2CF_2-\overset{O}{\overset{\|}{C}}-OH$

[C=O] 37

▶1675 VS
CCl_4 [0.3 g/l]

$\overset{O}{\overset{\|}{C}}-NH-CH_3$

[C=O] 2

▶1675 VS
$CHCl_3$ [1.39 g/l]

$\overset{O}{\overset{\|}{C}}-NH_2$

29

▶1675 S
B (0.0563)

H CH_2CH_3
C=C
H_3C CH_3
(trans)

428, see also 165

▶1675 SVB
B (0.728)

$(CH_3)_2CHCH_2CH(CH_3)CH_2CH_3$

138

▶1675 SB
B (0.728)

$(CH_3)_2CHC(CH_3)_2CH_2CH_2CH_3$

134

▶1675 S
C_6H_6
CaF_2

O

[C=O] 7

▶1675 S
CH_3CN
CaF_2

O

[C=O] 7

▶1675 S
$Cl_2C{=}CCl_2$
CaF_2
O
H_3C O CH_3
[Ring] 7

▶1675 S
G
H_3C
H_3C-N N =O
[C=C] 6

▶1675 MS
B (0.064)
H CH_2CH_3
C=C
H_3C CH_3
(trans)
165

▶1675
$(CH_3)_2C{=}N{-}OH$
[C=N] 570

▶1674 S
B (0.0576)
F F F F
222

▶1674 S
$CHCl_3$
CaF_2
O
O
[C=O] 7

▶1674 S
C_6H_6
CaF_2
O
H_3C O CH_3
[Ring] 7

▶1674 S
G
H O S H
N-C-C-N
C_6H_5 C_6H_{11}
[C=O] 31

▶1673 VS
CCl_4 [0.007 g/ml]
O
C-NH
CH_3
[C=O] 2

▶1673 W
B (0.2398)
CF_3
F
218

▶1673 S
CH_3CN
CaF_2
O
H_3C O CH_3
[Ring] 7

▶1673 M
A (50 mm Hg, 10 cm)
F F F F
223

▶1673
CH_3
[C=C] 542

▶1672 VS
$CHCl_3$ [7.6 g/l]
O
C-NH_2
29

▶1672 MS
B (0.064)
$(CH_3)_2C{=}CHCH_2CH_3$
166

▶1672 VSSp
CCl_4 [sat.] (0.025)
Br S $COCH_3$
348

▶1672 S
$CH_2Br_2CH_2Br_2$
CaF_2
O
O
[C=O] 7

▶1672 S
G
H O S H
N-C-C-N
C_6H_5 C_6H_5
[C=O] 31

▶1670 M
B (0.064)
H_3C CH_3
C=C
H CH_2CH_3
(cis)
164, see also 274

▶1670 S
B (0.169)
$CH_3CH{=}CHCH_2CH_3$
(trans)
207

▶1670 MS
B (0.169)
$CH_3(CH_2)_2CH{=}CH(CH_2)_2CH_3$
(trans)
205

▶1670 W
C_3H_7-n (pyridine)
431

▶1670 S
$[C_5H_4\overset{O}{\overset{\|}{C}}-CH_3]$ Os $[C_5H_5]$
[C=O]
450

▶1670 S
E, F
$HO-\overset{O}{\overset{\|}{C}}-H_2C$ (benzimidazole, N-H)
[C=O]
37

▶1670 S
$CHCl_3$
H_3C, CH_3 (pyranone)
[Ring]
7

▶1670 MSB
B (0.728)
$(CH_3)_3CCH_2CH_3$
145

▶1670
$CHCH_2\overset{O}{\overset{\|}{C}}-H$, C=O (two phenyl rings)
582

▶1669 S
A (200 mm Hg, 10 cm)
$CF_3CF_2CF_3$
235

▶1669 VS
CCl_4 [31%] (0.008)
Cl (thiophene, S) $COCH_3$
372

▶1669 W
B (0.064)
H_3CH_2C H
C=C
H CH_2CH_3
(trans)
168

▶1669
O= (cyclohexadiene) =O
[C=O]
690

▶1669 MW
B (0.064)
H_3C H
C=C
H $CH_2CH_2CH_3$
(trans)
169

▶1668 M
N
(1,3,5-triazine: N, N, N)
14

▶1668 S
C_6H_6
HNO_3
[NO_2 asymm. stretch]
642

▶1667 M
E
(1,3,5-triazine: N, N, N)
14

▶1667 WB
B (0.169)
$(CH_3)_3CCH_2\overset{CH_3}{CH}CH_2CH_3$
285

▶1667 SVB
B (0.728)
$CH_3CH_2\overset{CH_3}{\underset{CH_3}{C}}-\overset{CH_3}{CH_2CHCH_2CH_3}$
302

▶1667 SVB
B (0.728)
$(CH_3)_2CH\overset{CH_3}{CH}CH_2CH_2CH_3$
139

▶1667 S
$CH_2Br_2CH_2Br_2$
CaF_2
H_3C, CH_3 (pyranone)
[Ring]
7

▶1667 W
B (0°C)
$(CH_3)_2CH\underset{CH_3}{CH}CH(CH_3)_2$
284

▶1666 VS
$CHCl_3$
H_3C CH_3
H_3CO O N
[Ring] 15

▶1666 VVS
$CHCl_3$
H_3C CH_3
H_5C_2-O O N
[Ring] 15

▶1665 S
G
H O S H
N-C-C-N
C_6H_{11} C_6H_{11}
[C=O] 31

▶1665 VVS
$CHCl_3$
H_3C CH_3
HO O N
[Ring] 15

▶1665 S
E
O O_2N
NH-NH NO_2
S =O
N-H
580

▶1665 S
D (0.013)
$CH_2CH_2CH_2CH_3$
N
CH_3
574

▶1664 MS
B (0.0563)
$(CH_3)_2C{=}CHC(CH_3)_3$
204

▶1664 S
G
H O O H
N-C-C-N
C_6H_5 C_6H_5
[C=O] 31

▶1664 SB
B (0.728)
CH_3
$(CH_3)_2CHCHCH_2CH_3$
142

▶1664 S
$CH_3CH{=}N\text{-}N{=}CHCH_3$
[C=N] 567

▶1663 VS
B (0.2398)
CF_3
219

▶1663 MS
A (600 mm Hg, 10 cm)
$CF_3CF_2CF_2CF_2CF_3$
237

▶1662 S
CH_3CN
CaF_2
O
O
[C=O] 7

▶1662 M
C
CH_3
CH_3
185

▶1661 SB
B (0.728)
$(CH_3)_2CHCHCH(CH_3)_2$
CH_2
CH_3
301

▶1661 SB
B (0.728)
$(CH_3)_2CHCH(CH_3)_2$
144

▶1661 SB
B (0.728)
$(CH_3)_2CHCHCH(CH_3)_2$
CH_3
135

▶1661 S
$CHCl_3$
S
H_3C O CH_3
[Ring] 7

▶1661 SSh
$CHCl_3$
S
O
[Ring] 7

▶1661 W
B (0.2398)
F CF_3
F_2 F_2
F_2 F_2
F CF_3
230

▶1661-1646

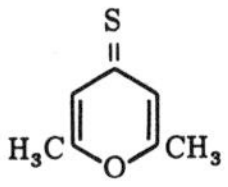

[Ring] 689

▶1660 VS
$CHCl_3$

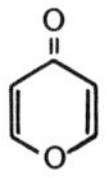

[Ring] 7

▶1660 S
C_6H_6, $CHBr_2CHBr_2$
CaF_2

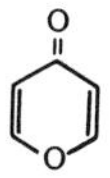

[C=O] 7

▶1660 M
$CHCl_3$

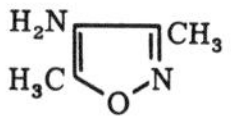

[Ring] 15

▶1660 M
E, F

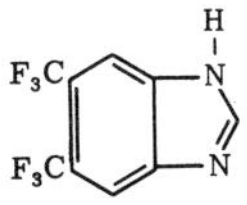

37

▶1660 M
CCl_4 [0.00658 m/1]

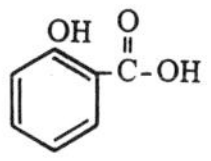

448

▶1660 VVS
$CHCl_3$

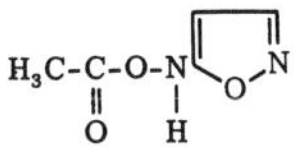

[Ring] 15

▶1659
B

$CH_2Cl-C(=O)-N(C_2H_5)_2$

674

▶1659 M
A

CHF_2CH_3

233

▶1659 VS

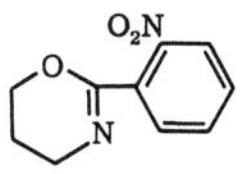

26

▶1658 VS
B (0.0563)

$CH_3CH=CHCH_2CH_3$
(cis)

208

▶1658 S

$[C_5H_4C(=O)-CH_3]Ru[C_5H_5]$

[C=O] 450

▶1658 S
n-C_6H_{14}, $Cl_2C=CCl_2$
CaF_2

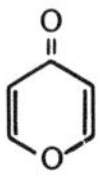

[C=O] 7

▶1658 SB
B (0.728)

$(CH_3)_2CHCH(CH_2CH_3)_2$

136

▶1658 S

$[C_5H_4C(=O)-CH_3]Fe[C_5H_5]$

[C=O] 450

▶1658 W
B

$(C_6H_5)_2SiCl_2$

86

▶1658 W
B

$C_6H_5SiCl_3$

86

▶1658 VVS
$CHCl_3$

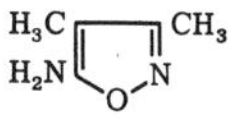

[Ring] 15

▶1657 S
G

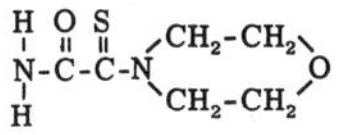

[C=O] 31

▶1657 S
G

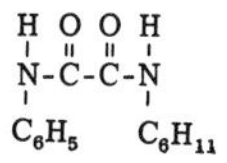

[C=O] 31

▶1657 VS $CHCl_3$ [0.0064 g/ml] [C=O] 2	▶1655 $(H_2N)_2C{=}O$ 589
▶1656 VS B (0.068) H_3C C_3H_7-n C=C H H (cis) 276	▶1655 C_6H_6 [0.0044 m/l] OH O C-OH 448
▶1656 VS B (0.127) $H_2C{=}C(CH_3){-}(CH_2)_4CH_3$ 98	▶1654 VS G H O O H; N-C-C-N; H C_6H_{11} [C=O] 31
▶1656 S B (0.064) $CH_3CH{=}CH(CH_2)_4CH_3$ (cis) 160	▶1654 M B (0.10) CH_3 192
▶1656 W CCl_4 $(C_6H_5)_3SiCl$ 86	▶1653 VS C_6H_{12} N-N= =O CH_3 [C=O] 35
▶1656 VSSh B (0.0576) CF_3 F F [C=O] 217	▶1653 VS $CHBr_3$ [0.0077 g/ml] O C-NH CH_3 [C=O] 2
▶1655 VS B H_3CO OCH_3 O N [C=N] 26	▶1653 VS C_6H_{12} N=N OH Cl [CO] 35
▶1655 VS C_2Cl_4 N=N OH NO_2 [CO] 35	▶1653 VS B (0.0563) $H_2C{=}C(CH_3)CH_2CH_2CH_3$ 206
▶1655 S G H_3C N- N =O CH_3 [C=C] 6	▶1653 W B $(C_6H_5)_2CH_3SiCl$ 86 ▶1653 M B $C_6H_5(CH_3)SiCl_2$ 86
▶1655 M C H_3C CH_3 179	▶1653 VS B (0.238) H_3CH_2C CH_2CH_3 C=C H H (cis) 275

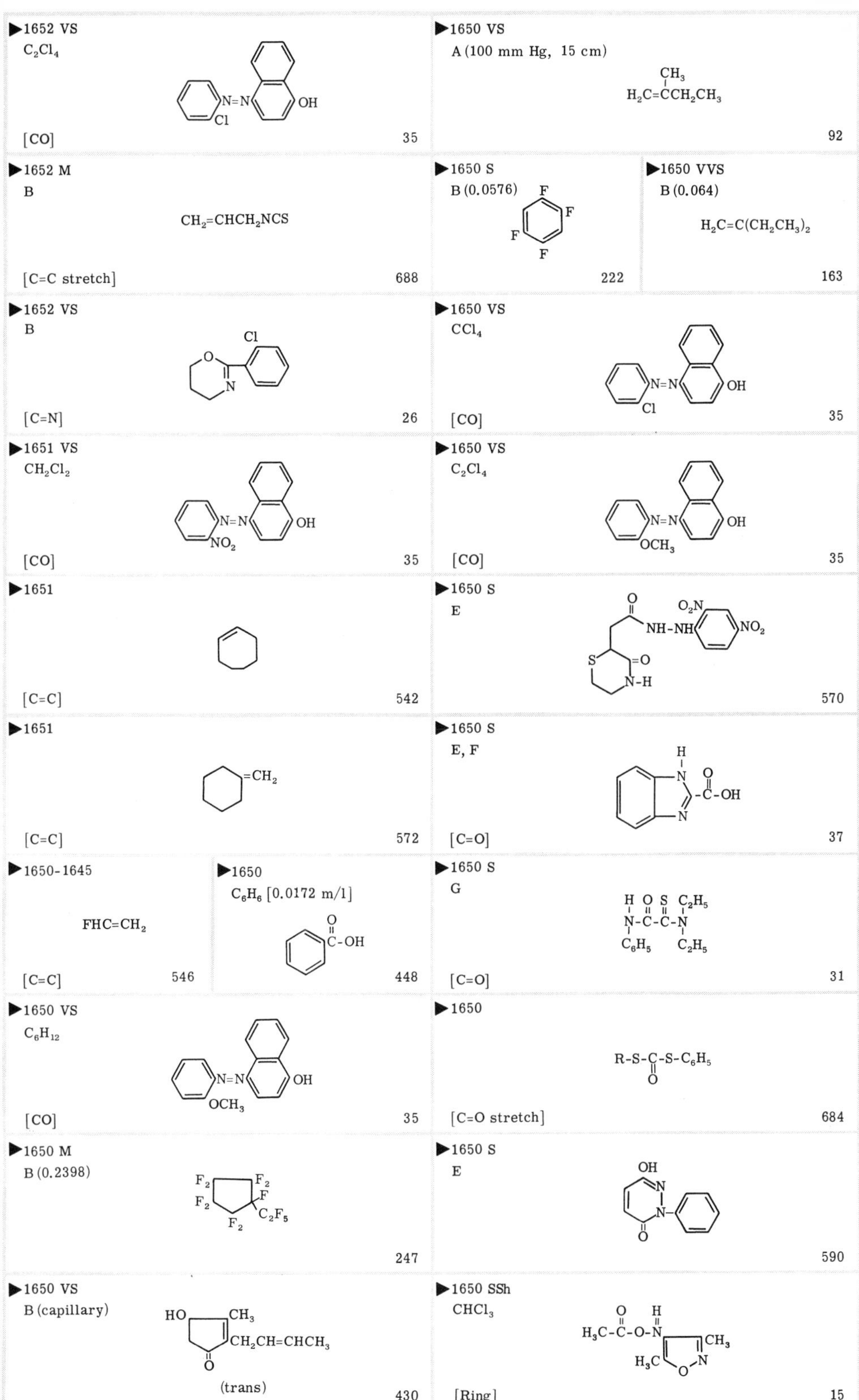
▶1652 VS
C2Cl4
N=N
OH
Cl
[CO]
35
▶1650 VS
A (100 mm Hg, 15 cm)
CH3
H2C=CCH2CH3
92
▶1652 M
B
CH2=CHCH2NCS
[C=C stretch]
688
▶1650 S
B (0.0576)
F
F
F
F
222
▶1650 VVS
B (0.064)
H2C=C(CH2CH3)2
163
▶1652 VS
B
Cl
O
N
[C=N]
26
▶1650 VS
CCl4
N=N
OH
Cl
[CO]
35
▶1651 VS
CH2Cl2
N=N
OH
NO2
[CO]
35
▶1650 VS
C2Cl4
N=N
OH
OCH3
[CO]
35
▶1651
[C=C]
542
▶1650 S
E
O
O2N
NH-NH
NO2
S
=O
N-H
570
▶1651
=CH2
[C=C]
572
▶1650 S
E, F
H
N
O
-C-OH
N
[C=O]
37
▶1650-1645
FHC=CH2
[C=C]
546
▶1650
C6H6 [0.0172 m/l]
O
C-OH
448
▶1650 S
G
H O S C2H5
N-C-C-N
C6H5 C2H5
[C=O]
31
▶1650 VS
C6H12
N=N
OH
OCH3
[CO]
35
▶1650
R-S-C-S-C6H5
O
[C=O stretch]
684
▶1650 M
B (0.2398)
F2
F2
F2
F
C2F5
F2
247
▶1650 S
E
OH
N
N-
O
590
▶1650 VS
B (capillary)
HO
CH3
CH2CH=CHCH3
O
(trans)
430
▶1650 SSh
CHCl3
O H
H3C-C-O-N
CH3
H3C
N
O
[Ring]
15

▶1650
B

$H_3C-C(=O)-NH-C_2H_5$

[Amide I, C=O] 529

▶1650
P

$La(NO_3)_3$

45

▶1650
E

$Mg(NO_3)_2$

45

▶1650 VS
B (0.15)

128

▶1650 VS
B (0.239)

$CH_3(CH_2)_2CH=CH(CH_2)_2CH_3$
(cis)

422

▶1650
E

$Sm(NO_3)_3$

45

▶1650
E

$Be(NO_3)_2$

45

▶1650

N
N

415

▶1650

$C(=O)-NH-CH_2-C(=O)-O-C_2H_5$

594

▶1650

$H_3C-C(=O)-CH(CH_3)-C(=O)-O-C_2H_5$

[C=O, enol form] 579

▶1650

O_2N $CH=CH-C(=O)-O-C_2H_5$

[C=C] 576

▶1649 VS
C_2Cl_4

N-N= OH
OCH_3

[CO] 35

▶1649 S
A (200 mm Hg, 10 cm)

F_2 F_2
F_2 F_2

243

▶1649 VS
B (0.13)

$H_2C=CHCH_2C(CH_3)_3$

102

▶1649 VS
B (0.127)

$H_2C=CH(CH_2)_3CH(CH_3)_2$

97

▶1649 S
A (10 mm Hg, 58 cm)

$H_2C=CH(CH_2)_5CH_3$

289

▶1649

$C(=O)-NH-CH_3$

[Amide I, C=O] 592

▶1648 VS
CH_2Cl_2

N=N OH
O_2N

[CO] 35

▶1648 VS
CCl_4

N=N OH
OCH_3

[CO] 35

▶1648 S
C_6H_{12}

N=N OH
CH_3

[CO] 35

▶1648 S
E, F

H
N
$C(=O)-OH$
N
F_3C

[CO] 37

▶1648
E or N

CH_2OH
N

41

▶1648 I $CH_3CH{=}CH{-}C({=}O){-}H$ [C=C] 545	▶1645 VS B (0.13) $H_2C{=}C(CH_3)CH_2C(CH_3)_3$ 96
▶1647 VVS $CHCl_3$ [Ring] 15	▶1645 S E, $CHCl_3$ 327
▶1647 MS A (36 mm Hg, 15 cm) $H_2C{=}CHCH(CH_3)_2$ 131	▶1645 VS B (0.13) $H_2C{=}CHC(CH_3)_3$ 129
▶1647 VS C_2Cl_4, CCl_4 [CO] 35	▶1645 SSh B (0.0563) $(CH_3)_2C{=}CHC(CH_3)_3$ 204
▶1647 $H_3C{-}C({=}O){-}N(C_2H_5)_2$ [C=O] 532	▶1645 S G $C_6H_{11}NH{-}C({=}O){-}C({=}O){-}NHC_6H_{11}$ 31
▶1646 VVS $CHCl_3$ 7	▶1645 S E, F [C=O] 37
▶1646 VS B (0.10) 175	▶1645 E $Al(NO_3)_2$ 45
▶1646 S C 181	▶1644 VS B [C=N] 26
▶1646 S E $\cdot HBr$ 596	▶1644 VS 26
▶1646 $CHCl_3$ [C=N] 597	▶1644 VS CH_2Cl_2 [CO] 35

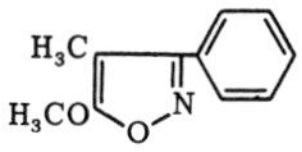

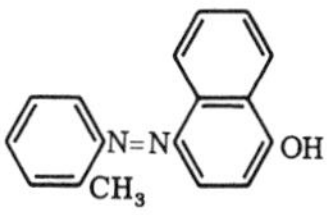

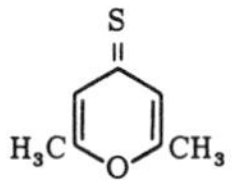

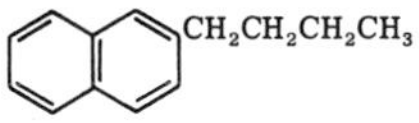

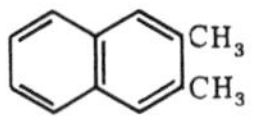

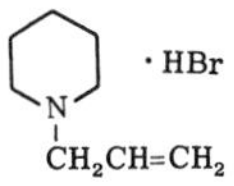

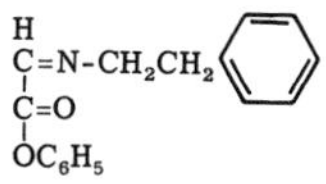

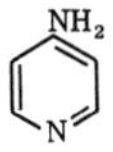

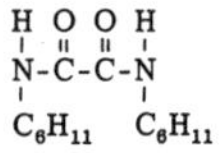

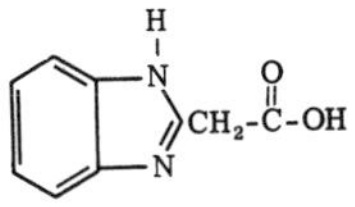

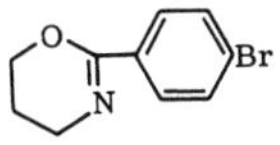

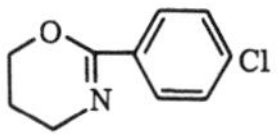

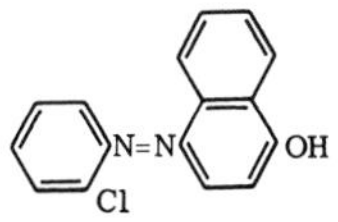

▶1644 S
D (0.013)
N
$CH_2CH=CH_2$
596

▶1643 S
B (0.10)
$CH_2CH_2CH_3$
177

▶1644 S
D (0.013)
N
$CH_2CH=CH_2$
596

▶1643 S
$CHCl_3$
S
O
[Ring] 7

▶1642 VS
B (0.064)
$H_2C=CHCHCH_2CH_3$
CH_3
167

▶1644 S
n-C_6H_{14}
CaF_2
O
H_3C O CH_3
[C=O] 7

▶1642 VS
E, G, CH_2Cl_2
N=N OH
H_3CO
[CO] 35

▶1644 MS
$CHCl_3$
H_3C CH_3
H_3C O N
[Ring] 15

▶1642 VS
B (0.066)
$H_2C=CH(CH_2)_6CH_3$
271

▶1643 VS
$CHCl_3$
N=N OH
O_2N
[CO] 35

▶1642 VS
CCl_4 [20 g/l]
$CH_2CH(CH_3)_2$
$CH_3CH=CH(CH_2)_2CH=CH(CH_2)_2CO-NH$
357

▶1643 VS
B
H_3C CH_3
183

▶1642 S
n-C_6H_{14}, CH_3CN
CaF_2
O
O
[C=O] 7

▶1643 VS
B (0.127)
$H_2C=CH(CH_2)_5CH_3$
290

▶1642 MSh
B (0.064)
H CH_2CH_3
C=C
H_3C CH_3
(trans) 165

▶1642 SSp
B (0.064)
$C_{17}H_{34}$
(1-heptadecene)
159

▶1643 VS
CH_2Cl_2
N=N OH
OCH_3
[CO] 35

▶1642 M
E, F
H
N
CF_3
N
H_3CO
37

▶1643 VS
$CHBr_3$
N=N OH
NO_2
[CO] 35

▶1642 VS
B (0.066)
$H_2C=CH(CH_2)_7CH_3$
270

▶1643 VS
CH_2Cl_2
N=N OH
Cl
[CO] 35

▶1642 M
C_6H_6 [0.0044 m/l]
OH O
C-OH
448

▶1642 W CH_2Cl_2 37	▶1640 E $Gd(NO_3)_3$ 45
▶1641 VS CH_2Cl_2 [CO] 35	▶1640 E $Ce(NO_3)_3$ 45
▶1641 VS $CHCl_3$ [CO] 35	▶1640 E $LiNO_3$ 45
▶1641 MS $(C_2H_5)_2O$ CaF_2 [NH_2] 19	▶1640 P $Y(NO_3)_3$ 45
▶1641 CH_2Cl_2 [CO] 35	▶1640 P $Ca(NO_3)_2$ 45
▶1640 VS E, G 35	▶1640 P $Pb(NO_3)_3$ 45
▶1640 VS CCl_4 [9.0 g/1] 3	▶1640 41
▶1640 VS CH_2Cl_2 [CO] 35	▶1639 VS $CHCl_3$ [CO] 35
▶1640 VS $CHCl_3$ [CO] 35	▶1639 VS E, G 35
▶1640 M C_6H_6 [0.0172 m/1] 448	▶1639 VS B (0.238) $H_2C{=}CC(CH_3)_3$ CH_3 272

▶1639 VVS
B (0.10)

200

▶1639 S
CH_2Cl_2

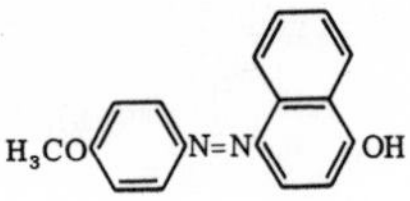

H_3CO N=N OH

[CO] 35

▶1639 SSh
C (0.15)

$(CH_3)_3CNO_2$

325

▶1639 W
B

$(CH_3)_2CHCHCH(CH_3)_2$ with CH_3 on the central CH

284

▶1638 VS
$CHBr_3$

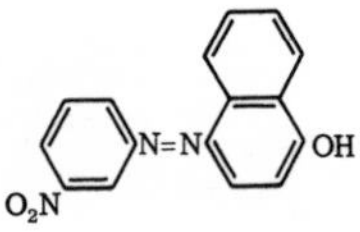

N=N OH O_2N

[CO] 35

▶1638 VS
$CHCl_3$

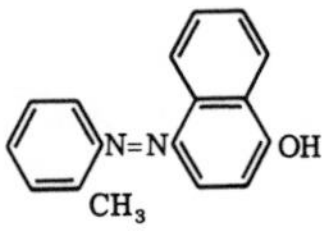

N=N OH CH_3

[CO] 35

▶1638 S
B

CH_3 CH_3

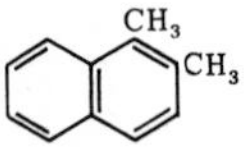

188

▶1638 SSh
$CHBr_2CHBr_2$
CaF_2

O O

[C=O] 7

▶1638
H

$CH_3CH{=}CH{-}\overset{O}{\overset{\|}{C}}{-}H$

[C=C] 545

▶1637 VVS
B

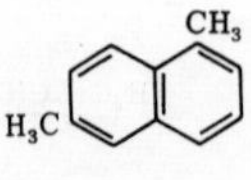

CH_3 H_3C

184

▶1637 VS
B (0.065)

$H_2C{=}CHCH_2C(CH_3)_3$

273

▶1637 S
B (0.0153)

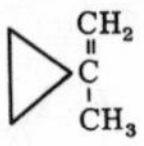

CH_2 = C − CH_3

288

▶1637 VSSp
B (0.0576)

CF_3 F F

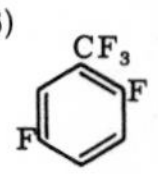

217

▶1637 M
B (0.955)

$CF_2Cl{-}CFCl_2$

255

▶1637 SSh
$CHCl_3$
CaF_2

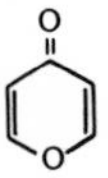

O O

[C=O] 7

▶1636 VVS
$CHCl_3$

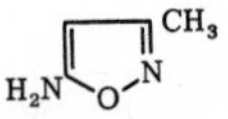

CH_3 H_2N O N

15

▶1636 VS
$CHBr_3$

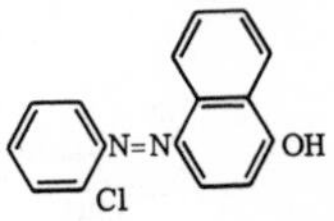

N=N OH Cl

[CO] 35

▶1636 VS
CH_2Cl_2

N−N= =O CH_3

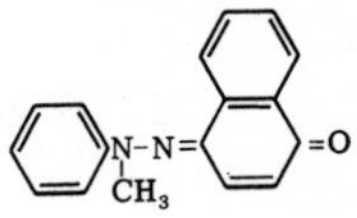

[C=O] 35

▶1636 VS
CS_2 (0.728)

$CH_2{=}CHCH{=}CHCH{=}CHCH{=}CH_2$

202

▶1636 VS
CCl_4 [25%]

$H_2C{=}\overset{CH_3}{\overset{|}{C}}CO_2CH_3$

193

▶1635 VVS
$CHCl_3$

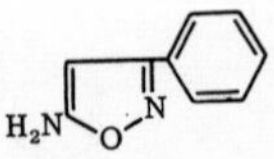

H_2N O N

[Ring] 15

▶1635 VVS
CH_2Cl_2 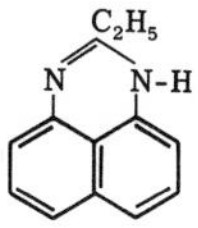
37

▶1635 VVS
CH_2Cl_2 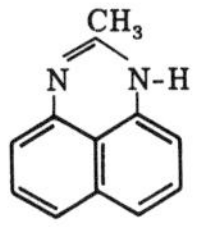
37

▶1635 VS
$CHBr_3$ 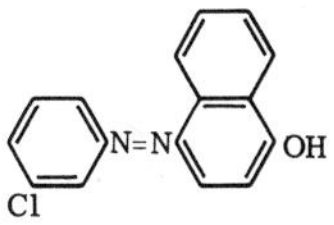
Cl
[CO]
35

▶1635 S
$Cl_2C=CCl_2$
CaF_2
H_3C CH_3
7

▶1635 S
C_6H_6, $CHBr_2CHBr_2$
CaF_2 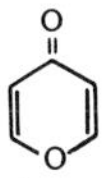
[C=O]
7

▶1635 S
$CHCl_3$
CH=N-N=CH
597

▶1635 M
E, F
CH_2CH_2-C-OH
37

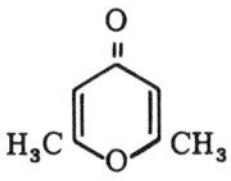

▶1635 M
E, F
CH_2OH
CF_3
37

▶1635 M
E, F
CH_2OH
F_3C
37

▶ 1635 VVS
C_2Cl_4
CaF_2
NO_2 NH_2 NO_2
[NH_2]
19

▶1635
E
$Y(NO_3)_3$
45

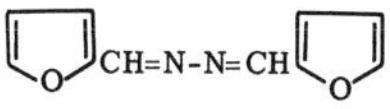

▶1635
P
$AgNO_3$
45

▶1635
P
$Ni(NO_3)_2$
45

▶1635
P
$Ba(NO_3)_2$
45

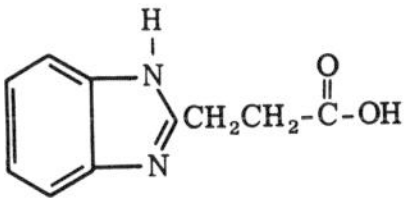

▶1634 VS
$CHCl_3$ 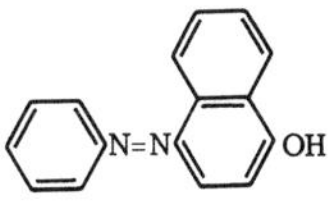
[CO]
35

▶1634 VS
E, CCl_4 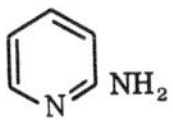
NH_2
327

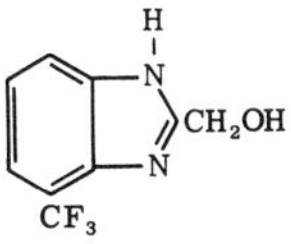

▶1634 VS
B (0.0576) 
225

▶ 1634 S
$CHCl_3$
CaF_2 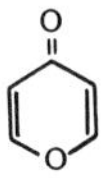
[C=O]
7

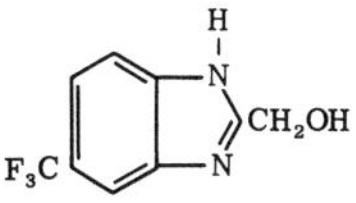

▶ 1634 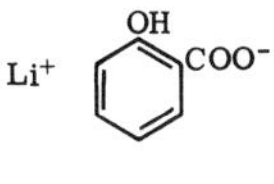
Li^+ OH COO^-
[Aromatic ring]
683

▶ 1634 S 
$[C_5H_4C(=O)-C_6H_5]Ru[C_5H_4C(=O)-C_6H_5]$
[C=O]
450

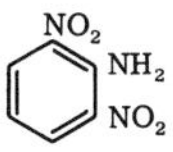

▶1633 VS
$CHCl_3$
N-N= =O
CH_3 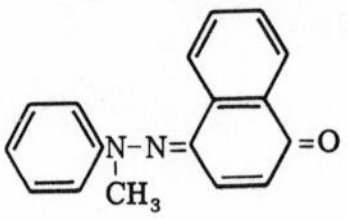
[C=O] 35

▶1633 VS
$CHCl_3$ 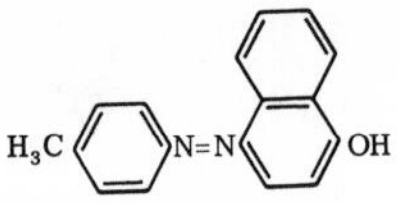
H_3C N=N OH
[CO] 35

▶1633
CH_2Cl_2 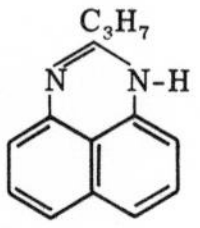
C_3H_7
N N-H
37

▶1633 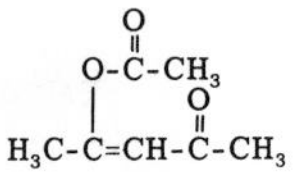
O
O-C-CH_3
O
H_3C-C=CH-C-CH_3
[C=C] 545

▶1633 VVS
C_5H_5N
CaF_2 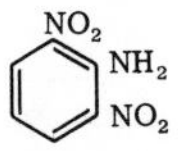
NO_2
NH_2
NO_2
[NH_2] 19

▶1632 VVS
CH_2Cl_2 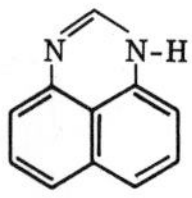
N N-H
37

▶1632 VS
E, G 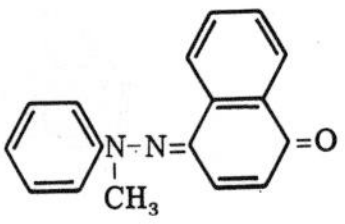
N-N= =O
CH_3
[C=O] 35

▶1632 S
CH_3CN 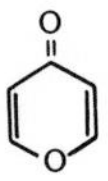
O
O
[C=O] 7

▶1632
P
$Gd(NO_3)_3$
45

▶1632 VS
C_5H_5N
CaF_2 
NH_2
O_2N NO_2
[NH_2] 19

▶1632
O O
H_3C-C-CH-C-O-C_2H_5
CH_3
[C=C] 595

▶1631 S
E, $CHCl_3$ 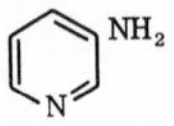
NH_2
N
327

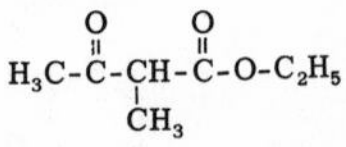

▶1631 S 
O O
[C_5H_4C-] Fe [C_5H_4C-]
450

▶1631 S 
O
[C_5H_4C-] Fe [C_5H_5]
[C=O] 450

▶1630 VS
E, G 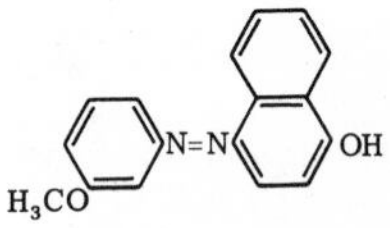
N=N OH
H_3CO
[CO] 35

▶1630 VS
CH_2Cl_2 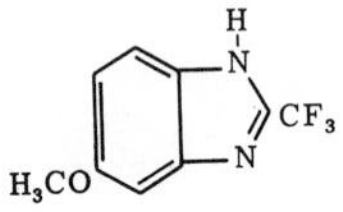
H
N
CF_3
N
H_3CO
37

▶1630 VS
E, G 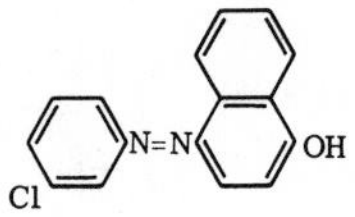
N=N OH
Cl
[CO] 35

▶1630 VS
$CHBr_3$ 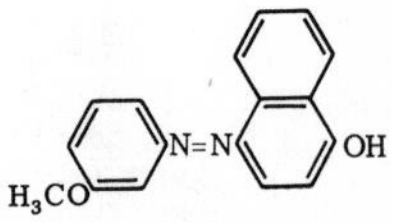
N=N OH
H_3CO
[CO] 35

▶1630 S
F 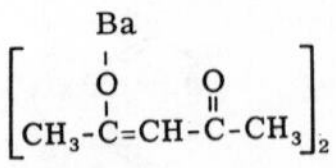
Ba
O O
[CH_3-C=CH-C-CH_3]$_2$
16

▶1630 S
B (0.10) 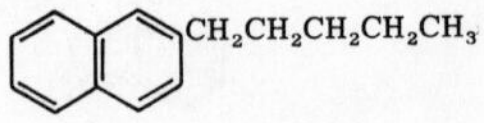
$CH_2CH_2CH_2CH_2CH_3$
173

▶1630 S
$CHCl_3$

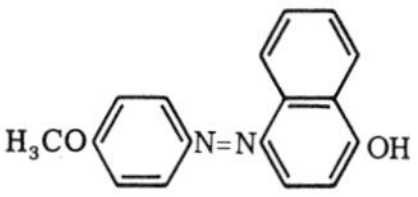

[CO] 35

▶1630 S
B (0.0184)

$CH_2=CHCH_2CH_2CH=CH_2$

374

▶1630 M
B (0.10)

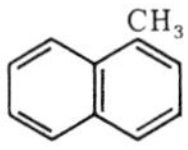

CH_3

192

▶1630

$-[CH_2-CH=CH-CH_2]_n-$
(polybutadiene)

[C=C] 628

▶1630
P

$LiNO_3$

45

▶1630
E or N

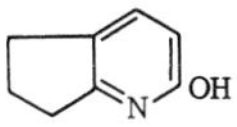

OH

41

▶1630
E or N

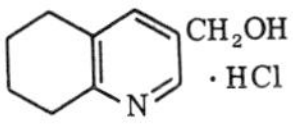

CH_2OH
$\cdot HCl$

41

▶1630 S
B (0.0576)

F F F F

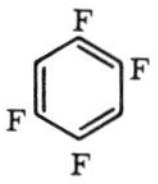

222

▶1630
K

$Ca(NO_3)_2$

45

▶1630
P

$Al(NO_3)_2$

45

▶1630
P

$Ce(NO_3)_3$

45

▶1630
P

$Cr(NO_3)_3$

45

▶1630
P

$Hg_2(NO_3)_2$

45

▶1630
P

$La(NO_3)_3$

45

▶1630
P

$Sm(NO_3)_3$

45

▶1630
P

$Zn(NO_3)_2$

45

▶1629 VVS
$CHCl_3$

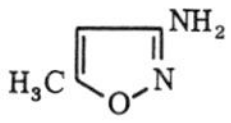

NH_2
H_3C O N

[Ring] 15

▶1629 VVS
$CHCl_3$

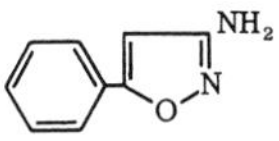

NH_2
O N

[Ring] 15

▶1629 S
B (0.0576)

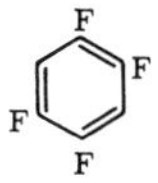

F F F F

222

▶1629 VS
$CHBr_3$

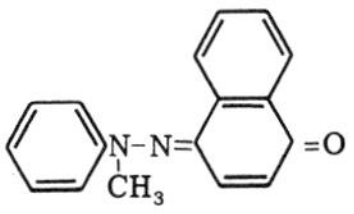

N-N= =O
CH_3

[C=O] 35

▶1629 VS
$CHBr_3$ 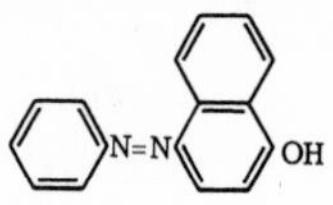
N=N OH
[CO] 35

▶1629 S
F 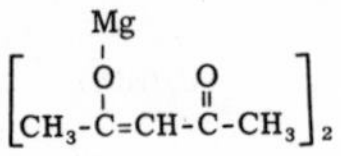
Mg
O O
$[CH_3-C=CH-C-CH_3]_2$
16

▶1629 SB
B (0.728)
$(CH_3)_2CHCH_2CHCH_2CH_3$
CH_3
138

▶1629 S
B (0.0104) 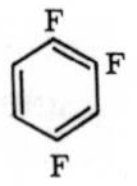
F F F
224

▶1629 S
G 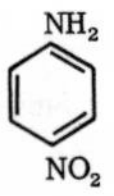
NH_2
NO_2
[NH_2] 21

▶1628 VS
E, G 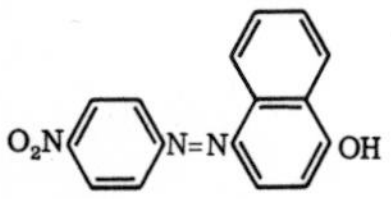
O_2N N=N OH
[CO] 35

▶1628 VS
$CHBr_3$ 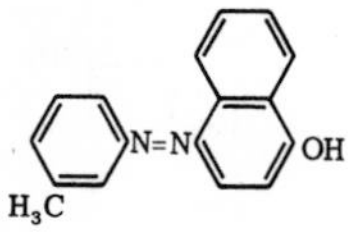
N=N OH
H_3C
[CO] 35

▶1628 VS
$CHBr_3$
N=N OH
OCH_3 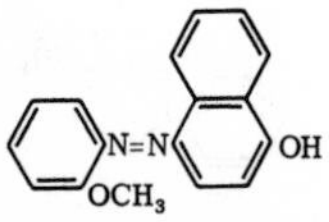
[CO] 35

▶1628 VS
C_2Cl_4 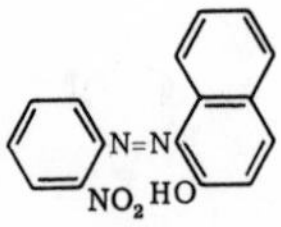
N=N
NO_2 HO
[CO] 35

▶1628-1618 S 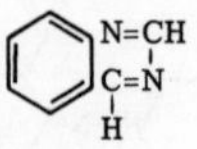
N=CH
C=N
H
[Ring] 598

▶1628 S
C_6H_6
CaF_2 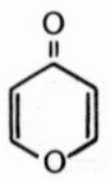
O
O
7

▶1628 S
$CHCl_3$ 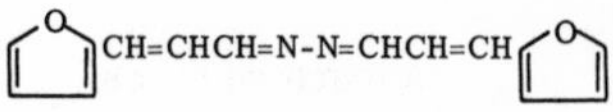
O CH=CHCH=N-N=CHCH=CH O
597

▶1628 M
$CHCl_3$ 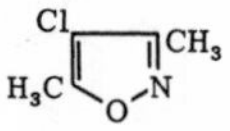
Cl CH_3
H_3C O N
[Ring] 15

▶1628 M
E, F 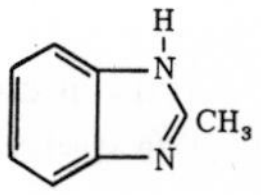
H
N
CH_3
N
37

▶1628
E or N 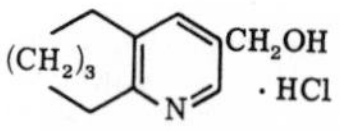
$(CH_2)_3$ CH_2OH
N · HCl
41

▶1627 VVS
$CHCl_3$ 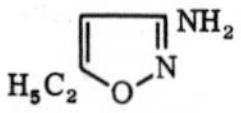
NH_2
H_5C_2 O N
15

▶1627 VS
$CHBr_3$ 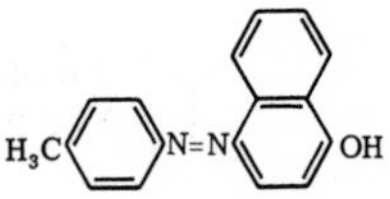
H_3C N=N OH
[CO] 35

▶1627 VS
E, G 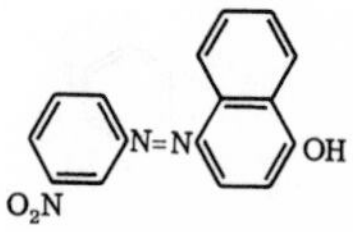
N=N OH
O_2N
[CO] 35

▶1627 VS
E, G 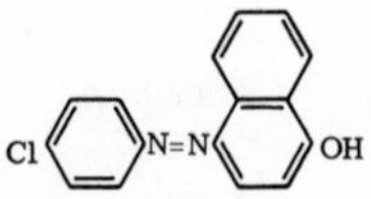
Cl N=N OH
[CO] 35

▶1627 S 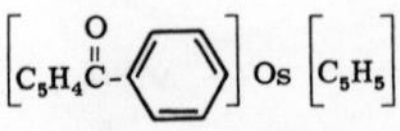
[$C_5H_4C(=O)$-] Os [C_5H_5]
[C=O] 450

▶1627 M CH_2Cl_2 H, N, CH_3, N, CF_3 37	▶1626 M A (200 mm Hg, 10 cm) F_2 F_2 F_2 F_2 243
▶1627 CN, =O, N, H 41	▶1626 O, CH=CH-C-OH [C=C] 545
▶1627 I CaF_2 NO_2, NH_2 [NH_2] 19	▶1626 CaF_2 NO_2, NH_2 [NH_2] 19
▶1627 $CHCl_3$ CaF_2 NH_2, O_2N, NO_2 [NH_2] 19	▶1626 CH=CH$_2$, CH=CH$_2$ [C=C] 599
▶1626 W B (0.2398) F, CF_3, F_2, F_2, F_2, F_2, F, CF_3 230	▶1625 VS E, G N=N, OH 35
▶1626 VS $CHBr_3$ N=N, OH, CH_3 [CO] 35	▶1625 VS E, G H_3CO, N=N, OH [CO] 35
▶1626 VS $CHBr_3$ H_3CO, N=N, OH [CO] 35	▶1625 VS E, G H_3C, N=N, OH [CO] 35
▶1626 S F K, O, O, CH_3-C=CH-C-CH_3 16	▶1625 VS E, G N=N, OH, Cl [CO] 35
▶1626 S [C_5H_4C(=O)-] Fe [C_5H_5] [C=O] 450	▶1625 VS E, G N=N, OH, CH_3 [CO] 35
▶1626 M E, F H, N, C_2H_5, N 37	▶1625 VS C_6H_{12} N=N, HO, OCH_3 [CO] 35

▶1625 VS
C_6H_{12} 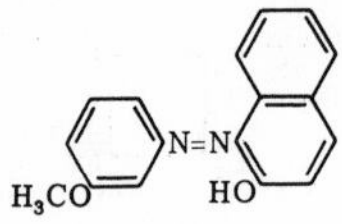
N=N
H_3CO HO
[CO] 35

▶1625 VS
C_6H_{12} 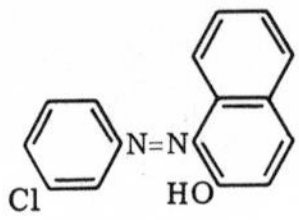
N=N
Cl HO
[CO] 35

▶1625 VS
C_6H_{12} 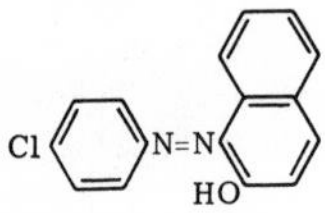
Cl N=N
HO
[CO] 35

▶1625 VS
C_2Cl_4 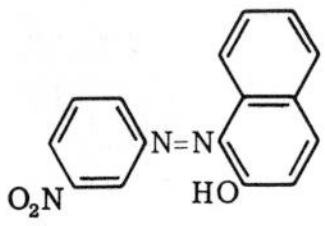
N=N
O_2N HO
[CO] 35

▶1625 VS
E, G 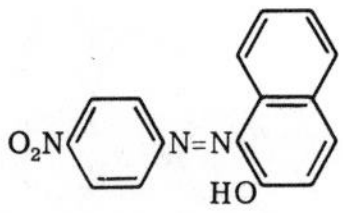
O_2N N=N
HO
[CO] 35

▶1625 S*
G
* A broad band appears at 3000-2100 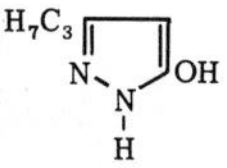
H_7C_3 N N OH H
[C=C] 6

▶1625 S*
* A broad band appears at 3000-2100 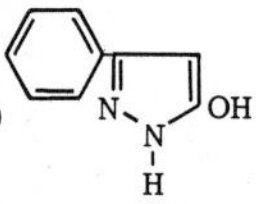
N N OH H
6

▶1625 S
E, F 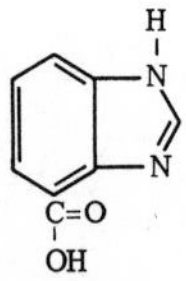
H N N
C=O OH
37

▶1625 M
$CHCl_3$ 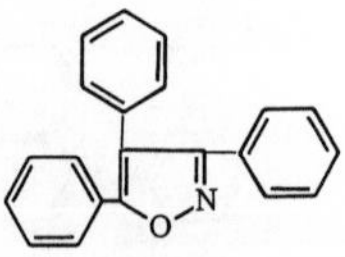
O N
[Ring] 15

▶1625 M
E, F 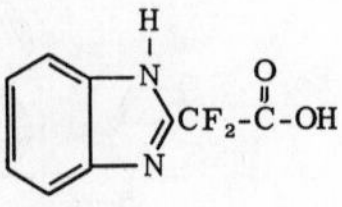
H N N CF_2-C(=O)-OH
37

▶1625 M
B
$H_2C{=}C(CH_3){-}B(OC_4H_9)_2$
[C=C] 4

▶1625 M
E, F 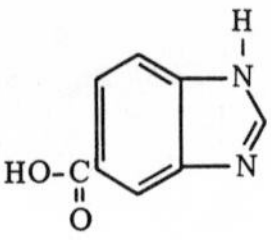
H N N
HO-C(=O)
37

▶1625
E
$Cr(NH_3)_5(NO_3)_3$
45

▶1625
P
$Mg(NO_3)_2$
45

▶1625
$F_3C{-}C(=O){-}ONa \cdot 2F_3C{-}C(=O){-}OH$
[C=O asymm. stretch] 525

▶1625 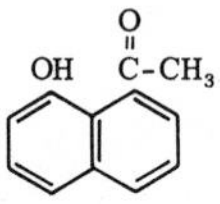
OH C(=O)-CH_3
629

▶1625
$Cl_3C{-}NO_2$
[NO_2] 600

▶1624 VVS
$CHCl_3$ 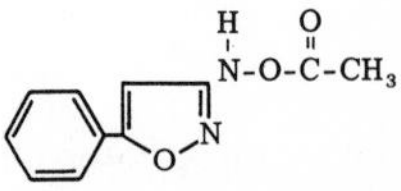
H O
N-O-C(=O)-CH_3
O N
15

▶1624 VS
C_6H_{12} 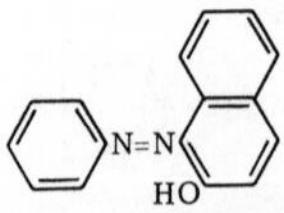
N=N
HO
[CO] 35

▶1624 VS
CH_2Cl_2, $CHCl_3$ 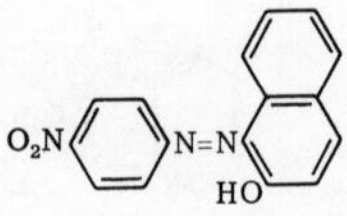
O_2N N=N
HO
[CO] 35

▶1624 VS
CH_2Cl_2 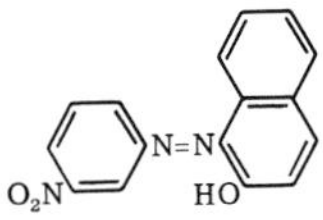
O_2N N=N HO
[CO] 35

▶1624 VS
CH_2Cl_2
N=N NO_2 HO
[CO] 35

▶1624 VS
C_6H_{12}
N=N CH_3 HO
[CO] 35

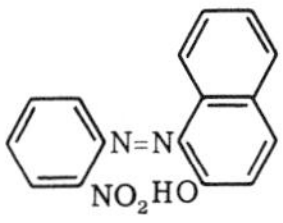

▶1624 VS
C_6H_{12}
H_3C N=N HO
[CO] 35

▶1624 VS
C_6H_{12}
H_3C N=N HO
[CO] 35

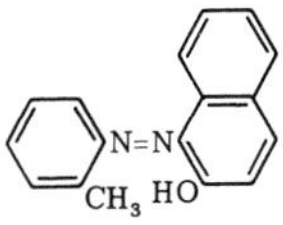

▶1624 VS
C_6H_{12}
H_3CO N=N HO
[CO] 35

▶1624 VS
C_2Cl_4
N=N HO OCH_3
[CO] 35

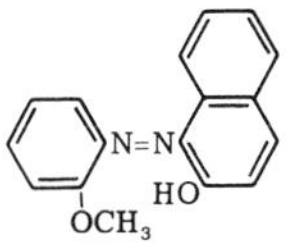

▶1624 VS
C_2Cl_4
H_3CO N=N HO
[CO] 35

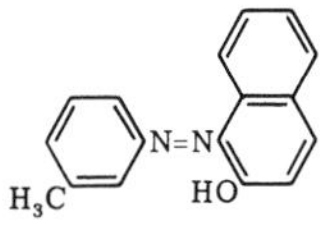

▶1624 SSh
B (0.0576)
F
226

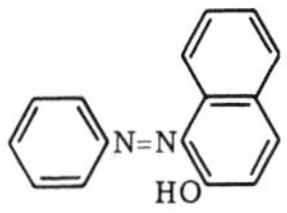

▶1624 VVS
$CHCl_3$
CaF_2
NH_2 NO_2
[NH_2] 19

▶1624 VVS
C_2Cl_4
CaF_2 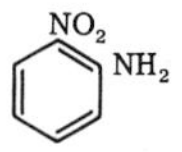
NO_2 NH_2
[NH_2] 19

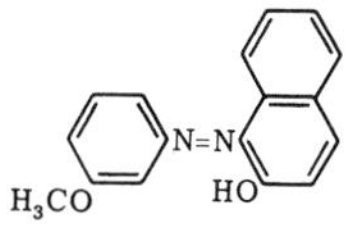

▶1623 VS
$CHCl_3$ 
S O
[Ring] 7

▶1623 VVS
B (0.056)
CH_2=CHCH=CHCH=CH_2
203

▶1623 VS
C_2Cl_4
N=N HO
[CO] 35

▶1623 VS
C_2Cl_4, CCl_4
H_3CO N=N HO
[CO] 35

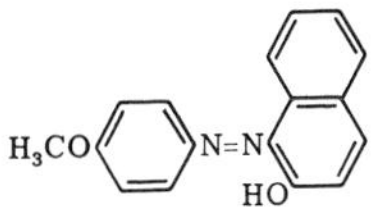

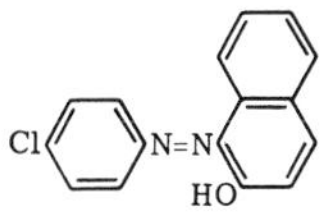

▶1623 VS
CCl_4
H_3CO N=N HO
[CO] 35

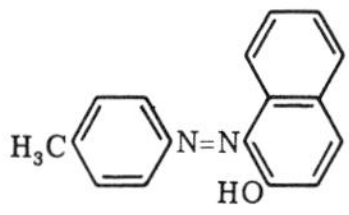

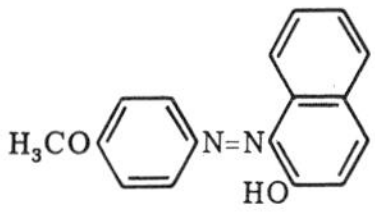

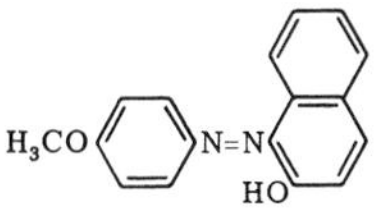

▶1623 VS
C_2Cl_4, CCl_4
Cl N=N HO
[CO] 35

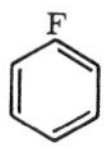

▶1623 VS
C_2Cl_4, CCl_4 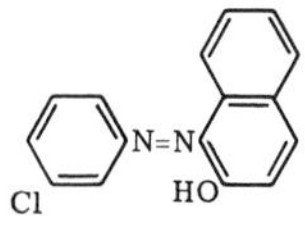
N=N Cl HO
[CO] 35

▶1623 VS
C_2Cl_4 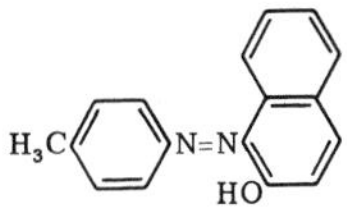
H_3C N=N HO
[CO] 35

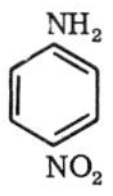

▶1623 VS
C_2Cl_4, CCl_4 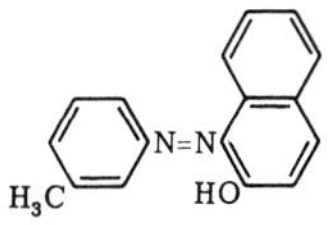
N=N H_3C HO
[CO] 35

▶1623 S
E, F 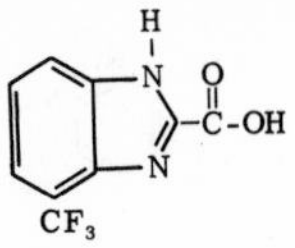
H / N / $C(=O)-OH$ / N / CF_3
37

▶1623 S
$[(C_5H_5)Fe(C_5H_4)]-C(=O)-[(C_5H_4)Ru(C_5H_5)]$
[C=O]
452

▶1623 S
C 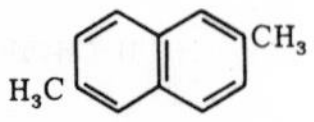
H_3C, CH_3
180

▶1623 M
$CHCl_3$ 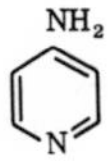
NH_2 / N
327

▶1622 VS
E, G, $CHCl_3$ 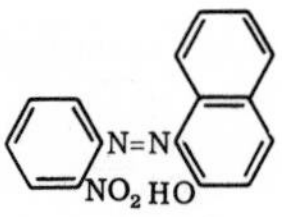
N=N, NO_2 HO
[CO]
35

▶1622 VS
E, G, CH_2Cl_2 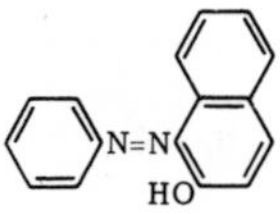
N=N, HO
[CO]
35

▶1622 VS
CCl_4 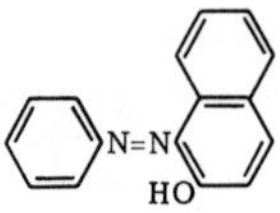
N=N, HO
[CO]
35

▶1622 VS
E, G 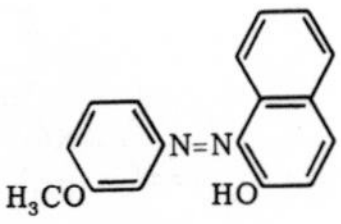
H_3CO, N=N, HO
[CO]
35

▶1622 VS
CCl_4, $CHCl_3$, CH_2Cl_2
N=N, HO, OCH_3 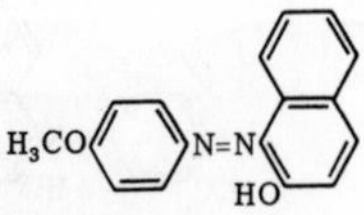
[CO]
35

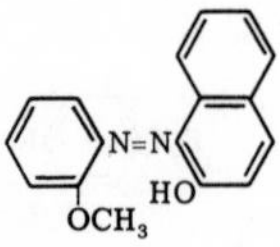

▶1622 VS
CH_2Cl_2, $CHCl_3$
H_3CO, N=N, HO
[CO]
35

▶1622 VS
CH_2Cl_2, $CHCl_3$ 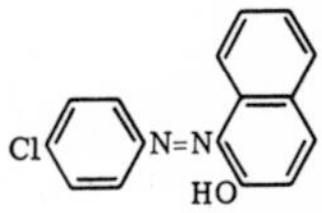
H_3C, N=N, HO
[CO]
35

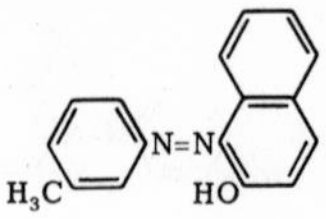

▶1622 VS
CCl_4, CH_2Cl_2
N=N, Cl HO
[CO]
35

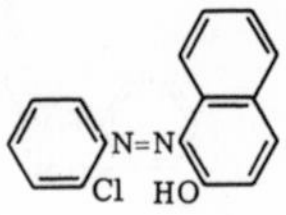

▶1622 VS
$CHCl_3$
N=N, Cl, HO
[CO]
35

▶ 1622 VS
CH_2Cl_2, $CHCl_3$
Cl, N=N, HO
[CO]
35

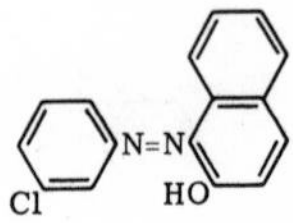

▶1622 VS
C_2Cl_4 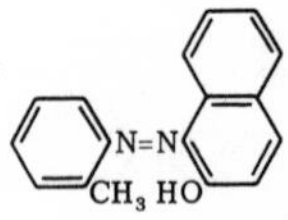
N=N, CH_3 HO
[CO]
35

▶1622 VS
$CHCl_3$ [8 g/l] 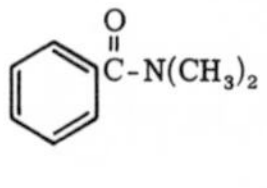
$C(=O)-N(CH_3)_2$
3

▶1622 M
E, F 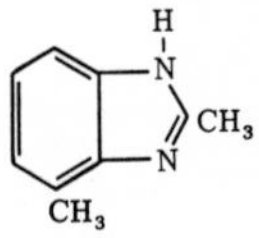
H / N / CH_3 / N / CH_3
37

▶1622 MW
CH_2Cl_2 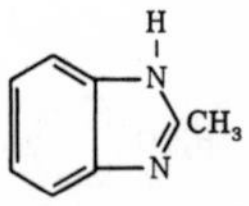
H / N / CH_3 / N
37

▶1622 MW
CH_2Cl_2 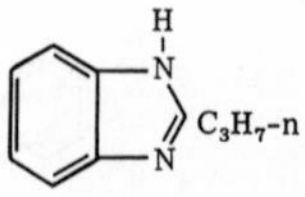
H / N / C_3H_7-n / N
37

▶1622 W
CH_2Cl_2 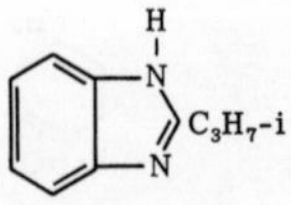
H / N / C_3H_7-i / N
37

▶1622

H_3C-S / $C=N-CH_3$ (phenyl)

606

▶1621 S
CH_3CN
CaF_2

O, H_3C, O, CH_3

[C=O]

7

▶1622
P

$Bi(NO_3)_3$

45

▶1621 S
F

Al, O, O, $[CH_3-C=CH-C-CH_3]_3$

16

▶1621 VS
CH_2Cl_2, $CHCl_3$

N=N, H_3CO, HO

[CO]

35

▶1621 S
$Cl_2C=CCl_2$, C_6H_6
CaF_2

O, O

7

▶1621 VS
$CHCl_3$

N=N, HO

[CO]

35

▶1621 M
$CHCl_3$

NH_2, N

327

▶1621 VS
CH_2Cl_2, $CHCl_3$

N=N, CH_3 HO

[CO]

35

▶1621 M
CCl_4

N, NH_2

327

▶1621 VS
CH_2Cl_2

H_3C, N=N, HO

[CO]

35

▶1621 MVB
B (0.40)

$CH_3CH_2-S-CH_2CH_3$

172

▶1621 VS
CH_2Cl_2

N=N, Cl, HO

[CO]

35

▶1621 VS
C_5H_5N
CaF_2

Cl, NH_2, NO_2

[NH_2]

19

▶1621 VS
$CHCl_3$

N=N, O_2N, HO

[CO]

35

▶1620 VVS
CH_2Cl_2

H, N, CH_3, N

37

▶1621 VS
CCl_4

N=N, CH_3 HO

[CO]

35

▶1620 VS
B

NH_2

[NH_2 bend]

687

▶1621 VS
A (775 mm Hg, 10 cm)

F_2, F_2, F_2, F_2

243

▶1620 VS
E, G

N=N, H_3C, HO

[CO]

35

▶1620 VS
E, G
[CO]
35

▶1620 VS
E, G
[CO]
35

▶1620 VS
$CHCl_3$
[CO]
35

▶1620 VS
B (0.065)
212

▶1620 S*
G
* A broad band appears at 3000-2200
[C=C]
6

▶1620 S*
G
* A broad band appears at 3000-2100
[C=C]
6

▶1620
B
[NH_2 bend]
687

▶1620 S*
G
* A broad band appears at 3200-2200
[C=C]
6

▶1620 SSh
$CHCl_3$
CaF_2
[C=O]
7

▶1620 M
CCl_4 [0.0082 m/l]
448

▶1620 MW
CH_2Cl_2
37

▶1620 MW
CH_2Cl_2
37

▶1620 W
CH_2Cl_2
37

▶1620
E
$CsNO_3$
45

▶1620
E
$Fe(NO_3)_3$
45

▶1620
E
$Ni(NO_3)_2$
45

▶1620
K
$La(NO_3)_3$
45

▶1620
P
$Be(NO_3)_2$
45

▶1620
P
$Cd(NO_3)_2$
45

▶1620
P
$Co(NO_3)_2$
45

▶1620
P

$Cr(NH_3)_5(NO_3)_6$

45

▶1620
P

$Pr(NO_3)_3$

45

▶1620
P

$RbNO_3$

45

▶1620
P

$Th(NO_3)_4$

45

▶1620
P

$Zn(NO_3)_2$

45

▶1620
E or N

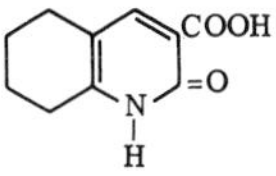

COOH
=O
N
H

41

▶1620
E or N

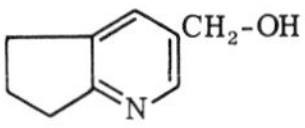

CH_2-OH
N

41

▶1620

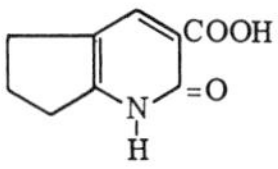

COOH
=O
N
H

41

▶1620
E, CH_2Cl_2

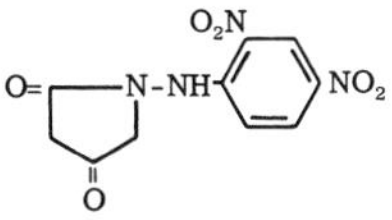

O_2N
O= N-NH NO_2
O

577

▶1620

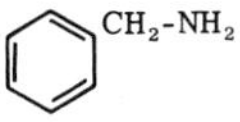

CH_2-NH_2

[NH_2 deformation] 603

▶1620

F_3C-NO_2

[NO_2] 600

▶1620

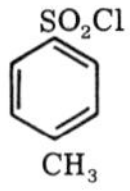

SO_2Cl
CH_3

[Ring] 389

▶1620

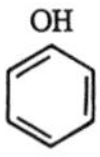

OH

632

▶1619 VVS
$CHCl_3$

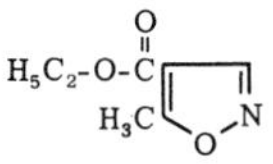

O
H_5C_2-O-C
H_3C N
O

[Ring] 15

▶1619 S
CH_2Cl_2

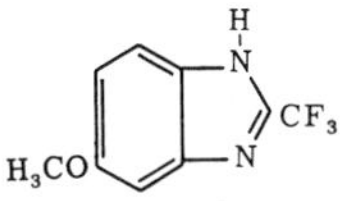

H
N
CF_3
H_3CO N

37

▶1619 W
CH_2Cl_2

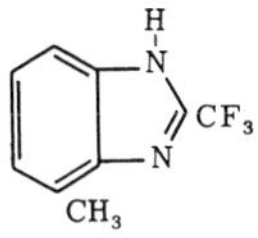

H
N
CF_3
N
CH_3

37

▶1618 VVS
$CHCl_3$

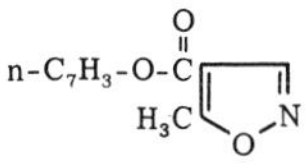

O
n-C_7H_3-O-C
H_3C N
O

[Ring] 15

▶1618 VVS
$CHCl_3$

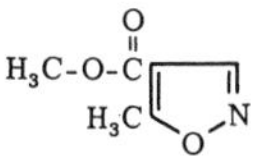

O
H_3C-O-C
H_3C N
O

[Ring] 15

▶1618 VS
E, G

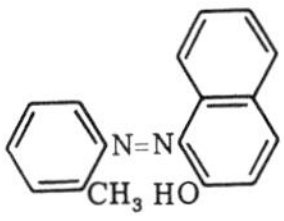

N=N
CH_3 HO

[CO] 35

▶1618 VSB
A (682 mm Hg, 10 cm)

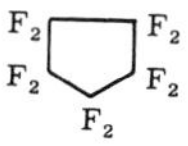

F_2 F_2
F_2 F_2
F_2

245

▶1618 VS B (0.08) [isoquinoline structure] 197	▶1617 VS $CHCl_3$ $H_3C-C(=O)-O-N(H)-$[3-methylisoxazolyl] [Ring] 15
▶1618 S F $[CH_3-C(OCa)=CH-C(=O)-CH_3]_2$ 16	▶1617 VS E, G $H_3C-C_6H_4-N=N-$[HO-naphthyl] [CO] 35
▶1618 M B (0.064) $(H_3C)(CH_3)C=C(H)(CH_2CH_3)$ (cis) 164, see also 274	▶1617 VS E, G [Cl-phenyl]$-N=N-$[HO-naphthyl] [CO] 35
▶1618 S F $CH_3-C(ONa)=CH-C(=O)-CH_3$ [CO?] 16	▶1617 S [1,3,5-triazine] 14
▶1618 M A (200 mm Hg, 10 cm) $CF_3CF_2CF_3$ 235	▶1617 CH_3NH_2Cl [NH_2 deformation] 259
▶1618 MW CH_2Cl_2 [benzotriazole, N-H] 37	▶1616 S CCl_4 [NH_2, NO_2 benzene] [NH_2] 21
▶1618 S C_2Cl_4, CCl_4 CaF_2 [Cl, NH_2, NO_2 benzene] [NH_2] 19	▶1616 S F $[CH_3-C(ONi)=CH-C(=O)-CH_3]_2$ 16
▶1618 $NH_2-C(=O)-OC_2H_5$ [Amide II] 604 ▶1618 S B (0.10) [tetrahydronaphthalene, CH_3] 269	▶1616 S F $CH_3-C(OLi)=CH-C(=O)-CH_3$ 16
▶1617 VVS $CHCl_3$ $H_3C-C(=O)-O-N(H)-$[3-phenylisoxazolyl] [Ring] 15	▶1616 MS $CHCl_3$ [3,5-diphenylisoxazole] [Ring] 15
▶1617 MS B (0.0576) [CF_3, F, F benzene] 217	▶1616 $[(CH_2)_3$-pyridine$-CH_2-NH]_2$ 41

▶1616
E or N 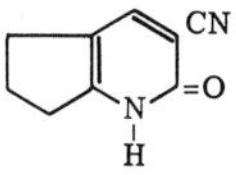
41

▶1615
$(CH_2)_3$ COOH =O N H
41

▶1615 VVS
CH_2Cl_2 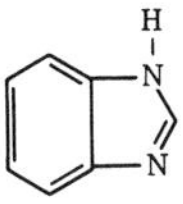
37

▶1615
CH_3OH
$H_3C-C(=O)-N(C_2H_5)_2$
[C=O]
532

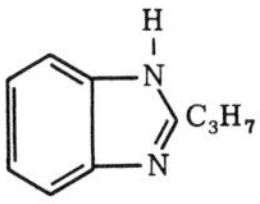

▶1615 VVS
CH_2Cl_2
C_3H_7
37

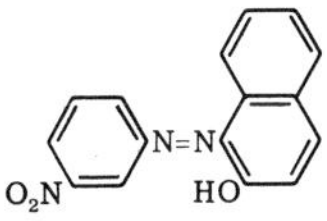

▶1615
E
$Cr(NO_3)_3$
45

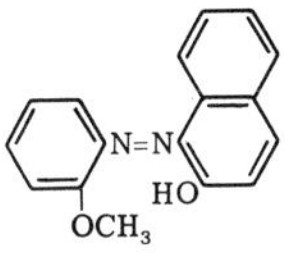

▶1615 VS
B (0.0576)
CH_3 F
220

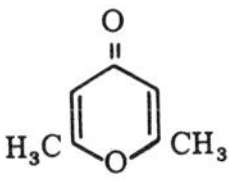

▶1615
P
$Ce(NH_4)_2(NO_3)_6$
45

▶1615 VS
E, G
O_2N N=N HO
[CO]
35

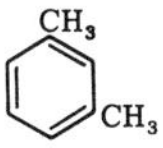

▶1615
P
$Ni(NO_3)_2$
45

▶1615 VS
E, G
N=N HO OCH_3
[CO]
35

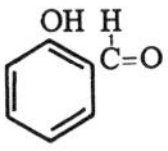

▶1614 VVS
CH_2Cl_2 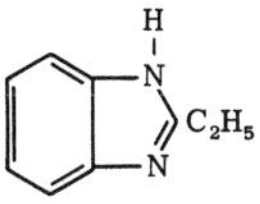
C_2H_5
37

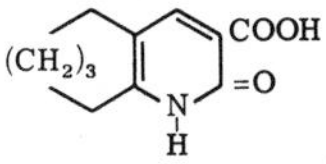

▶1615 S
$CHCl_3$
CaF_2
O H_3C O CH_3
7

▶1614 S
CH_3CN 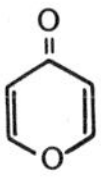
7

▶1615 SSp
A (6.3 mm Hg, 40 cm)
CH_3 CH_3
69

▶1613 VVS
$CHCl_3$ 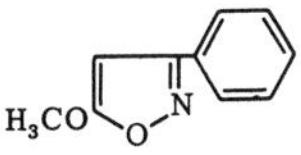
H_3CO O N
15

▶1615 M
$CHCl_3$
$H_2C=C(CH_3)-B(OH)_2$
[C=C]
4

▶1613 VS
$CHCl_3$ 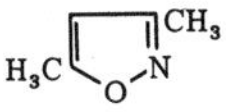
H_3C O N CH_3
[Ring]
15

▶1615
OH H C=O
605

▶1613 S
B (0.0104) 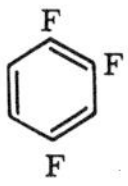
F F F
224

▶1613 SSh
B (0.10)
[cyclooctatetraene structure]
200

▶1613 S
[structure: NH_2, Cl benzene]
534

▶1613 S
B (0.728)
$(CH_3)_3CC(CH_3)_2CH_2CH_2CH_3$
304

▶1613 S
F
$[CH_3\text{-}C(O\text{-}Co)\text{=}CH\text{-}C(\text{=}O)\text{-}CH_3]_2$
16

▶1613 S
$CH_3\text{-}C(O\text{-}Cs)\text{=}CH\text{-}C(\text{=}O)\text{-}CH_3$
16

▶1613 S
F
$[CH_3\text{-}C(O\text{-}Sr)\text{=}CH\text{-}C(\text{=}O)\text{-}CH_3]_2$
16

▶1613 S
F
$[CH_3\text{-}C(O\text{-}Zr)\text{=}CH\text{-}C(\text{=}O)\text{-}CH_3]_4$
16

▶1613 M
CH_2Cl_2
$(CH_3)_3N \cdot BCl_2 \cdot CH\text{=}CH_2$
[C=C]
4

▶1613
[structure: NH_2, Cl benzene]
534

▶1613 W
B
$(C_6H_5)_2(CH_3)SiCl$
86

▶1613 W
B
$C_6H_5SiCl_3$
86

▶1613
E or N
[structure: tetrahydroquinoline with $C(\text{=}O)\text{-}OC_2H_5$]
41

▶1613
E or N
[structure: $(CH_2)_3$ fused pyridine with OH]
41

▶1612 VS
E, G
[structure: H_3CO-phenyl-N=N-naphthyl, HO]
[CO]
35

▶1612 VS
B (0.0576)
[structure: CF_3 benzene]
219

▶1612 S
$CHBr_2CHBr_2$
CaF_2
[structure: 4-pyrone]
7

▶1612 S
$CHBr_2CHBr_2$
CaF_2
[structure: 2,6-dimethyl-4-pyrone, H_3C, CH_3]
[C=O]
7

▶1612 S
E, F
[structure: benzimidazole, H-N, $CH_2\text{-}C(\text{=}O)\text{-}OH$]
37

▶1612 S
$[(C_5H_5)Fe(C_5H_4)]\text{-}C(\text{=}O)\text{-}[(C_5H_4)Fe(C_5H_5)]$
[C=O]
450

▶1612 S
$CH_3\text{-}C(O\text{-}Ag)\text{=}CH\text{-}C(\text{=}O)\text{-}CH_3$
16

▶1612 M
CCl_4 [0.00658 m/l]
448

▶1612
E or N
41

▶1612
E or N
41

▶1611 VVS
B (0.10)
187

▶1611 VS
B
183

▶1611 VVS
B (0.10)
188

▶1611
$CHCl_3$
606

▶1611
492

▶1611
E or N
41

▶1610 VVS
$CHCl_3$
[Ring]
15

▶1610 VVS
B (0.10)
184

▶1610 VS
B (0.029)
277

▶1610 S
$C_2H_2Cl_4$
[NH_2]
648

▶1610
E, $CHCl_3$
327

▶1610 S
E
[NH_2]
8

▶1610
CCl_4
327

▶1610 S*
G
* A broad band appears at 3000-2100
[C=C]
6

▶1610 S
E
16

▶1610 S
F
16

▶1610 S
16

▶1610 MSSh
B (0.2398)
$H_3C\text{-}O\text{-}CF_2CHFCl$
356

▶1610 M
B
$H_2C{=}CH \cdot B(OC_4H_9)_2$
[C=C]
4

▶1610 M
CH_2Cl_2 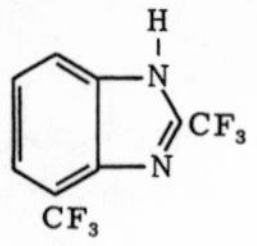
CF_3 37

▶1610 M
E, F 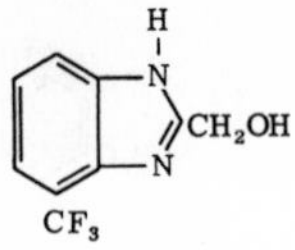
CH_2OH, CF_3 37

▶1610 M
E, F 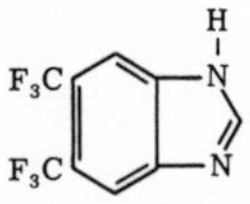
F_3C, F_3C 37

▶1610 MW
CH_2Cl_2 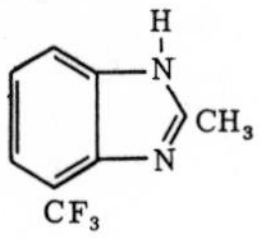
CH_3, CF_3 37

▶1610
G 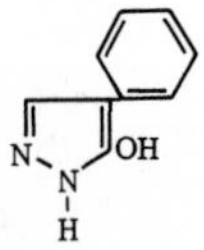
OH 6

▶1610
G 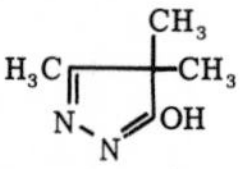
H_3C, CH_3, CH_3, OH 6

▶1610
I
$H_3C-NH-C(=O)-NH-CH_3$
[C=O] 609

▶1610
E or N 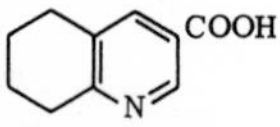
COOH 41

▶1610
E or N 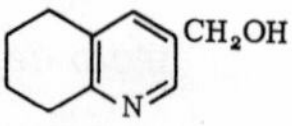
CH_2OH 41

▶1610
E or N
$(CH_2)_3$, CH_2NH_2 41

▶1610
O_2N, $CH=CH-C(=O)-O-C_2H_5$
[Ring] 576

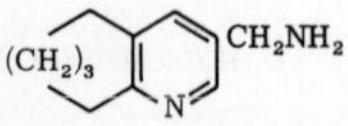

▶1610 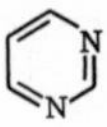
415

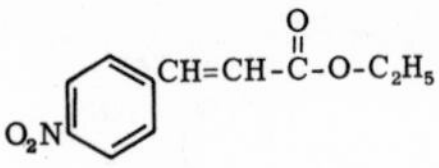

▶1610 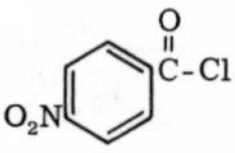
O_2N, $C(=O)-Cl$ 385

▶1610 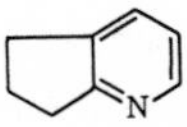
41

▶1610
E or N 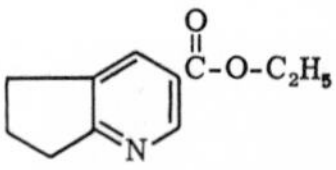
$C(=O)-O-C_2H_5$ 41

▶1610
E or N 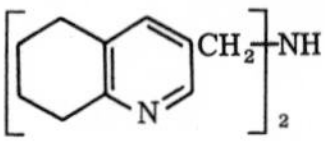
$[\ldots CH_2]_2NH$ 41

▶1609.6 VS
B (0.03) 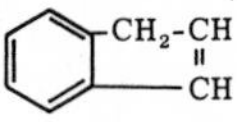
$CH_2-CH=CH$ 9

▶1609 S
C 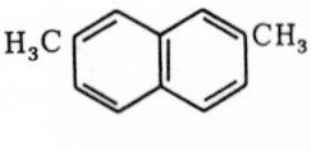
H_3C, CH_3 179

▶1609 S
E 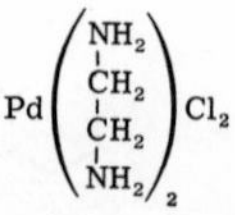
$Pd(NH_2-CH_2-CH_2-NH_2)_2Cl_2$
$[NH_2]$ 8

▶1608 VS
B (0.029)
$CH(CH_3)_2$, $(CH_3)_2HC$, $CH(CH_3)_2$ 278

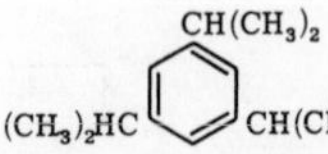

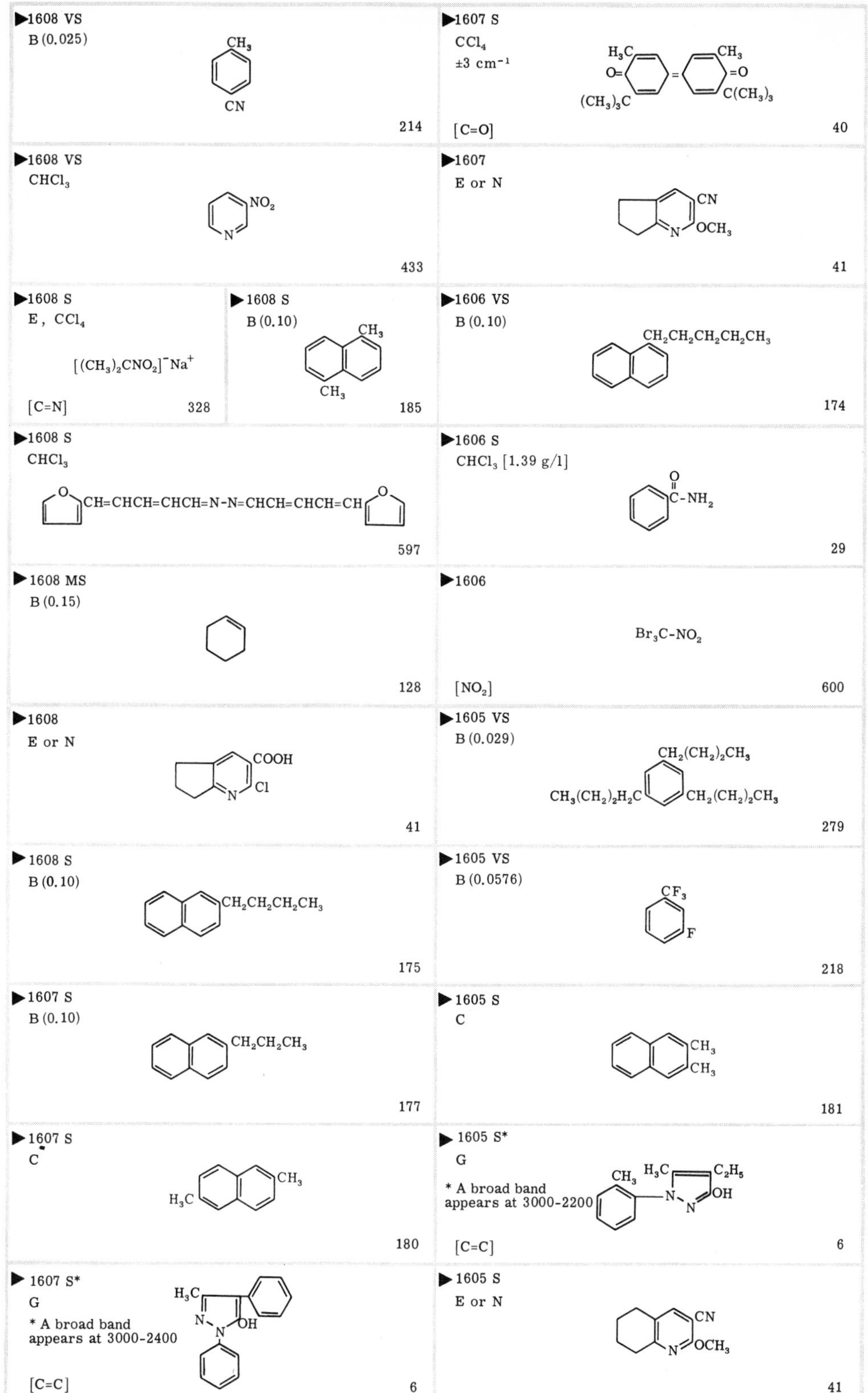
▶1608 VS
B (0.025)
CH3
CN
214
▶1607 S
CCl4
±3 cm-1
H3C
CH3
O=
=O
(CH3)3C
C(CH3)3
[C=O]
40
▶1608 VS
CHCl3
NO2
N
433
▶1607
E or N
CN
N
OCH3
41
▶1608 S
E, CCl4
[(CH3)2CNO2]-Na+
[C=N]
328
▶1608 S
B (0.10)
CH3
CH3
185
▶1606 VS
B (0.10)
CH2CH2CH2CH2CH3
174
▶1608 S
CHCl3
O
CH=CHCH=CHCH=N-N=CHCH=CHCH=CH
O
597
▶1606 S
CHCl3 [1.39 g/l]
O
C-NH2
29
▶1608 MS
B (0.15)
128
▶1606
Br3C-NO2
[NO2]
600
▶1608
E or N
COOH
N
Cl
41
▶1605 VS
B (0.029)
CH2(CH2)2CH3
CH3(CH2)2H2C
CH2(CH2)2CH3
279
▶1608 S
B (0.10)
CH2CH2CH2CH3
175
▶1605 VS
B (0.0576)
CF3
F
218
▶1607 S
B (0.10)
CH2CH2CH3
177
▶1605 S
C
CH3
CH3
181
▶1607 S
C
H3C
CH3
180
▶1605 S*
G
* A broad band appears at 3000-2200
CH3
H3C
C2H5
N
N
OH
[C=C]
6
▶1607 S*
G
* A broad band appears at 3000-2400
H3C
N
N
OH
[C=C]
6
▶1605 S
E or N
CN
N
OCH3
41

▶ 1605 M
$CHCl_3$

$H_2C{=}CH \cdot B(OH)_2$

[C=C] 4

▶1605
G

H_3C N N OCH_3

[C=C] 6

▶1605

O= =O
H_3C CH_3

[C=O] 518

▶1605
G

C_2H_5 H_3C C_2H_5 N N OH

[C=C] 6

▶1605
P

$Hg(NO_3)_2$

45

▶1605 S

C_3H_7-n
N

431

▶1604 VVS
B (0.10)

CH_3
CH_3

186

▶1604 S
$CHCl_3$

NH_2
N

327

▶1604 S

C_4H_9-i
N

431

▶ 1604 S
CH_2Cl_2

F H F N CF_3 F N F

37

▶1604 M
B

O_2N O N

26

▶1604 M
$CHCl_3$
±3 cm^{-1}

H_3C CH_3 O= =O $(CH_3)_3C$ $C(CH_3)_3$

[C=O] 40

▶1604
E or N

CN
N

41

▶1604 VS
B (0.10)

$CH_2CH_2CH_2CH_3$

176

▶1603 S
E

NO_2

[Ring] 20

▶1603 SSp
B (0.025)

CH_3 CN

312

▶1603 MW
B (0.2398)

F CF_3 F_2 F_2 F_2 F_2 F_2

232

▶1603 M
CH_2Cl_2

H N N

37

▶1603 S
CCl_4
±3 cm^{-1}

t-H_9C_4 C_4H_9-t O= =O t-H_9C_4 C_4H_9-t

[C=O] 40

▶1603 S
B (0.0104)

CH_3
F

220

▶1602 VS
B (0.029)

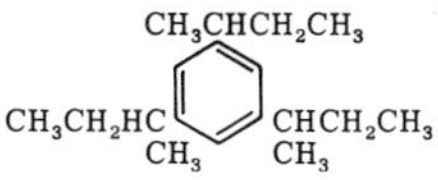

$CH_3CHCH_2CH_3$
CH_3CH_2HC $CHCH_2CH_3$
CH_3 CH_3

280

▶1602
E or N

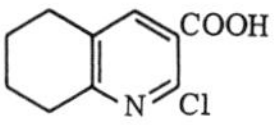

COOH
N Cl

41

▶1602
E or N

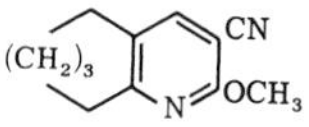

$(CH_2)_3$ CN
N OCH_3

41

▶1601 VS
$CHCl_3$

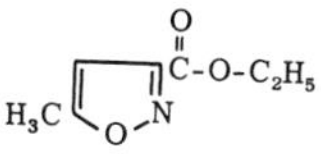

O
$C-O-C_2H_5$
H_3C O N

15

▶1601 VS
$CHCl_3$

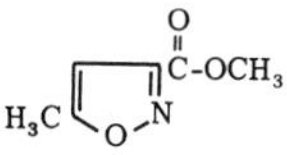

O
$C-OCH_3$
H_3C O N

15

▶1601 S
B (0.0576)

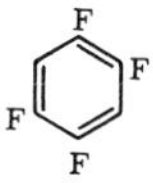

F
F
F
F

222

▶1601 S
B (0.10)

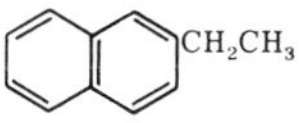

CH_2CH_3

189

▶1601 VS
B (0.065)

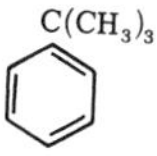

$C(CH_3)_3$

133

▶1601 M
G

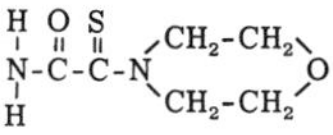

H O S
N-C-C-N CH_2-CH_2 O
H CH_2-CH_2

[NH]

31

▶1601 MW
CH_2Cl_2

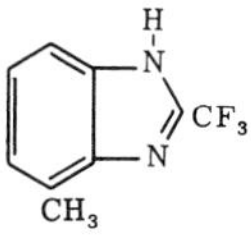

H
N
CF_3
N
CH_3

37

▶1601
CCl_4
±3 cm^{-1}

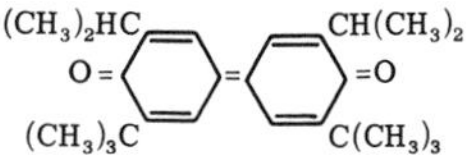

$(CH_3)_2HC$ $CH(CH_3)_2$
O= =O
$(CH_3)_3C$ $C(CH_3)_3$

[C=O]

40

▶1601 S
B (0.0576)

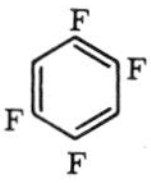

F
F
F
F

222

▶1600 VS
$CHCl_3$

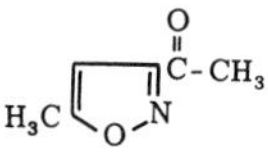

O
$C-CH_3$
H_3C O N

[Ring]

15

▶1600 VS
B (0.0104)

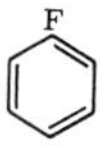

F

226

▶1600 VS
$CHCl_3$

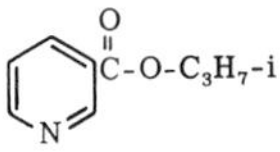

O
$C-O-C_3H_7$-i
N

433

▶1600 S
E, F

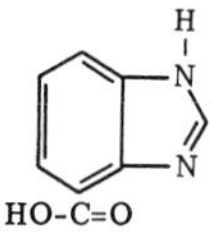

H
N
N
HO-C=O

37

▶1600 S

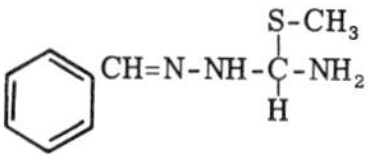

$S-CH_3$
$CH=N-NH-C-NH_2$
H

380

▶1600 S

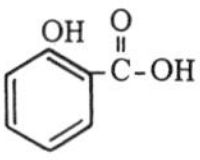

OH O
C-OH

611

▶1600 S

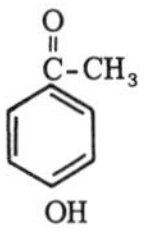

O
$C-CH_3$
OH

610

▶1600 S
CCl_4

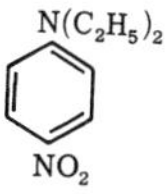

$N(C_2H_5)_2$
NO_2

21

▶1600 S F $[CH_3-C(O-Zn)=CH-C(=O)-CH_3]_2$ 16	▶1600 $C_6H_5-C(=O)-OH$ [Ring] 370
▶1600 S* G * A broad band appears at 3000-2200 [C=C] 6	▶1600 $H_2C=CH-C \equiv CH$ [C=C] 544
▶1600 S E NH_2 327	▶1599 VS B (0.10) CH_3 192
▶1600 M E, F $CH_2CH_2-C(=O)-OH$ 37	▶1599 VS $CHCl_3$ $C(=O)-O-C_4H_9-i$ 433
▶1600 M E CH_2OH 34	▶1599 VS $CHCl_3$ H_3C [Ring] 15
▶1600 M E CH_2OH 34	▶1599 S B F 26
▶1600 MW $CHCl_3$ CH_3 327	▶1599 S $CHCl_3$ $N(H)-C(=O)-CH_3$ 433
▶1600 MW $CHCl_3$ $CH_2CH_2-C(=O)-OC_2H_5$ 433	▶1599 S G NH_2 NO_2 21
▶1600 G H_3C, OH [C=C] 6	▶1599 CCl_4 $\pm 3\ cm^{-1}$ $(CH_3)_2HC$, $CH(CH_3)_2$ [C=O] 40
▶1600* G * Broad bands appear at 2700-2300 and 1900-1700 H_3C, OCH_3 [C=C] 6	▶1599 VS B (0.10) CH_3 192

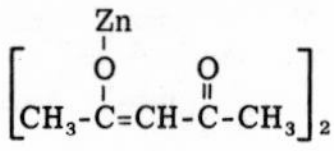

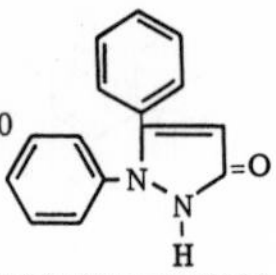

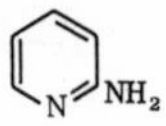

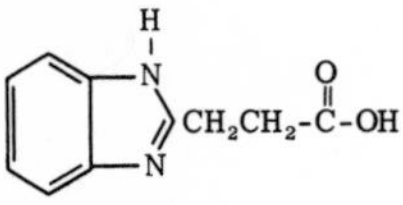

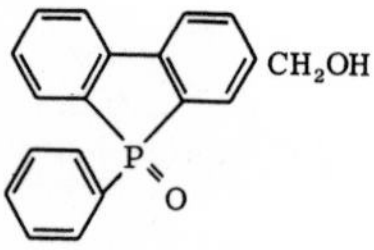

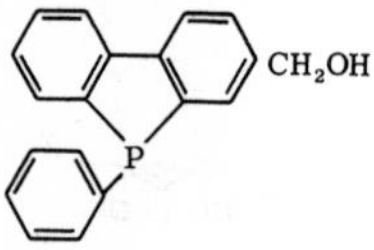

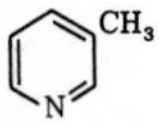

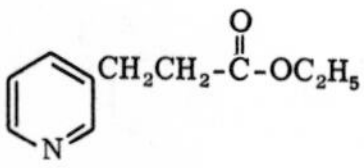

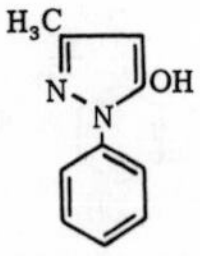

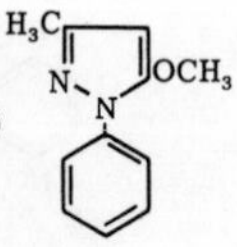

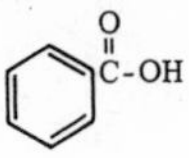

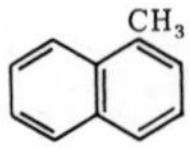

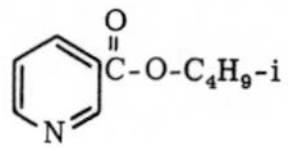

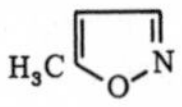

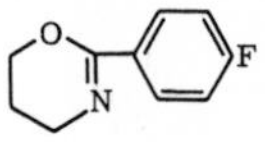

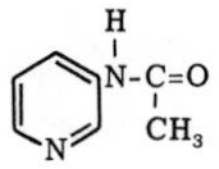

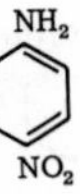

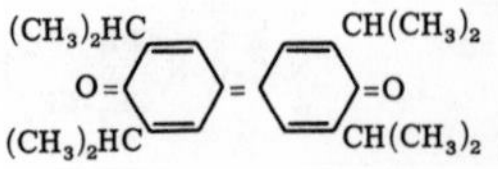

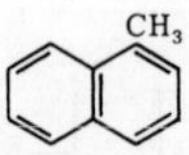

▶1598 VS
$CHCl_3$ 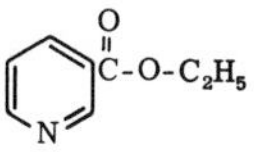
O
C-O-C_2H_5
N
433

▶1598 S
B (0.10) 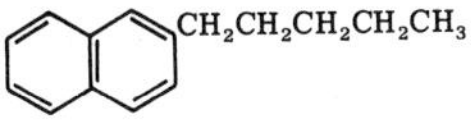
$CH_2CH_2CH_2CH_2CH_3$
173

▶1598 MW
$CHCl_3$ 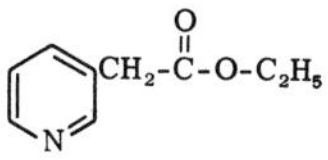
O
CH_2-C-O-C_2H_5
N
433

▶1597 VVS
CH_2Cl_2 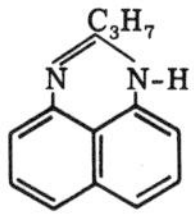
C_3H_7
N N-H
37

▶1597 VS
B (0.0104) 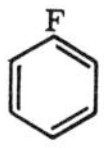
F
226

▶1597 S
$CHCl_3$ 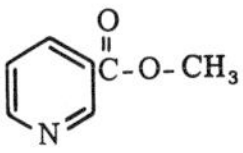
O
C-O-CH_3
N
433

▶1597 S
$CHCl_3$ 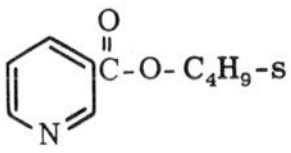
O
C-O-C_4H_9-s
N
433

▶1597 S
G 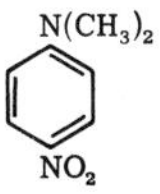
$N(CH_3)_2$
NO_2
21

▶1597 M
CH_2Cl_2 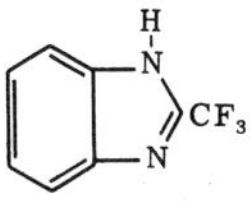
H
N
CF_3
N
37

▶1597 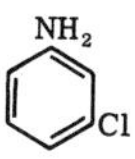
NH_2
Cl
534

▶1597 S
$CHCl_3$
±3 cm^{-1} 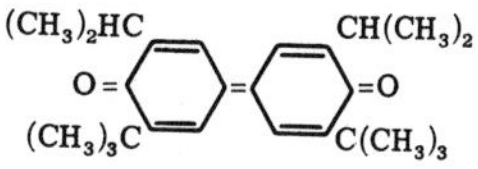
$(CH_3)_2HC$ $CH(CH_3)_2$
O= =O
$(CH_3)_3C$ $C(CH_3)_3$
[C=O]
40

▶1597
G 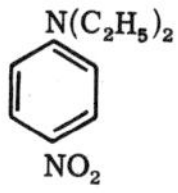
$N(C_2H_5)_2$
NO_2
21

▶1597 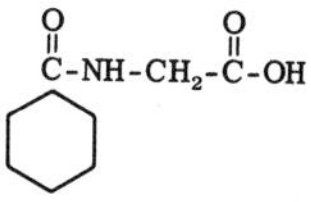
O O
C-NH-CH_2-C-OH
613

▶1597 VS
E 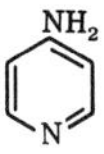
NH_2
N
327

▶1597 VS
B (0.051) 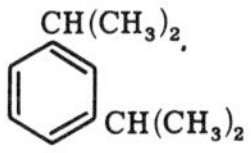
$CH(CH_3)_2$
$CH(CH_3)_2$
297

▶1596 S
E 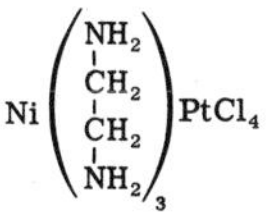
NH_2
CH_2
Ni CH_2 $PtCl_4$
NH_2 3
[NH_2]
8

▶1596 M
$CHCl_3$ 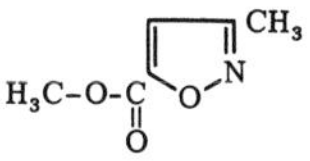
CH_3
N
H_3C-O-C O
O
[Ring]
15

▶1596 MW
$CHCl_3$ 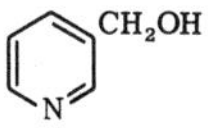
CH_2OH
N
433

▶1596 S
G
±3 cm^{-1} 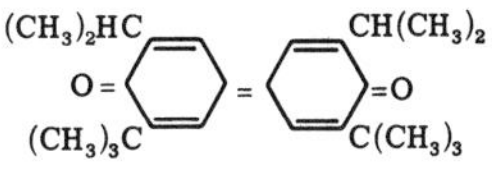
$(CH_3)_2HC$ $CH(CH_3)_2$
O= = =O
$(CH_3)_3C$ $C(CH_3)_3$
[C=O]
40

▶1596 VVS
CH_2Cl_2 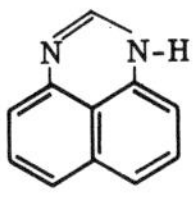
N N-H
37

▶1596 VVS CH_2Cl_2 37	▶1595 E or N $(CH_2)_3$ CN Cl N 41
▶1595 VS $CHCl_3$ C-O-C_3H_7-n 433	▶1595 OH HO HO OH =O =O [C=O] 624
▶1595 VS $CHCl_3$ NH-CH_3 433	▶1595 VS $CHCl_3$ C-H 433
▶1595 S C CH_3 191	▶1594 E or N Cl 41
▶1595 S CCl_4 $N(CH_3)_2$ NO_2 21	▶1594 E or N CN Cl 41
▶1595 S* G * A broad band appears at 3200-2200 [C=C] 6	▶1594 VS $CHCl_3$ C-O-C_4H_9-n 433
▶1595 VVS CH_2Cl_2 37	▶1593 S E or N CN Cl 41
▶1595 MS $CHCl_3$ H_5C_2-O-C 15	▶1593 M CCl_4 ±3 cm^{-1} [C=O] 40
▶1595 MB B (0.2398) 247	▶1593 W CH_2Cl_2 C_2F_5 37
▶1595 M E, F C_2H_5 37	▶1593 VS C CH_3 CH_3 182

▶1592 VVS
B (0.04)

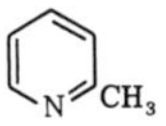

(2-methylpyridine structure: CH_3, N)

198

▶1592 S
B (0.0104)

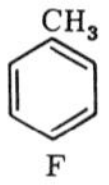

CH_3 / F (4-fluorotoluene structure)

220

▶1592 VS
B (0.0104)

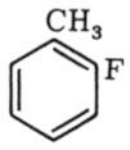

CH_3 / F (2-fluorotoluene structure)

221

▶1592 S
CCl_4

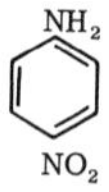

NH_2 / NO_2 (4-nitroaniline structure)

21

▶1592 S
F

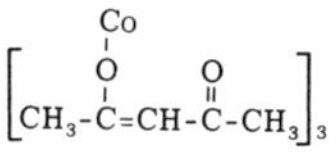

$[CH_3\text{-}C(\text{-O-Co})\text{=CH-}C(\text{=O})\text{-}CH_3]_3$

16

▶1592 S
F

$[CH_3\text{-}C(\text{-O-Mn})\text{=CH-}C(\text{=O})\text{-}CH_3]_3$

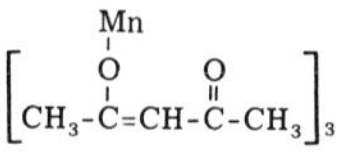

16

▶1592 S
$CHCl_3$

$CH_3CH\text{=}CHCH\text{=}N\text{-}N\text{=}CHCH\text{=}CHCH_3$

597

▶1592 M
B

$C_6H_5SiCl_3$

86

▶1592 M
CH_2Cl_2

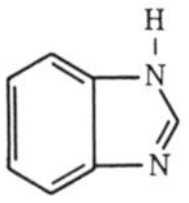

(benzimidazole structure: H, N, N)

37

▶1592 MW
$CHCl_3$

(pyridyl)$CH\text{=}CH\text{-}C(\text{=O})\text{-O-}C_2H_5$

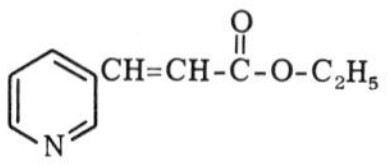

433

▶1592 W
CCl_4

$(C_6H_5)_3SiCl$

86

▶1592
$CHCl_3$
±3 cm^{-1}

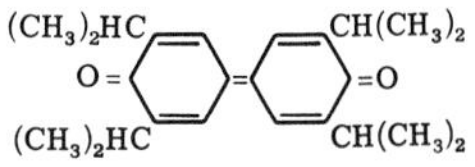

$(CH_3)_2HC$, $CH(CH_3)_2$, O=, =O, $(CH_3)_2HC$, $CH(CH_3)_2$ (diphenoquinone structure)

[C=O]

40

▶1591 S
B (0.055)

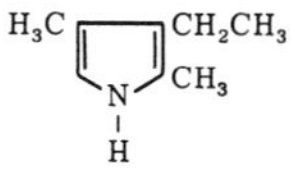

H_3C, CH_2CH_3, CH_3, N, H (pyrrole structure)

343

▶1591 S
B

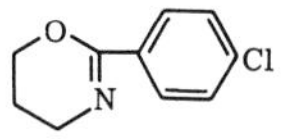

O, N, Cl (2-(4-chlorophenyl)-dihydro-oxazine structure)

26

▶1591 S
$CHCl_3$

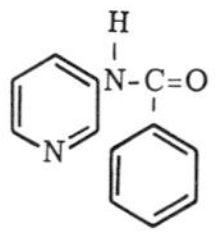

H, N-C=O, N (pyridyl, phenyl structure)

433

▶1591 M
$CHCl_3$
±3 cm^{-1}

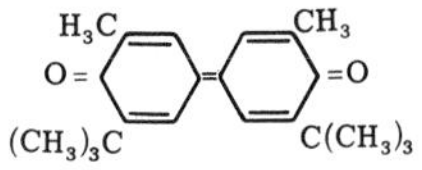

H_3C, CH_3, O=, =O, $(CH_3)_3C$, $C(CH_3)_3$ (diphenoquinone structure)

[C=O]

40

▶1591 M
E, F

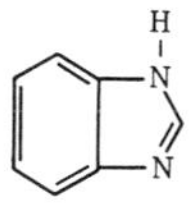

(benzimidazole structure: H, N, N)

37

▶1590 VS
B (0.10)

(thiophene structure: S)

199

▶1590 VS
$CHCl_3$

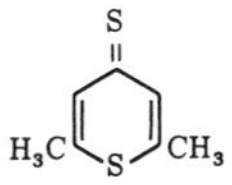

S, H_3C, S, CH_3 (thiopyranthione structure)

[Ring]

7

▶1590 VS
B (0.10)

CH_2CH_3 (1-ethylnaphthalene structure)

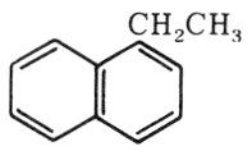

190

▶1590 S
CHCl$_3$ 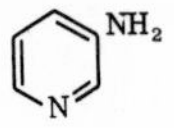
433

▶1590 S
CHCl$_3$ 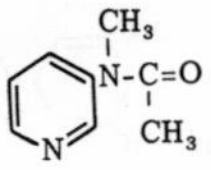
433

▶1590 S
E 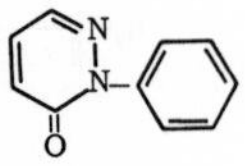
590

▶1590 S
E, CH$_2$Cl$_2$ 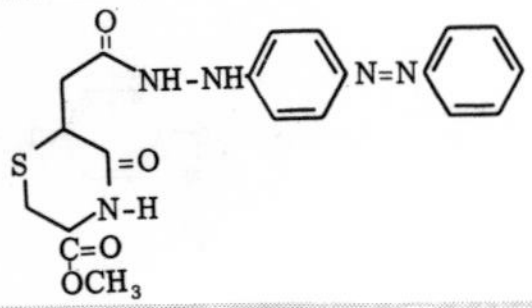
580

▶1590 S*
G
* A broad band appears at 3000-2200 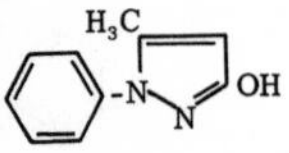
[C=C]
6

▶1590 S
$(C_6H_5)_2SiCl_2$
86

▶1590 M
B 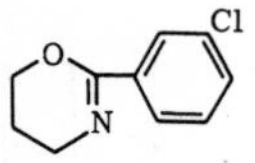
26

▶1590 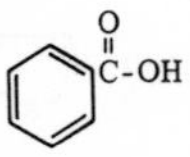
[Ring]
370

▶1590
E or N 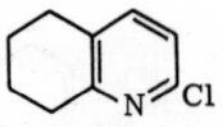
41

▶1590 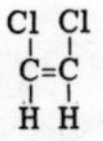
(cis)
[C=C]
543

▶1590
G 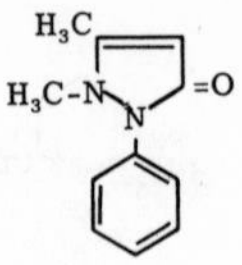
[C=C]
6

▶1589
E or N 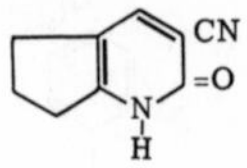
41

▶1589 VS
B (0.10) 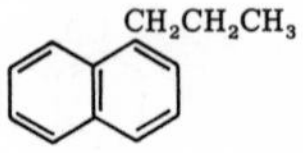
178

▶1589 VS
CHCl$_3$ 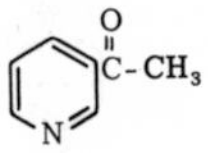
433

▶1589 S
G 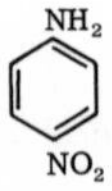
21

▶1589 S
CHCl$_3$ 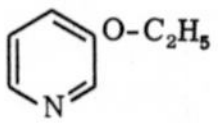
433

▶1589 M
B 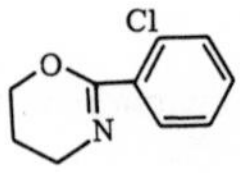
26

▶1589 MW
CHCl$_3$ 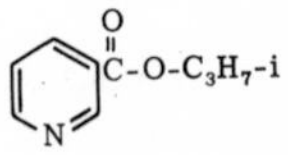
433

▶1589 MW
CHCl$_3$ 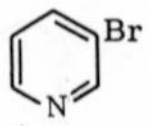
433

▶1588
E or N 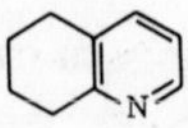
41

▶1588 VS
B (0.055)

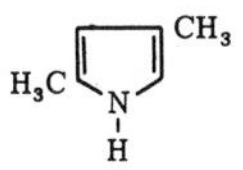

344

▶1588 S
E or N

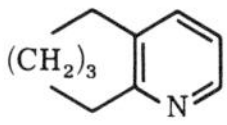

41

▶1588 MW
$CHCl_3$

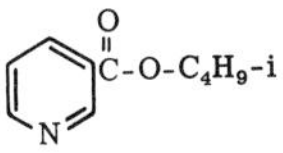

433

▶1588
E or N

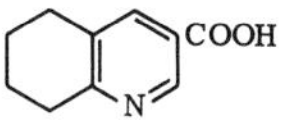

41

▶1587.7 M
B (0.2)

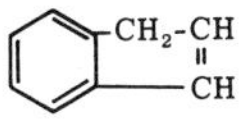

9

▶1587
E or N

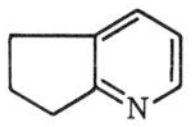

41

▶1587 S
E, $CHCl_3$

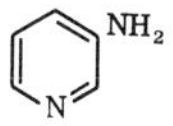

327

▶1587 S
F

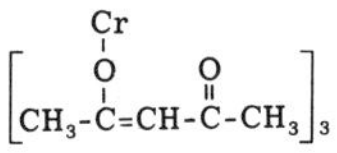

16

▶1587 S

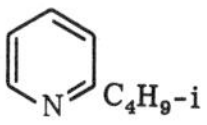

455

▶1587 M
B (0.2398)

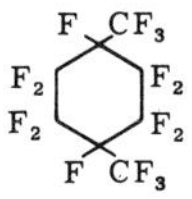

230

▶1586 VS
B (0.10)

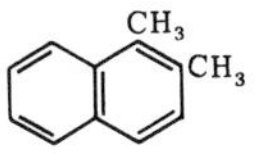

188

▶1585 VS (Raman only)
A

[C=C] 677

▶1585 VS
F

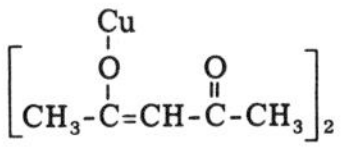

16

▶1585 S
B

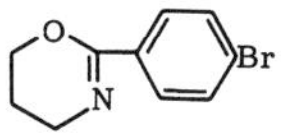

26

▶1585 VW
B (0.2398)

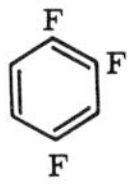

224

▶1585 S
E

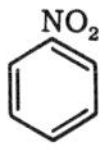

[Ring] 20

▶1585 MSB
B (0.2398)

$(CF_3)_2CFCF_2CF_3$

240

▶1585 SSh
B (0.10)

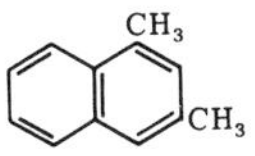

187

▶1585
G

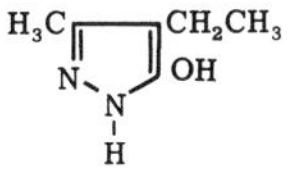

[C=C] 6

▶1585
E or N

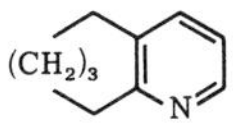

41

▶1584 VS $CHCl_3$ [7.6 g/l] C_6H_5-C(=O)-NH_2 29	▶1582 M $CHCl_3$ (pyridyl)CH_2CH_2-C(=O)-O-C_2H_5 433
▶1584 S $CHCl_3$ [1.39 g/l] C_6H_5-C(=O)-NH_2 29	▶1582 M $CHCl_3$ (pyridine)CH_3 433
▶1584 MW $CHCl_3$ (phenylpyridine) 433	▶1582 E or N $(CH_2)_3$ (pyridine)CH_2OH 41
▶1583 VS $CHCl_3$ (pyridine)NH-CH_3 433	▶1581 S E Ni (NH_2-CH_2-CH_2-NH_2)$_3$ $PtCl_4$ [NH_2] 8

▶1583 S
D
(furyl)CH=CHCH=CHCH=CHCH=N-N=CHCH=CHCH=CHCH=CH(furyl)
597

▶1582 VS A (200 mm Hg, 10 cm) $CF_3CF_2CF_3$ 235	▶1581 S $CHCl_3$ (pyridine)N(CH_3)-C(=O)-CH_3 433
▶1582 S F Fe [CH_3-C(-O-)=CH-C(=O)-CH_3]$_3$ 16	▶1581 S (pyridine)C_3H_7-n 431
▶1582 S CH_3N=O [N=O stretch] 614	▶1581-1566 S (benzo) N=CH, C=N, H [N=C ring] 598
▶1582 MSh B (0.10) (cyclooctatetraene) 200	▶1581 MS $CHCl_3$ (pyridine)C(=O)-H 433
▶1582 M E $[H_2CNO_2]^-Na^+$ [C=N] 328	▶1581 MW $CHCl_3$ (pyridine)Cl 433

▶1581
E or N 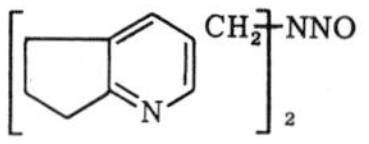
[(6,7-dihydro-5H-cyclopenta[b]pyridin-7-yl)CH_2]$_2$NNO
41

▶1580
$CHCl_3$ 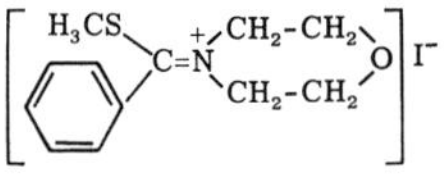
$[C_6H_5(H_3CS)C{=}N^+(CH_2-CH_2)_2O]\ I^-$
615

▶1580 S
E 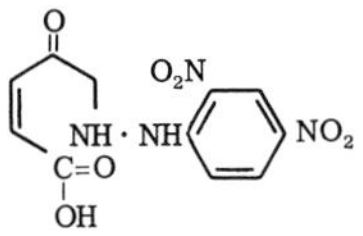
NH·NH–$C_6H_3(NO_2)_2$; C=O; OH
575

▶1580 S
E 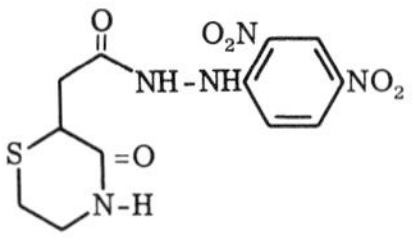
NH-NH–$C_6H_3(NO_2)_2$; S; =O; N-H
580

▶1580 S
G 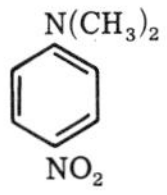
$N(CH_3)_2$; NO_2
21

▶1580 S
G 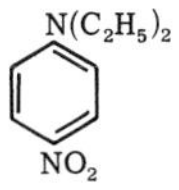
$N(C_2H_5)_2$; NO_2
21

▶1580 S
$CHCl_3$ 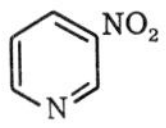
NO_2; N
433

▶1580 S
$CHCl_3$ 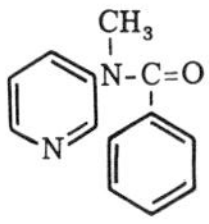
CH_3; N-C=O; N
433

▶1580 M
B 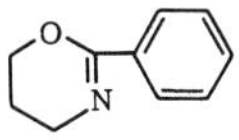
O; N
26

▶1580 M
C
KH_2PO_4
28

▶1580 M
D, CCl_4
$[H_2C\cdot NO_2]^-Na^+$
[C=N] 328

▶1580 M
$CHCl_3$ 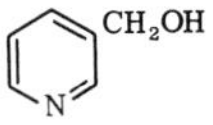
CH_2OH; N
433

▶1580 M
$CHCl_3$ 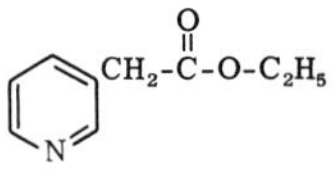
$CH_2-C(=O)-O-C_2H_5$; N
433

▶1580 MW
$CHCl_3$ 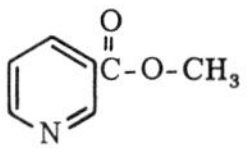
$C(=O)-O-CH_3$; N
433

▶1580 MW
$CHCl_3$ 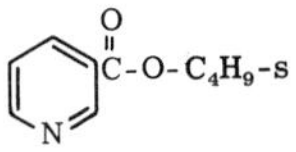
$C(=O)-O-C_4H_9$-s; N
433

▶1580 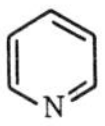
N
405

▶1580
E or N
N 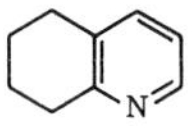
41

▶1580 M
CH_3NO_2
50

▶1579 S
$CHCl_3$ 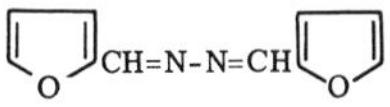
CH=N-N=CH; O; O
597

▶1579 MW
CCl_4 [0.007 g/ml] 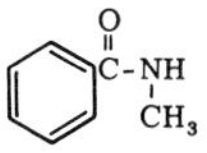
$C(=O)-NH-CH_3$
[Aromatic ring] 2

▶1578 S
$CHCl_3$ 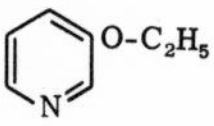
$O-C_2H_5$ N
433

▶1578 M
CCl_4 [0.04 g/ml] 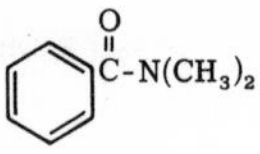
$C(=O)-N(CH_3)_2$
3

▶1578 MW
$CHCl_3$ 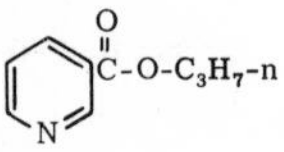
$C(=O)-O-C_3H_7$-n N
433

▶1578 MW
$CHCl_3$ 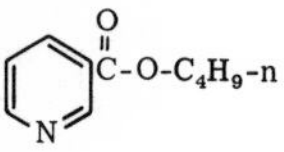
$C(=O)-O-C_4H_9$-n N
433

▶1578 MW
$CHCl_3$ [0.0064 g/ml] 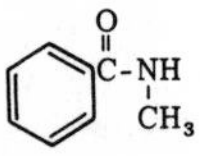
$C(=O)-NH-CH_3$
[Aromatic ring]
2

▶1578
E or N 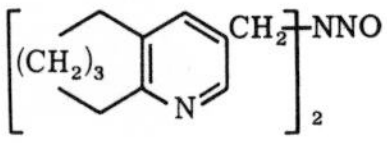
$[(CH_2)_3$ N $CH_2]_2$NNO
41

▶1577 M
B (0.0576)
$(C_4F_9)_3N$
346

▶1577 W
B (0.0576) 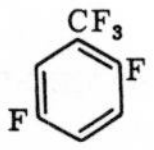
CF_3 F F
217

▶1577 SB
B (0.2398)
$CF_3(CF_2)_5CF_3$
241

▶1577 M
B
$CH_3CH_2C(CH_3)_2-CH_2CH_3$
336

▶1577
E or N 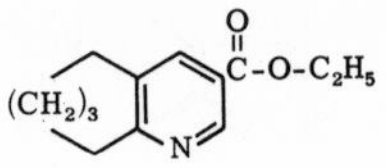
$(CH_2)_3$ N $C(=O)-O-C_2H_5$
41

▶1577 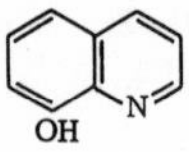
OH N
616

▶1576 S
$CHCl_3$ 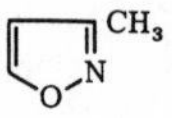
CH_3 N O
15

▶1576 MW
$CHCl_3$ 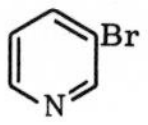
Br N
433

▶1576
E or N 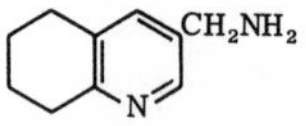
CH_2NH_2 N
41

▶1576
E or N
$(CH_2)_3$ N 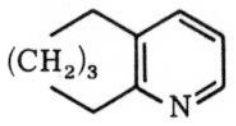
41

▶1575 VVS
B (0.04) 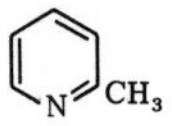
N CH_3
198

▶1575 VS
B (0.08) 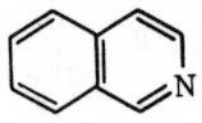
N
197

▶1575 VS
B
$[(CH_3)_2N-BO]_3$
32

▶1575 VS
E 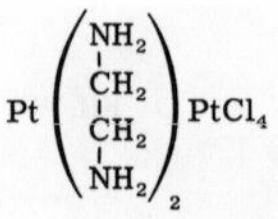
Pt $(NH_2-CH_2-CH_2-NH_2)_2$ $PtCl_4$
[NH_2]
8

▶1575 S
B 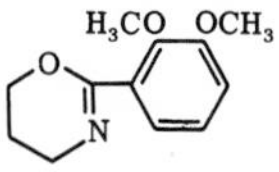
H_3CO OCH_3
O
N
26

▶1575 VW
B (0.2398) 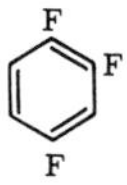
F
F
F
224

▶1575 S
G 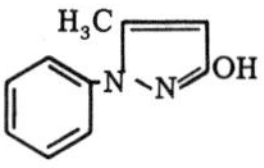
H_3C
N N OH
[C=C]
6

▶1575 S
$CHCl_3$ 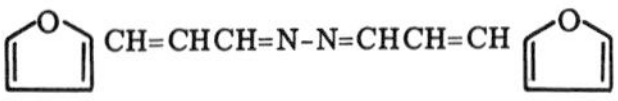
O CH=CHCH=N-N=CHCH=CH O
597

▶1575 W
B (0.0563)
$(CH_3)_2C=CHC(CH_3)_3$
204

▶1575 M
$CHCl_3$ 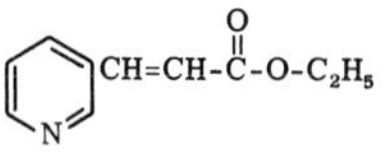
O
N CH=CH-C-O-C_2H_5
433

▶1575
G 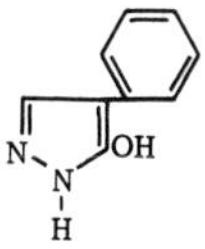
N OH
N
H
[C=C]
6

▶1575 MW
$CHBr_3$
O
C-NH
CH_3 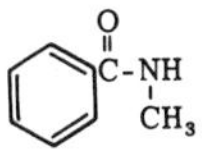
[Aromatic ring]
433

▶1575
E or N
$(CH_2)_3$ CH_2NH_2
N 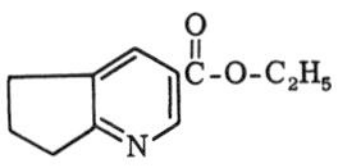
41

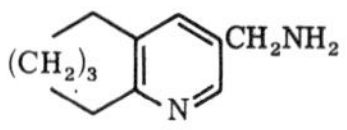

▶1575
E or N 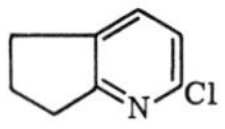
O
C-O-C_2H_5
N
41

▶1575
E or N
N Cl
41

▶1574.3 M
B (0.2) 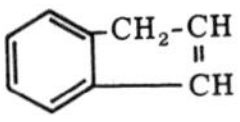
CH_2-CH
CH
9

▶1574 VVS
CH_2Cl_2 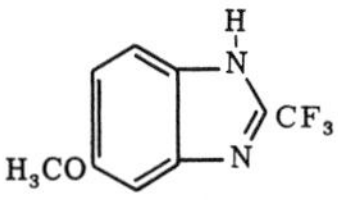
H
N
CF_3
N
H_3CO
37

▶1574 VS
E 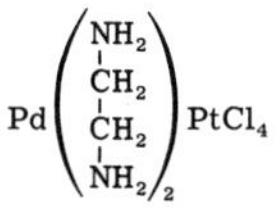
Pd $\left(\begin{array}{c}NH_2\\ CH_2\\ CH_2\\ NH_2\end{array}\right)_2$ $PtCl_4$
$[NH_2]$
8

▶1574 S
$CHCl_3$ 
S
O
7

▶1573 S
$CHCl_3$ 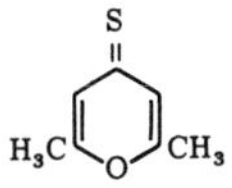
S
H_3C O CH_3
7

▶1573 M
C 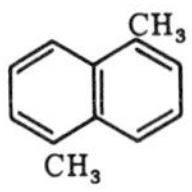
CH_3
CH_3
185

▶1573 MW
$CHCl_3$ 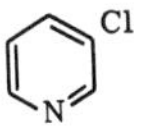
Cl
N
433

▶1573
G 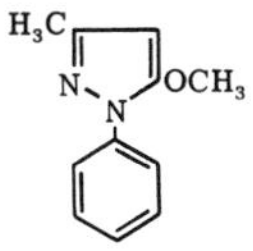
H_3C
N N OCH_3
[C=C]
6

▶1573
E or N 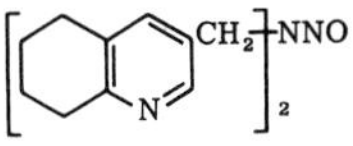
$[$ N $CH_2]_2$NNO
41

▶1572 S A (775 mm Hg, 10 cm) 243	▶1570 S E or N 41
▶1572 S E or N 41	▶1570 M E, F 37
▶1572 M B 26	▶1570 G [C=C] 6
▶1572 M CCl_4 327	▶1570 G [C=C] 6
▶1572 M $CHCl_3$ 327	▶1570 405
▶1572 MW $CHCl_3$ 433	▶1570 [C=C] 542
▶1572 E or N 41	▶1570 617
▶1571 S E $[NH_2]$ 8	▶1570 415
▶1571 E or N 41	▶1568 VS E $[NH_2]$ 8
▶1570 S I 618	▶1567 VS A (600 mm Hg, 10 cm) $CF_3CF_2CF_2CF_2CF_3$ 237

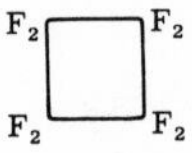

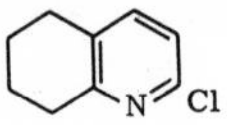

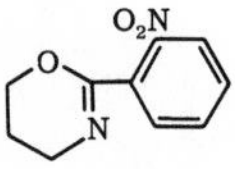

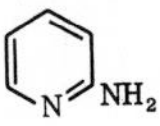

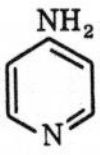

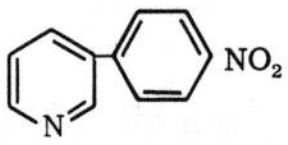

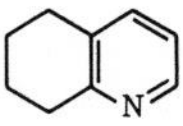

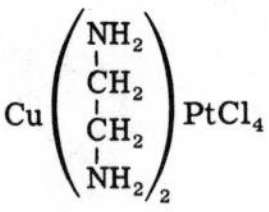

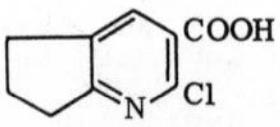

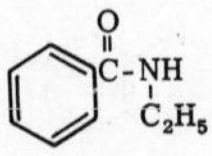

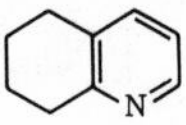

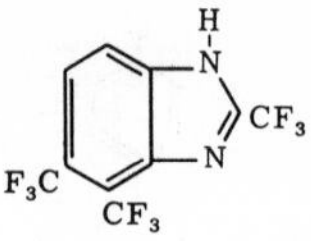

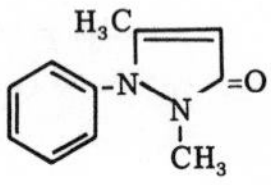

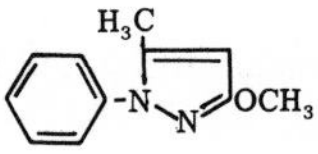

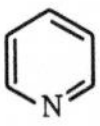

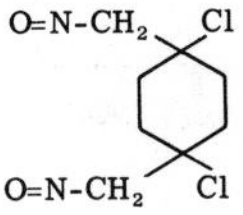

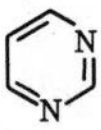

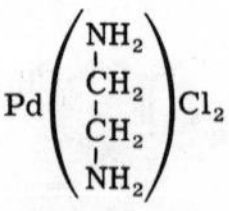

▶1567 VVS
CCl_4 [25%]
CH_3
$H_2C{=}CCO_2CH_3$
193

▶1567 W
B (0.2398) 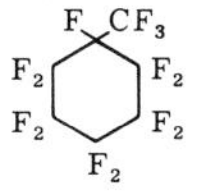
F CF_3 / F_2 F_2 / F_2 F_2 / F_2
232

▶1567 S
F
VO
O O
[CH_3-C=CH-C-CH_3]$_2$
16

▶1567 S
I
O
H_3C-C-NH-CH_3
46

▶1567 W
B (0.2398) 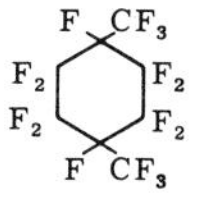
F CF_3 / F_2 F_2 / F_2 F_2 / F CF_3
230

▶1567 MW
$CHCl_3$ 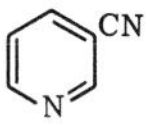
CN / N
433

▶1567 W
B
$(C_6H_5)_2SiCl_2$
86

▶1567 W
B
$(C_6H_5)_2(CH_3)SiCl$
86

▶1566 VVS
CH_2Cl_2 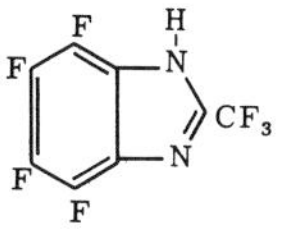
F H / F N / CF_3 / F N / F
37

▶1566 
[C=C]
542

▶1565 S
G 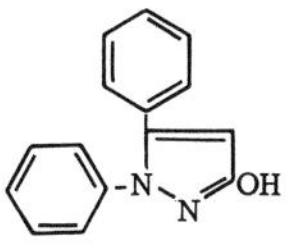
-N-N / OH
[C=C]
6

▶1565 S
G 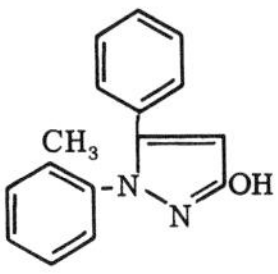
CH_3 / -N-N / OH
[C=C]
6

▶1565 S
G 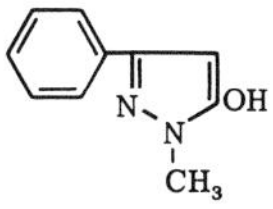
N-N / OH / CH_3
[C=C]
6

▶1565 S
Sc
O O
[CH_3-C=CH-C-CH_3]$_3$
16

▶1565 S 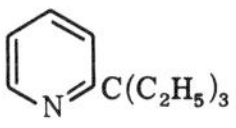
N $C(C_2H_5)_3$
455

▶1565 S 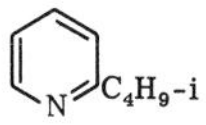
N C_4H_9-i
455

▶1565 M
E, F 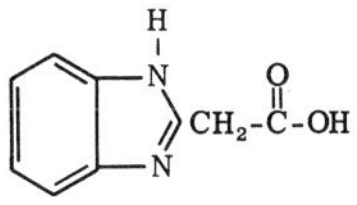
H / N / CH_2-C(=O)-OH / N
37

▶1565
B
O
H_3C-C-NH-CH_3
[Amide II]
533

▶1565
I
O
Cl-CH_2-C-NH-CH_3
[Amide II]
533

▶1565
CCl_4 [0.0128 m/l] 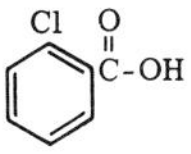
Cl O / C-OH
448

▶1564
E or N 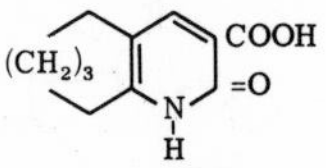
$(CH_2)_3$ COOH =O N H
41

▶1563 M
C 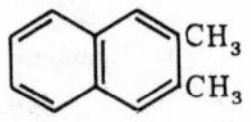
CH_3 CH_3
181

▶1563
E or N 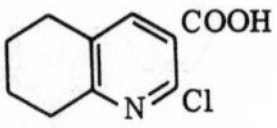
COOH N Cl
41

▶1563 S
A (400 mm Hg, 10 cm)
$(CH_3)_2CHCH_2CH_3$
333

▶1562 S
E or N 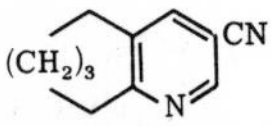
$(CH_2)_3$ CN N
41

▶1562 M
E, F 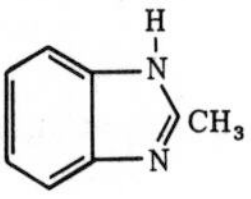
H N CH_3 N
37

▶1562
E or N 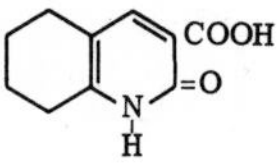
COOH =O N H
41

▶1562
E or N 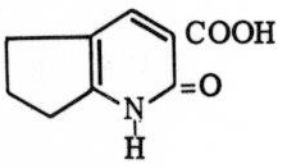
COOH =O N H
41

▶1562 S
E or N 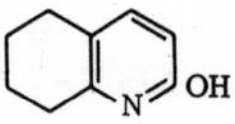
N OH
41

▶1561 VS
E 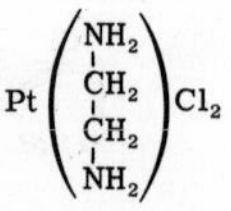
Pt (NH_2–CH_2–CH_2–NH_2) Cl_2
[NH_2]
8

▶1561 S
$CHCl_3$ 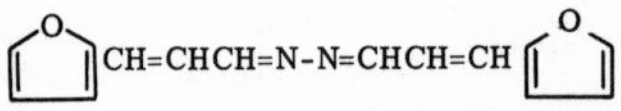
O CH=CHCH=N-N=CHCH=CH O
597

▶1561 S 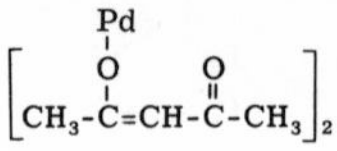
Pd O O
$[CH_3-C=CH-C-CH_3]_2$
16

▶1560 S
E 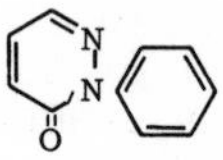
N N O
590

▶1560 S
F 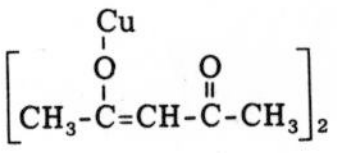
Cu O O
$[CH_3-C=CH-C-CH_3]_2$
16

▶1560 S
E or N 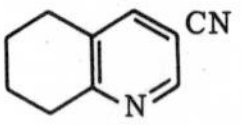
CN N
41

▶1560 M
C 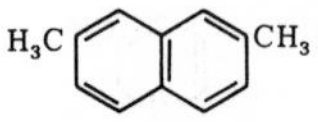
H_3C CH_3
179

▶1560 M
E 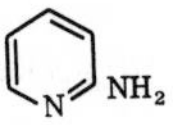
N NH_2
327

▶1560 M
E, F 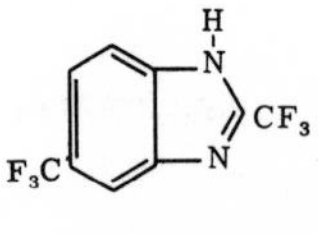
H N CF_3 N F_3C
37

▶1560
G 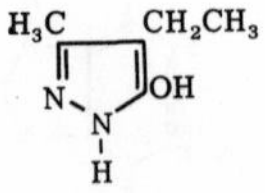
H_3C CH_2CH_3 N N OH H
[C=C]
6

▶1560 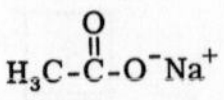
$H_3C-C(=O)-O^-Na^+$
[COO^-]
525

▶1560 NO_2 / NO_2 (dinitrobenzene structure) [NO_2 asymm. stretch] 620	▶1558 N=O [-N=O] 621
▶1560 E or N $(CH_2)_3$ N OH 41	▶1556 VS N N N 14
▶1558 VVS B (0.003) CH_3NO_2 311	▶1555 S A (775 mm Hg, 10 cm) $CF_2Cl\text{-}CF_2Cl$ 257
▶1558 S B (0.10) S 199	▶1555 S F Al, O, O $[CH_3\text{-}C\text{=}CH\text{-}C\text{-}CH_3]_3$ 16
▶1558 S G H_3C CH_3 N N OH [C=C] 6	▶1555 MS E, F H N H_3C H_3C CF_3 N 37
▶1558 M E, F H N CF_3 N Cl 37	▶1555 M C CH_3 CH_3 185
▶1558 M E, F H N CF_3 N O_2N 37	▶1555 M E, F H N CF_3 N 37
▶1558 G H_3C N N OH [C=C] 6	▶1555 M E, F H N CH_3 N F_3C CF_3 37
▶1558 E or N N 41	▶1555 G H_3C N N OH H [C=C] 6
▶1558 E or N CN N 41	▶1555 CH_3 $H_2C\text{=}CCH_2NO_2$ [NO_2] 600

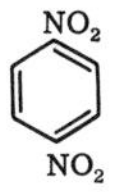

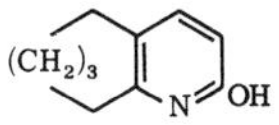

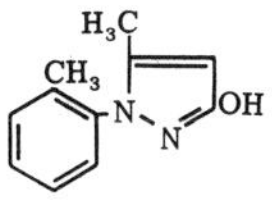

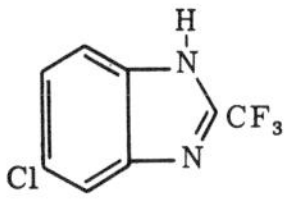

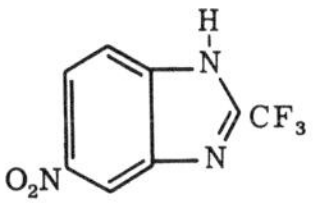

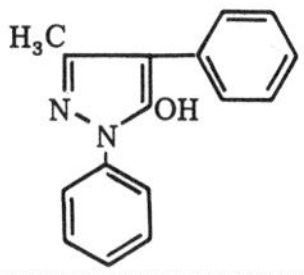

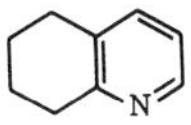

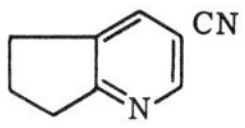

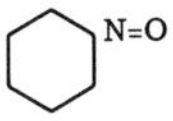

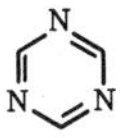

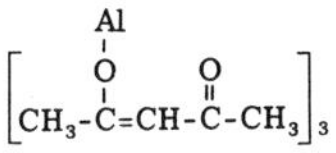

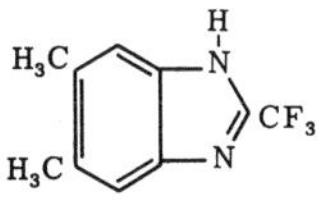

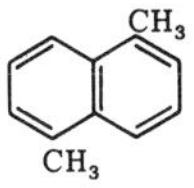

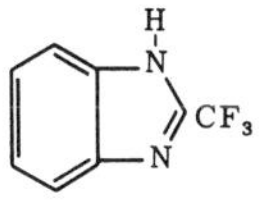

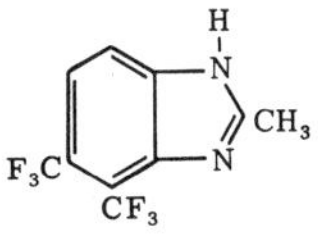

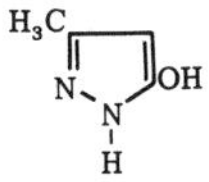

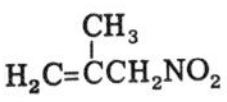

▶1554 S
CH_2Cl_2

H, N, CF_3, N, CF_3

37

▶1553.3 S
B (0.2)

CH_2-CH, =, CH

9

▶1553 W
B (0.2398)

F_2, F_2, F_2, F, C_2F_5, F_2

247

▶1553 SB
B (0.2398)

$(CF_3)_2CFCF_2CF_3$

240

▶1553 MS
E, F

H, N, CH_3, N, CH_3

37

▶1553 MW
CH_2Cl_2

H, N, CF_3, N, F_3C, CF_3

37

▶1553
E or N

N, · HCl

41

▶1552 S
$CHCl_3$

O, CH=N-N=CH, O

597

▶1552 MS
E, F

H, N, CF_3, N, H_3C

37

▶1551 M
E, F

H, N, CF_3, N, CH_3

37

▶1550 VS
N, E

N, N, N

14

▶1550 S
CH_2Cl_2

H, N, CF_3, N, F_3C

37

▶1550 MS
E, F

H, N, CF_3, N, H_3CO

37

▶1550 MS
E, F

H, N, O, CH_2CH_2-C-OH, N

37

▶1550 VS
B (0.003)

$CH_3(CH_2)_2CH_2NO_2$

310

▶1550 M
B (0.10)

CH_3, CH_3

188

▶1550 MW
CCl_4 [20 g/l]

$CH_2CH(CH_3)_2$
CH_3CH=$CH(CH_2)_2CH$=$CH(CH_2)_2CONH$

357

▶1550 M
E, F

H, N, CH_3, N, CF_3

37

▶1550 M
CCl_4

$[(CH_3)_2CNO_2]^-Na^+$

328

▶1550
B

O
Cl-CH_2-C-NH-CH_3

[Amide II]

533

▶1550 S
A (200 mm Hg, 10 cm)
$CF_3CF_2CF_3$
235

▶1548 VVS
CH_2Cl_2
F, F, F, F, H, N, CF_3, N
37

▶1548 S
A (682 mm Hg, 10 cm)
F_2, F_2, F_2, F_2, F_2
245

▶1548 VS
B (0.003)
$CH_3(CH_2)_2CH_2NO_2$
310

▶1548 SB
B (0.2398)
$CF_3(CF_2)_5CF_3$
241

▶1548 VWB
B (0.2398)
F_9C_4-O-C_4F_9
266

▶1548 S
CH_2Cl_2
H, N, CH_3, N, H_3C
37

▶1548 W
$C_6H_5(CH_3)SiCl_2$
86

▶1547 M
C
CH_3 CH_3
182

▶1547 M
E, F
H, N, CH_2CN, N
37

▶1547 M
E, F
H, N, C_2H_5, N
37

▶1546 VVS
B (0.003)
$C_3H_7NO_2$
392

▶1546 VS
B (0.003)
$CH_3CH_2CH(NO_2)CH_3$
324

▶1546 VS
B (0.003)
$(CH_3)_2CHCH_2NO_2$
326

▶1545 VS
B
O, H_3C, O_2N, N-C_6H_{13}-n
[NO_2 asymm. stretch]
26

▶1545 VS
CH_2Cl_2
H, N, CH_3, N, CF_3
37

▶1545 VS
CH_2Cl_2
H, N, CH_3, N, F_3C, CF_3
37

▶1545 VS
CH_2Cl_2
H, N, CF_3, N, CH_3
37

▶1545 MS
E, F
H, N, CF_3, N, H_3C, CH_3
37

▶1545 S
G
H_3C, N, N, OH, H_3C
[C=C]
6

▶1545 M E, F [2-(hydroxymethyl)-5-(trifluoromethyl)benzimidazole: F_3C, N-H, CH_2OH] 37	▶1543 MS CH_2Cl_2 [methylbenzimidazole: H_3C, N-H] 37
▶1545 M $CHCl_3$ [4H-pyran-4-thione: S, O] [Ring] 7	▶1543 M E, F [2-(trifluoromethyl)benzimidazole carboxylic acid: N-H, CF_3, HO-C=O] 37
▶1545 G [pyrazol-ol: H_5C_2, N, N-H, OH] [C=C] 6	▶1543 W B $C_6H_5HSiCl_2$ 86
▶1545 G [pyrazol-ol: H_7C_3, N, N-H, OH] [C=C] 6	▶1543 [phenylpyrazol-ol: N, N-H, OH] [C=C] 6
▶1544 VS B [tetrahydro-1,3-oxazine: O, H_3C, O_2N, $N-CH_3$] [NO_2] 26	▶1543 E or N [tetrahydroquinoline: CN, N, Cl] 41
▶1543 VS A (0.8 mm Hg, 10 cm) [tetrafluorobenzene: F, F, F, F] 223	▶1542 VS B [tetrahydro-1,3-oxazine: O, H_3C, O_2N, $N-CH_3$] [NO_2] 26
▶1543 VVS B (0.003) $(CH_3)_2CHNO_2$ 366	▶1542 VS B [tetrahydro-1,3-oxazine: O, H_5C_2, O_2N, $N-C_5H_{11}$-n] [NO_2] 26
▶1543 VS B [tetrahydro-1,3-oxazine: O, H_5C_2, O_2N, $N-C_3H_7$-i] [NO_2] 26	▶1542 VS B [tetrahydro-1,3-oxazine: O, H_5C_2, O_2N, $N-C_6H_{13}$-n] [NO_2] 26
▶1543 VS B [tetrahydro-1,3-oxazine: O, H_5C_2, O_2N, $N-C_4H_9$-n] [NO_2] 26	▶1542 VS CH_2Cl_2 [2-methylbenzimidazole: N-H, CH_3] 37
▶1543 S G [indazol-ol: N, N-CH_3, OH] [C=C] 6	▶1542 G [pyrazol-ol: $(CH_3)_3C$, N, N-H, OH] [C=C] 6

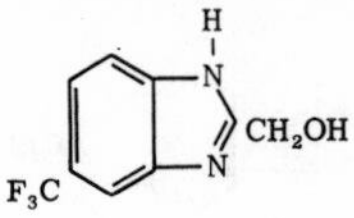

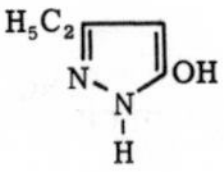

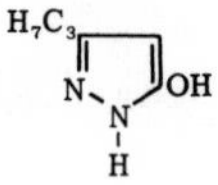

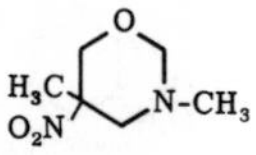

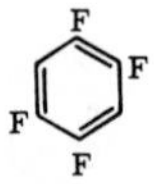

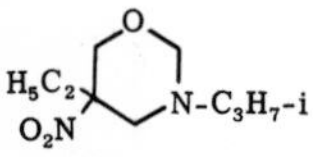

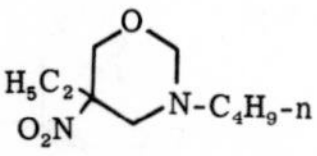

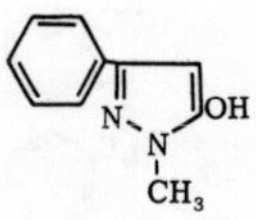

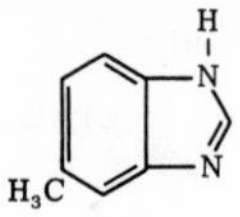

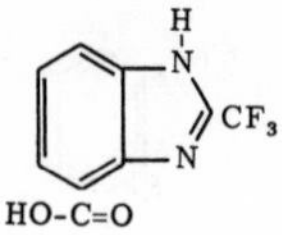

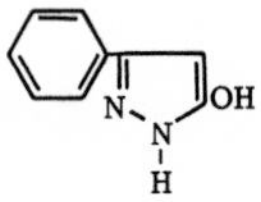

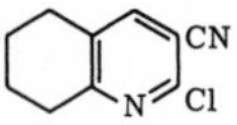

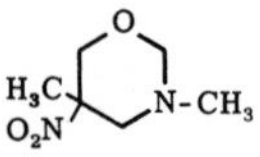

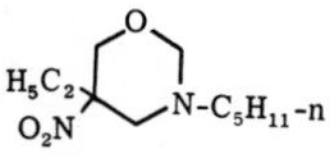

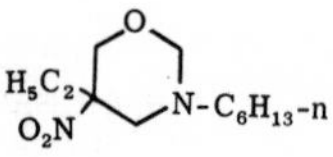

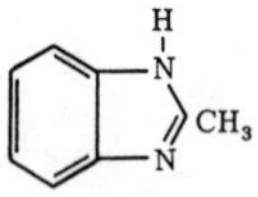

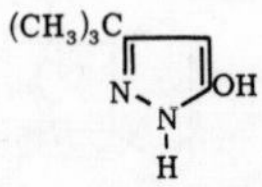

▶1541 VS
B 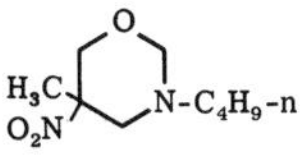
H_3C, O_2N, O, $N-C_4H_9-n$
[NO_2] 26

▶1541 S
C 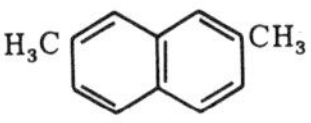
H_3C, CH_3
179

▶1540 VVS
CH_2Cl_2 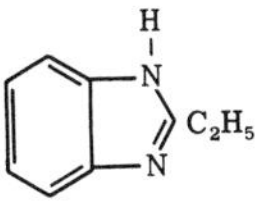
H, N, C_2H_5, N
37

▶1540 VS
B 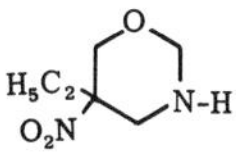
O, H_5C_2, O_2N, N-H
[NO_2] 26

▶1540 S 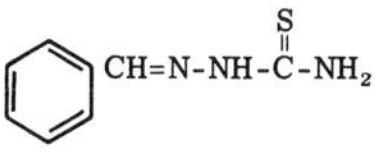
S
CH=N-NH-C-NH_2
380

▶1540 M
E, F 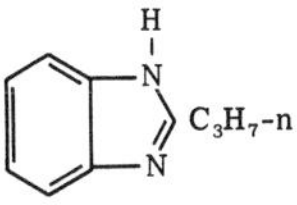
H, N, C_3H_7-n, N
37

▶1540 M
E, F 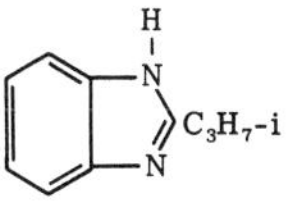
H, N, C_3H_7-i, N
37

▶1540 M
E, F 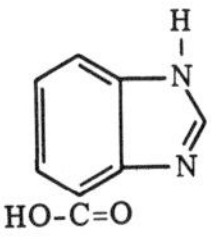
H, N, N, HO-C=O
37

▶1540 M
E 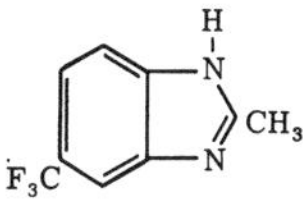
H, N, CH_3, N, F_3C
37

▶1540
G 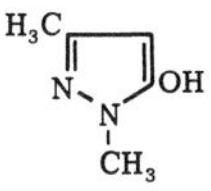
H_3C, N, N, OH, CH_3
[C=C] 6

▶1540 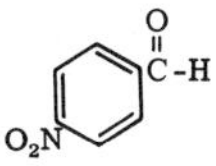
O
C-H, O_2N
387

▶1540
E or N
CN, Cl, N 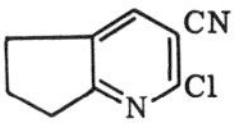
41

▶1538 VSSh
B (0.2398)
H_3C-O-CF_2CHClF
356

▶1538 M
E, F 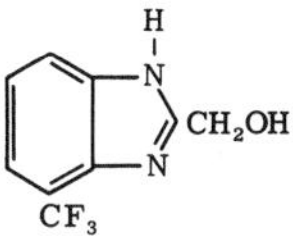
H, N, CH_2OH, N, CF_3
37

▶1538 M
CH_2Cl_2 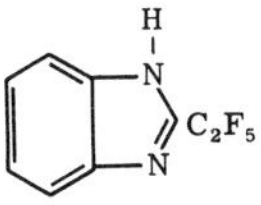
H, N, C_2F_5, N
37

▶1538 MW
CH_2Cl_2 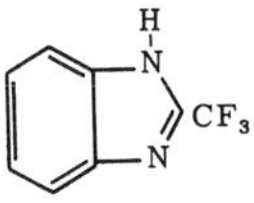
H, N, CF_3, N
37

▶1538 
H_3C, NO_2
[NO_2] 600

▶1537 VVS
$CHCl_3$ 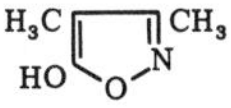
H_3C, CH_3, HO, O, N
[Ring] 15

▶1536 S
A 
F, F, F, F, F, F
623

▶1535 VS
CH_2Cl_2 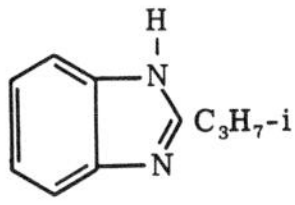
H, N, C_3H_7-i, N
37

▶1535 S G H O O H / N-C-C-N (H; C_6H_{11}) 31	▶1531 VS A (775 mm Hg, 10 cm) F_2 F_2 F_2 F_2 (cyclobutane) 243
▶1534 VS B $[(CH_3)_2N{-}BO]_3$ 32	▶1531 W B (0.0576) F CF_3, F_2 F_2, F_2 F_2, F CF_3 230
▶1534 VS B (film) F, F, F, F (benzene) 222	▶1531 MW B (0.136) $CH_3CHCH_2CH_3$, $CH_3CH_2HC(CH_3)$, $CHCH_2CH_3(CH_3)$ 280
▶1534 VVS C (0.003) $(CH_3)_3CNO_2$ 325	▶1531 MS B (0.136) $CH(CH_3)_2$, $(CH_3)_2HC$, $CH(CH_3)_2$ 278
▶1534 MS G H S S H / N-C-C-N (CH_2OH CH_2OH) [NH] 31	▶1530 S G H_3C, CH_3, C_2H_5, OH, N, N [C=C] 6
▶1534 MWSh B (0.238) $H_2C{=}CHCH_2C(CH_3)_3$ 273	▶1530 $Ce(NO_3)_6^{=}$ 631
▶1534 W A $CH_3CH_2CH(CH_3)CH_2CH_3$ 332	▶1529 S G H S S H / N-C-C-N (HO_2CH_2C, CH_2CO_2H) [NH] 31
▶1534 H $H_3C{-}C({=}O){-}NH{-}CH_3$ [Amide II] 533	▶1529 W B $CH_3CH_2CH(CH_3)(CH_2)_2CH_3$ 330
▶1533 S E, F H, N, CF_3, N, O_2N 37	▶1528 VVS $CHCl_3$ H_3C, CH_3, H_3CO, O, N [Ring] 15
▶1532 M CH_2Cl_2 H, N, C_3H_7, N, CH_3 37	▶1528 VS B (0.10) CH_2CH_3 (naphthalene) 189

▶1528 W
B 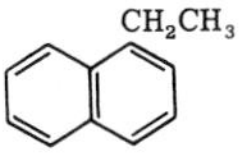
CH_2CH_3
190

▶1527 VS
E 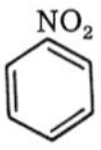
NO_2
[NO_2 asymm. stretch]
20

▶1527 VS
G 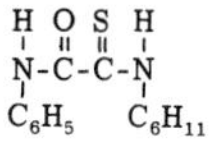
H O S H
N-C-C-N
C_6H_5 C_6H_{11}
31

▶1527 S
C 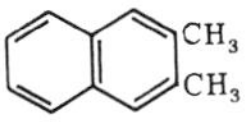
CH_3
CH_3
181

▶1527 S
G 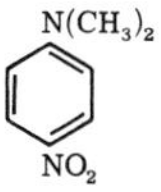
$N(CH_3)_2$
NO_2
21

▶1527 M
B (0.955)
$CF_2Cl\text{-}CFCl_2$
255

▶1527 
CH_3
NO_2
[NO_2 asymm. stretch]
620

▶1526 VS
G 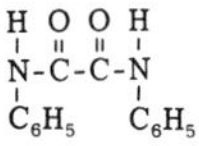
H O O H
N-C-C-N
C_6H_5 C_6H_5
31

▶1526 M
E, F 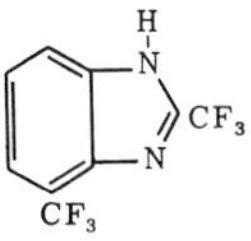
H
N
N
CF_3
CF_3
37

▶1526 W
$CHCl_3$
S
H_3C S CH_3 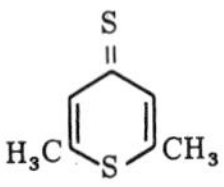
[Ring]
7

▶1525 S
A (6.9 mm Hg, 40 cm) 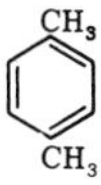
CH_3
CH_3
123

▶1525 M
CH_2Cl_2 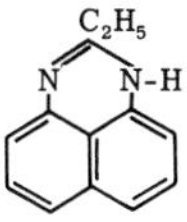
C_2H_5
N N-H
37

▶1525 W
$CHCl_3$ 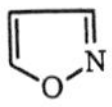
O N
15

▶1525 W
$CHCl_3$ 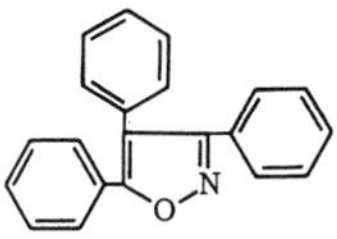
O N
15

▶1524 W
B (0.0576) 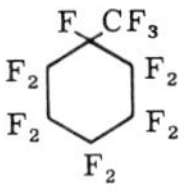
F CF_3
F_2 F_2
F_2 F_2
F_2
232

▶1524 M
$CHBr_3$
O
C-NH
CH_3 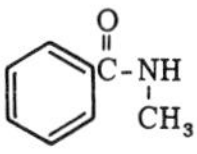
[Amide II]
2

▶1524 S
C 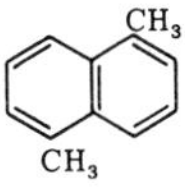
CH_3
CH_3
185

▶1524 MS
B (0.055) 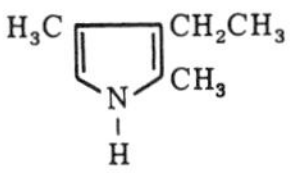
H_3C CH_2CH_3
CH_3
N
H
343

▶1524 M
CCl_4 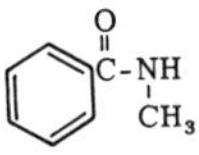
O
C-NH
CH_3
[Amide II]
2

▶1524 M
CH_2Cl_2 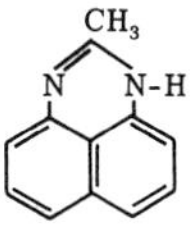
CH_3
N N-H
37

▶1524 M
CH_2Cl_2 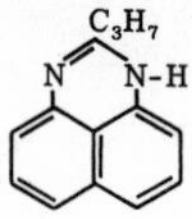
37

▶1523 VVS
B (0.10) 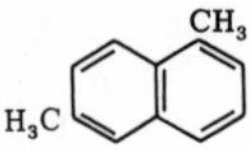
184

▶1523 M
C 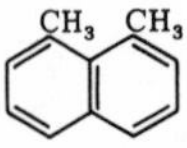
182

▶1522 VS
B 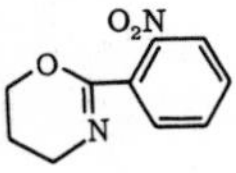
26

▶1522 VS
B (0.0104) 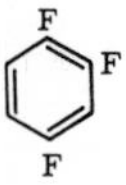
224

▶1522 M
CCl_4 [31%] (0.025) 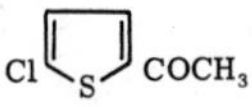
372

▶1521 VVS
$CHCl_3$ 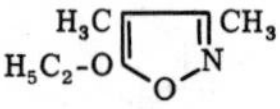
[Ring]
15

▶1520 VS
A (7.6 mm Hg, 40 cm) 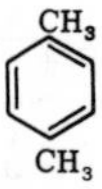
123, see also 68

▶1520 
406

▶1520 S
G 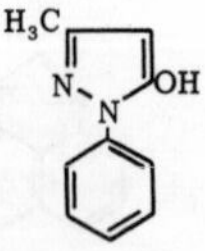
6

▶1520 S
G 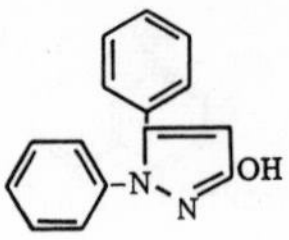
[C=C]
6

▶1520 MW
CCl_4 [sat.] (0.025) 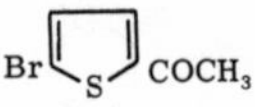
348

▶1520 M
E, F 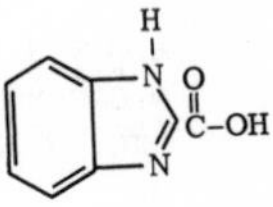
37

▶1520 M
E, F 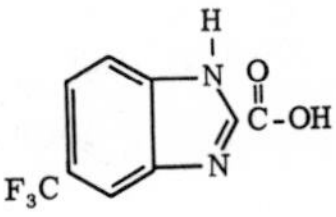
37

▶1520
P
$Th(NO_3)_4$
45

▶1518 S
E, F 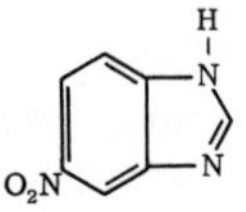
37

▶1518 VS
B (0.033) 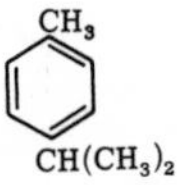
132

▶1518 VS
$CHCl_3$ 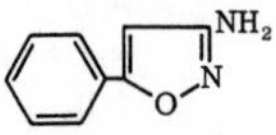
[Ring]
15

▶1518 S
B (0.10) 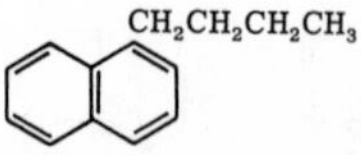
176

▶1517 M
B (0.065) 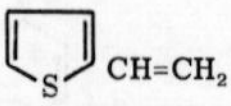
212

▶1517 VS
A (12.5 mm Hg, 10 cm) 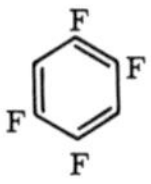
F, F, F, F
223

▶1517 VVSSp
B (0.028) 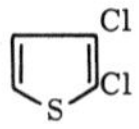
Cl, Cl, S
162

▶1517 VVS
B (0.10) 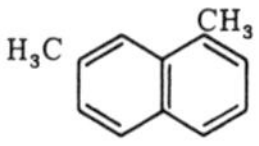
H_3C, CH_3
183

▶1517 S
G 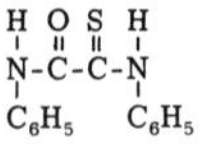
H O S H
N-C-C-N
C_6H_5 C_6H_5
[NH] 31

▶1517 S
G 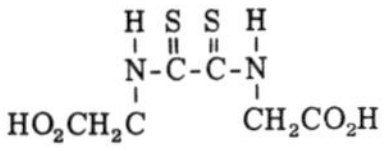
H S S H
N-C-C-N
HO_2CH_2C CH_2CO_2H
[NH] 31

▶1516
A 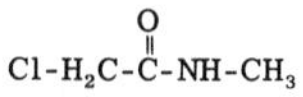
$Cl-H_2C-C(=O)-NH-CH_3$
[Amide II] 533

▶1516 VVS
B (0.10) 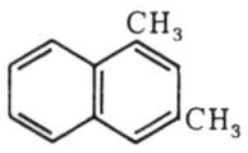
CH_3, CH_3
187

▶1516 S
G 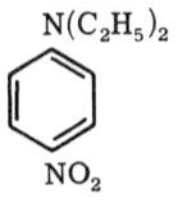
$N(C_2H_5)_2$
NO_2
21

▶1516 S
G 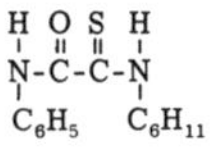
H O S H
N-C-C-N
C_6H_5 C_6H_{11}
[NH] 31

▶1515 
OH
632, see also 36

▶1515 S
A (385 mm Hg, 10 cm)
CF_3-CCl_3
253

▶1515 VS
B (0.028) 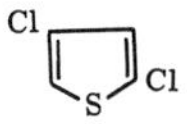
Cl, S, Cl
211

▶1515 S
A (6.9 mm Hg, 40 cm) 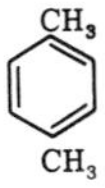
CH_3
CH_3
123

▶1515 S
G 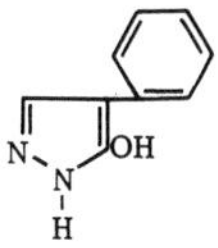
N, N, OH, H
[C=C] 6

▶1515 S
G 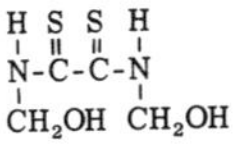
H S S H
N-C-C-N
CH_2OH CH_2OH
[NH] 31

▶1515 S
CCl_4 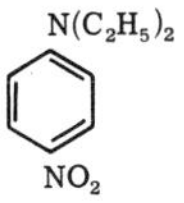
$N(C_2H_5)_2$
NO_2
21

▶1515 S
$CHCl_3$ 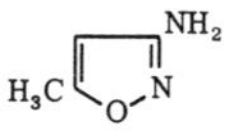
NH_2, H_3C, O, N
15

▶1515 S 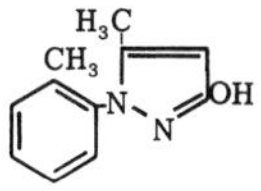
H_3C, CH_3, N, N, OH
[C=C] 6

▶1515 M
E, F 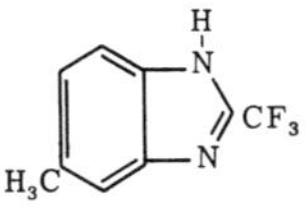
H, N, CF_3, N, H_3C
37

▶1515
$O_2N-CH=C(CH_3)_2$
[NO_2] 600

▶1514 S
B (0.055)
H_3C CH_3 N H
344

▶1513 VS
B (0.033)
CH_3 $CH(CH_3)_2$
132

▶1513 VS
A (12.5 mm Hg, 10 cm)
F F F F
223

▶1513 VS
B (film)
CH_3 F
220

▶1513 VS
B (0.10)
$CH_2CH_2CH_2CH_3$
175

▶1513 W
B (0.0576)
F CF_3 F_2 F_2 F CF_3 F_3C F F_2
228

▶1513 VS
G
$H_3C-\overset{O}{\overset{\|}{C}}-\overset{S}{\overset{\|}{C}}-N(CH_2-CH_2)_2O$
[C-N?]
31

▶1513 S
G
H O S
N-C-C-N(CH_2-CH_2)_2O
H
[C-N?]
31

▶1513 S
$CHCl_3$
NH_2 H_5C_2 O N
[Ring]
15

▶1512 VS
B (0.10)
$CH_2CH_2CH_2CH_2CH_3$
174

▶1512 VS
G
H O O H
N-C-C-N
C_6H_5 C_6H_{11}
[NH]
31

▶1512 VS
G
H O S H
N-C-C-N
H C_6H_{11}
31

▶1512 S
B (0.10)
CH_3
192

▶1512 S
C
H_3C CH_3
179

▶1511 VS
B (0.051)
$CH(CH_3)_2$ $CH(CH_3)_2$
298

▶1511 VS
B (film)
F F
225

▶1511 S
G
H O O H
N-C-C-N
C_6H_{11} C_6H_{11}
[NH]
31

▶1511 S
G
H O O H
N-C-C-N
C_6H_5 C_6H_{11}
[NH]
31

▶1511 S
G
H O S C_2H_5
N-C-C-N
C_6H_5 C_2H_5
[NH]
31

▶1511 MW
CCl_4 [20 g/l]
$CH_2CH(CH_3)_2$
$CH_3CH{=}CH(CH_2)_2CH{=}CH(CH_2)_2CO\text{-}NH$
357

▶1510 VS
G 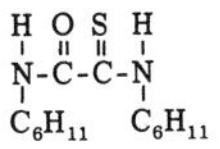
H O S H / N-C-C-N / C_6H_{11} C_6H_{11}
31

▶1510 S
C 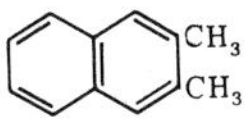
CH_3 CH_3
181

▶1510 S
G 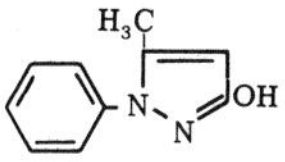
H_3C N N OH
[C=C]
6

▶1510 M
E, F 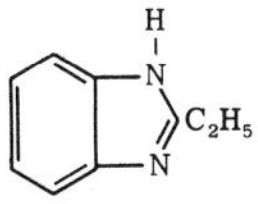
H N C_2H_5 N
37

▶1510 M
E, F 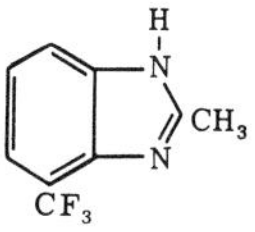
H N CH_3 N CF_3
37

▶1510
G 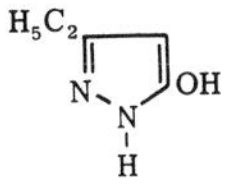
H_5C_2 N N OH H
[C=C]
6

▶1510 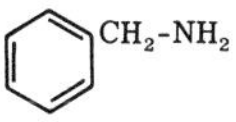
CH_2-NH_2
383

▶1509 S
G 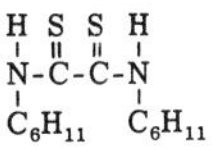
H S S H / N-C-C-N / C_6H_{11} C_6H_{11}
[NH]
31

▶1508 VS
B (0.10) 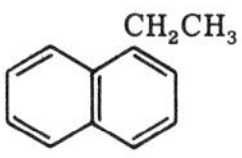
CH_2CH_3
190

▶1508 M (rotational line)
A (141 mm Hg, 15 cm)
H_3C-C≡CH
87

▶1508
CH_2F_2
[CH_2 deformation]
536

▶1507 VS
$CHCl_3$ 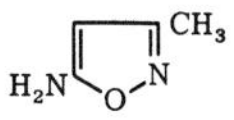
CH_3 H_2N O N
[Ring]
15

▶1506 VS
B (0.0104) 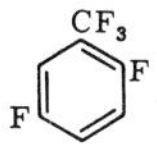
CF_3 F F
217

▶1506 VS
B (0.025) 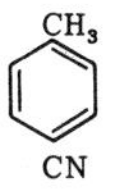
CH_3 CN
214

▶1506 MS
A (200 mm Hg, 10 cm)
CF_2Cl-CF_2Cl
257

▶1506 M
E 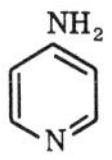
NH_2 N
327

▶1506 S 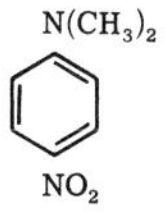
$N(CH_3)_2$ NO_2
[NO_2 asymm. stretch]
620, see also 21

▶1505 S
G 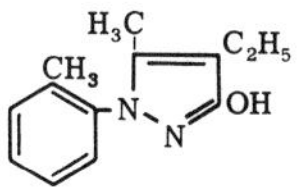
H_3C C_2H_5 CH_3 N N OH
[C=C]
6

▶1505 M
B (0.10) 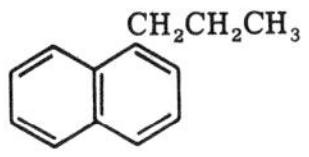
$CH_2CH_2CH_3$
178

▶1505 M
C 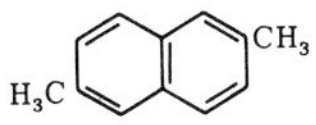
H_3C CH_3
180

Entry	Phase / conditions	Compound	Note	Ref.
▶1505	E	$Cr(NH_3)_5(NO_3)_3$		45
▶1505	G	H_3C, N, N–H, OH (pyrazole ring)	[C=C]	6
▶1505	G	H_3C, N–N, OCH_3 (N-phenyl pyrazole)	[C=C]	6
▶1505	G	N–N–H, OH (phenyl pyrazole)	[C=C]	6
▶1505		CH_3, N–N, OH (diphenyl pyrazole)	[C=C]	6
▶1504 VS	B (0.025)	CH_3 (tetrahydronaphthalene)		269
▶1504 VVS	$CHCl_3$	H_3C, CH_3, H_2N, O–N (isoxazole)	[Ring]	15
▶1504 M	CCl_4 [0.04 g/ml]	$C_6H_5C(=O)-N(CH_3)_2$		3
▶1504 M	$CHCl_3$	S, O (pyran-4-thione)		7
▶1503 S	G	NH_2, NO_2 (benzene ring)		21, see also 626
▶1503 S	G	H S S H / N–C–C–N / C_6H_{11} C_6H_{11}	[NH]	31
▶1502 S	A (200 mm Hg, 10 cm)	F_2 F_2 F_2 F_2 (cyclobutane)		243
▶1502 W	B (0.0576)	F, CF_3, F_2 F_2 F_2 F_2 F_2 (cyclohexane)		232
▶1501 VS	B (0.10)	$CH_2CH_2CH_2CH_2CH_3$ (naphthalene)		173
▶1500 VS	A (24.1 mm Hg, 40 cm)	(benzene)		309
▶1500 SB	G	N–N–H, OH (pyrazole)	[C=C]	6
▶1500 M	A (4.2 mm Hg, 40 cm)	$CH(CH_3)_2$ (benzene)		66
▶1500 S	A (6.3 mm Hg, 40 cm)	CH_3, CH_3 (benzene)		69
▶1500 S	A (8.3 mm Hg, 40 cm)	CH_2CH_3 (benzene)		70
▶1500 M	A (3.2 mm Hg, 40 cm)	$CH_2CH_2CH_3$ (benzene)		67

▶ 1500

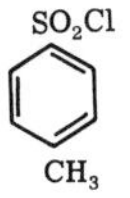

SO_2Cl

CH_3

[Ring] 389

▶ 1500

H_2C-CH_2 / O

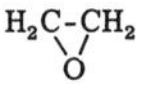

$[CH_2]$ 537, see also 64, 381

▶ 1499 VS

B (0.0104)

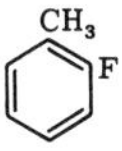

CH_3 F

221

▶1499 VS

B (0.0104)

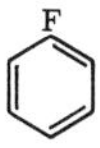

F

226

▶1499 S

A (4.2 mm Hg, 40 cm)

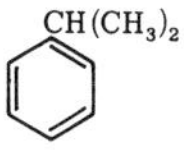

$CH(CH_3)_2$

66

▶1499 M

$CHCl_3$

NH_2 N

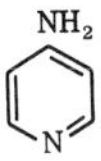

327

▶1497 W

A (4.2 mm Hg, 40 cm)

Grating

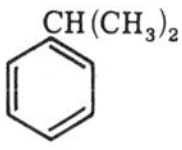

$CH(CH_3)_2$

121, see also 66

▶1497 VS

B (0.0104)

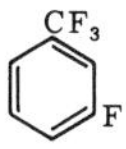

CF_3 F

218

▶1497 VS

E

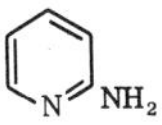

N NH_2

327

▶1497 S

A

F F F F F N

635

▶1497 MS

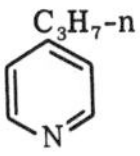

C_3H_7-n N

431

▶1496 MWSh

$CHCl_3$

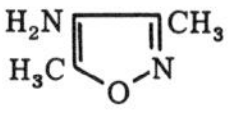

H_2N CH_3 H_3C O N

[Ring] 15

▶ 1496 S

$CHCl_3$

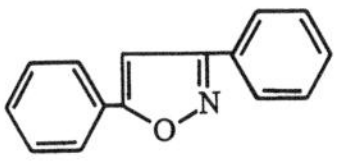

O N

[Ring] 15

▶1495 W

B (0.0576)

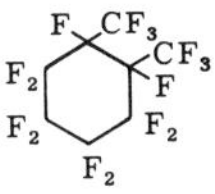

F CF_3 CF_3 F F_2 F_2 F_2 F_2

231

▶1495 VS

B (0.033)

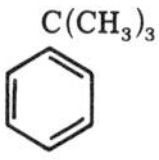

$C(CH_3)_3$

133

▶1495 S

H

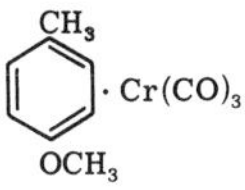

CH_3 · $Cr(CO)_3$ OCH_3

17

▶1495 M

C

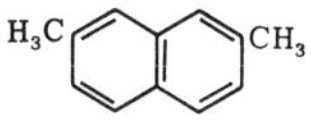

H_3C CH_3

179

▶ 1495

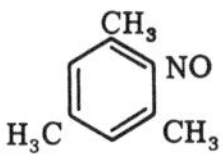

CH_3 NO H_3C CH_3

[CNO] 634

▶1495

OH

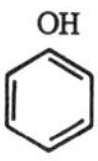

632, see also 36

▶1493 VS

B

$(H_5C_2-NH-B-N-C_2H_5)_3$

[Ring] 32

▶1493 VS
B (0.10)
CH_2CH_3
189

▶1493 MSh
B (0.0576)
$(CF_3)_2CFCF_2CF_3$
240

▶1493 W
$C_6H_5HSiCl_2$
86

▶1492 S
E, F
H
N
CF_3
N
O_2N
37

▶1492 M
B
O
N
26

▶1492 M
E, F
H
N
N
CF_3
37

▶1491 M (rotational lines)
A (141 mm Hg, 15 cm)
$H_3C\text{-}C{\equiv}CH$
87, see also 51

▶1490 VS
B (0.028)
Cl Cl
S
161

▶1490 S
B (0.955)
$CF_2Cl\text{-}CFCl_2$
255

▶1490 S
H
CH_3
$\cdot Cr(CO)_3$
CH_3
17

▶1490 WSh
$CHCl_3$
H_3C CH_3
H_3C N
O
[Ring] 15

▶1490 W
$C_6H_5SiCl_3$
86

▶1490
A
O
$H_3C\text{-}C\text{-}NH\text{-}CH_3$
46

▶1490
$CHCl_3$
O
C-H
O_2N
387

▶1489 VVS
$CHCl_3$
N
H_3CO O
[Ring] 15

▶1489 VS
$CHCl_3$
O
$H_5C_2\text{-}O\text{-}C$
H_3C N
O
[Ring] 15.

▶1489 S
$CHCl_3$
O
$n\text{-}H_7C_3\text{-}O\text{-}C$
H_3C N
O
[Ring] 15

▶1488 VVS
$CHCl_3$
H_2N N
O
[Ring] 15

▶1488 VS
A (682 mm Hg, 10 cm)
F_2 F_2
F_2 F_2
F_2
245

▶1488 VVSSp
B (0.025)
CH_3
CN
312

▶1488 VS
$CHCl_3$ 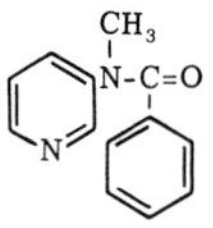
CH_3
N-C=O
N
433

▶1488 S
B (0.0576)
CH_3CH_2-O-CF_2CHFCl
261

▶1488 S
$CHCl_3$ 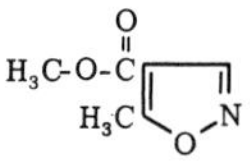
O
H_3C-O-C
H_3C N
O
[Ring] 15

▶1488 MS
E 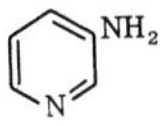
NH_2
N
327

▶1488 M
$C_6H_5(CH_3)SiCl_2$
86

▶1488 W
CCl_4
$(C_6H_5)_3SiCl$
86

▶1488 VVS
B (0.051) 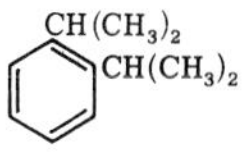
$CH(CH_3)_2$
$CH(CH_3)_2$
296

▶1487 VVS
$CHCl_3$ 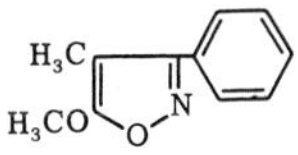
H_3C
H_3CO N
O
[Ring] 15

▶1486 VVS
B (0.051) 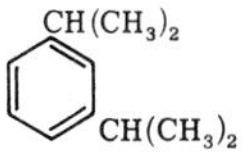
$CH(CH_3)_2$
$CH(CH_3)_2$
297

▶1486 S
A (24.1 mm Hg, 40 cm) 
309

▶1486 VS
B (0.08) 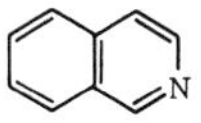
N
197

▶1486 S
$CHCl_3$ 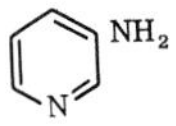
NH_2
N
433, see also 327

▶1486 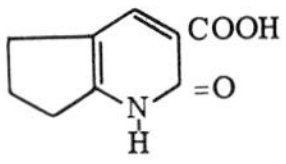
COOH
=O
N
H
41

▶1485 S
CCl_4 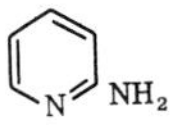
N NH_2
327

▶1485 M
C 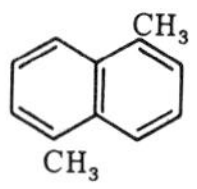
CH_3
CH_3
185, see also 42

▶1485
P
$Y(NO_3)_3$
45

▶1485 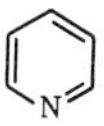
N
405

▶1484 SB
B (0.0576)
$CF_3(CF_2)_5CF_3$
241

▶1484 SB
B (0.008)
CH_3
$(CH_3)_3CC$-$CH(CH_3)_2$
CH_3
306

▶1484 S
G 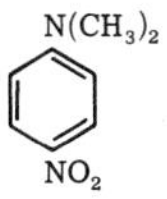
$N(CH_3)_2$
NO_2
[NO_2 stretch] 21

▶1484 S
$CHCl_3$ 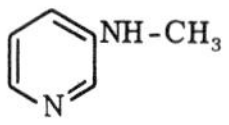
NH-CH_3
N
433

▶1484 MS
$CHCl_3$ 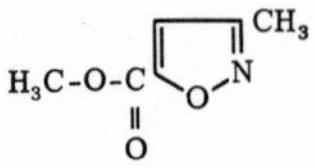
$H_3C-O-C(=O)$–[isoxazole]–CH_3 (O, N)
[Ring] 15

▶1484 M
$CHBr_3$, $CHCl_3$, CCl_4 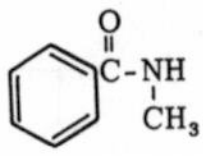
[benzene]–$C(=O)-NH-CH_3$
[CH_3] 2

▶1483.2 VS
B (0.03) 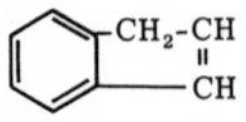
[benzene]–$CH_2-CH=CH$
9

▶1482 S
B 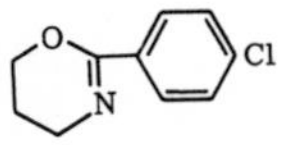
[ring O, N]–[benzene]–Cl
26

▶1482 SSh
$CHCl_3$ 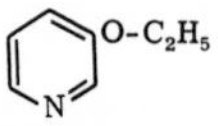
[pyridine, N]–$O-C_2H_5$
433

▶1482 M
E 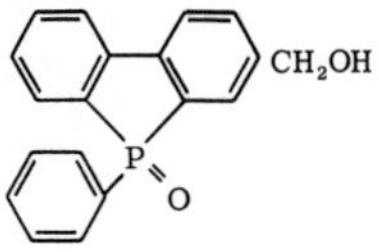
CH_2OH, P, O
34

▶1482 M
CCl_4 [0.00658 m/l] 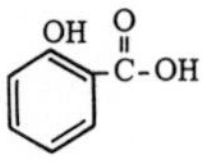
[benzene] OH, $C(=O)-OH$
448

▶1482
E or N 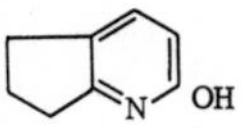
[ring] N, OH
41

▶1481 VS
A (20 mm Hg, 15 cm)
$(CH_3)_3CH$
373

▶1481 VS
B (0.016)
Grating
$(CH_3CH_2)_3CH$
82, see also 143, 156

▶1481 VSSh
B (0.0088).
$(CH_3)_3CCH_2CH(CH_3)CH(CH_3)_2$
303

▶1481 MSh
B (0.008)
$H_2C=C(CH_3)C(CH_3)_3$
272

▶1481 VS
$CHCl_3$ 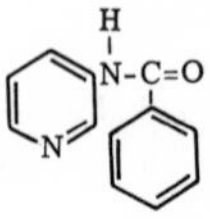
[pyridine, N]–$N(H)-C(=O)$–[benzene]
433

▶1481 M
$CHCl_3$ 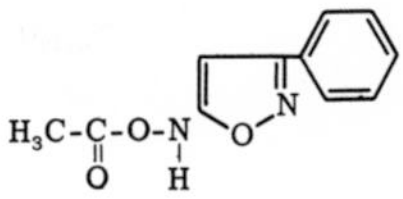
$H_3C-C(=O)-O-N(H)$–[isoxazole, O, N]–[benzene]
[Ring] 15

▶1481 M
$CHCl_3$ 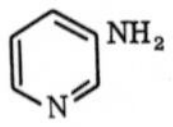
[pyridine, N]–NH_2
327

▶1480 VVS
$CHCl_3$ 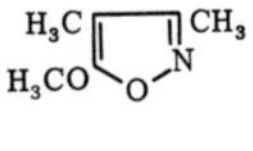
H_3C, H_3CO, CH_3 [isoxazole, O, N]
[Ring] 15

▶1480 VS
A (18 mm Hg, 40 cm)
$(CH_3)_2CH(CH_2)_4CH_3$
294, see also 120

▶1480 M
B (0.0104)
n-H_7C_3-O-CF_2CHFCl
263

▶1480 VS
$CHCl_3$ 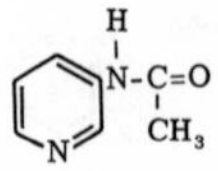
[pyridine, N]–$N(H)-C(=O)-CH_3$
433

▶1480 VS
$CHCl_3$ 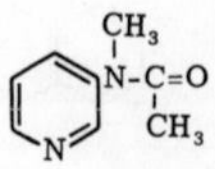
[pyridine, N]–$N(CH_3)-C(=O)-CH_3$
433

▶1480 VS
$CHCl_3$
[Ring]
15

▶1480 W
B (film)
228

▶1480 S
H
$\cdot Cr(CO)_3$
17

▶1480 MS
$CHCl_3$
[Ring]
15

▶1480 M
E
34

▶1480 M
CCl_4 [0.0128 m/l]
448

▶1480
K
$Gd(NO_3)_3$
45

▶1480
E or N
41

▶1479 S
B (0.0088)
$(CH_3)_2C{=}CHC(CH_3)_3$
204

▶1479 VS
B (0.0088)
$(CH_3)_3CCH_2CH_2C(CH_3)_3$
305

▶1479 VS
B (0.0088)
$(CH_3)_3CCH(CH_2CH_3)_2$
300, see also 78

▶1479 M
$CHCl_3$
433

▶1479 M
$CHCl_3$
433

▶1479 M
$CHCl_3$
433

▶1479 W
$C_6H_5HSiCl_2$
86

▶1478 VS
B (0.030)
$(CH_3)_3CCH_2CH_2CH_3$
155, see also 154

▶1478 VS
B (0.015)
$H_2C{=}C(CH_3)CH_2C(CH_3)_3$
96

▶1478 MW
$CHCl_3$
433

▶1478
E or N
41

▶1478
E or N
41

▶1477 SSh
A (36 mm Hg, 15 cm)
$H_2C{=}CHCH(CH_3)_2$
131

▶1477 MS
$CHCl_3$
[Ring]
15

▶1477 VS
B (0.015)
$H_2C{=}CHCH_2C(CH_3)_3$
102, see also 273

▶1477 M (rotational lines)
A (141 mm Hg, 15 cm)
$H_3C{-}C{\equiv}CH$
87, see also 51

▶1477 VS
B (0.015)
$H_2C{=}CHC(CH_3)_3$
129

▶1477 M
$CHCl_3$
$CH{=}CH{-}C(=O){-}O{-}C_2H_5$
433

▶1477 VS
B (0.0088)
$(CH_3)_3CC(CH_3)_2{-}CH_2CH_2CH_3$
304

▶1476 VS
A (9 mm Hg, 40 cm)
Grating
$(CH_3)_3C(CH_2)_3CH_3$
116

▶1477 VS
B
$CH_3CH_2C(CH_3)_2{-}CH_2CH_3$
336, see also 148, 149

▶1476 VS
B (0.018)
Grating
$(CH_3)_3CCH_2CH_2CH_3$
154, see also 155

▶1477 VS
B (0.033)
$(CH_3)_3CCH_2CH(CH_3)CH_2CH_3$
79

▶1476 S
G
$N(C_2H_5)_2$
NO_2
[NO_2 stretch]
21

▶1477 VS
B
$(CH_3)_3CCH(CH_3)CH_2CH_3$
340

▶1475 VVS
$CHCl_3$
[Ring]
15

▶1477 VW
A (200 mm Hg, 10 cm)
$CF_2Cl{-}CF_2Cl$
257

▶1475 VS
A
$(CH_3)_3CCH_2CH_3$
94

▶1477 S
D
CH=CHCH=CHCH=CHCH=N-N=CHCH=CHCH=CHCH=CH
597

▶1477 VS
B (0.0088)
$(CH_3)_3CCH(CH_3)C(CH_3)_3$
307

▶1475 VS
A
$CH_3CH_2CH(CH_3)CH_2CH_3$
332

▶1475

CH_3F

[CH_3 asymm. bend] 539

▶1475 VS

B

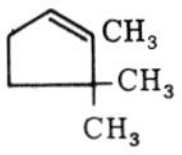

CH_3 CH_3 CH_3

126

▶1475 VS

B (0.0088)

$(CH_3)_2CHCHCH(CH_3)_2$
CH_2
CH_3

301

▶1475 VS

B

$(CH_3)_2CHCHCH(CH_3)_2$
CH_3

284, see also 135

▶1475 S

E

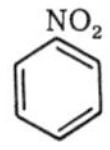

NO_2

[Ring] 20

▶1475 S

H

$CH_3\text{-}O\text{-}C_6H_5 \cdot Cr(CO)_3$

17

▶1475 S

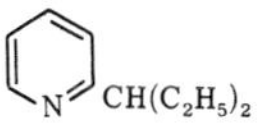

N $CH(C_2H_5)_2$

455

▶1475

E

$Hg_2(NO_3)_2$

45

▶1475

K

$La(NO_3)_3$

45

▶1474 VVS

$CHCl_3$

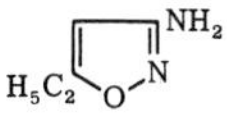

NH_2 H_5C_2 O N

[Ring] 15

▶1474 VVS

$CHCl_3$

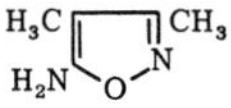

H_3C CH_3 H_2N O N

[Ring] 15

▶1474 MS

B (capillary)

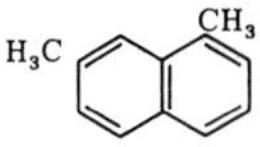

H_3C CH_3

183

▶1474 VS

B (0.033)

$(CH_3)_3CCH(CH_2CH_3)_2$

78, see also 300

▶1474 VS

$CHCl_3$

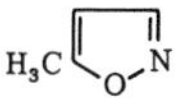

H_3C O N

[Ring] 15

▶1474 M

C

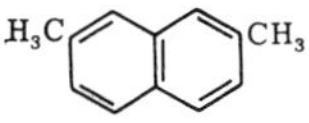

H_3C CH_3

179

▶1474

E or N

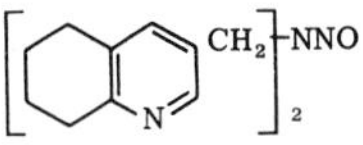

[N CH_2]$_2$ NNO

41

▶1473 VVS

$CHCl_3$

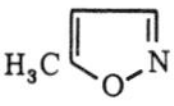

H_3C O N

[Ring] 15

▶1473 VVS

$CHCl_3$

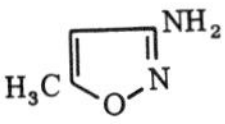

NH_2 H_3C O N

[Ring] 15

▶1473 VS

A (775 mm Hg, 10 cm)

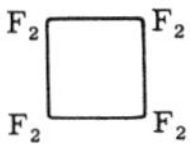

F_2 F_2 F_2 F_2

243

▶1473 VS

B (0.0088)

$(CH_3)_2CHCH_2CH_2CH(CH_3)_2$

137, see also 339

▶1473 VVS
B (0.04)

N CH$_3$

198. see also 438

▶1473 VS
C

$(CH_3)_3CNO_2$

325

▶1473 VSSh
$CHCl_3$

O
C-O-C$_2$H$_5$
H$_3$C O N

[Ring]

15

▶1473 S
$CHCl_3$

O CH=N-N=CH O

597

▶1473

CH$_2$NH$_2$
(CH$_2$)$_3$
N

41

▶1472 VS
B (0.033)

$(CH_3)_3CCHCH(CH_3)_2$
CH$_3$

293

▶1472 VS
B (capillary)

CH$_3$
CH$_3$

187

▶1472 S
A (18.8 mm Hg, 40 cm)
Grating

CH$_3$
$(CH_3)_3CCHCH_2CH_3$

110, see also 71, 90

▶1472 S
A (6 mm Hg, 40 cm)
Grating

$(CH_3)_2CHCH_2CH_2CH(CH_3)_2$

113

▶1472 S
$CHCl_3$

Cl
N

433

▶1472 MW
$CHCl_3$

NO$_2$
N

433

▶1472
E or N

COOH
(CH$_2$)$_3$
N =O
H

41

▶1472
E or N

O
C-OC$_2$H$_5$
(CH$_2$)$_3$
N

41

▶1471 VS
A

$(CH_3)_2CHCH_2CH_3$

333

▶1471 VS
B (0.0088)

$(CH_3)_2CHCH_2CH(CH_3)_2$

141, see also 150, 151

▶1471 VS
B (0.0088)

CH$_3$
$(CH_3)_2CHC-CH_2CH_2CH_3$
CH$_3$

134

▶1471 VS
B (0.0088)

$(CH_3)_2CHCHCH(CH_3)_2$
CH$_3$

135

▶1471 VS
B (0.0088)

CH$_3$
$(CH_3)_2CHCHCH_2CH_2CH_3$

139, see also 338

▶1471 VS
B (0.0088)

$(CH_3)_3CCH_2CHCH(CH_3)_2$
CH$_3$

303

▶1471 M
B (0.0104)

CH$_3$
F

221

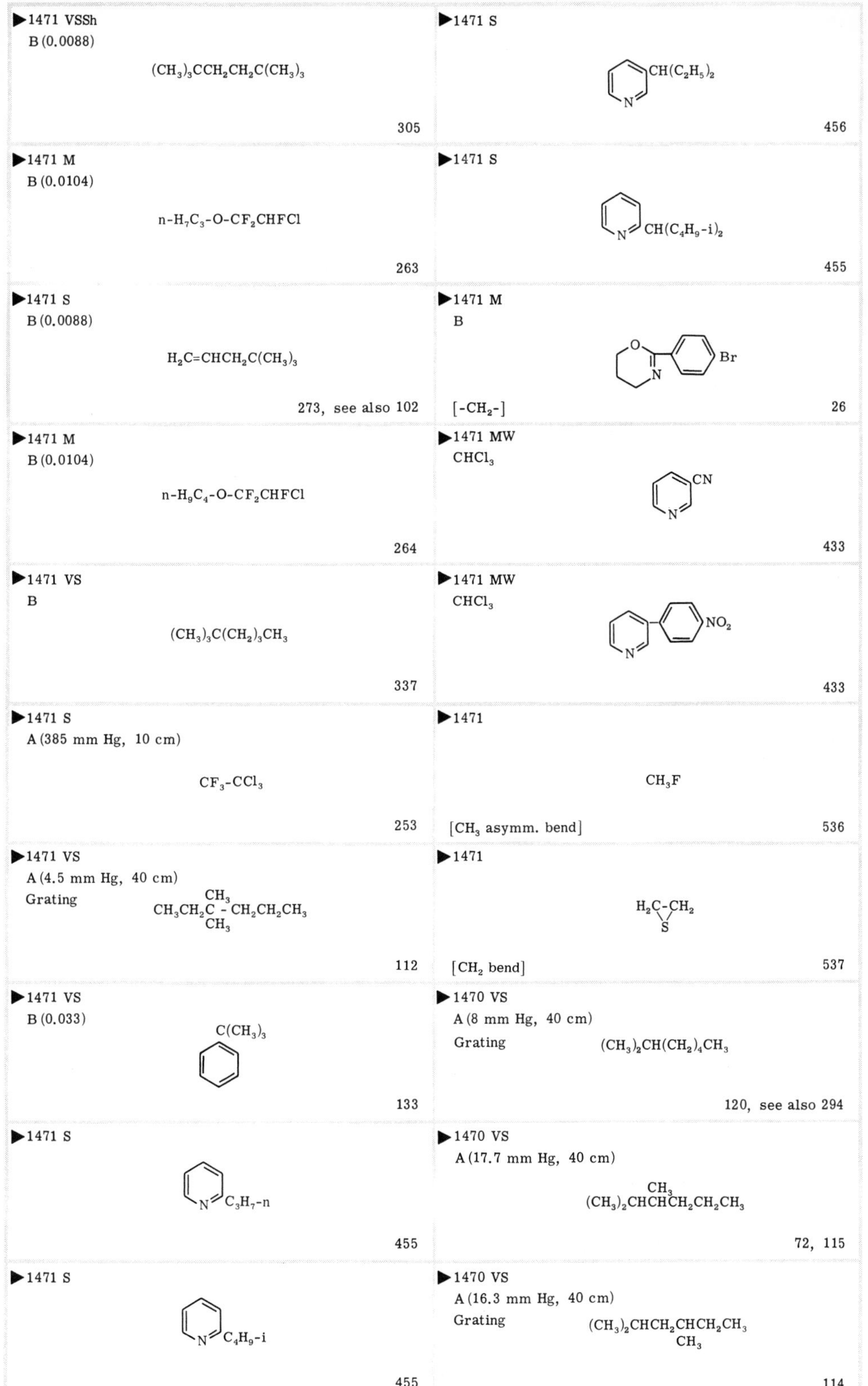

▶1471 VSSh
B (0.0088)
$(CH_3)_3CCH_2CH_2C(CH_3)_3$
305

▶1471 S
$CH(C_2H_5)_2$ (pyridine)
456

▶1471 M
B (0.0104)
$n\text{-}H_7C_3\text{-}O\text{-}CF_2CHFCl$
263

▶1471 S
$CH(C_4H_9\text{-}i)_2$ (pyridine)
455

▶1471 S
B (0.0088)
$H_2C{=}CHCH_2C(CH_3)_3$
273, see also 102

▶1471 M
B
O, N, Br
[$-CH_2-$]
26

▶1471 M
B (0.0104)
$n\text{-}H_9C_4\text{-}O\text{-}CF_2CHFCl$
264

▶1471 MW
$CHCl_3$
CN, N
433

▶1471 VS
B
$(CH_3)_3C(CH_2)_3CH_3$
337

▶1471 MW
$CHCl_3$
NO_2, N
433

▶1471 S
A (385 mm Hg, 10 cm)
$CF_3\text{-}CCl_3$
253

▶1471
CH_3F
[CH_3 asymm. bend]
536

▶1471 VS
A (4.5 mm Hg, 40 cm)
Grating
$CH_3CH_2C(CH_3)_2\text{-}CH_2CH_2CH_3$
112

▶1471
$H_2C\text{-}CH_2$, S
[CH_2 bend]
537

▶1471 VS
B (0.033)
$C(CH_3)_3$ (benzene)
133

▶1470 VS
A (8 mm Hg, 40 cm)
Grating
$(CH_3)_2CH(CH_2)_4CH_3$
120, see also 294

▶1471 S
$C_3H_7\text{-}n$ (pyridine)
455

▶1470 VS
A (17.7 mm Hg, 40 cm)
$(CH_3)_2CHCH(CH_3)CH_2CH_2CH_3$
72, 115

▶1471 S
$C_4H_9\text{-}i$ (pyridine)
455

▶1470 VS
A (16.3 mm Hg, 40 cm)
Grating
$(CH_3)_2CHCH_2CH(CH_3)CH_2CH_3$
114

▶1470 VS
A (12.6 mm Hg, 40 cm)
$CH_3(CH_2)_6CH_3$
89, see also 91

▶1470 VS
A (18.8 mm Hg, 40 cm)
$(CH_3)_3CCH(CH_3)CH_2CH_3$
71, see also 110

▶1470 VSB
A (16.6 mm Hg, 40 cm)
$(CH_3)_2CHC(CH_3)_2-CH_2CH_3$
90

▶1470 VS
A (40 mm Hg, 58 cm)
CH_3 (cyclopentane ring)
105

▶1470 VS
B (0.014)
$(CH_3)_2CHCH_2CH(CH_3)_2$
151, see also 141, 150

▶1470 VSB
B (0.029)
$CH(CH_3)_2$, $(CH_3)_2HC$, $CH(CH_3)_2$ (benzene ring)
278

▶1470 VS
$CHCl_3$
$H_3C-C(=O)-O-N(H)$ – ring with CH_3, H_3C, O, N
[Ring]
15

▶1470 VS
A (15.9 mm Hg, 40 cm)
$CH_3CH_2CH(CH_3)(CH_2)_3CH_3$
106, see also 119

▶1470 S
H
CH_3 (benzene ring) $\cdot Cr(CO)_3$
17

▶1470 S
CCl_4
$N(C_2H_5)_2$ (benzene ring) NO_2
[NO_2 stretch]
21

▶1470 M
G
H S S H
N-C-C-N
CH_2OH CH_2OH
[HCH]
31

▶1470 MW
$CHCl_3$
N (pyridine ring – phenyl)
433

▶1470 Sh
B
O, N, Cl (ring structure)
[$-CH_2-$]
26

▶1470
$H_2C-C(=O)-O-C_2H_5$
$H_2C-C(=O)-O-C_2H_5$
[CH_2 or CH_3]
478

▶1470
$H_5C_2-C(=O)$
O
$H_5C_2-C(=O)$
[CH_2]
638

▶1469 VVS
$CHCl_3$
NH_2, N, O (phenyl – isoxazole ring)
[Ring]
15

▶1469 VS
B (0.026)
$(CH_3)_2CH(CH_2)_3CH_3$
80, see also 158, 331

▶1469 VS
B
H_3CO, OCH_3, O, N (ring structure)
[$-CH_2-$]
26

▶1469 VS
B (0.028)
$CH_3(CH_2)_5CH_3$
85

▶1469 VS
B (0.033)
$(CH_3)_3C(CH_2)_2CH(CH_3)_2$
292

▶1469 VS
B (0.030)
$(CH_3)_3CCH_2CH_2CH_3$
155, see also 154

▶1469 S
B
Cl O N
[$-CH_2-$]
26

▶1469 M
A (4 mm Hg, 40 cm)
Grating
$CH_3CH_2CH(CH_2)_3CH_3$
CH_3
119, see also 106

▶1468 VSVB
A (200 mm Hg, 10 cm)
$CF_3CF_2CF_2CF_2CF_3$
237

▶1468 S
B (0.0088)
$(CH_3)_3CCH_2CH_3$
145

▶1468 VS
B (0.0088)
$(CH_3)_2CHCH(CH_2CH_3)_2$
136

▶1468 VS
B
$(CH_3)_2CHCH_2CHCH_2CH_3$
CH_3
334

▶1468 VS
B (0.0088)
CH_3
$(CH_3)_3CCH_2CHCH_2CH_3$
285

▶1468 VS
B (capillary)
CH_3
CH_3
186

▶1468 VS
B
Grating
$(CH_3)_2CH(CH_2)_3CH_3$
158, see also 80, 331

▶1468 VS
B (0.0085)
Grating
$CH_3(CH_2)_5CH_3$
84

▶1468 VS
B (0.018)
Grating
$(CH_3)_3CCH_2CH_2CH_3$
154, see also 155

▶1468 VS
B (0.036)
$H_2C{=}CH(CH_2)_3CH(CH_3)_2$
97

▶1468 VS
B (0.008)
$(CH_3)_2CHCH_2CHCH_2CH_3$
CH_3
138

▶1468 VS
$CHCl_3$
N O
[Ring]
15

▶1468 S
A (8 mm Hg, 40 cm)
$(CH_3CH_2)_2CHCH_2CH_2CH_3$
117

▶1468 VS
B (0.0153)
H_3C CH_3
CH_2CH_3
215

▶1468 M
A (100 mm Hg, 10 cm)
CH_3-CCl_3
251

▶1468 M
B
O N F
[$-CH_2-$]
26

▶1468
N Cl
41

▶1468
E or N

N Cl

41

▶1467 VS
A (6 mm Hg, 40 cm)
Grating

$CH_3CH_2CH(CH_3)CH(CH_3)CH_2CH_3$

111

▶1467 VS
A (12 mm Hg, 40 cm)

$CH_3(CH_2)_2CH(CH_3)-(CH_2)_2CH_3$

118

▶1467 VS
B (0.028)

$C_{26}H_{54}$
(5,14-di-n-butyloctadecane)

283

▶1467 VS
B (0.028)

$CH_3(CH_2)_5CH(C_3H_7)(CH_2)_5CH_3$

281

▶1467 S
B

$CH_3CH_2CH(CH_3)(CH_2)_2CH_3$

330, see also 81, 157

▶1467 M
M

$Pt(NH_2-CH_2-CH_2-NH_2)_2Cl_2$

[CH_2]

8

▶1467
E or N

N Cl

41

▶1466 VS
A (12.6 mm Hg, 40 cm)

$CH_3(CH_2)_6CH_3$

91

▶1466 VS
B (0.036)

$CH_3CH_2CH{=}CH(CH_2)_3CH_3$

99

▶1466 VS
B (0.033)

$(CH_3)_3CCH_2CH(CH_3)CH_2CH_3$

79

▶1466 VS
B (0.0088)

$(CH_3)_2CHCH(CH_3)CH_2CH_3$

142, see also 152, 153

▶1466 VS
B (0.018)

$CH_3CH_2CH(CH_3)(CH_2)_2CH_3$

81, see also 157

▶1466 VS
B (0.015)

$H_2C{=}CHC(CH_3)_3$

129

▶1466 VS
B (0.0088)

$CH_3CH_2C(CH_3)_2-CH_2CH_2CH_3$

360

▶1466 S
B (0.0088)

$(CH_3)_2C{=}CHC(CH_3)_3$

204

▶1466 S
B (0.0088)

$CH_3CH{=}CHCH_2CH_3$
(cis)

208

▶1466 VS
B (0.0088)

$CH_3CH_2CH(CH_3)-CH(CH_3)CH_2CH_3$

83

▶1466 VS
B (0.0088)

$CH_3CH_2C(CH_3)_2-CH_2CH(CH_3)CH_2CH_3$

302

▶1466 VVS
C

CH_3 CH_3

182

▶1466 S
A (10 mm Hg, 58 cm)
$CH_3CH=CH(CH_2)_4CH_3$
(cis and trans)
101

▶1466 S
C_4H_9-i (pyridine)
431

▶1466 S
(pyridine) $C(C_2H_5)_3$
455

▶1466 S
C_3H_7-n (pyridine)
431

▶1466 M
B
O_2N (oxazine–phenyl)
[$-CH_2-$]
26

▶1466 M
CCl_4 [20 g/l] (0.51)
$CH_3CH=CH(CH_2)_2CH=CH(CH_2)_2CO-NH-CH_2CH(CH_3)_2$
357

▶1466 W
$C_6H_5SiCl_3$
86

▶1466
$H_3C-O-CH_3$
[CH_3 symm. bend]
539, see also 47

▶1465 VVS
$CHCl_3$
$H_3C-C(=O)-O-N(H)$– (3-methyl-5-methylisoxazolyl; H_3C, CH_3)
[Ring]
15

▶1465 VS
B (0.011)
$(CH_3)_3CCH(CH_3)_2$
147, see also 146

▶1465 VS
B (0.028)
$(C_2H_5)_2CH(CH_2)_{20}CH_3$
282

▶1465 VS
B (0.018)
Grating
$CH_3CH_2CH(CH_3)(CH_2)_2CH_3$
157, see also 81

▶1465 S
B
Cl (oxazine–phenyl)
[$-CH_2-$]
26

▶1465 S
H
CH_3, H_3C, CH_3 (benzene) $\cdot Cr(CO)_3$
17

▶1465 M
$CHCl_3$
Br (pyridine)
433

▶1465
E or N
CN, =O, N–H (cyclopenta-pyridinone)
41

▶1465
E or N
CN, =O, N–H (tetrahydroquinolinone)
41

▶1464 S
A (36 mm Hg, 15 cm)
$H_2C=CHCH(CH_3)_2$
131

▶1464 VS
B
$(CH_3)_2CHOH$
460

▶1464 VVS
B (0.10)
$CH_2CH_2CH_2CH_3$ (naphthalene)
176

▶1464 VVS
B (0.064)
$H_3CH_2C\ \ H$
$C{=}C$
$H\ \ CH_2CH_3$
(trans)
168

▶1464 VSSh
B (0.015)
$H_2C{=}\underset{|}{C}CH_2C(CH_3)_3$
CH_3
96

▶1464 VS
B (0.0088)
$(CH_3CH_2)_3CH$
143, see also 156

▶1464 VS
B
CH_3
$(CH_3)_2CHCHCH_2CH_2CH_3$
338, see also 139

▶1464 VS
B (0.0088)
$(CH_3CH_2)_2CHCH_2CH_2CH_3$
140

▶1464 VS
B (0.005)
Grating
$(CH_3)_2CHCH(CH_3)_2$
210, see also 144

▶1464 VS
B (0.0088)
CH_3
$CH_3CH_2C - CHCH_2CH_3$
$CH_3\ CH_3$
299

▶1464 VVS
CCl_4 [25%] (0.10)
$(CH_3)_2CHCH_2CH_2OH$
195

▶1464 VS
B (0.0576)
n-H_9C_4-O-CF_2CHFCl
264

▶1464 VS
CCl_4 [25%] (0.10)
$CH_3CH_2C(CH_3)_2$
OH
194

▶1464 S
B (0.008)
$H_2C{=}CH(CH_2)_7CH_3$
270

▶1464 VS
B (0.036)
$H_2C{=}CH(CH_2)_5CH_3$
290

▶1464 VVS
B (0.064)
$C_{17}H_{34}$
(1-heptadecene)
159

▶1464 VS
$CHCl_3$
H_3C–[isoxazole ring, O, N]–$\overset{O}{\overset{\|}{C}}$-O-$C_2H_5$
[Ring]
15

▶1464 VS
B (0.016)
Grating
CH_3
$CH_3CH_2C - CH_2CH_3$
CH_3
148, 149

▶1464 S
B
CH_3
$(CH_3)_2CHC - CH_2CH_3$
CH_3
341

▶1464 S
B (0.0088)
$CH_3(CH_2)_2CH{=}CH(CH_2)_2CH_3$
(trans)
205

▶1464 VS
B (0.065)
[thiophene ring, S]CH_2CH_3
213

▶1464 S
B (0.008)
$H_2C{=}\underset{|}{C}C(CH_3)_3$
CH_3
272

▶1464 S
C
[naphthalene ring] CH_3
CH_3
185

▶1464 MS
$CHCl_3$
[Ring]
15

▶1464 M
B (0.008)
$H_2C{=}CH(CH_2)_8CH_3$
352

▶1464 W
$CHCl_3$
7

▶1464 W
B
$C_6H_5HSiCl_2$
86

▶1463 VS
B (0.018)
Grating
$(CH_3)_3CCH(CH_3)_2$
146, see also 147

▶1463 VS
B (0.016)
$(CH_3CH_2)_3CH$
82

▶1463 VS
$CHCl_3$
[Ring]
15

▶1463 VS
B (0.055)
344

▶1463 M
M
$Ni(NH_2CH_2CH_2NH_2)_2PtCl_4$
$[CH_2]$
8

▶1463 MSh
$CHCl_3$
[Ring]
15

▶1463
E or N
41

▶1462 VVS
CCl_4 [25%] (0.10)
$CH_3(CH_2)_4OH$
393

▶1462 VS
CCl_4 [25%] (0.10)
$CH_3(CH_2)_7OH$
317

▶1462 M
B (0.0576)
230

▶1462 VS
B (0.016)
$(CH_3CH_2)_3CH$
156, see also 143

▶1462 S
B (0.008)
$H_3CH_2C(H)C{=}C(H)CH_2CH_3$
275

▶1462 VS
B (0.033)
132

▶1462 VS
B (0.013)
$CH_3CH{=}CHCH(CH_3)_2$
130

▶1462 S
B (0.008)
$CH_3(CH_2)_2CH{=}CH(CH_2)_2CH_3$
(trans)
205

▶1462 SSp
B (0.003)
$(CH_3)_2CHNO_2$
366

▶1462 VVS
B (0.10)
$CH_2CH_2CH_2CH_2CH_3$
174

▶1462 VSB
B (0.029)
$CH_2(CH_2)_2CH_3$
$CH_3(CH_2)_2H_2C$ $CH_2(CH_2)_2CH_3$
279

▶1462 S
B (0.010)
Grating
CH_3
$(CH_3)_2CHCHCH_2CH_3$
152, see also 142, 153

▶1462 S
B
O
H_3C N-CH_3
O_2N
[CH_2] 26

▶1462 S
B
O
H_5C_2 N- CH_3
O_2N
[CH_2] 26

▶1462 VVS
B (0.051)
$CH(CH_3)_2$
$CH(CH_3)_2$
296

▶1462 S
B
$(CH_3)_2CH(CH_2)_3CH_3$
331, see also 80, 158

▶1462 S
C_3H_7-n
N
431

▶1462 S
$CH(C_2H_5)_2$
N
431

▶1462 MS
$CHCl_3$
H_3C
H_3CO O N
[Ring] 15

▶1462 M
CCl_4 [0.00658 m/l]
OH O
C-OH
448

▶1462
E or N
COOH
N
41

▶1461 VS
B (0.029)
$CH_3CHCH_2CH_3$
CH_3CH_2HC $CHCH_2CH_3$
CH_3 CH_3
280

▶1461 VS
B (0.015)
CH_3
$(CH_3)_2CHCHCH_2CH_3$
153, see also 142, 152

▶1461 VS
B (0.036)
CH_3
104, see also 633

▶1461 S
B
O
H_5C_2 N-C_3H_7-n
O_2N
[CH_2] 26

▶1461 M
CCl_4
CH_3
N
Cl_3P PCl_3
N
CH_3
384

▶1460 M
B (0.0104)
F
F
F
224

▶1460 VS
B (0.0104)
CF_3
219

▶1460 S
B (0.008)
H_3C C_3H_7-n
C=C
H H
(cis)
276

▶1460 VVS
CCl_4 [25%] (0.10)
$CH_3CH_2CHCH_2CH_3$
OH
196

▶1460 VS
CCl_4 [25%] (0.10)
$CH_3(CH_2)_5OH$
320

▶1460 VS
B (0.036)
$H_2C=C(CH_3)(CH_2)_4CH_3$
98

▶1460 VVS
B (0.10)
CH_3
192

▶1460 VS
C
CH_3 CH_3
181

▶1460 VVS
B (0.10)
CH_2CH_3
190

▶1460 VS
$CHCl_3$
H_2N O N CH_3
[Ring]
15

▶1460 VVS
B (0.051)
$CH(CH_3)_2$
$CH(CH_3)_2$
298

▶1460 VS
B (0.0104)
$H_3C\text{-}O\text{-}CF_2CHClF$
356

▶1460 VS
B
H_3C CH_3
CH_3
124

▶1460 S
E or N
$(CH_2)_3$ N COOH Cl
41

▶1460 SSp
B (0.0104)
F
226

▶1460 VSSp
B (0.025)
CH_3 CN
312

▶1460 S
B (0.0104)
CH_3 F
221

▶1460 S
B
O H_5C_2 O_2N $N\text{-}C_2H_5$
[CH_2]
26

▶1460 VSVB
B (0.029)
$CH(CH_3)_2$
H_3C CH_3
277

▶1460 S
B
O H_5C_2 O_2N $N\text{-}C_3H_7\text{-}i$
[CH_2]
26

▶1460 VS
B (0.036)
$CH_3CH=CH(CH_2)_4CH_3$
(cis)
100

▶1460 S
B
O H_5C_2 O_2N $N\text{-}C_4H_9\text{-}n$
[CH_2]
26

▶1460 S
B (0.003)
$CH_3(CH_2)_2CH_2NO_2$
310

▶1460 S
B
O H_5C_2 O_2N $N\text{-}C_5H_{11}\text{-}n$
[CH_2]
26

▶1460 S
B
H_5C_2 O_2N (morpholine ring, O) N-C_6H_{13}-n
[CH_2] 26

▶1460 S
H
CH_3 · $Cr(CO)_3$ CH_3
17

▶1460 S
H
O ‖ C-OCH_3 · $Cr(CO)_3$
17

▶1460 S
E or N
O ‖ C-OH N Cl
41

▶1460 S
E or N
N OH
41

▶1460 M
A (8.3 mm Hg, 40 cm)
CH_2CH_3
70

▶1460 M
A (4.2 mm Hg, 40 cm)
$CH(CH_3)_2$
66, see also 121

▶1460 M
A (100 mm Hg, 10 cm)
CHF_2CH_3
233

▶1460 S
B (0.0088)
$(CH_3)_3CCHC(CH_3)_3$ with CH_3
307

▶1460
G
H_3C CH_3 CH_3 N N OH
[C=C] 6

▶1460
C_2H_5SCN
[CH_2] 639

▶1460 VS (Raman only)
A
CH_3OH
[CH_3 bend] 678

▶1459 VSSh
B (0.028)
$CH_3(CH_2)_5CH_3$
85

▶1459 VS
B (0.028)
$CH_3(CH_2)_5CH(C_3H_7)(CH_2)_5CH_3$
281

▶1459 VS
B (0.028)
$C_{26}H_{54}$
(5,14-di-n-butyloctadecane)
283

▶1459 S
B
H_3C O_2N (morpholine ring, O) N-C_3H_7-i
[CH] 26

▶1459 S
B
H_3C O N-H H_3C CH_3
[CH] 26

▶1459 S
B
H_5C_2 O_2N O N-H
26

▶1459 M
$CHCl_3$
H_3C CH_3 O N
[Ring] 15

▶1459 M
A (4.2 mm Hg, 40 cm)
Grating
$CH(CH_3)_2$
121, see also 66

▶1458 SSh
B (0.088)

$(CH_3)_3CC(CH_3)_2-CH_2CH_2CH_3$

304

▶1458 S
A (100 mm Hg, 10 cm)

CH_3-CCl_3

251, see also 51

▶1458 VS
B (0.0104)

CF_3, F (fluorobenzotrifluoride ring)

218

▶1458 VS
B (0.0153)

H_3C, CH_3, CH_2CH_3 (cyclohexane ring)

215

▶1458 VVS
CCl_4 [25%] (0.10)

$CH_3(CH_2)_3OH$

353

▶1458 S
B (0.0088)

$H_2C=C(CH_3)CH_2CH_2CH_3$

206

▶1458 VS
B (0.0085)
Grating

$CH_3(CH_2)_5CH_3$

84

▶1458 S
B

O, H_3C, O_2N, $N-C_3H_7$-n (ring)

[CH_2]

26

▶1458 W
B (0.0104)

CH_3, F (benzene ring)

220

▶1458 S
B (0.0088)

$CH_3CH=CHCH_2CH_3$
(cis)

208

▶1458 MS
B(?)[9.5%] (0.003)

$(CH_3)_2CHCH_2NO_2$

326

▶1458 S
B (0.2398)

$CF_2Cl-CFCl_2$

255

▶1458 S
B (0.0088)

$CH_3CH=CHCH_2CH_3$
(trans)

207

▶1458 M
M

$Pd(NH_2-CH_2-CH_2-NH_2)_2Cl_2$

[CH_2]

, 8

▶1458 VS
B (0.018)
Grating

$CH_3CH_2CH(CH_3)(CH_2)_2CH_3$

157, see also 81

▶1458 M
A (3.2 mm Hg, 40 cm)

$CH_2CH_2CH_3$ (benzene ring)

67, see also 122

▶1458 VVS
B (0.10)

$CH_2CH_2CH_2CH_3$ (naphthalene ring)

175

▶1457.8 VS
B (0.03)

$CH_2-CH=CH$ (fused to benzene ring)

9

▶1458 S
B(?)[9%] (0.003)

$CH_3CH_2CH(NO_2)CH_3$

324

▶1457 S
A (10 mm Hg, 58 cm)

$H_2C=CH(CH_2)_5CH_3$

289

▶1457 S B (ring structure: H_3C, O_2N, N-C_6H_{13}-n) [CH_2] 26	▶1456 VVS CCl_4 [25%] (0.10) $CH_3CHCH_2CH_3$ OH 319
▶1457 $F_3C-C(=O)-O^-Na^+$ [O=C-O] 525	▶1456 VVS B (0.064) $(CH_3)_2C=CHCH_2CH_3$ 166
▶1457 E or N $(CH_2)_3$ (fused pyridine ring, N) 41	▶1456 VS B $[(CH_3)_2N-BO]_3$ 32
▶1456 VVS B (0.064) $H_2C=C(CH_2CH_3)_2$ 163	▶1456 VVS B (0.064) $H_2C=CHCHCH_2CH_3$ CH_3 167
▶1456 S B (0.008) H_3C CH_3 C=C H CH_2CH_3 (cis) 164, 274	▶1456 VVS B (0.064) H_3C H C=C H $CH_2CH_2CH_3$ (trans) 169
▶1456 S B CH_3-C-R [CH_3 asymm. bend] 691	▶1456 VS $CHCl_3$ H_3C (isoxazole ring, O, N) C(=O)-CH_3 [Ring] 15
▶1456 VS CCl_4 [25%] (0.10) $CH_3CH_2CHCH_2OH$ CH_3 321	▶1456 VVS B (0.20) H_3C (cyclopentane) CH_3 (trans) 308
▶1456 VS B (0.064) $CH_3CH=CH(CH_2)_4CH_3$ (cis) 160, see also 100	▶1456 S B (0.025) CH_3 (benzene ring) CN 214
▶1456 S B (0.0153) $CH_3(CH_2)_2CH=CH(CH_2)_2CH_3$ (cis) 422	▶1456 VS B (0.0153) (cyclopropyl) C(=CH_2)-CH_3 288
▶1456 VVS CCl_4 [25%] (0.10) $CH_3CH(CH_2)_2CH_3$ OH 318	▶1456 M $CHCl_3$ Cl, CH_3, H_3C (isoxazole ring, O, N) [Ring] 15

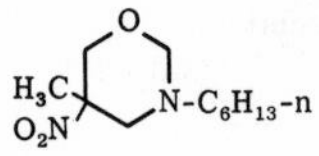

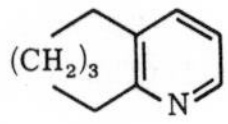

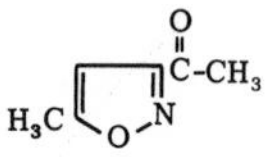

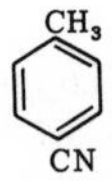

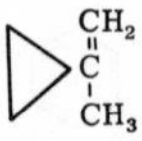

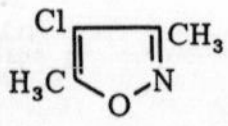

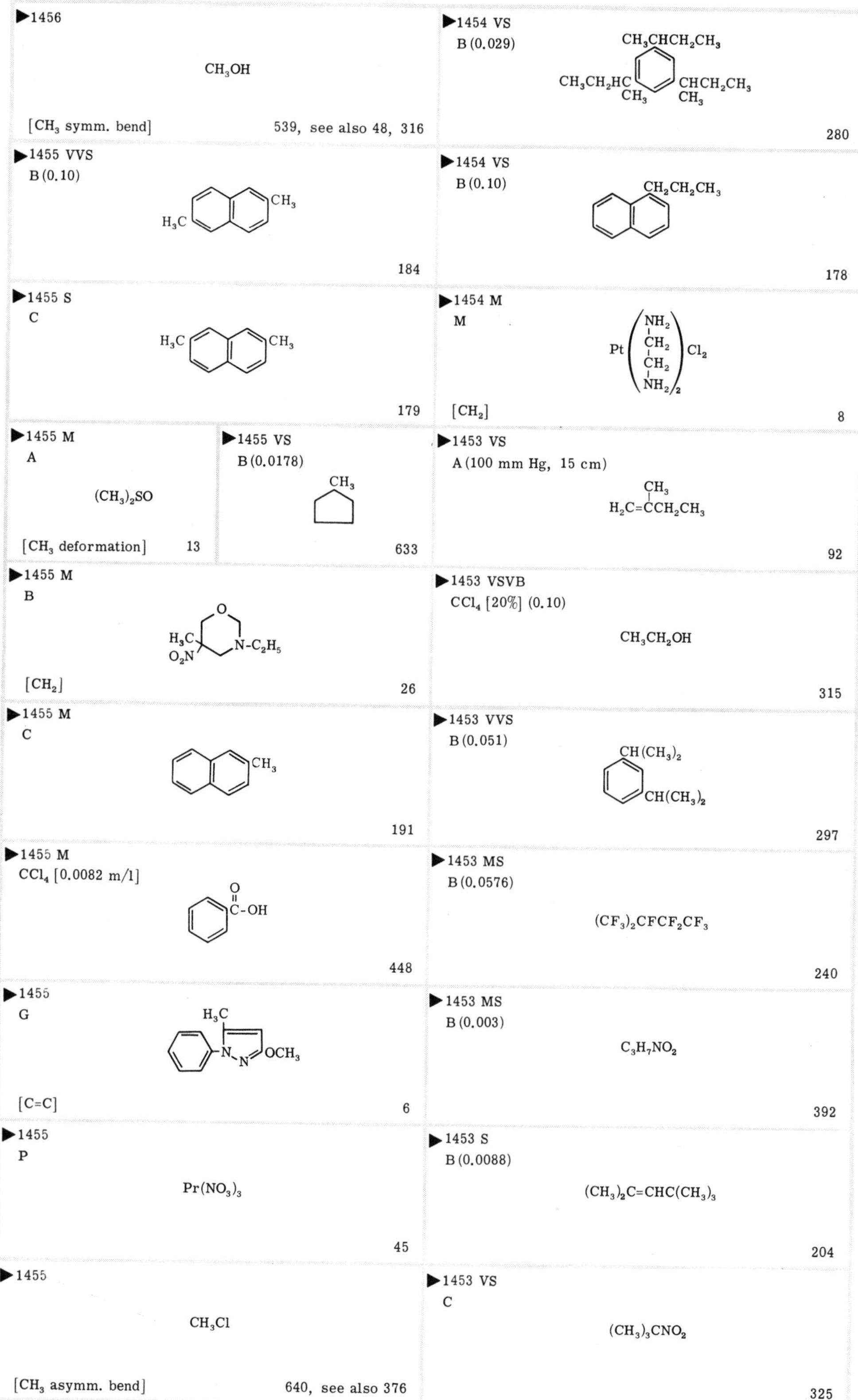
▶1456
CH3OH
[CH3 symm. bend]
539, see also 48, 316
▶1454 VS
B (0.029)
CH3CHCH2CH3
CH3CH2HC
CHCH2CH3
CH3
CH3
280
▶1455 VVS
B (0.10)
H3C
CH3
184
▶1454 VS
B (0.10)
CH2CH2CH3
178
▶1455 S
C
H3C
CH3
179
▶1454 M
M
Pt
NH2
CH2
CH2
NH2
Cl2
[CH2]
8
▶1455 M
A
(CH3)2SO
[CH3 deformation]
13
▶1455 VS
B (0.0178)
CH3
633
▶1453 VS
A (100 mm Hg, 15 cm)
CH3
H2C=CCH2CH3
92
▶1455 M
B
O
H3C
O2N
N-C2H5
[CH2]
26
▶1453 VSVB
CCl4 [20%] (0.10)
CH3CH2OH
315
▶1455 M
C
CH3
191
▶1453 VVS
B (0.051)
CH(CH3)2
CH(CH3)2
297
▶1455 M
CCl4 [0.0082 m/l]
O
C-OH
448
▶1453 MS
B (0.0576)
(CF3)2CFCF2CF3
240
▶1455
G
H3C
N
N
OCH3
[C=C]
6
▶1453 MS
B (0.003)
C3H7NO2
392
▶1455
P
Pr(NO3)3
45
▶1453 S
B (0.0088)
(CH3)2C=CHC(CH3)3
204
▶1455
CH3Cl
[CH3 asymm. bend]
640, see also 376
▶1453 VS
C
(CH3)3CNO2
325

▶1453 S
B

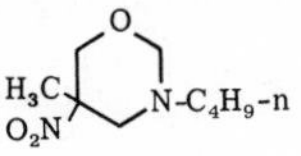

H_3C, O_2N — morpholine ring (O), N-C_4H_9-n

26

▶1453 S
B

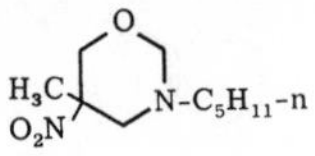

H_3C, O_2N — morpholine ring (O), N-C_5H_{11}-n

[CH_2]

26

▶1453 S
G

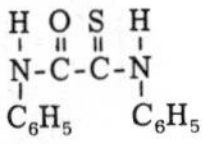

H O S H
N-C-C-N
C_6H_5 C_6H_5

[NH]

31

▶1453 M
G

H O S H
N-C-C-N
C_6H_{11} C_6H_{11}

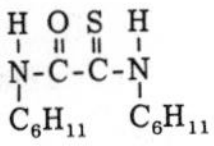

[HCH]

31

▶1452 VS
B (0.036)

103

▶1452 VS
B

$(C_2H_5\text{-NH-B-N-}C_2H_5)_3$

[CH]

32

▶1452 VVS
B (0.10)

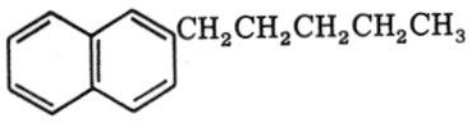

$CH_2CH_2CH_2CH_2CH_3$

173

▶1452 VS
B (0.0576)

$CH_3CH_2\text{-O-}CF_2CHFCl$

261

▶1452 S
C

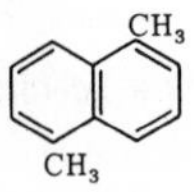

CH_3
CH_3

185

▶1452 VS
$CHCl_3$

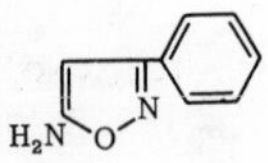

H_2N O N

[Ring]

15

▶1452 VS
$CHCl_3$

H_3CO O N

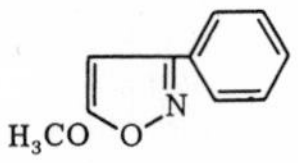

[Ring]

15

▶1451 VS
B (0.025)

$CH_3CH_2SCH_3$

209

▶1451 VS
B (0.064)

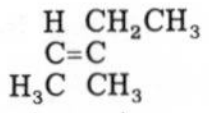

H CH_2CH_3
C=C
H_3C CH_3
(trans)

165

▶1451 VSB
B (0.025)

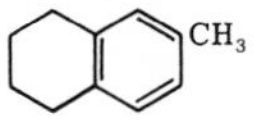

CH_3

269

▶1451 VSVB
CCl_4 [20%] (0.10)

CH_3OH

316, see also 48

▶1451 VS
B (0.0576)

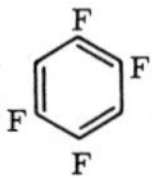

F F F F

222

▶1451 SVB
B (0.0576)

$(C_4F_9)_3N$

346

▶1451 VVS
B (0.025)

$CH_3CH_2\text{-S-}CH_2CH_3$

172

▶1451 S
$CHCl_3$

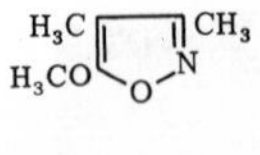

H_3C CH_3
H_3CO O N

[Ring]

15

▶1451 S
$CHCl_3$

H_3C
H_3CO O N

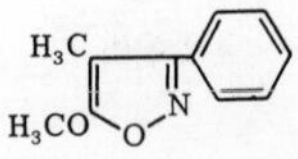

[Ring]

15

▶1450 VVS
B (0.10) 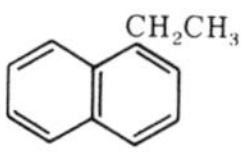
2-ethylnaphthalene (CH_2CH_3)
190

▶1450 VS
B 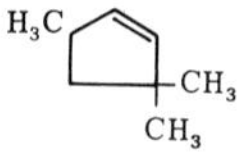
H_3C, CH_3, CH_3
125

▶1450 VS
$CHCl_3$ 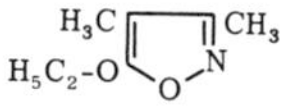
H_3C, CH_3, H_5C_2-O, O, N
[Ring]
15

▶1450 VS
$CHCl_3$ 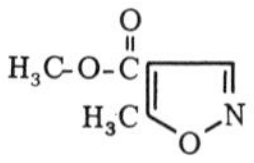
O, H_3C-O-C, H_3C, O, N
[Ring]
15

▶1450 S
C 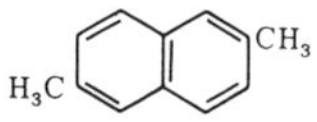
H_3C, CH_3
180

▶1450 S
G
$Na_4P_2O_6$
28

▶1450 S
H 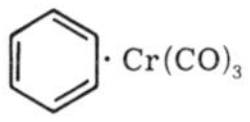
· $Cr(CO)_3$
17

▶1450 S
H 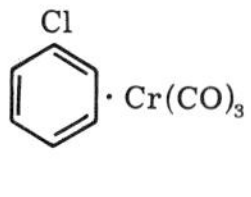
Cl
· $Cr(CO)_3$
17

▶1450 S
H 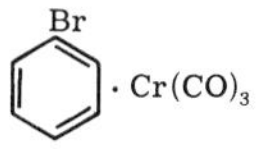
Br
· $Cr(CO)_3$
17

▶1450 VVS
B 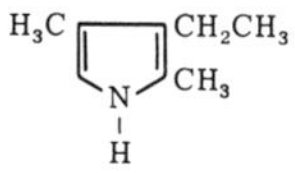
H_3C, CH_2CH_3, CH_3, N, H
343

▶1450
E
$Pr(NO_3)_3$
45

▶1450 S 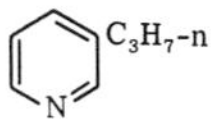
C_3H_7-n, N
431

▶1450
G 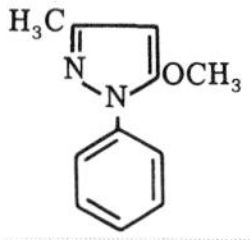
H_3C, N, N, OCH_3
[C=C]
6

▶1450
G 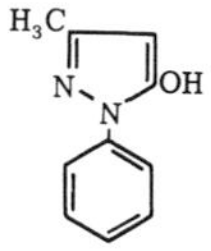
H_3C, N, N, OH
[C=C]
6

▶1450 S 
CH_3
CN
[CH_3 asymm. bend]
641

▶1449 VVSB
CCl_4 [25%] (0.10)
$CH_3CH_2CH_2OH$
314

▶1449 VS
B (0°C)
$(CH_3)_2CHCHCH(CH_3)_2$
CH_3
284, see also 135

▶1449 S
B 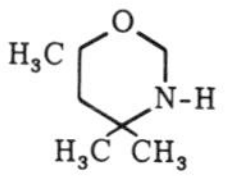
H_3C, O, N-H, H_3C, CH_3
[CH]
26

▶1449 S
B 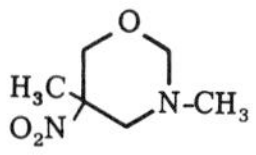
O, H_3C, O_2N, N-CH_3
26

▶1449 S
B 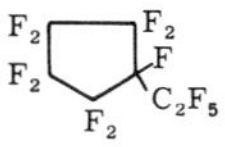
F_2, F_2, F_2, F_2, F, C_2F_5
247

▶1449 MS
G 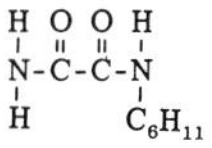
H O O H
N-C-C-N
H C_6H_{11}
[HCH]
31

▶1449 MB
A (100 mm Hg, 15 cm)
CH_3CH_2Cl
88

▶1449 M
G
H O S H
N-C-C-N
C_6H_{11} C_6H_{11}
[HCH] 31

▶1449 M
$CHCl_3$
CH_3 N O
[Ring] 15

▶1448 VS
B (capillary)
CH_3
CH_3
186

▶1448 VS
B (capillary)
CH_3
CH_3
187

▶1448 VSSh
B (0.036)
$CH_3CH{=}CH(CH_2)_4CH_3$
(cis)
100

▶1448 VS
B (0.055)
H_3C CH_3 N H
344

▶1448 VS
$CHCl_3$
H_3CO-C O N CH_3 O
[Ring] 15

▶1448 M
B
O N
[-CH_2-] 26

▶1448 M
G
H O O H
N-C-C-N
C_6H_{11} C_6H_{11}
[HCH] 31

▶1447 VS
B (0.034)
$(CH_3)_2C{=}C(CH_3)_2$
459

▶1447 VS
B (capillary)
H_3C CH_3
183

▶1447 S
G
H O S H
N-C-C-N
C_6H_5 C_6H_{11}
[HCH] 31

▶1447 S
G
H O S H
N-C-C-N
C_6H_5 C_6H_5
[HCH] 31

▶1447 S
$CHCl_3$
S
H_3C O CH_3
[CH_3] 7

▶1447 S
$CHCl_3$
O
CH=CH-C-O-C_2H_5
O_2N
596

▶1447 M
G
H S S H
N-C-C-N
C_6H_{11} C_6H_{11}
[HCH] 31

▶1447 M
$CHCl_3$
O
H_3C O CH_3
[CH_3] 7

▶1446 VS
B (0.033)
$C(CH_3)_3$
133

▶1446 S
$CHCl_3$
H_5C_2O-C O N CH_3 O
[Ring] 15

▶1446 M
$CHCl_3$
$H_3C-C(=O)-O-N(H)-$ (isoxazole, phenyl)
[Ring] 15

▶1445
CH_3Br
[CH_3 asymm. bend] 642

▶1445 VS
A
$CH_3CH_2CH(CH_3)CH_2CH_3$
332

▶1445 VS
A (12.5 mm Hg, 10 cm)
F, F, F, F (tetrafluorobenzene)
223

▶1445 W
B (0.0576)
F, CF_3, CF_3, F, F_2, F_2, F_2, F_2
231

▶1445 S
B (0.008)
H_3C C_3H_7-n
C=C
H H
(cis)
276

▶1445 VS
$CHCl_3$, E
NH_2, N
327

▶1445 S
B
$CH_3CH=CHCH=CHCH_3$
354

▶1445 S
B
HO
379

▶1445 VS
B (0.065)
S, CH_2CH_3
213

▶1445 S
B
O, H_3C, O_2N, N-CH_3
[CH_2] 26

▶1445 S
G
H O O H
N-C-C-N
C_6H_{11} C_6H_{11}
[HCH] 31

▶1445 MS
$CHCl_3$
H_3C, CH_3, H_3C, O, N
[Ring] 15

▶1445 MS
B (0.08)
N
197

▶1445 M
M
$[H_2C \cdot NO_2]^- Na^+$
[CH_2 deformation] 328

▶1445 M
CCl_4 [0.04 g/ml]
O
C-$N(CH_3)_2$
3

▶1445 M
$CHCl_3$
S, H_3C, S, CH_3
[CH_3] 7

▶1445 W
$CHCl_3$
H_3C, O, N
[Ring] 15

▶1445 WSh
$CHCl_3$
NH_2, H_3C, O, N
[Ring] 15

▶1445
G
N, N, OH, CH_3
[C=C] 6

▶1445
G
H_3C, OH, N, N, H_3C
[C=C]
6

▶1445
G
H_3C, OH, N, N, H
[C=C]
6

▶1444 S
B (0.0088)
$CH_3(CH_2)_2CH{=}CH(CH_2)_2CH_3$
(trans)
205

▶1444 VS
CCl_4, E
N, NH_2
327

▶1444 VS
C
CH_3 CH_3
182

▶1444 S
G
CH=N-NH-C(=S)-NH_2
N, H
380

▶1443 VS
A (775 mm Hg, 10 cm)
F_2 F_2 F_2 F_2
243

▶1443 VS
B (0.015)
$H_2C{=}CHCH_2C(CH_3)_3$
102, see also 273

▶1443 VS
B (0.0104)
CF_3, F, F
217

▶1443 VS
B (0.0104)
F, F, F
224

▶1443 S
B
O, H_5C_2, O_2N, N-C_3H_7-i
[CH_2]
26

▶1443 S
B
O, H_5C_2, O_2N, N-C_5H_{11}-n
[CH_2]
26

▶1443 SSp
B (0.003)
$(CH_3)_2CHNO_2$
366

▶1442 VSSh
B (0.036)
$H_2C{=}CH(CH_2)_5CH_3$
290

▶1442 VS
$CHCl_3$
NH_2, N
433, see also 327

▶1442 S
B
O, H_5C_2, O_2N, N-C_2H_5
26

▶1442 VS
B (0.10)
$CH_2CH_2CH_3$
177

▶1442 S
C
CH_3, H_3C
180

▶1442 S
G
H O S C_2H_5
N-C-C-N
C_6H_5 C_2H_5
[HCH]
31

▶1442 M
CCl_4 [0.00658 m/l]
OH, C(=O)-OH
448

▶1442
CCl_4 [0.0128 m/l]
Cl, C(=O)-OH
448

▶1442 M
$CHCl_3$ 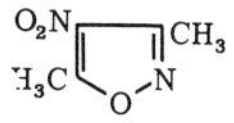
O_2N, CH_3, H_3C, O, N
[Ring] 15

▶1442 WSh
$CHCl_3$ 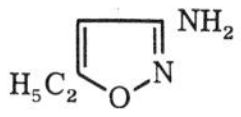
NH_2, H_5C_2, O, N
[Ring] 15

▶1441 VSSh
B (0.015)
$H_2C{=}C(CH_3)CH_2C(CH_3)_3$
96

▶1441 VSB
B (0.025) 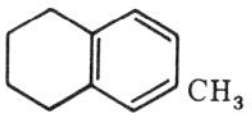
CH_3
269

▶1441 S
B 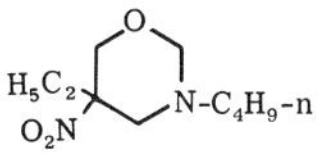
O, H_5C_2, O_2N, $N{-}C_4H_9{-}n$
[CH_2] 26

▶1441 SSh
CCl_4 [20 g/l] (0.51)
$CH_3CH{=}CH(CH_2)_2CH{=}CH(CH_2)_2CO{-}NH{-}CH_2CH(CH_3)_2$
357

▶1441 VS
B (0.10) 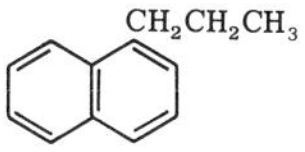
$CH_2CH_2CH_3$
178

▶1441
CH_3I
[CH_3 asymm. bend] 536

▶1440 VSSh
B (0.036)
$H_2C{=}CH(CH_2)_3CH(CH_3)_2$
97

▶1440 VVS
CCl_4 [25%] (0.10)
$H_2C{=}C(CH_3)CO_2CH_3$
193

▶1440 S
B 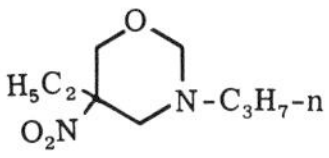
O, H_5C_2, O_2N, $N{-}C_3H_7{-}n$
[CH_2] 26

▶1440 S
E 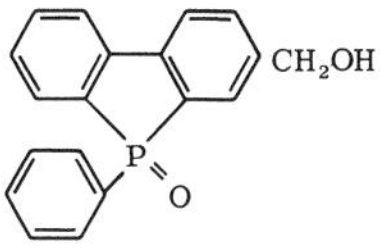
CH_2OH, P, O
34

▶1440 S
E 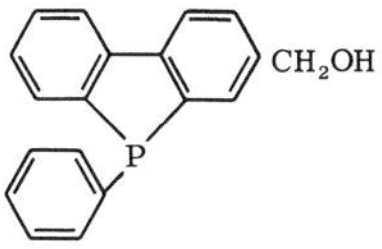
CH_2OH, P
34

▶1440 S
G 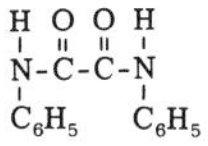
$C_6H_5NH{-}CO{-}CO{-}NHC_6H_5$
31

▶1440 S
G 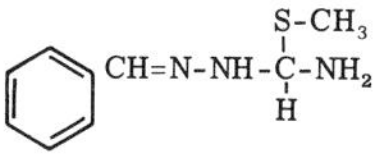
$C_6H_5CH{=}N{-}NH{-}CH(S{-}CH_3){-}NH_2$
380

▶1440 S
G 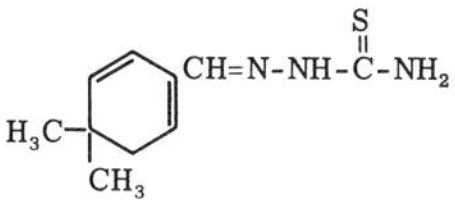
H_3C, CH_3, $CH{=}N{-}NH{-}C(=S){-}NH_2$
380

▶1440 MS
A
$(CH_3)_2SO$
[CH_3 deformation] 13

▶1440 M
G 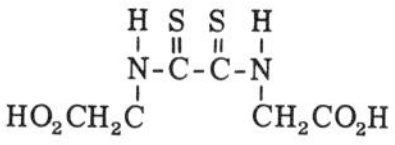
$HO_2CH_2C{-}NH{-}CS{-}CS{-}NH{-}CH_2CO_2H$
[HCH] 31

▶1440 MW
$CHCl_3$ 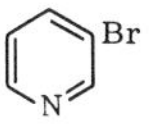
Br, N
433

▶1440 MW
$CHCl_3$ 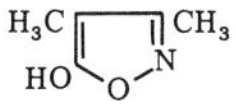
H_3C, CH_3, HO, O, N
[Ring] 15

▶1440
G

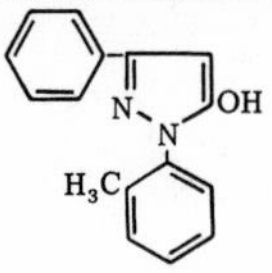

OH N N H_3C

[C=C] 6

▶1440
G

H_3C N N OH CH_3

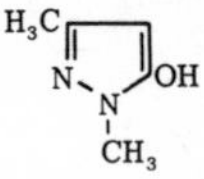

[C=C] 6

▶1439 M
B (0.0088)

$H_2C{=}CH(CH_2)_8CH_3$

352

▶1439 VS
A (12.5 mm Hg, 10 cm)

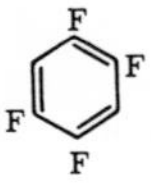

F F F F

223

▶1439 VS
B (0.0576)

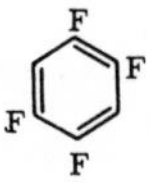

F F F F

222

▶1439 S
B

H_3CO OCH_3 O N

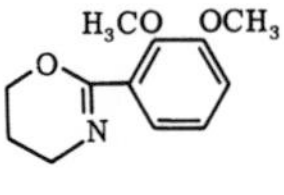

[$-CH_2-$] 26

▶1439 S
$CHCl_3$

$CH_3CH{=}CHCH{=}N{-}N{=}CHCH{=}CHCH_3$

597

▶1439 MS
G

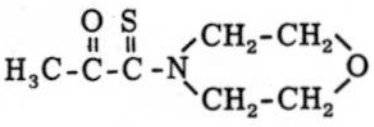

$H_3C{-}C(=O){-}C(=S){-}N(CH_2{-}CH_2)_2O$

[HCH] 31

▶1439 MSh
B (0.0088)

$H_2C{=}CH(CH_2)_6CH_3$

271

▶1438 VSSp
B (0.065)

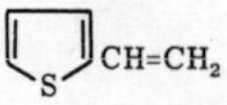

S $CH{=}CH_2$

212

▶1438 M
$CHCl_3$, E

NH_2 N

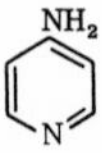

327

▶1437 M
B (0.0088)

$H_2C{=}CHCH_2C(CH_3)_3$

273, see also 102

▶1437 S
B (0.0576)

$CF_3(CF_2)_5CF_3$

241

▶1437 VS
B (0.0576)

F F

225

▶1437 VS
B (0.06)

128

▶1437 VVS
B (0.51)

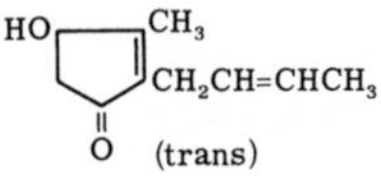

HO CH_3 $CH_2CH{=}CHCH_3$ O (trans)

430

▶1437 VS
B (0.025)

$CH_3CH_2SCH_3$

209

▶1437 S
C

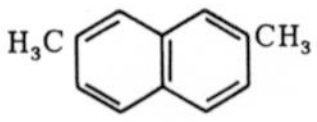

H_3C CH_3

179

▶1437 S
$CHCl_3$

H_3C CH_3 H_2N O N

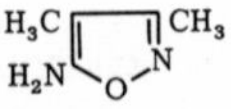

[Ring] 15

▶1437 S
A (200 mm Hg, 10 cm)

$CF_3CF_2CF_3$

235

▶1437 SSh
C 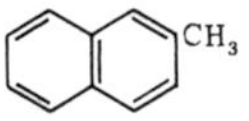
CH_3
191

▶1437 MW
$CHCl_3$
O
H_3C O CH_3 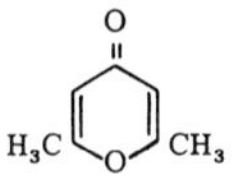
7

▶1437 W
CCl_4
$[(CH_3)_2NO_2]^- Na^+$
[CH_3 asymm. bend] 328

▶1436 VVS
B (0.10) 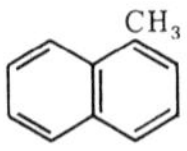
CH_3
192, see also 42

▶1435 S
B (0.003)
$C_3H_7NO_2$
392

▶1435 VVS
B (0.10) 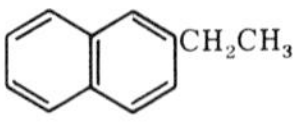
CH_2CH_3
189

▶1435 VS
$CHCl_3$ 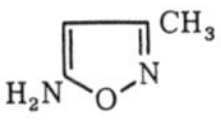
CH_3
H_2N O N
[Ring] 15

▶1435 MSSh
B (0.0576)
$(CF_3)_2CFCF_2CF_3$
240

▶1435 MW
B (0.0104) 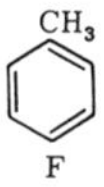
CH_3
F
220

▶1435 M
C 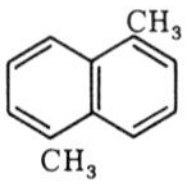
CH_3
CH_3
185

▶1435 M
G 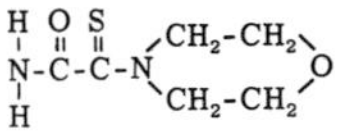
H O S
N-C-C-N CH_2-CH_2 O CH_2-CH_2
H
[HCH] 31

▶1435
G 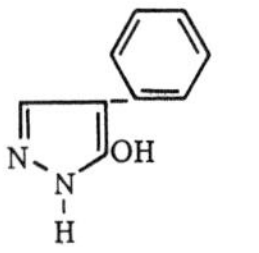
N N OH
H
[C=C] 6

▶1435
CH_2Br-CH_2Br
[CH_2 bend] 536

▶1435 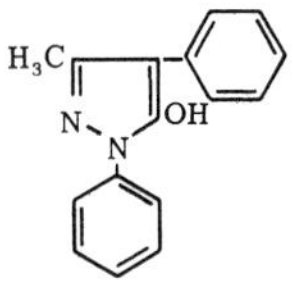
H_3C
N N OH
[C=C] 6

▶1434 VVS
B (0.10) 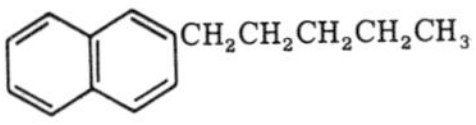
$CH_2CH_2CH_2CH_2CH_3$
173

▶1434 M (rotational line)
A (141 mm Hg, 15 cm)
H_3C-C≡CH
87

▶1434 M
B 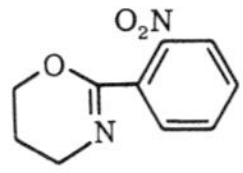
O_2N
O N
[-CH_2-] 26

▶1433 VS
B 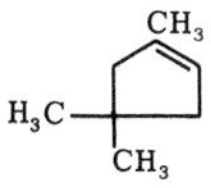
CH_3
H_3C
CH_3
127

▶1433 VS
B (capillary) 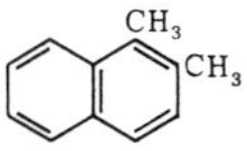
CH_3
CH_3
188

▶1433 VVS
B (0.10) 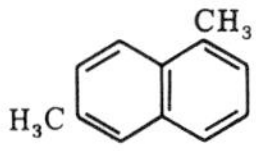
CH_3
H_3C
184

▶1433 VVS
B (0.04)

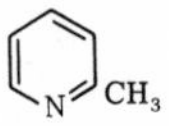

N CH$_3$

198, see also 438

▶1433 VS
C

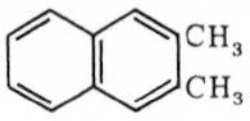

CH$_3$
CH$_3$

181

▶1433 VS
$CHCl_3$

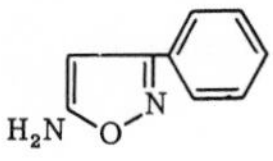

H$_2$N O N

[Ring] 15

▶1433 S
CCl_4

$(C_6H_5)_3SiCl$

86

▶1433 S

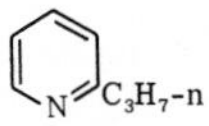

N C$_3$H$_7$-n

455

▶1433 S

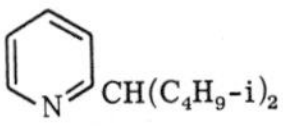

N CH(C$_4$H$_9$-i)$_2$

455

▶1433 M
B

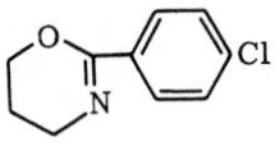

O N Cl

[-CH$_2$-] 26

▶1432 VVS
$CHCl_3$

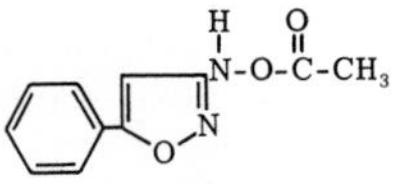

H O
N-O-C-CH$_3$
O N

[Ring] 15

▶1432 VS
C

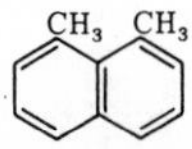

CH$_3$ CH$_3$

182

▶1431 VS
B (0.003)

$CH_3(CH_2)_2CH_2NO_2$

310

▶1431 M
B (?)[9.5%] (0.003)

$(CH_3)_2CHCH_2NO_2$

326

▶1431 VS
$CHCl_3$

S
O

[Ring] 7

▶1431 VS
$CHCl_3$

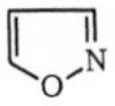

O N

[Ring] 15

▶1431 VS

$(C_6H_5)_2SiCl_2$

86

▶1431 S
B

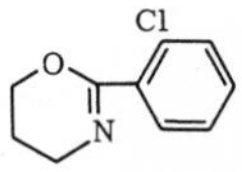

Cl
O N

[-CH$_2$-] 26

▶1431 M
A (385 mm Hg, 10 cm)

CF_3-CCl_3

253

▶1431 M
A (100 mm Hg, 10 cm)

CH_3-CCl_3

251, see also 51

▶1431 M
$CHCl_3$

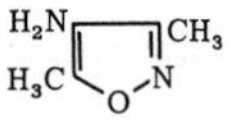

H$_2$N CH$_3$
H$_3$C O N

[Ring] 15

▶1431

O
H$_3$C-C-CH$_3$

[CH$_3$ bend] 537, see also 49

▶1430 VS
$CHCl_3$

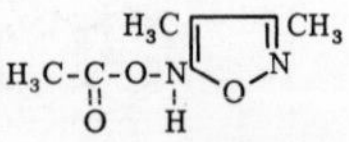

H$_3$C CH$_3$
H$_3$C-C-O-N O N
O H

[Ring] 15

▶1430 MWSh $CHCl_3$ [pyridine]-NH-CH_3 433	▶1428 S $CHCl_3$ H_3C, CH_3, H_3C [isoxazole] [Ring] 15
▶1429 CH_2Cl_2 376	▶1428 S $CHCl_3$ [pyridine]-C(=O)-H 433
▶1429 VVS B (0.056) CH_2=CHCH=CHCH=CH_2 203	▶1428 M $CHCl_3$ [pyridine]-[phenyl]-NO_2 433
▶1429 VS $CHCl_3$ H_3C-C(=O)-O-N(H)-[isoxazole], CH_3, H_3C [Ring] 15	▶1427 VS $CHCl_3$ n-H_7C_3-O-C(=O)-[isoxazole], H_3C [Ring] 15
▶1429 VS B (0.10) [naphthalene]-$CH_2CH_2CH_3$ 177	▶1427 VS $CHCl_3$ H_5C_2-O-C(=O)-[isoxazole], H_3C [Ring] 15
▶1429 S [pyridine]-$C(C_2H_5)_3$ 455	▶1427 MS B (0.08) [isoquinoline] 197
▶1429 MS [pyridine]-$CH(C_2H_5)_2$ 431	▶1426 MS $CHCl_3$ [pyridine]-CH_2CH_2-C(=O)-O-C_2H_5 433
▶1429 CH_2Cl_2 [CH_2 bend] 536	▶1426 MS $CHCl_3$ [pyridine]-CH_2-C(=O)-O-C_2H_5 433
▶1428 VS B (capillary) [naphthalene], CH_3, CH_3 186	▶1426 M B (0.015) H_2C=CH$CH_2C(CH_3)_3$ 102
▶1428 S $CHCl_3$ [pyridine]-NO_2 433	▶1426 M $CHCl_3$ H_3C-[isoxazole]-C(=O)-O-C_2H_5 [Ring] 15

▶1425 VS
A

CHF_2CH_3

233

▶1425 VVS
CCl_4 [31%] (0.008)

Cl S $COCH_3$

372

▶1425 VS
B (0.064)

$H_2C{=}CHCHCH_2CH_3$
CH_3

167

▶1425 S
$CHCl_3$

$O{-}C_2H_5$
N

433

▶1425 W
B (0.0576)

F CF_3
F_2 F_2
F CF_3
F_3C F
F_2

228

▶1425
P

$Ca(NO_3)_2$

45

▶1424 S
$CHCl_3$

O
$C{-}O{-}C_3H_7{-}i$
N

433

▶1424 M
$CHCl_3$

CH_2OH
N

433

▶1423 VS
$CHCl_3$

NH_2
H_3C O N

[Ring]

15

▶1423 VS
$CHCl_3$

CH_3
$N{-}C{=}O$
N

433

▶1423 S
C

CH_3

191

▶1423 S
G

N N OH
H

[C=C]

6

▶1422 MS
A (36 mm Hg, 15 cm)

$H_2C{=}CHCH(CH_3)_2$

131

▶1422 S
B

O Cl
N

[$-CH_2-$]

26

▶1422 S
$CHCl_3$

O
$C{-}O{-}C_2H_5$
N

433

▶1422 M
B

$[(C_2H_5)_2P\cdot BH_2]_3$

[CH_2]

30

▶1422 W
B

F CF_3
CF_3
F_2 F
F_2 F_2
F_2

231

▶1422 M
$CHCl_3$

O
$C{-}CH_3$
H_3C O N

[Ring]

15

▶1422

$H_2C(C{\equiv}N)_2$

466

▶1421 S
$CHCl_3$

O
$C{-}O{-}CH_3$
N

433

▶1421 MW
$CHCl_3$ [0.0064 g/ml]

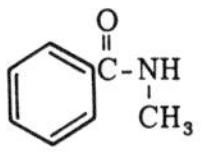

O
C-NH
CH_3

[CH_3]
2

▶1420 VVS
$CHCl_3$

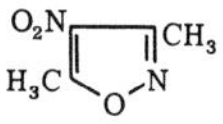

O_2N CH_3
H_3C O N

[Ring]
15

▶1420 VS
$CHCl_3$

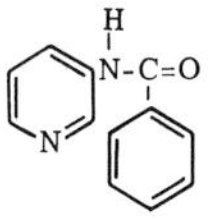

H
N-C=O
N

433

▶1420 VS
$CHCl_3$

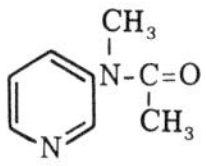

CH_3
N-C=O
CH_3
N

433

▶1420 VS
$CHCl_3$

H
N-C=O
CH_3
N

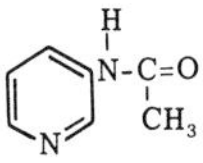

433

▶1420 SSh
B (0.003)

CH_3NO_2

311

▶1420 S
G

$Na_4P_2O_6$

28

▶1420 S
$CHCl_3$

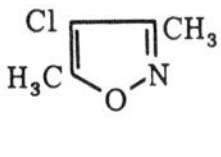

Cl CH_3
H_3C O N

[Ring]
15

▶1420 S
$CHCl_3$

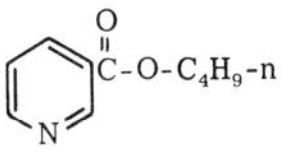

O
C-O-C_4H_9-n
N

433

▶1420 S
$CHCl_3$

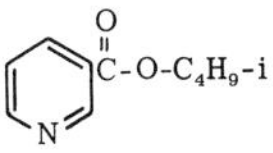

O
C-O-C_4H_9-i
N

433

▶1420 S
$CHCl_3$

O
C-O-C_4H_9-s
N

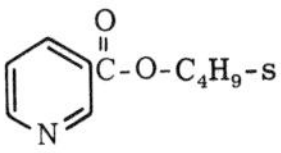

433

▶1420 M (rotational line)
A (141 mm Hg, 15 cm)

$H_3C-C{\equiv}CH$

87

▶1420 M
C

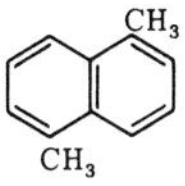

CH_3
CH_3

185

▶1420 M
CCl_4 [0.0082 m/l]

O
C-OH

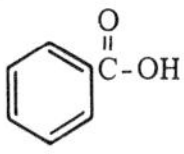

448, see also 38

▶1420

$Ce(NO_3)_6^{=}$

631

▶1419 VS
$CHCl_3$

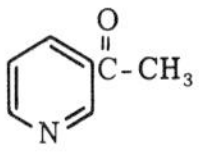

O
C-CH_3
N

433

▶1419 VS
B (0.055)

CH_3
H_3C N
H

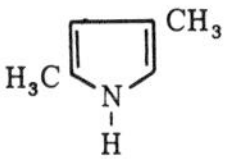

344

▶1419 S
B (0.0184)

$H_2C{=}CHCH_2CH_2CH{=}CH_2$

374

▶1419 S
$CHCl_3$

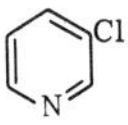

Cl
N

433

▶1419 S
$CHCl_3$

O
C-O-C_3H_7-n
N

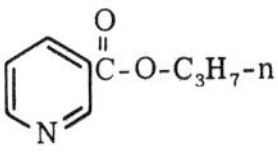

433

▶1419 M
A

$(CH_3)_2SO$

[CH_3 deformation] 13

▶1418 VVS
A (200 mm Hg, 10 cm)

$CF_3CF_2CF_2CF_2CF_3$

237

▶1418 VVS
B (0.051)

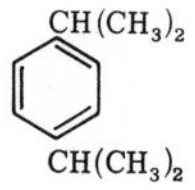

$CH(CH_3)_2$
$CH(CH_3)_2$

298

▶1418

CH_3NH_2

[CH_3 symm. bend] 539

▶1418 SSp
B (0.066)

$H_2C=CH(CH_2)_8CH_3$

352

▶1418 S
B (0.028)

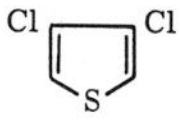

Cl Cl
S

161

▶1418 S
$CHCl_3$

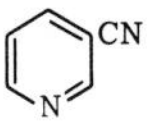

CN
N

433

▶1418 S
$CHCl_3$

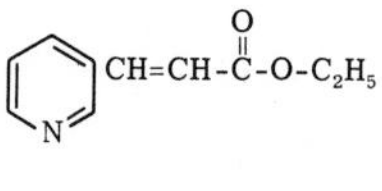

O
CH=CH-C-O-C_2H_5
N

433

▶1418

O
H_3C-C-OH

[CH_3 bend] 537, see also 50

▶1417 VS
$CHCl_3$

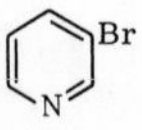

Br
N

433

▶1417 VS
B

$(C_2H_5\text{-NH-B-N-}C_2H_5)_3$

[CH] 32

▶1417 M
C_6H_6 [0.0044 m/l]

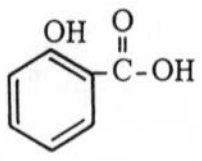

OH O
C-OH

448

▶1416 VS
B (capillary)

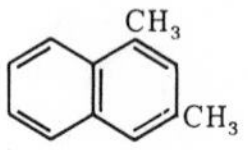

CH_3
CH_3

187

▶1416 VS
B (0.028)

Cl
Cl
S

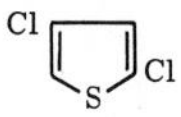

211

▶1416 M
B (0.015)

$H_2C=CHC(CH_3)_3$

129

▶1416 MW
$CHCl_3$

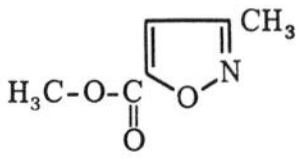

CH_3
H_3C-O-C O N
O

[Ring] 15

▶1415 S
B

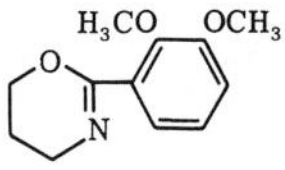

H_3CO OCH_3
O
N

26

▶1415 MW
CCl_4 [0.007 g/ml]

O
C-NH
CH_3

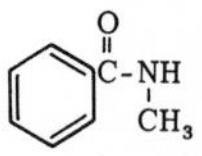

[CH_3] 2

▶1415 MW
$CHBr_3$ [0.0077 g/ml]

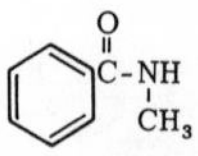

O
C-NH
CH_3

2

▶1415 MW
$CHCl_3$

CH_3
H_5C_2-O-C O N
O

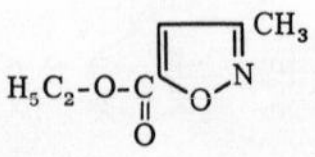

[Ring] 15

▶1415
E
$Ba(NO_3)_2$
45

▶1414 VVS
2,5-dichlorothiophene (Cl, Cl, S)
211

▶1414 VSSh
A (200 mm Hg, 10 cm)
$CF_3CF_2CF_3$
235

▶1413 MW
I
$H_3C-C(=O)-NH-CH_3$
46

▶1414 VS
A
CHF_2CH_3
233

▶1413 MW
CCl_4 [0.3 g/l]
$C_6H_5-C(=O)-NH-CH_3$
[CH_3]
2

▶1414 SSh
B (0.0576)
p-difluorobenzene (F, F)
225

▶1412 VS
B
$[(CH_3)_2N-BO]_3$
[CH_3]
32

▶1414 W
B (0.0576)
perfluoromethylcyclohexane (F, CF_3, F_2, F_2, F_2, F_2, F_2)
232

▶1412 VS
B (film)
1,3,5-trioxane (O, O, O)
342

▶1414 VS
$CHCl_3$
3,5-dimethylisoxazole (H_3C, CH_3, O, N)
[Ring]
15

▶1412 S
pyridine, C_3H_7-n (N)
431

▶1414 S
$CHCl_3$
4-pyrone (O, O)
[Ring]
7

▶1412 S
pyridine, C_4H_9-i (N)
431

▶1414 VVSSp
CCl_4 [sat.] (0.025)
2-acetyl-5-bromothiophene (Br, S, $COCH_3$)
348

▶1412 M
E
nitrobenzene (NO_2)
[Ring]
20

▶1414 M
$CHCl_3$
methyl 5-methylisoxazole-3-carboxylate (H_3C, O, N, $C(=O)-O-CH_3$)
[Ring]
15

▶1412 M
P
$[(C_2H_5)_2P \cdot BCl_2]_3$
[CH_2]
30

▶1414 MW
$CHCl_3$
3-methylpyridine (CH_3, N)
433

▶1411 VS
$CHCl_3$
3-methylisoxazole (CH_3, O, N)
[Ring]
15

▶1411 M
$CHCl_3$ 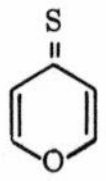
S
O
7

▶1410 VS
N, E 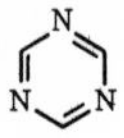
N N N
14

▶1410 S
H 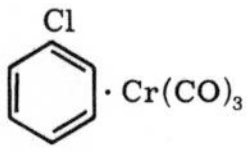
Cl
$\cdot Cr(CO)_3$
17

▶1410 S
P
$[(C_2H_5)_2P \cdot BBr_2]_3$
$[CH_2]$ 30

▶1410 S
P
$[(C_2H_5)_2P \cdot BI_2]_3$
$[CH_2]$ 30

▶1410 S
B (0.025) 
CH_3
CN
214

▶1410 M
CCl_4 [0.0128 m/l] 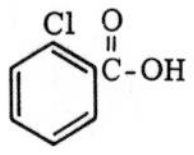
Cl O
C-OH
448

▶1410
O
H_5C_2-C-NH_2
465

▶1410 S
CH_3CHCOO^-
NH_3^+
[COO^- symm. stretch] 685

▶1408 W
B (0.008)
H_3C C_3H_7-n
C=C
H H
(cis)
276

▶1408 VVS
B (0.025) 
S
199

▶1408 MS
B (0.0104)
n-H_9C_4-O-CF_2CHFCl
264

▶1408 S
$CHCl_3$ 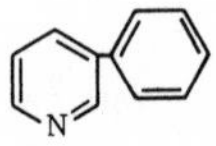
N
433

▶1408 S
$CHCl_3$ 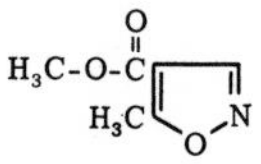
O
H_3C-O-C
H_3C O N
[Ring] 15

▶1408 S
$CHCl_3$ 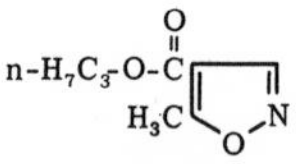
O
n-H_7C_3-O-C
H_3C O N
[Ring] 15

▶1408 S 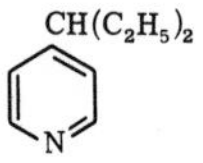
$CH(C_2H_5)_2$
N
431

▶1408 W
B
$(C_6H_5)_2(CH_3)SiCl$
86

▶1407 VVS
$CHCl_3$ 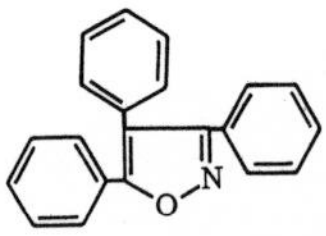
O N
15

▶1407 W
B (0.008)
H_3CH_2C CH_2CH_3
C=C
H H
275

▶1407 M (rotational line)
A (141 mm Hg, 15 cm)
H_3C-C≡CH
87

▶1407 M
$CHCl_3$
$H_3C-C(=O)-O-N(H)-$ (3-phenylisoxazolyl)
[Ring] 15

▶1407 MW
$CHCl_3$
Cl, N (chloropyridine)
433

▶1406 M
B (0.0088)
$CH_3CH=CHCH_2CH_3$
(cis)
208

▶1406 S
B (0.0576)
F, F, F (trifluorobenzene)
224

▶1406 VS
$CHCl_3$
(3,5-diphenylisoxazole) O, N
[Ring] 15

▶1406 S
B (0.064)
$CH_3CH=CH(CH_2)_4CH_3$
(cis)
160, see also 100

▶1406 MSh
B (0.0088)
$(CH_3)_3CC(CH_3)_2-CH(CH_3)_2$
306

▶1406 VSVSp
CS_2 (0.728)
$CH_2=CHCH=CHCH=CHCH=CH_2$
202

▶1406 VS
B (0.10)
$CH_2CH_2CH_2CH_2CH_3$ (naphthalene)
174

▶1405 S
$CHCl_3$
$H_5C_2-O-C(=O)-$, H_3C (isoxazole) O, N
15

▶1405 MSh
C
CH_3 (methylnaphthalene)
191

▶1404 VS
B (0.051)
$CH(CH_3)_2$ / $CH(CH_3)_2$ (diisopropylbenzene)
298

▶1405
A
$(CH_3)_2SO$
[CH_3 deformation] 13

▶1404 VS
C
$(CH_3)_3CNO_2$
325

▶1404 VS
B (0.10)
$CH_2CH_2CH_2CH_3$ (naphthalene)
176

▶1404 SSp
B (0.065)
S, $CH=CH_2$ (vinylthiophene)
212

▶1404 M
B
$C_6H_5(CH_3)SiCl_2$
86

▶1403 VS
A (100 mm Hg, 10 cm)
F_2, F_2, F_2, F_2 (octafluorocyclobutane)
243

▶1403 VS
A
CHF_2CH_3
233

▶1403 MS
B (0.0088)
$(CH_3)_3CC(CH_3)_2-CH_2CH_2CH_3$
304

▶1403 VS
CCl_4 [25%] (0.10)
$CH_3CH(OH)CH_2CH_3$
319

▶1403 S
B (0.028)
Cl, Cl, S (dichlorothiophene)
161

▶1402 VS
$CHCl_3$ 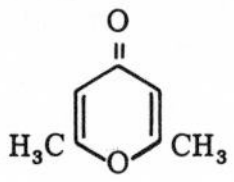
O; H_3C ... CH_3 (ring)
[Ring]
7

▶1401 S
B (0.015)
$H_2C{=}CHCH_2C(CH_3)_3$
102, see also 273

▶1401 VS
B (0.003)
CH_3NO_2
311

▶1401 M
C_6H_6
HNO_3
[NO_3^- ?]
642

▶1400 M
B 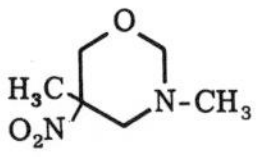
O; H_3C; O_2N; $N{-}CH_3$ (ring)
26

▶1400 M
B 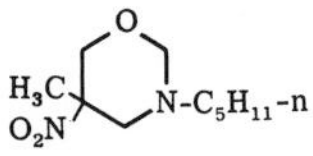
O; H_3C; O_2N; $N{-}C_5H_{11}{-}n$ (ring)
26

▶1400 M
B 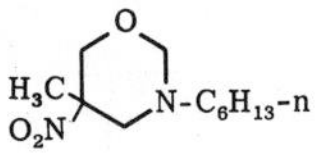
O; H_3C; O_2N; $N{-}C_6H_{13}{-}n$ (ring)
26

▶1400 M
G 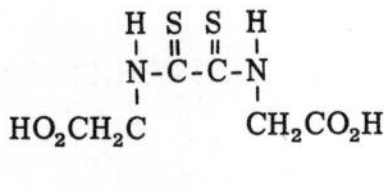
H S S H; N-C-C-N; HO_2CH_2C CH_2CO_2H
[HCH]
31

▶1399 MSh
B (0.011)
$(CH_3)_3CCH(CH_3)_2$
147, see also 146

▶1399 S
B (0.10) 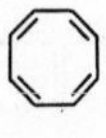
200

▶1399 M
B 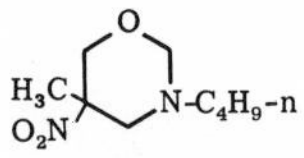
O; H_3C; O_2N; $N{-}C_4H_9{-}n$ (ring)
26

▶1399 M
A (100 mm Hg, 15 cm)
CH_3CH_2Cl
88

▶1399 W
B (0.055) 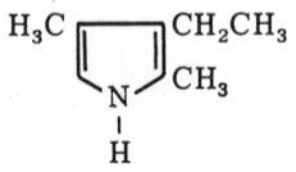
H_3C; CH_2CH_3; CH_3; N; H (ring)
343

▶1398 MSSh
B (0.018)
Grating
$(CH_3)_3CCH(CH_3)_2$
146, see also 147

▶1398 MS
A
$HC{\equiv}C{-}\overset{O}{\overset{\|}{C}}{-}H$
[HCO bend]
18

▶1397 VS
A (100 mm Hg, 10 cm)
$CH_3{-}CCl_3$
251, see also 51

▶1397 S
B (0.0088)
$(CH_3)_3CCH(CH_2CH_3)_2$
300, see also 78

▶1397 VVS
B (0.10) 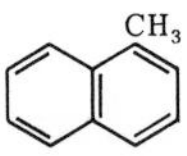
CH_3 (naphthalene ring)
192, see also 42

▶1397 MW
B (0.008)
$H_2C{=}CHCH_2C(CH_3)_3$
273, see also 102

▶1397 VSSp
B (0.003)
$(CH_3)_2CHNO_2$
366

▶1397 M
B (0.0576)

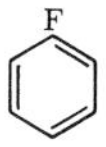

F

226

▶1396 M
G

H O S H
N-C-C-N
C_6H_{11} C_6H_{11}

31

▶1396

CH_3CN

[CH_3 symm. bend]

538

▶1395 MSSh
B (0.0088)

$(CH_3)_3CCH_2CH_3$

145

▶1395 S
B (0.0088)

$(CH_3)_3CCH_2CH_2C(CH_3)_3$

305

▶1395 SSh
B (0.0088)

$(CH_3)_3CCH_2CHCH(CH_3)_2$
CH_3

303

▶1395 S
B (?)[9.5%] (0.003)

$(CH_3)_2CHCH_2NO_2$

326

▶1395 M
B (0.008)

CH_3
$(CH_3)_3CCH_2CHCH_2CH_3$

285

▶1395 MS
B (0.0088)

$(CH_3)_2C{=}CHC(CH_3)_3$

204

▶1395 S
B (0.033)

$(CH_3)_3CCH(CH_2CH_3)_2$

78, see also 300

▶1395 VS
B (0.0088)

CH_3
$(CH_3)_3CCHC(CH_3)_3$

307

▶1395
E

$Zn(NO_3)_2$

45

▶1395
P

$Cr(NO_3)_3$

45

▶1394 VS
B (capillary)

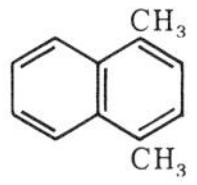

CH_3
CH_3

186

▶1394 S
B

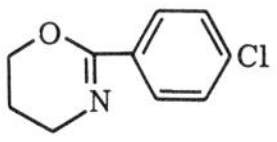

O
N
Cl

26

▶1394 MS
B (0.033)

$C(CH_3)_3$

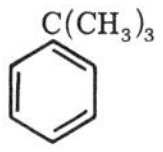

133

▶1394 MS
B (0.033)

CH_3
$(CH_3)_3CCH_2CHCH_2CH_3$

79

▶1393.2 VS
B (0.03)

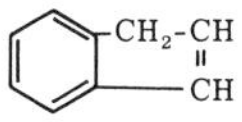

CH_2-CH
CH

9

▶1393 SSh
B (0.0088)

CH_3
$(CH_3)_2CHC - CH_2CH_2CH_3$
CH_3

134

▶1393 S
B (0.018)
Grating

$(CH_3)_3CCH_2CH_2CH_3$

154, 155

▶1393 MSh
B (0.0104)
n-H_7C_3-O-CF_2CHFCl
263

▶1393 MS
G
H(H)N-C(=O)-C(=S)-N(H)C_6H_{11}
31

▶1393 MSh
B (0.033)
$(CH_3)_3C(CH_2)_2CH(CH_3)_2$
292

▶1392 VS
B (0.055)
H_3C, CH_3, N-H (pyrrole ring)
344

▶1392
P
$La(NO_3)_3$
45

▶1392
P
$Nd(NO_3)_3$
45

▶1391 SSh
B (0.015)
$H_2C=C(CH_3)CH_2C(CH_3)_3$
96

▶1391 S
B (?) [9%] (0.003)
$CH_3CH_2CH(NO_2)CH_3$
324

▶1391 MSh
B (0.0576)
CF_3 (benzene ring)
219

▶1391 M
B
H_5C_2, O_2N, O, N-H (ring)
26

▶1391 M
M
Pt(NH_2-CH_2-CH_2-NH_2)$_2$$PtCl_4$
[CH_2]
8

▶1390 M
B
H_5C_2, O_2N, O, N-CH_3 (ring)
26

▶1390 S
B
H_5C_2, O_2N, O, N-C_3H_7-n (ring)
26

▶1390 M
B
H_5C_2, O_2N, O, N-C_6H_{13}-n (ring)
26

▶1390 M
B
O, N (ring), Br (benzene ring)
[-CH_2-]
26

▶1390 M
G
C_6H_5(H)N-C(=O)-C(=S)-N(H)C_6H_{11}
[CN]
31

▶1390
E
$Cu(NO_3)_2 \cdot 5H_2O$
45

▶1390
E
$Fe(NO_3)_3$
45

▶1390
G
H_3C, H_3C-N, N, =O (ring), benzene ring
[C=C]
6

▶1390
P
$Be(NO_3)_2$
45

▶1390
P
$Ni(NO_3)_2$
45

▶1390
P
$Sm(NO_3)_3$
45

▶1389 S
$CHCl_3$
S
H_3C O CH_3
7

▶1390
$CH_2{=}C{=}CH_2$
[CH_2 bend] 536

▶1390 M
A (10 mm Hg, 58 cm)
$H_2C{=}CH(CH_2)_5CH_3$
289

▶1389 M
A (10 mm Hg, 58 cm)
$CH_3CH{=}CH(CH_2)_4CH_3$
(cis and trans)
101

▶1390
P
$Th(NO_3)_4$
45

▶1389 M
CCl_4 [20 g/l] (0.51)
$CH_2CH(CH_3)_2$
$CH_3CH{=}CH(CH_2)_2CH{=}CH(CH_2)_2\text{-}CONH$
357

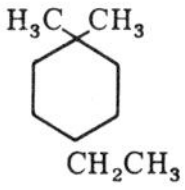

▶1389 S
B (0.0088)
CH_3
$CH_3CH_2C\text{-}CH_2CH_2CH_3$
CH_3
360

▶1388 VS
B (0.033)
$(CH_3)_3CCHCH(CH_3)_2$
CH_3
293

▶1389 SSp
B (0.0153)
H_3C CH_3
CH_2CH_3
215

▶1388 S
CCl_4 [0.04 g/ml] 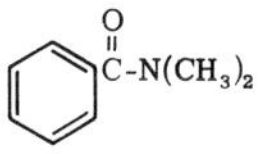
O
$C\text{-}N(CH_3)_2$
3

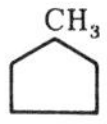

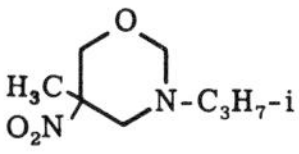

▶1389 S
B (0.0088)
CH_3
$(CH_3)_2CHCHCH_2CH_2CH_3$
139, see also 338

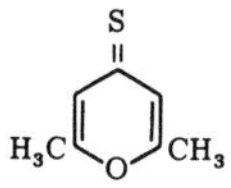

▶1388 MS
G
H S S H
N-C-C-N
C_6H_{11} C_6H_{11}
[CN?] 31

▶1389 VS
B (0.0088)
CH_3
$(CH_3)_3CC\text{-}CH(CH_3)_2$
CH_3
306

▶1388 M
B
$[(C_2H_5)_2P \cdot BH_2]_3$
[CH_3 symm. deformation] 30

▶1389 SSh
B (0.0088)
$(CH_3)_3CCH_2CHCH(CH_3)_2$
CH_3
303

▶1388 M
B 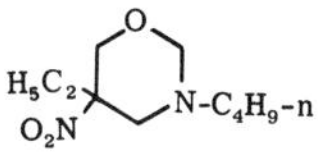
O
H_5C_2
O_2N
$N\text{-}C_4H_9\text{-}n$
26

▶1389 S
A (40 mm Hg, 58 cm)
CH_3
105

▶1388 M
C 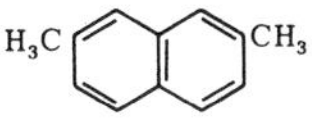
H_3C CH_3
179

▶1389 S
B
O
H_3C
O_2N
$N\text{-}C_3H_7\text{-}i$
26

▶1388 M
P
$[(C_2H_5)_2P \cdot BCl_2]_3$
[CH_3 symm. deformation] 30

▶1387 VS A $(CH_3)_2CHCH_2CH_3$ 333	▶1387 S (pyridine ring)$CH(C_4H_9\text{-}i)_2$ 455
▶1387 VS A (100 mm Hg, 10 cm) $CH_3\text{-}CCl_3$ 251, see also 51	▶1387 M (rotational line) A (141 mm Hg, 15 cm) $H_3C\text{-}C{\equiv}CH$ 87, see also 51
▶1387 S B (0.0088) $(CH_3)_2CHCH(CH_3)CH_2CH_3$ 142, 153, see also 152	▶1387 MW A (6.3 mm Hg, 40 cm) (benzene ring) CH_3, CH_3 69
▶1387 SSp B (0.0088) $(CH_3)_2CHCH_2CH(CH_3)_2$ 141, 151, see also 150, 335	▶1387 CH_2Br_2 [CH_2] 646
▶1387 VSB B (0.065) $H_2C{=}C(CH_3)C(CH_3)_3$ 272	▶1386 VS B (0.10) (naphthalene ring)$CH_2CH_2CH_2CH_2CH_3$ 174
▶1387 S B (0.0088) $(CH_3)_2CHCH(CH_3)CH(CH_3)_2$ 135	▶1386 VS B (0.008) Grating $(CH_3)_2CHCH_2CH(CH_3)_2$ 150, see also 141, 151
▶1387 S B (0.036) $H_2C{=}CH(CH_2)_3CH(CH_3)_2$ 97	▶1386 S B (0.010) Grating $(CH_3)_2CHCH(CH_3)CH_2CH_3$ 152, see also 142, 153
▶1387 S B (0.0088) $(CH_3)_2CHCH_2CH(CH_3)_2$ 141, 151, see also 150	▶1386 S B (0.033) $(CH_3)_3C(CH_2)_2CH(CH_3)_2$ 292
▶1387 W B (0.0576) (benzene ring) CF_3, F, F 217	▶1386 S P $[(C_2H_5)_2P\cdot BI_2]_3$ [CH_3 symm. deformation] 30
▶1387 S (pyridine ring)$C_4H_9\text{-}i$ 455	▶1386 M B (morpholine ring) H_5C_2, O_2N, $N\text{-}C_3H_7\text{-}i$ 26

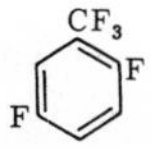

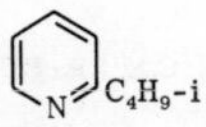

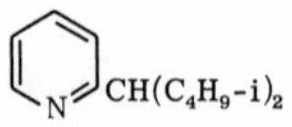

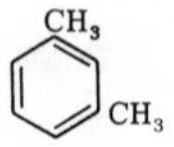

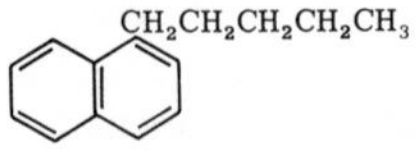

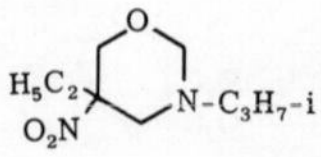

▶1386 M
G
H S S H / N-C-C-N / CH_2OH CH_2OH (C=S, C=S)
[HCH]
31

▶1385 S
B (0.0153)
cyclopropyl–C(=CH_2)–CH_3
288

▶1385 VS
A
$CH_3CH_2CH(CH_3)CH_2CH_3$
332

▶1385 S
B (0.0088)
$(CH_3)_2CHCH(CH_2CH_3)_2$
136

▶1385 VS
CCl_4 [25%] (0.10)
$(CH_3)_2CHCH_2CH_2OH$
195

▶1385 MSh
B (0.0104)
n-H_7C_3-O-CF_2CHFCl
263

▶1385 VS
B (0.003)
$C_3H_7NO_2$
392

▶1385 MS
B (0.0088)
$(CH_3)_2C{=}CHC(CH_3)_3$
204

▶1385 VS
B
CH_3, CH_3, CH_3 (cyclopentene)
126

▶1385 VSSp
B (0.033)
CH_3, $CH(CH_3)_2$ (benzene)
132

▶1385 VSSp
B (0.029)
$CH(CH_3)_2$, $(CH_3)_2HC$, $CH(CH_3)_2$ (benzene)
278

▶1385 S
B
H_3C, O, N-H, H_3C CH_3
[CH]
26

▶1385 S
B (0.0088)
$(CH_3)_2CHCH_2CH_2CH(CH_3)_2$
137, see also 339

▶1385 SSp
B (0.025)
CH_3, CN (benzene)
312

▶1385 S
B (0.025)
CH_3, CN (benzene)
214

▶1385 VS
B (0.064)
H CH_2CH_3 / C=C / H_3C CH_3
(trans)
165

▶1385
G
H_3C, N, N, =O, CH_3 (phenyl)
[C=C]
6

▶1385 S
B (0.0088)
$(CH_3)_2CHCH_2CH(CH_3)CH_2CH_3$
138, see also 334

▶1385
G
H_3C, OH, N, N, CH_3
[C=C]
6

▶1385 S
B (0.0088)
$(CH_3)_2CHCH(CH_2CH_3)CH(CH_3)_2$
301

▶1385
P
$Ce(NH_4)_2(NO_3)_6$
45

▶1385 M

H_2N-C(=O)-C(=S)-NH-C_6H_{11}

[C=S] 31

▶1385

SO_2Cl / CH_3 (benzene ring)

[S-O] 389

▶1384 VS
B

$[(CH_3)_2N \cdot BO]_3$

[CH] 32

▶1384 S
B (0.026)

$(CH_3)_2CH(CH_2)_3CH_3$

80, see also 158, 331

▶1384 VS
B (0.055)

H_3C, CH_2CH_3, CH_3, N-H (pyrrole ring)

343

▶1384 VS
B (0.10)

$CH_2CH_2CH_2CH_3$ (naphthalene)

176

▶1384 M
B

H_5C_2, O_2N, O, N-C_2H_5

26

▶1383 S
B (0.0576)

CH_3, F (benzene ring)

221

▶1383 VVS
B (0.51)

HO, CH_3, CH_2CH=$CHCH_3$, O (trans)

430

▶1383 S
A (100 mm Hg, 15 cm)

H_2C=$C(CH_3)CH_2CH_3$

92

▶1383 S
B (0.015)

H_2C=$CHC(CH_3)_3$

129

▶1383 VS
B (0.051)

$CH(CH_3)_2$, $CH(CH_3)_2$ (benzene ring)

296

▶1383 S
B (0.0576)

CH_3 / F (benzene ring)

220

▶1383
G

H_3C, N, N, OH (pyrazole ring with phenyl)

[C=C] 6

▶1382 VVS
B (0.10)

CH_2CH_3 (naphthalene)

190

▶1382 VS
B (0.005)

$(CH_3)_2CHCH(CH_3)_2$

144, 210

▶1382 S
B (0.018)
Grating

$(CH_3)_3CCH(CH_3)_2$

146, see also 147

▶1382 S
P

$[(C_2H_5)_2P \cdot BBr_2]_3$

[CH_3 symm. deformation] 30

▶1382 M
CCl_4 [0.00658 m/l]

OH, C(=O)-OH (benzene ring)

448, see also 471

▶1382 W
$CHCl_3$

S, H_3C, S, CH_3

[CH_3] 7

▶1382
P

$Hg_2(NO_3)_2$

45

▶1381 VS
B

$CH_3CH_2C(CH_3)_2-CH_2CH_3$

336, see also 148, 149

▶1381 W
CCl_4

$[(CH_3)_2NO_2]^-Na^+$

[CH_3 symm. deformation] 328

▶1381 S
B (0.029)

$CH(CH_3)_2$ / H_3C / CH_3 (benzene ring)

277

▶1381 VVS
B (0.003)

$CH_3(CH_2)_2CH_2NO_2$

310

▶1381 VS
B (0.0088)

$(CH_3)CHCH(CH_3)CH_2CH_2CH_3$

139

▶1381 VS
B (0.018)

$CH_3CH_2CH(CH_3)(CH_2)_2CH_3$

81, see also 157

▶1381 VS
B (0.030)

$(CH_3)_3CCH_2CH_2CH_3$

155, see also 154

▶1381 VS
B (0.011)
Grating

$(CH_3)_3CCH(CH_3)_2$

147, see also 146

▶1381 MSSh
B (0.0088)

$(CH_3)_3CCH_2CH_3$

145

▶1381 S
B (0.008)

$(CH_3)_2CHCH(CH_3)CH_2CH_2CH_3$

338

▶1381 VS
B (0.008)

$CH_3CH_2CH(CH_3)-CH(CH_3)CH_2CH_3$

83

▶1381 VS
B (0.0088)

$(CH_3)_3CC(CH_3)_2-CH_2CH_2CH_3$

304

▶1381 VSVB
CCl_4 [20%] (0.10)

CH_3CH_2OH

315

▶1381 VSVB
CCl_4 [25%] (0.10)

$CH_3CH_2CH_2OH$

314

▶1381 VVS
B (0.10)

CH_3 (naphthalene ring)

192, see also 42

▶1381 MSh
B (0.0104)

$F_9C_4-O-C_4F_9$

266

▶1381 S
B (0.0088)

$(CH_3CH_2)_3CH$

143, see also 156

▶1381 S
B (0.0088)

$(CH_3)_2CHCH(CH_3)CH_2CH_3$

142, 153, see also 152

▶1381

$H_3C-C(=O)-OH$

538, see also 51

▶1381 SSp
B (0.0153)

$CH_3(CH_2)_2CH{=}CH(CH_2)_2CH_3$
(cis)

422

▶1381 S
B (0.0088)

$(CH_3CH_2)_2CHCH_2CH_2CH_3$

140

▶1381 VS
B (0.0088)

$CH_3CH_2C(CH_3)(CH_2CH_3)-CHCH_2CH_3$ (CH₃ above and CH₃CH₃ below the C – CH bond)

299

▶1381 VS
B (0.0104)

$CH_3CH_2\text{-}O\text{-}CF_2CHFCl$

261

▶1381 VS
B (0.068)

H_3C H
C=C
H $CH_2CH_2CH_3$
(cis and trans)

276 (cis), 169 (trans)

▶1381 VVS
B (0.064)

$H_2C{=}CHCHCH_2CH_3$
CH_3

167

▶1381 VS
B (0.064)

$(CH_3)_2C{=}CHCH_2CH_3$

166

▶1381 VS
B (0.064)

$CH_3CH{=}CH(CH_2)_4CH_3$
(cis)

160, see also 100

▶1381 VSSh
B (0.0088)

CH_3
$(CH_3)_3CC-CH(CH_3)_2$
CH_3

306

▶1381 VS
B (0.068)

H_3C C_3H_7-n
C=C
H H
(cis)

276

▶1381 S
B (0.036)

$H_2C{=}CH(CH_2)_5CH_3$

290

▶1381 S
B (0.064)

$C_{17}H_{34}$
(1-heptadecene)

159

▶1381 S
CCl_4 [25%] (0.10)

CH_3
$H_2C{=}CCO_2CH_3$

193

▶1381 VS
C

CH_3 CH_3 (1,8-dimethylnaphthalene structure)

182

▶1381 S
B (0.065)

(thiophene ring, S) CH_2CH_3

213

▶1381 M
B

O (ring)
H_3C
O_2N N-C_3H_7-n

26

▶1381 M
B

O (ring)
H_3C
O_2N N-C_2H_5

26

▶1380 VS (Raman only)
A

$CH_3C{\equiv}CCH_3$

[CH_3 bend] 679

▶1380

CH_3CH_3

[CH_3 symm. bend] 538

▶1380 VS
A (17.7 mm Hg, 40 cm)

CH_3
$(CH_3)_2CHCHCH_2CH_2CH_3$

72, see also 115

▶1380 VS
A (15.9 mm Hg, 40 cm)

$CH_3CH_2CH(CH_2)_3CH_3$
CH_3

106, see also 119

▶1380 VS
B (0.016)

$(CH_3CH_2)_3CH$

156, see also 143

▶1380 VS
B (0.036)

CH_3
$H_2C{=}C(CH_2)_4CH_3$

98

▶1380 S
A (18 mm Hg, 40 cm)

$(CH_3)_2CH(CH_2)_4CH_3$

294, see also 120

▶1380 VS
B (0.033)

$(CH_3)_3CCH(CH_2CH_3)_2$

78, see also 300

▶1380 VS
B (0.033)

CH_3
$(CH_3)_3CCH_2CHCH_2CH_3$

79

▶1380 M
A (0.028)

$CH_3(CH_2)_5CH_3$

85

▶1380
E

$CsNO_3$

45

▶1380
G

C_2H_5
H_3C — C_2H_5
N N OH

[C=C]

6

▶1380
P

$AgNO_3$

45

▶1380
P

$Al(NO_3)_3$

45

▶1380
P

$Cd(NO_3)_2$

45

▶1380
P

KNO_3

45

▶1380 S
A (12.6 mm Hg, 40 cm)

$CH_3(CH_2)_6CH_3$

89

▶1379 VS
A (100 mm Hg, 10 cm)

$CH_3\text{-}CCl_3$

251, see also 51

▶1379 S
A (36 mm Hg, 15 cm)

$H_2C{=}CHCH(CH_3)_2$

131

▶1379 VSSp
B (0.028)

$CH_3(CH_2)_5CH(C_3H_7)(CH_2)_5CH_3$

281

▶1379 M
B (0.0088)

CH_3
$(CH_3)_3CCH_2CHCH_2CH_3$

285

▶1379 VS
B (0.033)

$(CH_3)_3CCH(CH_2CH_3)_2$

78, see also 300

▶1379 S
B (0.0088)

CH_3 CH_3
$CH_3CH_2C - CH_2CHCH_2CH_3$
CH_3

302

▶1379 MS B (0.008) $H_3C\ CH_3$ C=C $H\ CH_2CH_3$ (cis) 274	▶1379 VS B (0.013) $CH_3CH{=}CHCH(CH_3)_2$ 130
▶1379 SSh B (0.0088) $(CH_3)_3CCH_2CHCH(CH_3)_2$ CH_3 303	▶1379 S B (0.0088) $CH_3(CH_2)_2CH{=}CH(CH_2)_2CH_3$ (trans) 205
▶1379 VS B Grating $CH_3CH_2CH(CH_2)_2CH_3$ CH_3 157, see also 81	▶1379 VS B (0.064) $H_3CH_2C\ H$ C=C $H\ CH_2CH_3$ (trans) 168
▶1379 VS CCl_4 [25%] (0.10) CH_3 $CH_3CH_2CHCH_2OH$ 321	▶1379 VS B (0.028) Grating $CH_3(CH_2)_5CH_3$ 84
▶1379 S B (0.036) $CH_3CH_2CH{=}CH(CH_2)_3CH_3$ 99	▶1379 MS B (0.0088) $CH_3CH{=}CHCH_2CH_3$ (trans) 207
▶1379 VS B (0.003) CH_3NO_2 311, see also 50	▶1379 VSB CCl_4 [25%] (0.10) $CH_3(CH_2)_4OH$ 393
▶1379 VS B (0.028) $C_{26}H_{54}$ (5, 14-di-n-butyloctadecane) 283	▶1379 VS B (0.10) $CH_2CH_2CH_3$ (on naphthalene ring) 178
▶1379 VS B $(CH_3)_3C(CH_2)_3CH_3$ 337	▶1379 S B (0.10) CH_3 (on tetrahydronaphthalene ring) 269
▶1379 VSSp B (0.028) $(C_2H_5)_2CH(CH_2)_{20}CH_3$ 282	▶1379 SSp B (0.066) $CH_3(CH_2)_8CH{=}CH_2$ 352
▶1379 S B (?) [9.5%] (0.003) $(CH_3)_2CHCH_2NO_2$ 326	▶1379 S B (0.010) Grating CH_3 $(CH_3)_2CHCHCH_2CH_3$ 152, see also 142, 153

▶1379 S
B (0.016)
Grating

$CH_3CH_2C(CH_3)_2CH_2CH_3$

148, see also 149, 336

▶1379 VS
B (0.051)

$CH(CH_3)_2$ / $CH(CH_3)_2$ (benzene ring)

297

▶1379 S
B (0.051)

$CH(CH_3)_2$ / $CH(CH_3)_2$ (benzene ring)

298

▶1379 S
B (0.066)

$H_2C{=}CH(CH_2)_6CH_3$

271

▶1379 S
B (0.066)

$H_2C{=}CH(CH_2)_7CH_3$

270

▶1379 M
B

H_3C, O_2N, O, $N{-}C_6H_{13}{-}n$ (ring)

26

▶1379 M
C

CH_3 / CH_3 (naphthalene)

185

▶1379 W
B

$C_6H_5HSiCl_2$

86

▶1379 W

$H_3C{-}C{\equiv}CH$

[CH_3 symm. bend]

538, see also 51

▶1378 S
B (0.018)
Grating

$(CH_3)_3CCH_2CH_2CH_3$

154, see also 155

▶1378 S
B

H_5C_2, O_2N, O, $N{-}C_3H_7{-}n$ (ring)

26

▶1378 MS
B (0.007)

$CH_3CH_2C(CH_3)_2CH_2CH_3$

149, see also 148, 336

▶1378 S
B (0.036)

$CH_3CH{=}CH(CH_2)_4CH_3$
(cis)

100

▶1378 S
G

H_3C, CH_3, $CH{=}N{-}NH{-}C({=}S){-}NH_2$ (ring)

380

▶1378 M
B

H_5C_2, O_2N, O, $N{-}C_5H_{11}{-}n$ (ring)

26

▶1378 M
C

H_3C, CH_3 (naphthalene)

180

▶1378 S
B (0.025)

$CH_3CH_2SCH_3$

209

▶1377 VS
B (0.029)

$CH_3CHCH_2CH_3$, $CH_3CH_2HC(CH_3)$, $CHCH_2CH_3(CH_3)$ (benzene ring)

280

▶1377 S
B (0.0088)

$H_2C{=}C(CH_3)CH_2CH_2CH_3$

206

▶1377 VS
B (0.025)

$CH_3CH_2{-}S{-}CH_2CH_3$

172

▶1377 VS
B
$(CH_3)_2CHOH$
460

▶1377 VS
B (0°C)
$(CH_3)_2CHCHCH(CH_3)_2$
CH_3
284, see also 135

▶1377 VVS
B (0.10)
$CH_2CH_2CH_2CH_3$
175

▶1377 M
B (0.0576)
F
F
F
F
222

▶1377 VVS
CCl_4 [25%] (0.10)
$CH_3(CH_2)_3OH$
353

▶1377 S
B (0.036)
CH_3
104, see also 633

▶1377 SSp
B (0.029)
$CH_2(CH_2)_2CH_3$
$CH_3(CH_2)_2H_2C$ $CH_2(CH_2)_2CH_3$
279

▶1377 S
B (0.068)
H_3CH_2C CH_2CH_3
C=C
H H
(cis)
275

▶1377 S
B (capillary)
CH_3
CH_3
187

▶1377 S
$C_2H_2Cl_4$
NH_2
$S{=}C{-}CH_3$
648, see also 54, 55

▶1377 S
$CH(C_2H_5)_2$
N
431

▶1377 S
B (0.0088)
CH_3
$(CH_3)_3CCHC(CH_3)_3$
307

▶1377 M
B
Cl
O
N
26

▶1377 W
B
$(C_6H_5)_2SiCl_2$
86

▶1377 VVS
B (0.064)
$H_2C{=}C(C_2H_5)_2$
163

▶1376 VS
A
$(CH_3)_3CCH_2CH_3$
94

▶1376 VS
B (0.04)
N CH_3
198, see also 438

▶1376 VVS
CCl_4 [25%] (0.10)
$CH_3CHCH_2CH_3$
OH
319

▶1376 VVS
B (0.064)
H_3C CH_3
C=C
H CH_2CH_3
(cis)
164, see also 274

▶1376 VVS
B (0.10)
CH_3
H_3C
184

▶1376 W
B (0.0576)
F, F, F (trifluorobenzene ring)
224

▶1376 VS
CCl_4 [25%] (0.10)
$CH_3(CH_2)_7OH$
317

▶1376 S
B (0.0088)
$(CH_3)_2CHCHCH(CH_3)_2$
CH_3
135

▶1376 VS
B
$(C_2H_5\text{-}NH\text{-}B\text{-}N\text{-}C_2H_5)_3$
[CH]
32

▶1376 VS
CCl_4 [25%] (0.10)
$CH_3CH_2C(CH_3)_2$
OH
194

▶1376 S
B
O
H_3C, O_2N, $N\text{-}C_5H_{11}\text{-}n$
26

▶1376 VS
CCl_4 [25%] (0.10)
$CH_3CH_2CHCH_2CH_3$
OH
196

▶1376 S
B
$(CH_3)_2CH(CH_2)_3CH_3$
331, see also 80, 158

▶1376 SSh
B (0.033)
$(CH_3)_3CCHCH(CH_3)_2$
CH_3
293

▶1376 M
B
O
H_3C, O_2N, $N\text{-}C_4H_9\text{-}n$
26

▶1376 M
M
$Cu(NH_2\text{-}CH_2\text{-}CH_2\text{-}NH_2)_2PtCl_4$
$[CH_2]$
8

▶1376 M
$CHCl_3$
O
H_3C, O, CH_3
$[CH_3]$
7

▶1376 W
CCl_4
$(C_6H_5)_3SiCl$
86

▶1376
P
$Hg(NO_3)_2$
45

▶1375 S
B (0.0088)
CH_3
$(CH_3)_2CHC\text{-}CH_2CH_2CH_3$
CH_3
134

▶1375 S
C
CH_3
CH_3
181

▶1375 S
G
S
$CH{=}N\text{-}NH\text{-}C\text{-}NH_2$
380

▶1375 M
B
O
H_5C_2, O_2N, $N\text{-}C_4H_9\text{-}n$
26

▶1375
E
$NaNO_3$
45

▶1375
E
$Sm(NO_3)_3$
45

▶1375
G

(phenyl-pyrazole ring structure) OH, N, N, CH_3

[C=C] 6

▶1375
G

H_3C, CH_3, CH_3, N, N, OH

[C=C] 6

▶1375
P

$RbNO_3$

45

▶1375
P

$Zn(NO_3)_2$

45

▶1374 MS
B (0.056)

$CH_2=CHCH=CHCH=CH_2$

203

▶1374 M
B (0.0088)

$CH_3CH=CHCH_2CH_3$
(cis)

208

▶1374 VS
B

$(CH_3)_2CHCH_2CHCH_2CH_3$
CH_3

334, see also 138

▶1374 VS
B (0.015)

$H_2C=CHCH_2C(CH_3)_3$

102, see also 273

▶1374 S
B (0.003)

$(CH_3)_2CHNO_2$

366

▶ 1374 VVS
CCl_4 [25%] (0.10)

$CH_3CH(CH_2)_2CH_3$
OH

318

▶1374 VS
B (0.065)

$H_2C=CC(CH_3)_3$
CH_3

272

▶ 1374 VSSh
B (0.0088)

CH_3
$(CH_3)_3CC - CH(CH_3)_2$
CH_3

306

▶ 1374 MS
B (0.0088)

$(CH_3)_2C=CHC(CH_3)_3$

204

▶1374 VS
C

$(CH_3)_3CNO_2$

325

▶1374 S
B

$(CH_3)_2CHCH_2CH(CH_3)_2$

335, see also 141, 150, 151

▶1374 S

N, C_3H_7-n

455

▶1374 MS
G

H O S H
N-C-C-N
C_6H_5 C_6H_5

[C=S] 31

▶ 1374 MSh
C

CH_3

191

▶1374 M
$CHCl_3$

S
H_3C, O, CH_3

7

▶1374

CH_3CH_3

[CH_3 symm. bend] 538, see also 56

▶1373 VVS
C

182

▶1373 S
B

H_5C_2, O_2N, N-C_6H_{13}-n

26

▶1373 VVS
B (0.20)

H_3C, CH_3
(trans)

308

▶1373 M
M

Pt $\left(NH_2\text{-}CH_2\text{-}CH_2\text{-}NH_2\right)_2 Cl_2$

[CH_2] 8

▶1373 M
M

Pd $\left(NH_2\text{-}CH_2\text{-}CH_2\text{-}NH_2\right)$ $PtCl_4$

[CH_2] 8

▶1372 VSSh
B (0.015)

$H_2C{=}C(CH_3)CH_2C(CH_3)_3$

96

▶1372 VVS
B (0.08)

N

197

▶1372 VS
B (0.0088)

$(CH_3)_3CC(CH_3)_2\text{-}CH_2CH_2CH_3$

304

▶1372 S
B

$CH_3CH_2CH(CH_3)(CH_2)_2CH_3$

330, see also 81, 157

▶1372 VS
B (0.055)

H_3C, CH_2CH_3, CH_3, N, H

343

▶1372 S
B (0.036)

$H_2C{=}CH(CH_2)_3CH(CH_3)_2$

97

▶1372 VSSh
B (0.10)

CH_2CH_3

190

▶1372 M
CCl_4 [20 g/1] (0.51)

$CH_3CH{=}CH(CH_2)_2CH{=}CH(CH_2)_2\text{-}CO\text{-}NH\text{-}CH_2CH(CH_3)_2$

357

▶1372 M
M

Pd $\left(NH_2\text{-}CH_2\text{-}CH_2\text{-}NH_2\right)_2 Cl_2$

[CH_2] 8

▶1372
P

$Mg(NO_3)_2$

45

▶1371 VS
B (capillary)

CH_3, CH_3

188

▶1371 S
B (0.005)

$(CH_3)_2CHCH(CH_3)_2$

210, see also 144

▶1371 S
G

$CH{=}N\text{-}NH\text{-}C({=}S)\text{-}NH_2$, N, H

380

▶1370 VS
A (16.6 mm Hg, 40 cm)

$(CH_3)_3CCH(CH_3)CH_2CH_3$

71, see also 110

▶1370 VS
B

$(CH_3)_3CCH(CH_3)CH_2CH_3$

340

▶1370 VS
B (0.10)
$CH_2CH_2CH_2CH_2CH_3$
173

▶1370 VS
B
H_3C CH_3
CH_3
124

▶1370 S
B (0.018)
Grating
$(CH_3)_3CCH(CH_3)_2$
146, 147

▶1370 S
B (0.0088)
CH_3
$(CH_3)_2CHCHCH_2CH_2CH_3$
139, see also 338

▶1370 VS
B (0.0104)
$n\text{-}H_9C_4\text{-}O\text{-}CF_2CHFCl$
264

▶1370 S
B (0.0088)
CH_3
$(CH_3)_2CHCHCH_2CH_3$
142, 153, see also 152

▶1370 S
B (0.0088)
CH_3
$(CH_3)_2CHC - CH_2CH_2CH_3$
CH_3
134

▶1370 VS
B
$(CH_3)_2C=C(CH_3)_2$
459

▶1370 SSh
B (0.10)
$CH_2CH_2CH_2CH_3$
176

▶1370
E
$CsNO_3$
45

▶1370
E
KNO_3
45

▶1370
E
$RbNO_3$
45

▶1370
G
N N OH
H_3C
[C=C]
6

▶1370
P
$LiNO_3$
45

▶1370
$(CH_3)_4C$
655, see also 57

▶1369 VS
B (0.014)
Grating
$(CH_3)_2CHCH_2CH(CH_3)_2$
150, see also 141, 151, 335

▶1369 S
B
O
H_3C
N-H
H_3C CH_3
[CH]
26

▶1369 S
B (0.026)
$(CH_3)_2CH(CH_2)_3CH_3$
80, see also 158, 331

▶1369 M
M
$Pd\left(\begin{matrix} NH_2 \\ CH_2 \\ CH_2 \\ NH_2 \end{matrix}\right)Cl_2$
8

▶1368 VS
B (0.033)
$C(CH_3)_3$
133

▶1368 VS
B (0.033)

$(CH_3)_3CCHCH(CH_3)_2$
CH_3

293

▶1368 VS
B (0.033)

$(CH_3)_3C(CH_2)_2CH(CH_3)_2$

292

▶1368 S
B (0.0088)

$(CH_3)_2CHCH_2CH_2CH(CH_3)_2$

137, 339

▶1368 VS
B (0.008)

$H_2C{=}CHCH_2C(CH_3)_3$

273, see also 102

▶1368 S
B (0.0088)

$(CH_3)_2CHCHCH(CH_3)_2$
CH_3

135

▶1368 VS
CCl_4 [25%] (0.10)

$(CH_3)_2CHCH_2CH_2OH$

195

▶1368 S
B (0.0088)

$(CH_3)_2CHCHCH(CH_3)_2$
CH_2
CH_3

301

▶1368 S
B
Grating

$(CH_3)_2CH(CH_2)_3CH_3$

158, see also 80, 331

▶1368 S
B (0.0088)

$(CH_3)_2CHCH_2CH(CH_3)_2$

141, 151, see also 150, 335

▶1368 S

(pyridine ring, N) $CH(C_4H_9\text{-}i)_2$

455

▶1368 S
B (0.0088)

$(CH_3)_2CHCH_2CHCH_2CH_3$
CH_3

138, see also 334

▶1368 VS
B (0.0088)

CH_3
$(CH_3)_3CCHC(CH_3)_3$

307

▶1368 VSSh
B (0.0088)

CH_3
$CH_3CH_2C - CHCH_2CH_3$
CH_3CH_3

299

▶1368 M
B (0.010)
Grating

CH_3
$(CH_3)_2CHCHCH_2CH_3$

152, see also 142, 153

▶1368 VVS
B (0.0088)

$(CH_3)_3CCH_2CH_2C(CH_3)_3$

305

▶1368 SSp
B (0.0153)

H_3C CH_3 (cyclohexane ring) CH_2CH_3

215

▶1368 VS
B (0.0104)

$n\text{-}H_7C_3\text{-}O\text{-}CF_2CHFCl$

263

▶1367 MS
G

H S S H
N-C-C-N
HO_2CH_2C CH_2CO_2H

[C=S?]

31

▶1368 S
B (0.0088)

$(CH_3)_2CHCH(CH_2CH_3)_2$

136

▶1367 MW
$CHCl_3$

(isoxazole ring: O, N)

15

▶1366 VS
B (0.033)
$(CH_3)_3CCH_2CH(CH_3)CH_2CH_3$
79

▶1366 VS
B (0.033)
$(CH_3)_3CCH(CH_2CH_3)_2$
78, 300

▶1366 S
B (0.0088)
$(CH_3)_3CCH_2CH_3$
145

▶1366 VS
B (0.051)
$CH(CH_3)_2$ / $CH(CH_3)_2$ (benzene ring)
296

▶1366 S
$C_2H_2Cl_4$
$S{=}C(NH_2){-}CH_3$
651, see also 54, 55

▶1366 M
C
H_3C (naphthalene) CH_3
179

▶1366 M
M
$Pt(NH_2CH_2CH_2NH_2)Cl_2$
[CH_2]
8

▶1366
$H_2C{=}C(CH_3){-}CH_2NO_2$
[NO_2 symm. stretch]
600

▶1365 VSSp
B (0.029)
$CH(CH_3)_2$, $(CH_3)_2HC$, $CH(CH_3)_2$ (benzene ring)
278

▶1365 VS
B (0.030)
$(CH_3)_3CCH_2CH_2CH_3$
155, see also 154

▶1365 S
B (capillary)
H_3C, CH_3 (naphthalene)
183

▶1365 VSSp
B (0.033)
CH_3 / $CH(CH_3)_2$ (benzene ring)
132

▶1365 S
B (0.016)
Grating
$CH_3CH_2C(CH_3)_2{-}CH_2CH_3$
148, see also 149, 336

▶1365 S
B
H_3C, O_2N (morpholine ring, O) $N{-}C_3H_7{-}i$
26

▶1365
E
$LiNO_3$
45

▶1365
E
$Mg(NO_3)_2$
45

▶1365
G
H_3C (pyrazole ring, N, N) OH; H_3C (benzene ring)
[C=C]
6

▶1365
P
$Ca(NO_3)_2$
45

▶1364 VS
B (0.0104)
$CH_3CH_2{-}O{-}CF_2CHFCl$
261

▶1364 VS
B (0.015)
$H_2C{=}C(CH_3)CH_2C(CH_3)_3$
96

▶1364 VS
B (0.018)
Grating
$(CH_3)_3CCH_2CH_2CH_3$
154, see also 155

▶1364 VS
B (0.10)
$CH_2CH_2CH_3$
178

▶1364 S
C_4H_9-i
455

▶1362 S
B (0.029)
$CH(CH_3)_2$, H_3C, CH_3
277

▶1362 MB
A (200 mm Hg, 10 cm)
CF_2Cl-CF_2Cl
257

▶1362 SSp
B (0.0088)
$(CH_3)_2C=CHC(CH_3)_3$
204

▶1362 S
B (?) [9%] (0.003)
$CH_3CH_2CH(NO_2)CH_3$
324

▶1362 SSh
B (0.015)
$H_2C=CHC(CH_3)_3$
129

▶1362 S
B
H_5C_2, O_2N, O, $N-C_3H_7$-i
26

▶ 1362 S
B (0.10)
$CH_2CH_2CH_3$
177

▶1362 VS
B (0.051)
$CH(CH_3)_2$, $CH(CH_3)_2$
297

▶1362 MS
B (0.056)
$CH_2=CHCH=CHCH=CH_2$
203

▶1362 M
G
H S S H / N-C-C-N / C_6H_{11} C_6H_{11}
[C=S?]
31

▶1361.3 S
B (0.03)
$CH_2-CH=CH$
9

▶1361 SSh
B (0.013)
$CH_3CH=CHCH(CH_3)_2$
130

▶1361 VS
B (0.10)
CH_2CH_3
189

▶1361 SSp
CS_2 [sat.] (0.025)
Cl, S, $COCH_3$
348

▶1361 S
B
O, H_5C_2, O_2N, $N-CH_3$
26

▶1361 S
B (0.051)
$CH(CH_3)_2$, $CH(CH_3)_2$
298

▶1361 MS
CCl_4 [31%] (0.008)
Br, S, $COCH_3$
372

▶1360 W
B (0.136) 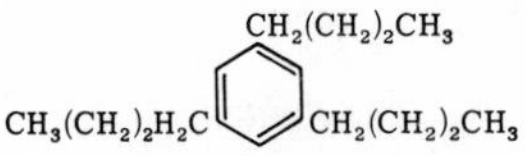
$CH_2(CH_2)_2CH_3$
$CH_3(CH_2)_2H_2C$ $CH_2(CH_2)_2CH_3$
279

▶1360 M
B 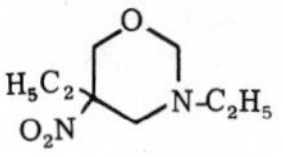
O
H_5C_2 N-C_2H_5
O_2N
26

▶1360
P
$Ba(NO_3)_2$
45

▶1360
P
$Cu(NO_3)_2 \cdot 5H_2O$
45

▶1360
$CHCl_3$ 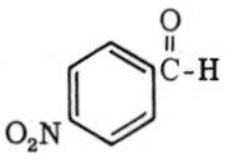
O
C-H
O_2N
[NO_2]
387

▶1359 VS
CCl_4 [25%] (0.10)
$CH_3(CH_2)_5OH$
320

▶1359 S
B 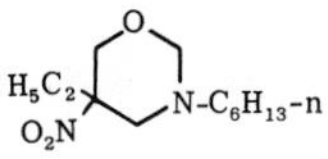
O
H_5C_2 N-C_6H_{13}-n
O_2N
26

▶1359 M
B 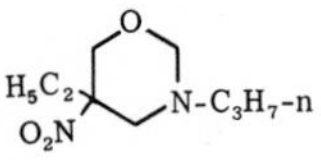
O
H_5C_2 N-C_3H_7-n
O_2N
26

▶1359 M
B 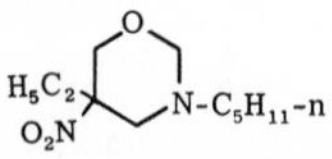
O
H_5C_2 N-C_5H_{11}-n
O_2N
26

▶1358 S
CCl_4 [25%] (0.10)
CH_3
$H_2C{=}CCO_2CH_3$
193

▶1358 S
CCl_4 [sat.] 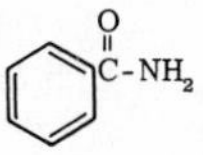
O
C-NH_2
29

▶1357 VSSp
B (0.003)
$(CH_3)_2CHNO_2$
366

▶1357 VS
CCl_4 [0.138 g/l] 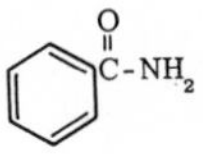
O
C-NH_2
29

▶1357 S
B 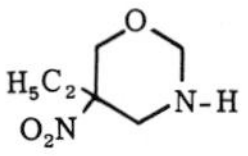
O
H_5C_2 N-H
O_2N
26

▶1357
P
$Co(NO_3)_2$
45

▶1357 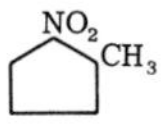
NO_2
CH_3
[NO_2 asymm. stretch]
600

▶1357 MS
B (film)
$(C_4F_9)_3N$
346

▶1355 S
B (0.0104)
$(CF_3)_2CFCF_2CF_3$
240

▶1355 VS
B (0.10) 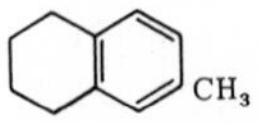
CH_3
269

▶1355 S
B 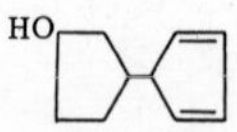
HO
379

▶1355 MWSh
B (0.066)

$H_2C{=}CH(CH_2)_6CH_3$

271

▶1355 MSh
B (0.068)

$H_3C\ \ CH_3$
$C{=}C$
$H\ \ CH_2CH_3$
(cis)

274

▶1355 M
C_6H_6 [0.0042 m/l]

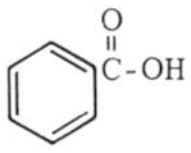

448, see also 38

▶1355 WSh
B

$CH_3CH_2CH(CH_2)_2CH_3$
CH_3

330

▶1355 WSh
B (0.033)

CH_3
$(CH_3)_3CCH_2CHCH_2CH_3$

79

▶1355
E

$NaNO_3$

45

▶1355
P

$Y(NO_3)_3$

45

▶1355

CH_3Cl

[CH_3 symm. bend] 539

▶1354 SSh
B (0.10)

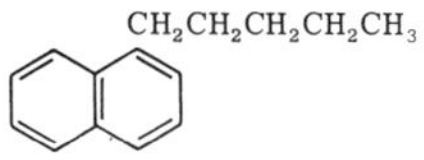

174

▶1353 VS
B (0.028)

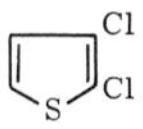

162

▶1353 SSh
B (0.10)

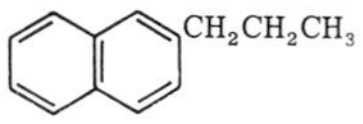

177

▶1353 S
B (?) [9%] (0.003)

NO_2
$CH_3CH_2CHCH_3$

324

▶1352 M
CCl_4 [0.0082 m/l]

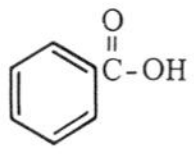

448, see also 38

▶1352
E

$AgNO_3$

45

▶1352
P

$Sr(NO_3)_2$

45

▶1351 VVS
A (25 mm Hg, 10 cm)

$CF_3CF_2CF_3$

235

▶1351 VS
E

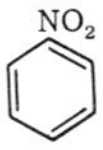

[NO_2 symm. stretch] 20

▶1351 S
C

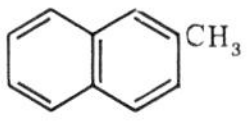

191

▶1351

CH_3N_3

[N=N=N symm. stretch] 650, see also 58, 59

▶1350 SSh
B (0.0104)

$F_9C_4\text{-}O\text{-}C_4F_9$

266

▶1350 S B 385	▶1349 S B [-CH_2-] 26
▶1350 S B H_3C, O_2N, N-C_3H_7-n 26	▶1349 S B H_3C, O_2N, N-C_6H_{13}-n 26
▶1350 S E, F F_3C, F_3C, H, N, N 37	▶1349 VVS B (0.20) H_3C, CH_3 (trans) 308
▶1350 MSh A (15.9 mm Hg, 40 cm) $CH_3CH_2CH(CH_3)(CH_2)_3CH_3$ 106, see also 119	▶1349 M E, F H, N, N, CH_2OH 37
▶1350 M B (0.036) CH_3 104, see also 633	▶1348 VVS B (0.028) Cl, Cl, S 161
▶1350 M E CH_2OH, P, O 34	▶1348 VS C $(CH_3)_3CNO_2$ 325
▶1350 OH 632, see also 36	▶1348 S B (0.068) H_3CH_2C H C=C H CH_2CH_3 (trans) 168
▶1350 O_2N, CH=CH-C(=O)-O-C_2H_5 [NO_2] 652	▶1348 S B F [-CH_2-] 26
▶1350 CH_3, NO_2 [NO_2 symm. stretch] 620	▶1348 S B H_3C, O_2N, N-C_4H_9-n [NO_2 symm. stretch] 26
▶1349 W B (0.097) $(CH_3)_2CH(CH_2)_3CH_3$ 80, see also 158, 331	▶1348 S B H_3C, O_2N, N-C_5H_{11}-n [NO_2 symm. stretch] 26

▶1348 S
E, F
H N CF_3 N O_2N
37

▶1346 MS
B (0.003)
$C_3H_7NO_2$
392

▶1346 VS
B
O_2N O N
[-CH_2-]
26

▶1346 SSh
B (0.151)
H_3C CH_3
CH_2CH_3
215

▶1346 S
E, F
H N C_3H_7 N
37

▶1346 SSh
B (0.10)
CH_2CH_3
190

▶1346 SSh
B (0.064)
H CH_2CH_3
C=C
H_3C CH_3
(trans)
165

▶1346 W
A (12.5 mm Hg, 40 cm)
$CH_3(CH_2)_6CH_3$
89

▶1346
P
$La(NO_3)_3$
45

▶1345 VS
B
O N Cl
[-CH_2-]
26

▶1345 VS
B
Cl O N
[-CH_2-]
26

▶1345 S
E, F
H N CF_3 N F_3C CF_3
37

▶1345 MS
E, F
H N C_2F_5 N
37

▶1345
E
$Pb(NO_3)_2$
45

▶1344 VS
B (0.003)
$(CH_3)_2CHCH_2NO_2$
326

▶1344 MSh
B (0.0563)
$(CH_3CH_2)_2CHCH_2CH_2CH_3$
140

▶1344 MS
B (0.036)
$CH_3CH_2CH{=}CH(CH_2)_3CH_3$
99

▶1344 S
B (film)
$CF_3(CF_2)_5CF_3$
241

▶1344 S
E, F
H N N O_2N
37

▶1344 MS
B (0.10)
CH_3
192, see also 42

▶1344 VSSh
B (0.10)
CH_3
H_3C
184

▶1344 M E, F (5-methoxy-2-trifluoromethylbenzimidazole structure; H_3CO, CF_3) 37	▶1342 S B $(C_2H_5\text{-NH-B-N-}C_2H_5)_3$ 32
▶1344 MW B (0.065) (2-thienyl)-$CH=CH_2$ 212	▶1342 S C_6H_6 [0.0044 m/l] (salicylic acid structure; OH, C(=O)-OH) 448
▶1343 VS B (0.239) $CH_3(CH_2)_2CH=CH(CH_2)_2CH_3$ (cis) 422	▶1342 M A (18 mm Hg, 40 cm) $(CH_3)_2CH(CH_2)_4CH_3$ 294, see also 120
▶1343 S B (2-(p-bromophenyl)dihydro-oxazine structure; O, N, Br) [$-CH_2-$] 26	▶1341 VSB B (0.300) $C_{26}H_{54}$ (5, 14-di-n-butyloctadecane) 283
▶1343 M B (morpholine structure; H_3C, O, N-H, H_3C, CH_3) [CH] 26	▶1341 WSh B (0.028) $CH_3(CH_2)_5CH(C_3H_7)(CH_2)_5CH_3$ 281
▶1343 M B (oxazine structure; H_5C_2, O_2N, O, $N\text{-}CH_3$) [NO_2 symm. stretch?] 26	▶1341 S B (0.10) (naphthalene)-$CH_2CH_2CH_3$ 178
▶1342 VS B (0.0563) $CH_3(CH_2)_2CH=CH(CH_2)_2CH_3$ (trans) 205	▶1341 S B (oxazine structure; H_5C_2, O_2N, O, N-H) [NO_2 symm. stretch?] 26
▶1342 VS B (0.157) $(CH_3)_3C(CH_2)_2CH(CH_3)_2$ 292	▶1341 M B (oxazine structure; H_5C_2, O_2N, O, $N\text{-}C_4H_9\text{-n}$) [NO_2 symm. stretch?] 26
▶1342 VS B (cyclopentene structure; CH_3, H_3C, CH_3) 127	▶1340 VVS B (0.028) (dichlorothiophene structure; Cl, S, Cl) 211
▶1342 VS B $[(CH_3)_2N\text{-}BO]_3$ [Ring] 32	▶1340 VSVB CCl_4 [25%] (0.10) $CH_3CH_2CH_2OH$ 314

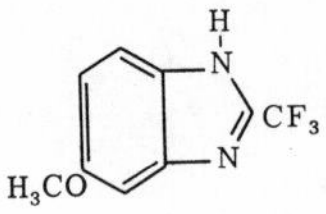

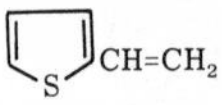

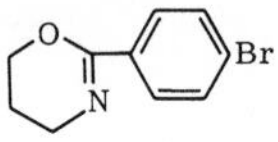

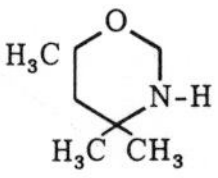

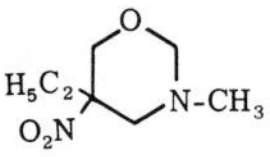

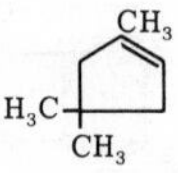

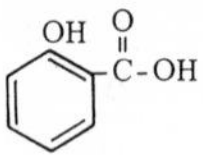

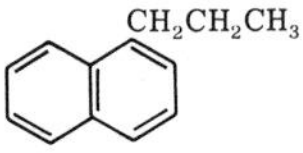

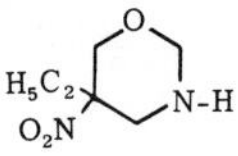

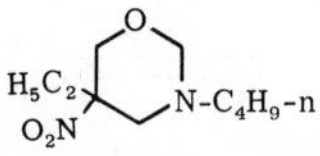

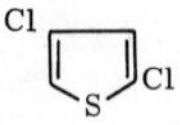

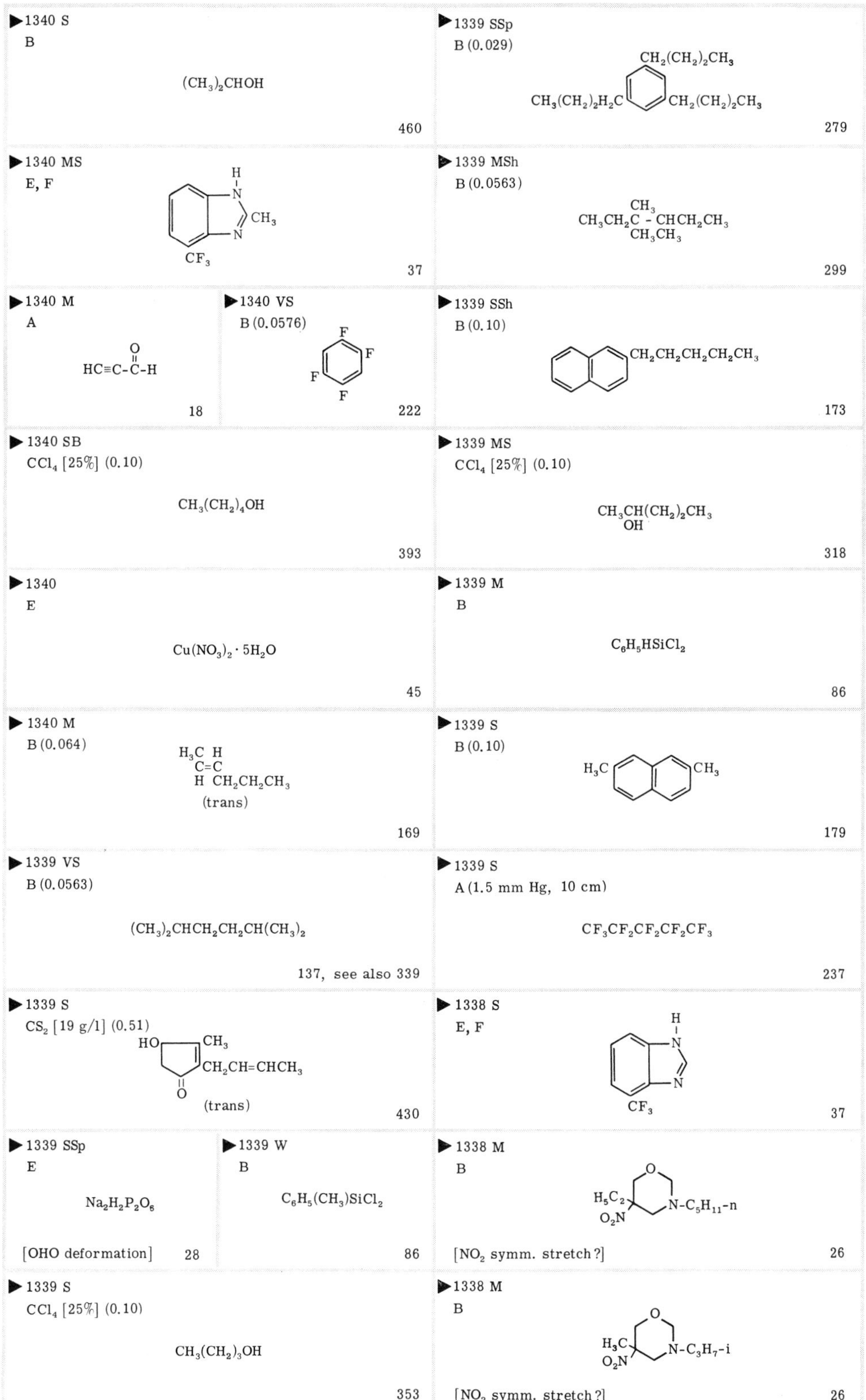

▶1340 S
B
$(CH_3)_2CHOH$
460

▶1339 SSp
B (0.029)
$CH_2(CH_2)_2CH_3$
$CH_3(CH_2)_2H_2C$ $CH_2(CH_2)_2CH_3$
279

▶1340 MS
E, F
H N CH_3 N CF_3
37

▶1339 MSh
B (0.0563)
CH_3
$CH_3CH_2C - CHCH_2CH_3$
CH_3CH_3
299

▶1340 M
A
HC≡C-C(=O)-H
18

▶1340 VS
B (0.0576)
F F F F
222

▶1339 SSh
B (0.10)
$CH_2CH_2CH_2CH_2CH_3$
173

▶1340 SB
CCl_4 [25%] (0.10)
$CH_3(CH_2)_4OH$
393

▶1339 MS
CCl_4 [25%] (0.10)
$CH_3CH(CH_2)_2CH_3$
OH
318

▶1340
E
$Cu(NO_3)_2 \cdot 5H_2O$
45

▶1339 M
B
$C_6H_5HSiCl_2$
86

▶1340 M
B (0.064)
H_3C H
C=C
H $CH_2CH_2CH_3$
(trans)
169

▶1339 S
B (0.10)
H_3C CH_3
179

▶1339 VS
B (0.0563)
$(CH_3)_2CHCH_2CH_2CH(CH_3)_2$
137, see also 339

▶1339 S
A (1.5 mm Hg, 10 cm)
$CF_3CF_2CF_2CF_2CF_3$
237

▶1339 S
CS_2 [19 g/l] (0.51)
HO CH_3
$CH_2CH=CHCH_3$
O
(trans)
430

▶1338 S
E, F
H N N CF_3
37

▶1339 SSp
E
$Na_2H_2P_2O_6$
[OHO deformation] 28

▶1339 W
B
$C_6H_5(CH_3)SiCl_2$
86

▶1338 M
B
O
H_5C_2 O_2N N-C_5H_{11}-n
[NO_2 symm. stretch?] 26

▶1339 S
CCl_4 [25%] (0.10)
$CH_3(CH_2)_3OH$
353

▶1338 M
B
O
H_3C O_2N N-C_3H_7-i
[NO_2 symm. stretch?] 26

▶1338 M
C 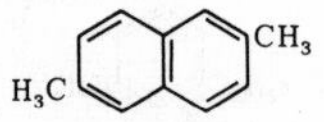
H_3C CH_3
180

▶1337 S
B (0.169)
$(CH_3CH_2)_3CH$
143, see also 156

▶1337 S
B 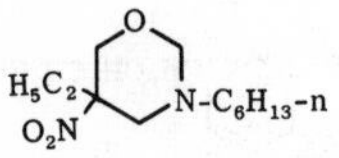
H_5C_2 O_2N O N-C_6H_{13}-n
[NO_2 symm. stretch ?]
26

▶1337 S
E, F 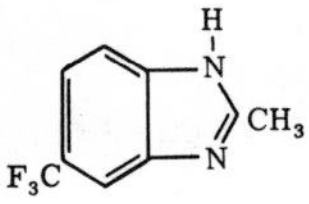
H N CH_3 N F_3C
37

▶1337 M
B
$C_6H_5SiCl_3$
86

▶1337 W
B
$(C_6H_5)_2(CH_3)SiCl$
86

▶1336 S
E, F 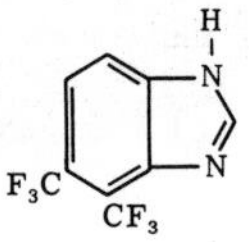
H N N F_3C CF_3
37

▶1336 MSh
B (0.10) 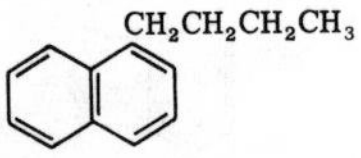
$CH_2CH_2CH_2CH_3$
176

▶1335 M 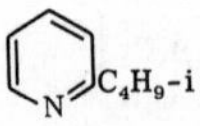
N C_4H_9-i
431

▶1335 M
B (capillary) 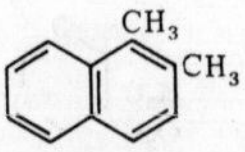
CH_3 CH_3
188

▶1335 S
C 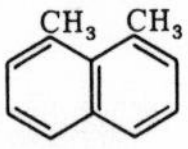
CH_3 CH_3
182

▶1335 MS 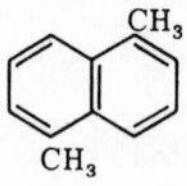
CH_3 CH_3
185

▶1335 S
E, F 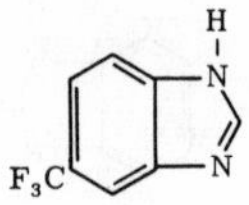
H N N F_3C
37

▶1335 MS
E 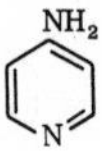
NH_2 N
327

▶1335 M 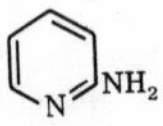
N NH_2
327

▶1335 MS
B (0.10) 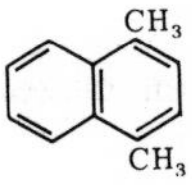
CH_3 CH_3
186

▶1335 M
$CHCl_3$
O H_3C O CH_3 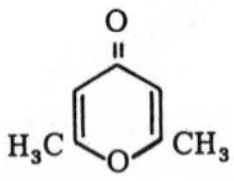
7

▶1335 M
B
$(C_6H_5)_2SiCl_2$
86

▶1335 MW
$CHCl_3$
S H_3C S CH_3 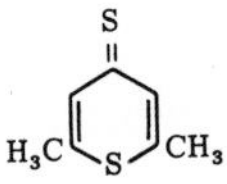
[Ring]
7

▶1335 M
B (0.013)
$CH_3CH{=}CHCH(CH_3)_2$
130

▶1335 VSVB
CCl_4 [20%] (0.10)
CH_3CH_2OH
315

▶1334 M
B (0.029) 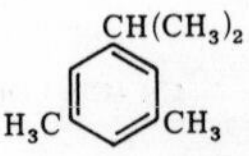
$CH(CH_3)_2$ H_3C CH_3
277

▶1334 S
B

H_3C CH_3

183

▶1334 S
B

H_3C CH_3 CH_3

125

▶1334 S
E, F

H N CH_2OH N CF_3

37

▶1333 S
E, F

H N CH_3 N F_3C CF_3

37

▶1334 M
B (0.10)

$(CH_3CH_2)_3CH$

156, see also 143

▶1333 VS
B (film)

CF_3 F

218

▶1332.5 M
B (0.03)

CH_2-CH =CH

9

▶1332 S
E, F

H N CH_2OH N F_3C

37

▶1332 S
M

$Ni(NH_2CH_2CH_2NH_2)_3PtCl_4$

[NH_2]

8

▶1332 MS
E, F

H_3C H_3C H N CF_3 N

37

▶1332 M
C_6H_6 [0.0128 m/l]

Cl O C-OH

448

▶1332 W
CS_2

$(C_6H_5)_3SiCl$

86

▶1332 MS
B (0.0563)

$(CH_3)_2CHCH(CH_2CH_3)_2$

136

▶1332 S

$N(CH_3)_2$ NO_2

[NO_2 asymm. stretch]

620

▶1330 VS
B (0.157)

$(CH_3)_3C(CH_2)_2CH(CH_3)_2$

292

▶1330 S
B (0.0563)

$(CH_3)_3CCH_2CHCH(CH_3)_2$ (with CH_3 on CH)

303

▶1330 S
B (0.169)

$(CH_3)_2CHCH(CH_3)CH_2CH_3$

142, see also 152, 153

▶1330 S
E, F

H N CF_3 N CF_3

37

▶1330 MS
B (0.051)

$CH(CH_3)_2$ $CH(CH_3)_2$

298

▶1330 M
C

CH_3 CH_3

181

▶1330 M
E

CH_2OH P

34

▶1330 M E, F 2-ethylbenzimidazole (structure) 37	▶1328 S B (0.0563) $(CH_3)_3CC(CH_3)_2-CH(CH_3)_2$ 306
▶1330 M C_6H_6 [0.0044 m/l] salicylic acid (structure: OH, C(=O)-OH) 448	▶1328 M B (0.014) $(CH_3)_2CHCH_2CH(CH_3)_2$ 151, see also 141, 150
▶1330 E $Be(NO_3)_2$ 45	▶1328 S E, F H_3C, CH_3, CF_3 benzimidazole (structure) 37
▶1330 K $Ca(NO_3)_2$ 45	▶1328 S CCl_4 [25%] (0.10) $CH_3CH_2C(CH_3)_2OH$ 194
▶1330 $Gd(NO_3)_3$ 45	▶1328 P $La(NO_3)_3$ 45
▶1329 MS E, F 2-CF_3 benzimidazole (structure) 37	▶1327 SSh B (0.10) $CH_2CH_2CH_3$ naphthalene (structure) 178
▶1329 MS E, F F_3C, CF_3 benzimidazole (structure) 37	▶1327 W B (0.055) H_3C, CH_2CH_3, CH_3 pyrrole, N-H (structure) 343
▶1329 M B (0.101) $(CH_3)_2CHCH(CH_3)CH_2CH_3$ 153, see also 142, 152	▶1326 VS B (0.2398) F benzene (structure) 226
▶1328 MB B (0.029) $CH_3CHCH_2CH_3$, $CH_3CH_2HC(CH_3)$, $CHCH_2CH_3(CH_3)$ benzene (structure) 280	▶1326 S B (0.064) H_5C_2, O_2N, N-C_4H_9-n oxazine (structure) 26
▶1328 VS B (0.014) $(CH_3)_2CHCH_2CH(CH_3)_2$ 141, 151, see also 150, 335	▶1326 S B $H_3CH_2C(H)C=C(H)CH_2CH_3$ (trans) 168

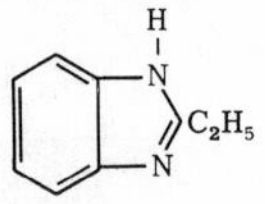

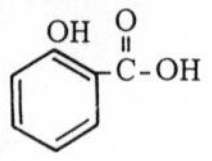

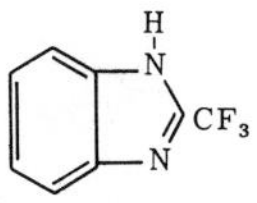

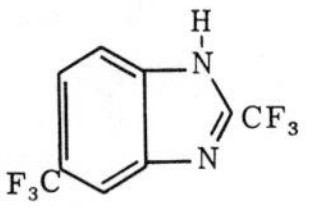

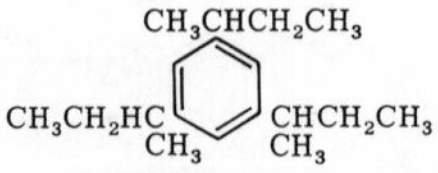

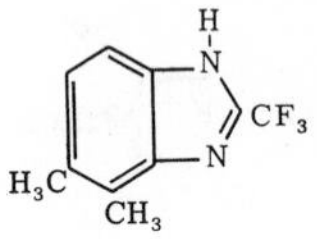

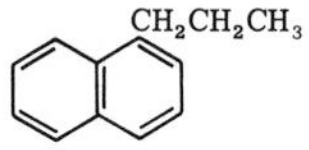

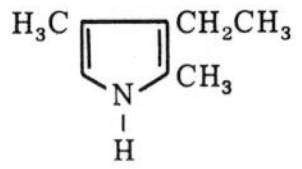

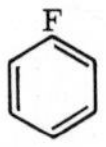

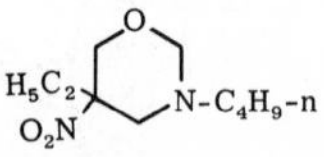

▶1326 SSh
B (0.028)

2,5-dichlorothiophene structure: Cl, Cl, S

161

▶1326 SSh
B (0.064)

H CH$_2$CH$_3$
C=C
H$_3$C CH$_3$
(trans)

165

▶1326 M
E

NH$_2$ (2-aminopyridine)

327

▶1326 M
M

Pt (NH$_2$–CH$_2$–CH$_2$–NH$_2$)$_2$ Cl$_2$

[NH$_2$ wag]

8

▶1325 VS
CCl$_4$ [31%] (0.025)

Cl–(thiophene)–COCH$_3$

372

▶1325 S
B

H$_5$C$_2$, O$_2$N, O, N-C$_6$H$_{13}$-n

26

▶1325 S
B

H$_5$C$_2$, O$_2$N, O, N-C$_5$H$_{11}$-n

26

▶1325 S
E, F

H, N, CF$_3$, N, H$_3$C

37

▶1325 S
E, F

H, N, CF$_3$, N, CH$_3$

37

▶1325 S
E, F

H, N, CF$_3$, N, Cl

37

▶1325 MS
E, F

H, N, C$_3$F$_7$, N, CH$_3$

37

▶1325 M
B (0.0563)

CH$_3$ CH$_3$
CH$_3$CH$_2$CH-CHCH$_2$CH$_3$

83

▶1325 M
E, F

H, N, CH$_2$CN, N

37

▶1325 M
B (0.015)

H$_2$C=CCH$_2$C(CH$_3$)$_3$
CH$_3$

96

▶1325
G

H$_3$C, N, N, OH

[C=C]

6

▶1325
P

Ce(NO$_3$)$_3$

45

▶1324 VVS
CCl$_4$ [25%] (0.10)

CH$_3$
H$_2$C=CCO$_2$CH$_3$

193

▶1324 S
B

H$_5$C$_2$, O$_2$N, O, N-H

26

▶1324 M
B

H$_5$C$_2$, O$_2$N, O, N-C$_3$H$_7$-i

26

▶1324 MS
B (0.10)

CH$_2$CH$_2$CH$_3$ (naphthalene)

177

▶1324 M
E, F

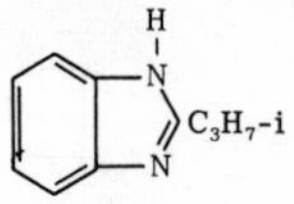

H
N
C_3H_7-i
N

37

▶1324 M
M

Pd $\left(NH_2 - CH_2 - CH_2 - NH_2 \right)_2$ Cl_2

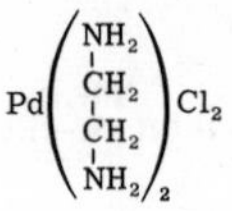

[NH_2] 8

▶1323 M
B (0.0563)

$(CH_3CH_2)_2CHCH_2CH_2CH_3$

140

▶1323 M
B (0.06)

128

▶1323

H_3C-S-CH_3

[CH_3 symm. bend] 539, see also 60

▶1322 S
E, F

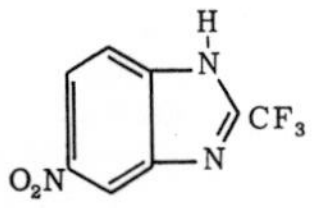

H
N
CF_3
O_2N
N

37

▶1322

$H_2C(C{\equiv}N)_2$

466

▶1321 MSh
B (film)

F_9C_4-O-C_4F_9

266

▶1321 S
B (0.0563)

$(CH_3)_2CHCHCH(CH_3)_2$
CH_3

135

▶1321 VS
B (0.0563)

$(CH_3)_2CHCHCH(CH_3)_2$
CH_2
CH_3

301

▶1321 M
B (0.0576)

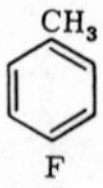

CH_3
F

220

▶1321 M
M

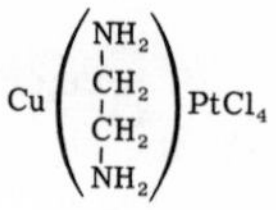

Cu $\left(NH_2 - CH_2 - CH_2 - NH_2 \right)$ $PtCl_4$

[NH_2] 8

▶1321 M
M

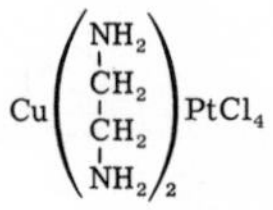

Cu $\left(NH_2 - CH_2 - CH_2 - NH_2 \right)_2$ $PtCl_4$

[NH_2] 8

▶1320 S
E, F

H
N
N
O_2N

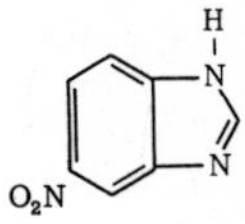

37

▶1320 M
B (0.030)

$(CH_3)_3CCH(CH_3)_2$

147, see also 146

▶1320 M
CCl_4 [0.0082 m/1]

O
C-OH

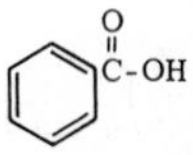

448, see also 38

▶1320
E

$Sm(NO_3)_3$

45

▶1320
G

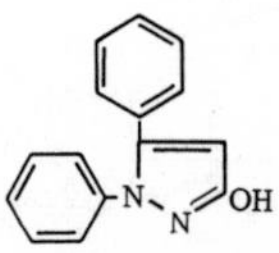

N
N
OH

[C=C] 6

▶1320
K

$Mn(NO_3)_2 \cdot xH_2O$

45

▶1320
P

$Sm(NO_3)_3$

45

▶1320
P
$Y(NO_3)_3$
45

▶1319 S
B
H_3C, CH_3, CH_3
125

▶1319 S
B (0.169)
$(CH_3CH_2)_3CH$
143, see also 156

▶1319 M
B (0.0563)
$(CH_3)_3CC(CH_3)_2-CH_2CH_2CH_3$
304

▶1319 SSp
CS_2 [sat.] (0.025)
Br, S, $COCH_3$
348

▶1319 M
B (0.064)
$H_2C{=}CHCH(CH_3)CH_2CH_3$
167

▶1318 S
B (?) [9%] (0.003)
$CH_3CH_2CH(NO_2)CH_3$
324

▶1318 S
B (0°C)
$(CH_3)_2CHCH(CH_3)CH(CH_3)_2$
284, see also 135

▶1318 S
B (0.065)
S, CH_2CH_3
213

▶1318 M
CCl_4
N, NH_2
327

▶1318 VSVB
CCl_4 [20%] (0.10)
CH_3CH_2OH
315

▶1317 VS
$CHCl_3$
O, O
[Ring]
7

▶1317 M
B (0.101)
$(CH_3CH_2)_3CH$
156, see also 143

▶1316 S
B (0.029)
$CH(CH_3)_2$, $(CH_3)_2HC$, $CH(CH_3)_2$
278

▶1316 S
E
NO_2
[Ring]
20

▶1315 S
B
H_3C, CH_3, CH_3
124

▶1315 S
B (0.157)
$(CH_3)_3C(CH_2)_2CH(CH_3)_2$
292

▶1315 MVB
B (0.036)
$H_2C{=}C(CH_3)(CH_2)_4CH_3$
98

▶1315
E
$Ce(NO_3)_3$
45

▶1315 VSB
B (0.0104)
H_3C-O-CF_2CHFCl
356

▶1315
F_3C-NO_2
[NO_2 symm. stretch]
600

▶1314 VS CCl_4 [25%] (0.10) $CH_3CHCH_2CH_3$ (OH) 319	▶1312 M B (0.064) H_3CH_2C H C=C H CH_2CH_3 (trans) 168
▶1314 MW B (0.003) CH_3NO_2 311 / ▶1314 VS B (0.051) $CH(CH_3)_2$, $CH(CH_3)_2$ (benzene ring) 297	▶1312 S B (0.10) CH_2CH_3 (naphthalene) 190
▶1314 M B (0.0563) $(CH_3)_2CHC(CH_3)_2-CH_2CH_2CH_3$ 134	▶1311 MS B (0.0563) $CH_3CH_2C(CH_3)_2-CH_2CH(CH_3)CH_2CH_3$ 302
▶1314 S $C_2H_2Cl_4$ NH_2 / S=C-CH_3 651, see also 54, 55 / ▶1314 M P $[(CH_3)_2P\cdot BCl_2]_3$ 30	▶1311 SB B (0.056) CH_2=CHCH=CHCH=CH_2 203
▶1314 S CS_2 [19 g/l] (0.51) HO, CH_3, CH_2CH=$CHCH_3$, O (trans) 430	▶1311 M B (0.0563) $(CH_3CH_2)_2CHCH_2CH_2CH_3$ 140
▶1314 M $CHCl_3$ NH_2, N (pyridine) 327	▶1311 MS B (film) n-H_7C_3-O-CF_2CHFCl 263
▶1312.5 S B (0.03) CH_2-CH = CH (benzene ring) 9	▶1311 S B (0.169) $(CH_3)_2CHCH(CH_3)CH_2CH_2CH_3$ 139, see also 338
▶1312 VS B (0.0104) CH_3CH_2-O-CF_2CHFCl 261	▶1311 M M Pt (NH_2-CH_2-CH_2-NH_2)$_2$ Cl_2 [CH_2 twist] 8
▶1312 M B (0.0563) $(CH_3)_2CHCH(CH_2CH_3)_2$ 136	▶1311 Cl_3C-NO_2 [NO_2 asymm. stretch] 600
▶1312 MSSh B (0.0563) $(CH_3)_3CCH_2CH(CH_3)CH(CH_3)_2$ 303	▶1310 M B (0.157) $(CH_3)_3CCH(CH_3)CH(CH_3)_2$ 293

▶1310 M
P

$[(CH_3)_2P \cdot BH_2]_3$

30

▶1310
G

H_3C CH_3 N N OH

[C=C]

6

▶1310
G

CH_3 N N OH

[C=C]

6

▶1310
G

H_3C CH_3 N N OH

[C=C]

6

▶1310

OH

632, see also 36

▶1310

O

$CH=CH-\overset{\|}{C}-O-C_2H_5$

O_2N

[C-O-C]

652

▶1309 S
B (film)

O O O

342

▶1309 MS
A (36 mm Hg, 15 cm)

$H_2C=CHCH(CH_3)_2$

131

▶1309 S
B (0.169)

CH_3
$(CH_3)_2CHCHCH_2CH_3$

142, see also 152, 153

▶1309 M
B

CH_3 CH_3 CH_3

126

▶1308 S
A

CH_3
$CH_3CH_2CHCH_2CH_3$

332

▶1308 S
B (0.10)

CH_2CH_3

189

▶1308 MSSp
B (0.0153)

$CH_3(CH_2)_2CH=CH(CH_2)_2CH_3$
(cis)

422

▶1308 VS
B (0.20)

H_3C CH_3

(trans)

308

▶1308 MW
B (0.104)

$CH_3CH_2CH(CH_2)_2CH_3$
CH_3

81, see also 157

▶1308 VSB
B (0.0104)

$n-H_9C_4-O-CF_2CHFCl$

264

▶1308 VSB
B (film)

$(C_4F_9)_3N$

346

▶1307 VS
B (0.0563)

$CH_3CH=CHCH_2CH_3$
(cis)

208

▶1307 S
B (0.169)

$(CH_3)_3CCH(CH_2CH_3)_2$

300, see also 78

▶1307 M
B (0.0563)

$(CH_3)_3CCH_2CH_3$

145

▶1307 MB
B (0.036)

$CH_3CH=CH(CH_2)_4CH_3$
(cis)

100, see also 160

▶1307 W
A

$(CH_3)_3CCH_2CH_3$

94

▶1307 W
B

$C_6H_5HSiCl_2$

86

▶1306 S
B

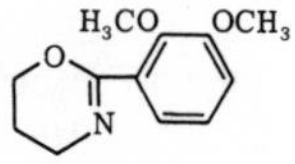

H_3CO OCH_3 O N

26

▶1306
P

$Pr(NO_3)_3$

45

▶1306 S
$CHCl_3$

S O

[Ring] 7

▶1305 S
B (0.003)

$(CH_3)_2CHNO_2$

366

▶1305 M
B (0.068)

H_3C CH_3
C=C
H CH_2CH_3
(cis)

164, 274

▶1305 M
B (0.003)

$CH_3(CH_2)_2CH_2NO_2$

310

▶1305 M
B (0.066)

$H_2C{=}CH(CH_2)_8CH_3$

352

▶1305 SB
CCl_4 [25%] (0.10)

$CH_3CH_2CHCH_2CH_3$
OH

196

▶1305 W
A (12.6 mm Hg, 40 cm)

$CH_3(CH_2)_6CH_3$

89

▶1305
P

$La(NO_3)_3$

45

▶1305

CH_3Br

[CH_3 symm. bend] 539, see also 61, 62, 63, 376

▶1305 VSSh
B (0.0576)

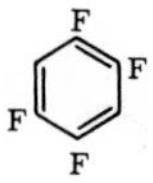

F F F F

222

▶1304 VSB
B (film)

H_9C_4-O-C_4F_9

266

▶1304 S
B (0.0157)

$(CH_3)_3CCH(CH_2CH_3)_2$

78, see also 300

▶1304 S
B (0.055)

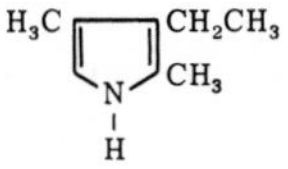

H_3C CH_2CH_3 CH_3 N H

343

▶1304 M
A

$(CH_3)_2SO$

[CH_3 symm. deformation] 13

▶1304 M
B

$C_6H_5SiCl_3$

86

▶1304 M
B

$(C_6H_5)_2SiCl_2$

86

▶1304 S
B (0.064)

$(CH_3)_2C{=}CHCH_2CH_3$

166

▶1304 MW
A (15.9 mm Hg, 40 cm)

$CH_3CH_2CH(CH_2)_3CH_3$
CH_3

106, see also 119

▶1303 M
P

$[(CH_3)_2P \cdot BCl_2]_3$

[CH_3 symm. deformation] 30

▶1303 VSB
B (0.0104)

$H_3C\text{-}O\text{-}CF_2CHFCl$

356

▶1302 VS
B

$(CH_3)_2CHOH$

460

▶1302 SSp
B (0.065)

CH_3
$CH(CH_3)_2$

132

▶1302 M
B (0.013)

$CH_3CH{=}CHCH(CH_3)_2$

130

▶1302 MSSh
B (film)

$CF_3(CF_2)_5CF_3$

240

▶1302 S
B (0.169)

CH_3
$CH_3CH_2C - CHCH_2CH_3$
CH_3 CH_3

299

▶1301 W
B (0.028)

$CH_3(CH_2)_5CH(C_3H_7)(CH_2)_5CH_3$

281

▶1300 W
A (1.5 mm Hg, 10 cm)

$CF_3CF_2CF_3$

235

▶1300 VSB
B (0.0104)

$CH_3CH_2\text{-}O\text{-}CF_2CHFCl$

261

▶1300 VS
B (0.0563)

$(CH_3)_3CCH_2CH_2C(CH_3)_3$

305

▶1300 S
B (0.0576)

CH_3
F

220

▶1300 SB
B (0.056)

$CH_2{=}CHCH{=}CHCH{=}CH_2$

203

▶1300 M
C

KH_2PO_4

28

▶1300
E

$Pr(NO_3)_3$

45

▶1300
E

$Y(NO_3)_3$

45

▶1299 VSSh
B (0.300)

$C_{26}H_{54}$
(5,14-di-n-butyloctadecane)

283

▶1299 VS
B (film)

F CF_3
F_2 F_2
F_2 F_2
F CF_3

230

▶1299 VS
B (0.068)

H_3CH_2C CH_2CH_3
C=C
H H
(cis)

275

▶1299 W B (0.0576) [structure: toluene ring with CH_3 and F substituents] 221	▶1299 M CCl_4 $[(CH_3)_2CNO_2]^- Na^+$ [NO_2 deformation + $C{=}NO_2$ wag] 328
▶1299 VS C_6H_6 HNO_3 [NO_2 symm. stretch] 642	▶1299 W B (0.036) $H_2C{=}CH(CH_2)_5CH_3$ 290
▶1299 S B (0.051) [benzene ring with $CH(CH_3)_2$ and $CH(CH_3)_2$] 298	▶1298 VS A (100 mm Hg, 15 cm) CH_3CH_2Cl 88
▶1299 S G [benzene ring]$CH{=}N{-}NH{-}C({=}S){-}NH_2$ 380	▶1298 MS B (0.036) $CH_3CH_2CH{=}CH(CH_2)_3CH_3$ 99
▶1299 S G H_3C, CH_3-[benzene ring]$CH{=}N{-}NH{-}C({=}S){-}NH_2$ 380	▶1298 MS B (0.10) [naphthalene ring]$CH_2CH_2CH_2CH_2CH_3$ 173
▶1299 MS [pyridine ring (N)]$C(C_2H_5)_3$ 455	▶1298 M P $[(CH_3)_2P \cdot BCl_2]_3$ [CH_3 symm. bend] 30
▶1299 MS [pyridine ring (N)]C_4H_9-i 455	▶1297 VSB B (0.0104) n-H_9C_4-O-CF_2CHFCl 264
▶1299 MS B (0.0088) $(CH_3)_2CHCH_2CH(CH_3)CH_2CH_3$ 138, see also 334	▶1297 VS B (0.10) [tetrahydronaphthalene ring]CH_3 269
▶1299 MWSh B (0.0563) $(CH_3)_2CHC(CH_3)_2{-}CH_2CH_2CH_3$ 134	▶1297 S A $(CH_3)_2CHCH_2CH_3$ 333
▶1299 MB E, $CHCl_3$ [pyridine ring (N)]NH_2 327	▶1297 SB B (0.136) $CH_3(CH_2)_2H_2C$-[benzene ring]-$CH_2(CH_2)_2CH_3$, $CH_2(CH_2)_2CH_3$ 279

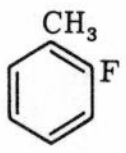

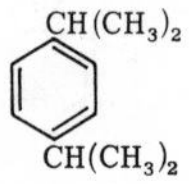

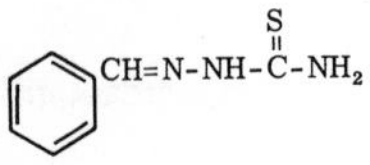

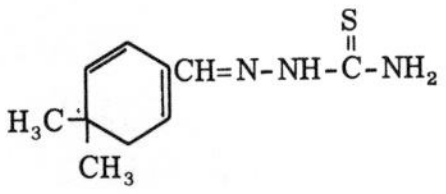

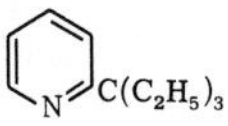

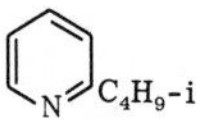

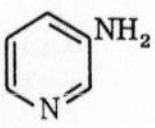

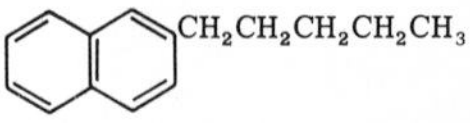

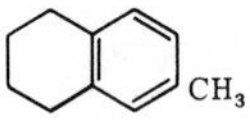

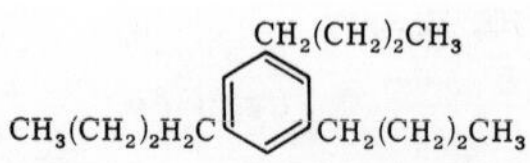

▶1297 VSB
CCl_4 [25%] (0.10)
$CH_3CH(CH_2)_2CH_3$
OH
318

▶1297 S
G
S
‖
CH=N-NH-C-NH_2
N
H
380

▶1297 M
B (0.0563)
CH_3
$(CH_3)_3CC$ - $CH_2CH_2CH_3$
CH_3
304

▶1297 M
B
$CH_3CH_2CH_2CHCH_2CH_3$
CH_3
330, see also 81, 157

▶1297 M
CCl_4 [0.00658 m/l]
OH O
‖
C-OH
448, see also 27

▶1297 M
CCl_4 [0.0128 m/l]
Cl O
‖
C-OH
448

▶1296 M
B (0.097)
$(CH_3)_2CH(CH_2)_3CH_3$
80, see also 158, 331

▶1296 M
M
Pt $\left(\begin{array}{c} NH_2 \\ CH_2 \\ CH_2 \\ NH_2 \end{array}\right)_2$ $PtCl_4$
[NH_2]
8, see also 43

▶1296 M
B (0.157)
$(CH_3)_3CCHCH(CH_3)_2$
CH_3
293

▶1295 S
A (6.2 mm Hg, 10 cm)
F_2 F_2
F_2 F_2
F_2
245

▶1295 VS
B (0.0104)
n-H_7C_3-O-CF_2CHFCl
263

▶1295 MS
B (0.003)
$C_3H_7NO_2$
392

▶1295 MS
B (0.0563)
CH_3 CH_3
CH_3CH_2C - $CH_2CHCH_2CH_3$
CH_3
302

▶1295 S
B (0°C)
$(CH_3)_2CHCHCH(CH_3)_2$
CH_3
284, see also 135

▶1295 S
B (0.003)
$(CH_3)_2CHCH_2NO_2$
326

▶1295 M
B (0.0563)
$(CH_3)_2CHCH(CH_2CH_3)_2$
136

▶1295 MS
N $CH(C_2H_5)_2$
455

▶1295 MW
B (0.104)
$CH_3CH_2CH(CH_2)_2CH_3$
CH_3
81, see also 157

▶1295 M
B (0.0563)
$(CH_3)_2CHCHCH(CH_3)_2$
CH_3
135

▶1295
P
$Gd(NO_3)_3$
45

▶1294 VSB
B (0.0104)

$H_3C\text{-}O\text{-}CF_2CHFCl$

356

▶1294 M
B (0.0563)

$(CH_3CH_2)_2CHCH_2CH_2CH_3$

140

▶1294 M
B

$CH_3CH_2C(CH_3)_2\text{-}CH_2CH_3$

336, see also 148, 149

▶1294 MS
B (0.169)

$(CH_3)_2CHCH_2CH_2CH(CH_3)_2$

137, see also 339

▶1294 M
P

$(CH_3)_2P\cdot BH_2$

[CH_3 symm. deformation] 30

▶1293 MS
B (0.0153)

CH_2 = C – CH_3 (cyclopropyl)

288

▶1292 VS
B (0.063)

CH_3, CN

312

▶1292 VVS
B (0.04)

N, CH_3

198, see also 438

▶1292 S
A (1.8 mm Hg, 10 cm)

F_2 F_2 F_2 F_2

243

▶1292 MS
B (0.238)

$H_2C{=}C(CH_3)C(CH_3)_3$

272

▶1292 M
B (0.064)

$H_2C{=}CHCH(CH_3)CH_2CH_3$

167

▶1292 S
B (0.063)

CH_3, CN

214

▶1292 M
B

H_3C, O, N-H, H_3C CH_3

26

▶1292 MW
B (0.169)

$CH_3CH_2CH(CH_3)\text{-}CH(CH_3)CH_2CH_3$

83

▶1291 VS
B (0.055)

CH_3, H_3C, N, H

344

▶1290
B

CH_3CF_3

[CF] 447

▶1290 VS
B

Cl, O, N

26

▶1290 VSB
CCl_4 [25%] (0.10)

$CH_3CH(OH)CH_2CH_3$

319

▶1290 W
B (0.0576)

F

226

▶1290 S
M

Pt (NH_2–CH_2–CH_2–NH_2) Cl_2

[NH_2] 8

▶1290 S
B (0.0563)

$CH_3CH{=}CHCH_2CH_3$
(trans)

207

▶1290 M
B (0.003)

$CH_3(CH_2)_2CH_2NO_2$

310

▶1290 MS
CCl_4 [25%] (0.10)

$CH_3(CH_2)_3OH$

353

▶1290 M
C_6H_6 [0.0172 m/l]

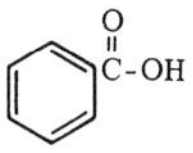

$C_6H_5C(=O){-}OH$

448, see also 38

▶1290 M
C_6H_6 [0.0044 m/l]

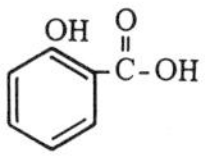

o-$HOC_6H_4C(=O){-}OH$

448

▶1290 MW
B (0.015)

$H_2C{=}CHCH_2C(CH_3)_3$

102, see also 273

▶1290

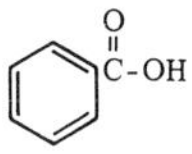

$C_6H_5C(=O){-}OH$

370, see also 38, 448

▶1290

$H_3C{-}C(=O){-}OH$

50

▶1289 SSh
A (1.5 mm Hg, 10 cm)

$CF_3CF_2CF_2CF_2CF_3$

237

▶1289 VS
B (0.003)

$CH_3CH_2CH(NO_2)CH_3$

324

▶1289 S

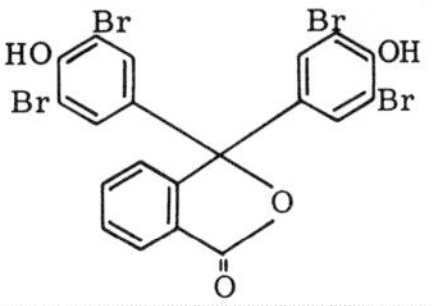

HO, Br, Br, OH, Br, Br, O, O (structure: phthalide bearing two dibromohydroxyphenyl groups)

39

▶1289 M
B (0.029)

$CH_3CH_2C(CH_3)_2{-}CH_2CH_3$

149, see also 148, 336

▶1289 M
B (0.169)

$(CH_3)_2CHCH(CH_3)CH_2CH_3$

142, see also 152, 153

▶1287.8 MS
B (0.03)

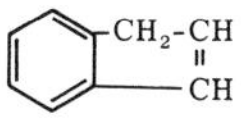

C_6H_4 fused $CH_2{-}CH{=}CH$ (indene)

9

▶1287 S
B (0.0104)

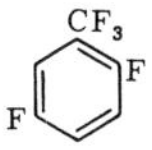

CF_3, F, F (structure: difluorobenzotrifluoride)

217

▶1287 W
B (0.0576)

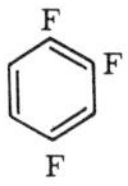

F, F, F (structure: trifluorobenzene)

224

▶1287 S
B (0.136)

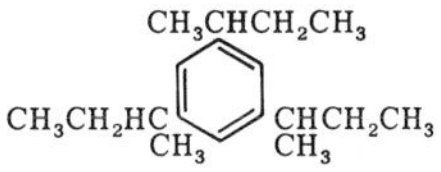

$CH_3CHCH_2CH_3$, $CH_3CH_2HC(CH_3)$, $CHCH_2CH_3$ CH_3 (structure: tri-sec-butylbenzene)

280

▶1287 S
E

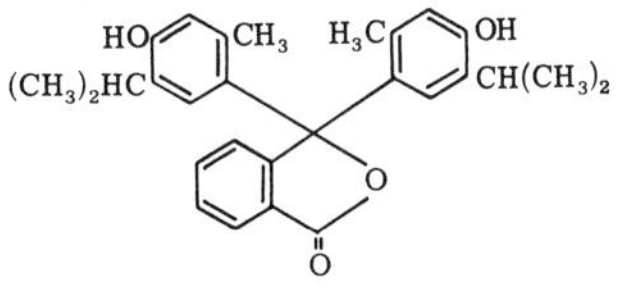

HO, CH_3, H_3C, OH, $(CH_3)_2HC$, $CH(CH_3)_2$, O, O (structure: phthalide bearing two hydroxy-methyl-isopropylphenyl groups)

39

▶1287 MS
E

HO, OH, O, O (structure: phthalide bearing two hydroxyphenyl groups)

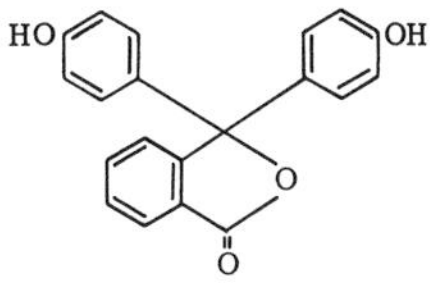

39

▶1287 M
M

$Pd(NH_2{-}CH_2{-}CH_2{-}NH_2)_2PtCl_4$

[NH_2]

8

▶1287 M
CCl_4 [0.0082 m/l]
O ‖ C-OH
448, see also 38

▶1286 MS
B (0.065)
$H_2C{=}CHCH_2C(CH_3)_3$
273, see also 102

▶1286 S
B
O_2N O N
26

▶1286 MB
A (100 mm Hg, 69 cm)
$CH_3\text{-}CCl_3$
251

▶1286 M
E
O O
39

▶1285 S
B (0.0563)
$(CH_3)_3CCH_2CH_2C(CH_3)_3$
305

▶1285 VS
B (0.10)
CH_3
269

▶1285 S
B (0.064)
H_3CH_2C H C=C H CH_2CH_3
(trans)
168

▶1285 MSSh
B (0.025)
S
199

▶1285 M
B (0.151)
H_3C CH_3 CH_2CH_3
215

▶1284 VSVB
B (film)
$(C_4F_9)_3N$
346

▶1284 S
B
Cl O N
26

▶1284 S
B (0.0563)
$(CH_3)_3CCH_2CH(CH_3)CH(CH_3)_2$
303

▶1284 M
B (0.169)
$(CH_3)_2CHCH_2CH(CH_3)CH_2CH_3$
138, see also 334

▶1284 M
B (0.169)
$(CH_3)_2CHCH(CH_3)CH_2CH_3$
142, see also 152, 153

▶1284 M
M
$Pd(NH_2\text{-}CH_2\text{-}CH_2\text{-}NH_2)Cl_2$
[NH_2 wag]
8

▶1284
$H_3C\text{-}C(=O)\text{-}CH_2\text{-}C(=O)\text{-}CH_3$
654

▶1284 MS
E
O O
39

▶1283 M
B (0.157)
$(CH_3)_3C(CH_2)_2CH(CH_3)_2$
292

▶1282 W
B (0.0576)
CH_3 F
221

▶1282 VS
B (0.0104)

$C_6H_4(CF_3)F$ [structure: CF3, F on benzene ring]

218

▶1282 VS
$CHCl_3$

[structure: 2,6-dimethyl-4H-pyran-4-thione; S, H_3C, CH_3, O]

[Ring]

7

▶1282 S
B (0.157)

$(CH_3)_3CCH_2CHCH_2CH_3$
CH_3

79, 285

▶1282 S
B (film)

[structure: perfluorocyclohexane ring with F, CF_3, F_2, F, CF_3, F_3C, F, F_2, F_2]

228

▶1282 M
B

[structure: H_3C, O_2N, O, N-CH_3 ring]

26

▶1282 M
C

[structure: naphthalene with CH_3 CH_3]

182

▶1282 M
M

$Cu(NH_2\text{-}CH_2\text{-}CH_2\text{-}NH_2)_2PtCl_4$

[CH_2]

8

▶1282 M
M

$Ni(NH_2\text{-}CH_2\text{-}CH_2\text{-}NH_2)_3PtCl_4$

[CH_2]

8

▶1282 M
$CHCl_3$ [0.0064 g/ml]

$C_6H_5C(=O)\text{-}NH\text{-}CH_3$

[Amide III]

2

▶1282
P

$Hg_2(NO_3)_2$

45

▶1282 M
E

[structure: $(H_3C)_2N$, $N(CH_3)_2$, O, O]

39

▶1281 VS
A (100 mm Hg, 15 cm)

CH_3CH_2Cl

88

▶1281 S
C

[structure: naphthalene with H_3C, CH_3]

179

▶1280 S
B (0.0563)

$(CH_3)_2CHCH(CH_3)_2$

144, see also 210

▶1280 W
B (0.0576)

[structure: benzene ring with CH_3, F]

220

▶1280 S
B

[structure: O, N ring attached to phenyl with Cl]

26

▶1280 S
B

[structure: O, N ring attached to phenyl with F]

26

▶1280 MW
B (0.101)

$CH_3(CH_2)_5CH_3$

85

▶1280 M
M

$Pd(NH_2\text{-}CH_2\text{-}CH_2\text{-}NH_2)_2Cl_2$

[CH_2]

8

▶1280

$(CH_3)_4C$

655, see also 57

▶1280

O_2N, C-H, O

387

▶1280 S
E

Cl, O, O

39

▶1279 S
B (film)

F, CF_3, F_2, F_2, F_2, F_2, F_2

232

▶1279 MSSp
B (0.065)

CH_3, $CH(CH_3)_2$

132

▶1279 MWSh
B (0.0088)

$(CH_3)_2CHCH_2CH(CH_3)_2$

141, see also 150, 151, 335

▶1278 VS
E

$[H_2CNO_2]^-Na^+$

[NO_2 asymm. stretch] 328

▶1278

CH_3CF_3

[CF] 447

▶1278 S
B (0.157)

$(CH_3)_3CCH_2CH(CH_3)CH_2CH_3$

79, see also 285

▶1278 S
B (0.10)

CH_3, CH_3

187

▶1278 M
C

SrH_2GeO_4

690

▶1278 M
E

N, NH_2

327

▶1277 VS
A (12.5 mm Hg, 10 cm)

F, F, F, F

223

▶1277 WSh
A (100 mm Hg, 10 cm)

CF_3-CCl_3

253

▶1277 VSSp
B (0.0104)

F, F, F, F

222

▶1277 VS
D, CCl_4

$[H_2CNO_2]^-Na^+$

[NO_2 asymm. stretch] 328

▶1277 S
B (0.10)

$CH_2CH_2CH_2CH_3$

176

▶1277 M
B (0.157)

$(CH_3)_3C(CH_2)_2CH(CH_3)_2$

292

▶1277 VW
B (0.0576)

CF_3

219

▶1277 M
B

O, H_3C, O_2N, N-C_3H_7-n

26

▶1277 M
B

O, H_3C, O_2N, N-C_4H_9-n

26

▶1277 M
B (0.051)

$CH(CH_3)_2$
$CH(CH_3)_2$

298

▶1277 MW
B (0.0563)

$(CH_3CH_2)_2CHCH_2CH_2CH_3$

140

▶1277 M
CCl_4

$[(CH_3)_2CNO_2]^-Na^+$

328

▶1276 S
B

H_3C $-CH_3$ CH_3

125

▶1276 S
B (0.0563)

$(CH_3)_2CHCH(CH_2CH_3)_2$

136

▶1276 S
B (0.169)

$(CH_3CH_2)_3CH$

143, see also 156

▶1276 S
B

O N Br

26

▶1276 M
CCl_4 [0.007 g/ml]

O C-NH CH_3

[Amide III]

2

▶1275 M
A

$H-C{\equiv}C-C(=O)-H$

18

▶1275
E

$Cr(NH_3)_5(NO_3)_3$

45

▶1275

$H_2C{=}CH{-}O{-}CH{=}CH_2$

666

▶1274 M
B (film)

O O O

342

▶1274 S
A (6.2 mm Hg, 10 cm)

$CF_2Cl{-}CF_2Cl$

257

▶1274 S
CCl_4 [25%] (0.10)

$CH_3CH_2C(CH_3)_2$ OH

194

▶1274 S
B (0.0563)

$(CH_3)_2CHCHCH(CH_3)_2$ CH_2 CH_3

301

▶1274 S
B (0.0563)

CH_3 CH_3
$CH_3CH_2C - CH_2CHCH_2CH_3$
CH_3

302

▶1274 S
B

O N

26

▶1274 S
C

CH_3 CH_3

181

▶1274 M
B (0.101)

$(CH_3CH_2)_3CH$

156, see also 143

▶1273 M
B

O H_5C_2 O_2N $N{-}C_3H_7{-}n$

26

▶1272 VS
B (0.238)
H_3C C_3H_7-n
C=C
H H
(trans)
276

▶1272 M
C_6H_6 [0.0172 m/1]
(benzoic acid: C_6H_5-C(=O)-OH)
448

▶1272 MS
B (0.003)
$C_3H_7NO_2$
392

▶1272 M
B (0.068)
H_3C CH_3
C=C
H CH_2CH_3
(cis)
274

▶1272 VVS
CCl_4 [31%] (0.025)
Cl–(thiophene)–$COCH_3$
372

▶1272 W
$CHCl_3$
S=(thiopyran) H_3C, CH_3
7

▶1272 S
B (0.10)
(naphthalene)–$CH_2CH_2CH_2CH_3$
175

▶1271 VS
A (12.5 mm Hg)
(benzene) F, F, F, F
223

▶1272 M
B (0°C)
$(CH_3)_2CHCHCH(CH_3)_2$
CH_3
284, see also 135

▶1271 VVS
B (0.08)
(isoquinoline) N
197

▶1272 M
CS_2 [20 g/l] (0.51)
$CH_2CH(CH_3)_2$
$CH_3CH=CH(CH_2)_2CH=CH(CH_2)_2CO-NH$
357

▶1271 VVS
CS_2 [sat.] (0.025)
Br–(thiophene)–$COCH_3$
348

▶1272 SSp
B (0.151)
H_3C CH_3
(cyclohexane)
CH_2CH_3
215

▶1271 VSB
B (film)
F CF_3 CF_3 F
F_2 F_2 F_2 F_2
231

▶1272 MW
B (0.169)
CH_3 CH_3
CH_3CH_2CH-$CHCH_2CH_3$
83

▶1271 S
B (0.0563)
$(CH_3)_2CHCHCH(CH_3)_2$
CH_3
135

▶1272 S
B (0.063)
CH_3
(benzene)
CN
214

▶1271 VSVB
CCl_4 [20%] (0.10)
CH_3CH_2OH
315

▶1272 S
C
(naphthalene)–CH_3
191

▶1271 MS
E
NH_2
(pyridine) N
327

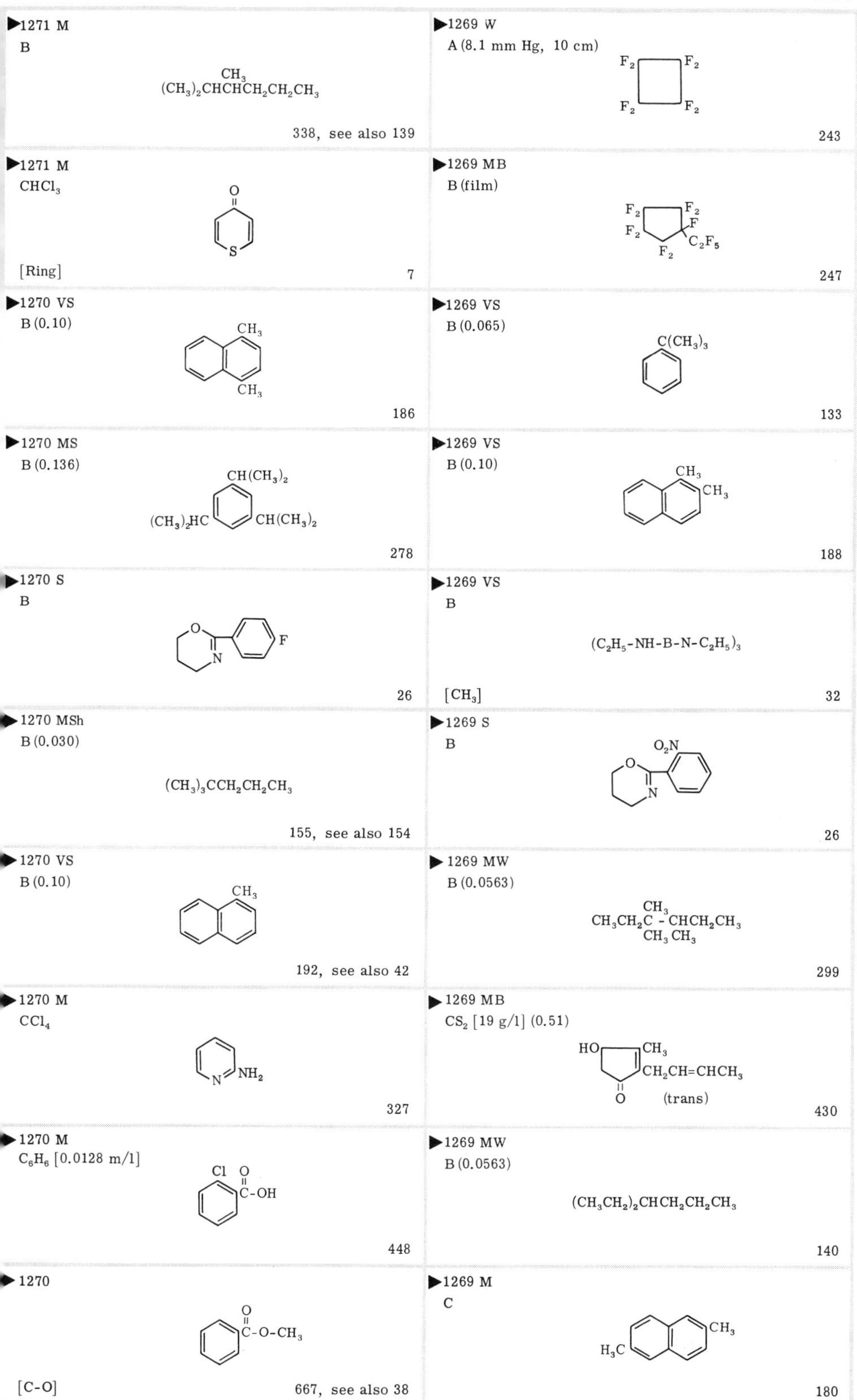

1271 M
B
(CH3)2CHCH(CH3)CH2CH2CH3
338, see also 139
1271 M
CHCl3
[Ring]
7
1270 VS
B (0.10)
186
1270 MS
B (0.136)
CH(CH3)2
(CH3)2HC
CH(CH3)2
278
1270 S
B
26
1270 MSh
B (0.030)
(CH3)3CCH2CH2CH3
155, see also 154
1270 VS
B (0.10)
CH3
192, see also 42
1270 M
CCl4
NH2
327
1270 M
C6H6 [0.0128 m/l]
448
1270
C-O-CH3
[C-O]
667, see also 38
1269 W
A (8.1 mm Hg, 10 cm)
243
1269 MB
B (film)
247
1269 VS
B (0.065)
C(CH3)3
133
1269 VS
B (0.10)
188
1269 VS
B
(C2H5-NH-B-N-C2H5)3
[CH3]
32
1269 S
B
O2N
26
1269 MW
B (0.0563)
299
1269 MB
CS2 [19 g/l] (0.51)
(trans)
430
1269 MW
B (0.0563)
(CH3CH2)2CHCH2CH2CH3
140
1269 M
C
180

▶1269 W B $C_6H_5HSiCl_2$ 86	▶1267 M B (0.0563) $(CH_3)_3CC(CH_3)_2-CH_2CH_2CH_3$ 304
▶1268 VSSp B (0.025) $CH_3CH_2SCH_3$ 209	▶1267 M B (0.015) $H_2C{=}CHC(CH_3)_3$ 129
▶1268 SSp B (0.065) $CH_3(CH_2)_2CH{=}CH(CH_2)_2CH_3$ (cis) 422	▶1267 M B (0.064) $H_3C(H)C{=}C(H)CH_2CH_2CH_3$ (trans) 169
▶1268 VS B (0.10) 2-naphthyl-$CH_2CH_2CH_2CH_2CH_3$ 173	▶1267 M CCl_4 [0.04 g/ml] $C_6H_5C(=O)-N(CH_3)_2$ 3
▶1268 M P $[(CH_3)_2P\cdot BH_2]_3$ [CH_3 symm. deformation] 30	▶1267 W B $C_6H_5SiCl_3$ 86
▶1268 MW P $[(C_2H_5)_2P\cdot BI_2]_3$ 30	▶1266 S B (0.0563) $CH_3(CH_2)_2CH{=}CH(CH_2)_2CH_3$ (trans) 205
▶1267 MW B (0.0576) $C_6H_4(CH_3)F$ (CH₃, F ortho) 221	▶1266 MS B (0.051) $C_6H_4(CH(CH_3)_2)_2$ (ortho) 296
▶1267 VS CS_2 CF_4 418	▶1266 MS B (0.06) cyclohexene 128
▶1267 S B (0.10) H_3C-naphthalene-CH_3 183	▶1266 M B CH_3, H_3C, H_3C cyclopentene 127
▶1267 MS B (0.169) $(CH_3)_2CHCH(CH_3)CH_2CH_2CH_3$ 139, see also 338	▶1266 M B (0.169) $(CH_3)_2CHCH(CH_3)CH_2CH_3$ 142, see also 152, 153

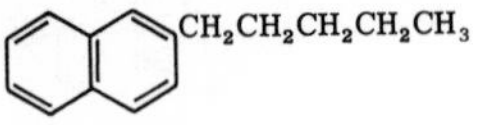

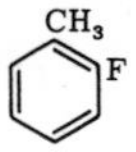

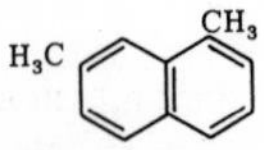

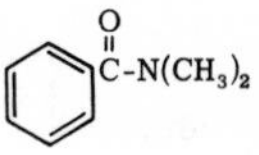

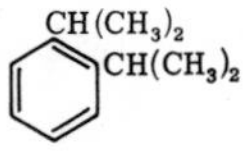

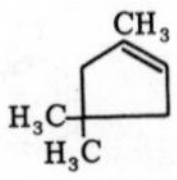

▶1266

CH_3CF_3

[CF] 447

▶1266

CH_2Cl_2

376

▶1265 S
B (0.10)

CH_3
H_3C

184

▶1265

$[(CH_2)_3 \ CH_2 \text{—}]_2 NNO$
N

41

▶1264 VS
B (film)

F CF_3
F_2 F_2
F_2 F_2
F_2

232

▶1264 S
B (film)

F CF_3
F_2 F_2
F_2 F_2
F CF_3

230

▶1264 VS
B

O
N
Cl

26

▶1264 M
B (film)

CF_3
F
F

217

▶1264 S
A (775 mm Hg, 10 cm)

CHF_2CH_3

233

▶1264 VS
B (0.10)

CH_2CH_3

189

▶1264 S
B (0.003)

NO_2
$CH_3CH_2CHCH_3$

324

▶1264 S
C_6H_6 [0.004 m/l]

OH O
C-OH

448, see also 38

▶1264 S
B (0.10)

$CH_2CH_2CH_2CH_2CH_3$

174

▶1264 M
A

CH_3
$CH_3CH_2CHCH_2CH_3$

332

▶1264 M
$CHCl_3$

NH_2
N

327

▶1263 VS
E

$[H_2C \cdot NO_2]^- Na^+$

[NO_2 asymm. stretch] 328

▶1263 S
B

O
N
Br

26

▶1263 WVB
B (0.036)

CH_3
$H_2C{=}C(CH_2)_4CH_3$

98

▶1263 S
A

CH_3SiH_2I

[CH_3 symm. bend] 669

▶1263 MS
G

O S
$H_3C\text{-}C\text{-}C\text{-}N$ $\langle CH_2\text{-}CH_2, CH_2\text{-}CH_2 \rangle$ O

[$N \langle CH_2\text{-}CH_2, CH_2\text{-}CH_2 \rangle O$]

31

▶1263 M B (0.015) $H_2C{=}C(CH_3)CH_2C(CH_3)_3$ 96	▶1261 M B (0.169) $(CH_3CH_2)_3CH$ 143, see also 156
▶1263 W B $(C_6H_5)_2SiCl_2$ 86	▶1261 S B $C_6H_5(CH_3)SiCl_2$ 86
▶1262 VVS A (1.5 mm Hg, 10 cm) $CH_3CH_2CH_3$ 235	▶1261 MS B (0.169) $(CH_3)_2CHCH_2CH_2CH(CH_3)_2$ 137, see also 339
▶1262 VS C CH_3 CH_3 (naphthalene) 182	▶1261 W B (0.163) $H_2C{=}CH(CH_2)_6CH_3$ 271
▶1262 VS D $[H_2C \cdot NO_2]^- Na^+$ [NO_2 asymm. stretch] 328	▶1261 W CS_2 $(C_6H_5)_3SiCl$ 86
▶1262 S B (0.10) (naphthalene) $CH_2CH_2CH_3$ 177	▶1260 VS B H_3C, O, N-H, H_3C CH_3 (morpholine ring) 26
▶1262 E $Hg_2(NO_3)_2$ 45	▶1260 S B H_5C_2, O_2N, O, N-CH_3 (ring) 26
▶1261 VS B H_3CO OCH_3, O, N (ring) 26	▶1260 MS B (0.10) (naphthalene) $CH_2CH_2CH_2CH_3$ 176
▶1261 VS CCl_4 $[H_2C \cdot NO_2]^- Na^+$ [NO_2 asymm. stretch] 328	▶1260 M B $[(C_2H_5)_2P \cdot BH_2]_3$ 30
▶1261 MS B (0.0563) $(CH_3)_2CHC(CH_3)_2-CH_2CH_2CH_3$ 134	▶1260 M B H_3C, O_2N, O, N-CH_3 (ring) 26

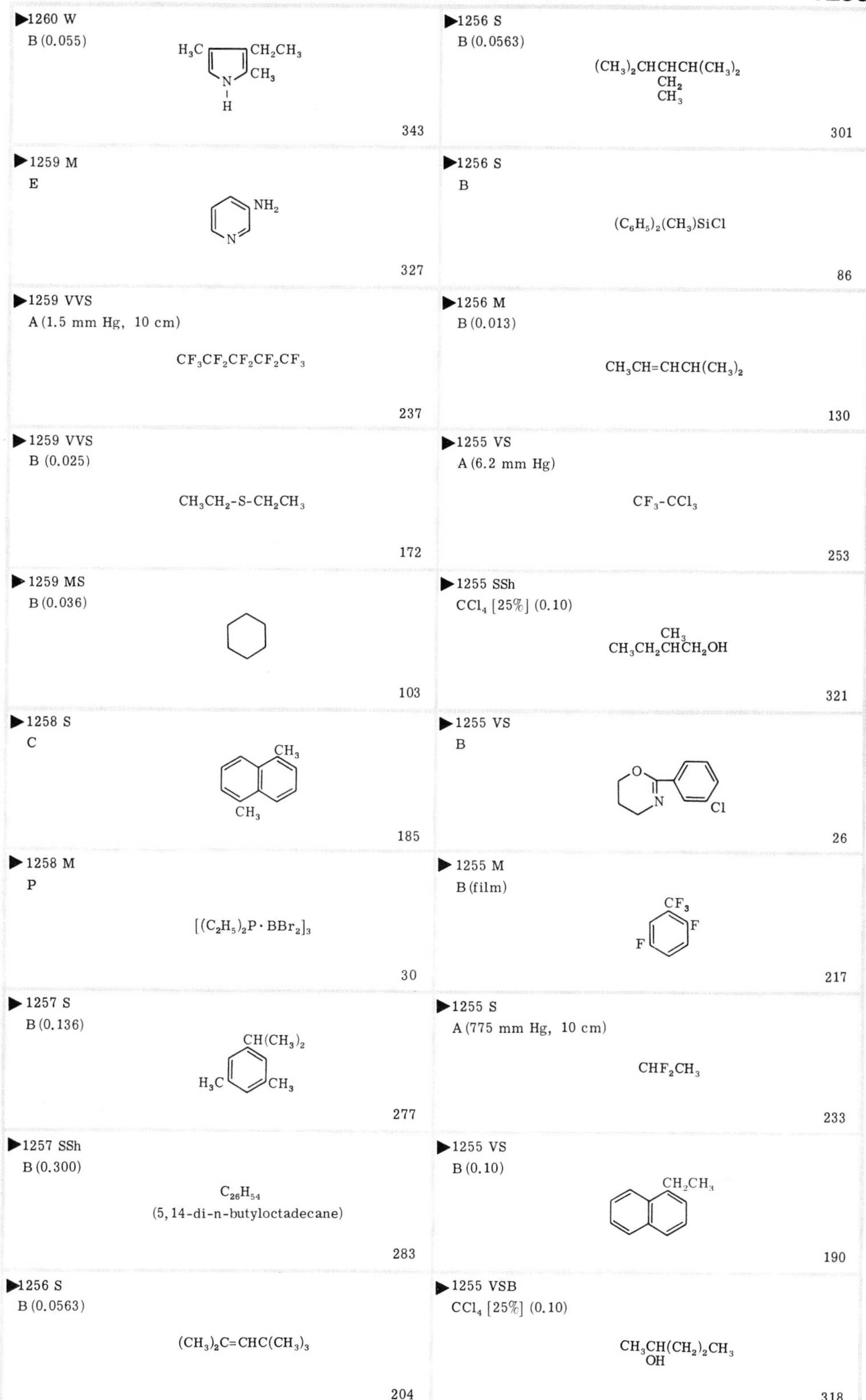

▶1260 W
B (0.055)
H_3C, CH_2CH_3, CH_3, N, H
343

▶1259 M
E
NH_2, N
327

▶1259 VVS
A (1.5 mm Hg, 10 cm)
$CF_3CF_2CF_2CF_2CF_3$
237

▶1259 VVS
B (0.025)
CH_3CH_2-S-CH_2CH_3
172

▶1259 MS
B (0.036)
103

▶1258 S
C
CH_3, CH_3
185

▶1258 M
P
$[(C_2H_5)_2P \cdot BBr_2]_3$
30

▶1257 S
B (0.136)
$CH(CH_3)_2$, H_3C, CH_3
277

▶1257 SSh
B (0.300)
$C_{26}H_{54}$
(5, 14-di-n-butyloctadecane)
283

▶1256 S
B (0.0563)
$(CH_3)_2C{=}CHC(CH_3)_3$
204

▶1256 S
B (0.0563)
$(CH_3)_2CHCHCH(CH_3)_2$
CH_2
CH_3
301

▶1256 S
B
$(C_6H_5)_2(CH_3)SiCl$
86

▶1256 M
B (0.013)
$CH_3CH{=}CHCH(CH_3)_2$
130

▶1255 VS
A (6.2 mm Hg)
CF_3-CCl_3
253

▶1255 SSh
CCl_4 [25%] (0.10)
CH_3
$CH_3CH_2CHCH_2OH$
321

▶1255 VS
B
O, N, Cl
26

▶1255 M
B (film)
CF_3, F, F
217

▶1255 S
A (775 mm Hg, 10 cm)
CHF_2CH_3
233

▶1255 VS
B (0.10)
CH_2CH_3
190

▶1255 VSB
CCl_4 [25%] (0.10)
$CH_3CH(CH_2)_2CH_3$
OH
318

▶1255 S B (0.056) $CH_2=CHCH=CHCH=CH_2$ 203	▶1253 M H_2O $[H_2C \cdot NO_2]^- Na^+$ 328
▶1255 MSh B (0.169) $(CH_3)_2CHCH(CH_3)CH_2CH_2CH_3$ 139, see also 338	▶1252 S B (0.0563) $(CH_3)_3CCH_2CH_3$ 145
▶1255 M C [2-methylnaphthalene structure, CH_3] 191	▶1252 S B [structure: O_2N, O, N] 26
▶1255 H_2C-CH_2 / O 537, see also 64, 381	▶1252 W B (film) [structure: F, CF_3, F_2, F_2, F, CF_3, F_3C, F, F_2] 228
▶1255 CH_3I [CH_3 symm. bend] 539	▶1252 MS B (film) $H_3C-O-CF_2CHFCl$ 356
▶1254 VS CS_2 CF_4 418	▶1252 S D $[CH_2CH(CN)]_n$ [CH tert.] 23
▶1253 VVS B (0.025) [thiophene structure, S] 199	▶1251 [1,3,5-triazine structure, N, N, N] 14
▶1253 S C $(CH_3)_3CNO_2$ 325	▶1250 VS B (0.157) $(CH_3)_3C(CH_2)_2CH(CH_3)_2$ 292
▶1253 S G [structure: indole ring, N-H] $CH=N-NH-C(=S)-NH_2$ 380	▶1250 VS B (0.0104) [structure: benzene ring, F, F, F] 224
▶1253 M B (0.169) $(CH_3)_2CHCH(CH_3)CH_2CH_3$ 142, see also 152, 153	▶1250 S A $(CH_3)_3CCH_2CH_3$ 94

▶1250 S
B

$(CH_3)_3C(CH_2)_3CH_3$

337

▶1250 S
B (0.055)

CH_3 H_3C N H

344

▶1250 MW
B (0.0563)

$(CH_3CH_2)_2CHCH_2CH_2CH_3$

140

▶1250 M
CCl_4

N NH_2

327

▶1250 MW
P

$[(C_2H_5)_2P \cdot BI_2]_3$

[CH_2 wag] 30

▶1250

O C-OH

[OH] 370

▶1249 S
B (0.030)

$(CH_3)_3CCH_2CH_2CH_3$

155, see also 154

▶1249 VS
G

H S S H
N-C-C-N
HO_2CH_2C CH_2CO_2H

[CCO] 31

▶1249 M
B

O
H_3C N-CH_3
O_2N

26

▶1249 M
CCl_4

CH_3
N
Cl_3P PCl_3
N
CH_3

[CH_3] 384

▶1248 VVS
B (0.0104)

H_5C_2-O-CF_2CHFCl

261

▶1248 VSSh
B (0.0576)

CF_2Cl-$CFCl_2$

255

▶1248 VS
B (0.136)

$CH(CH_3)_2$
$(CH_3)_2HC$ $CH(CH_3)_2$

278

▶1248 VS
B (0.08)

N

197

▶1248 M
B (0.003)

$CH_3(CH_2)_2CH_2NO_2$

310

▶1248 M
B (0.0563)

$(CH_3)_2CHCH(CH_2CH_3)_2$

136

▶1248

[CH_2]$_2$ NNO
N

41

▶1247 W
B (film)

F CF_3
F_2 F_2
F_2 F_2
F_2

232

▶1247 VS
B (0.169)

CH_3
$(CH_3)_3CCH_2CHCH_2CH_3$

285

▶1247 VS
B (0.157)

CH_3
$(CH_3)_3CCH_2CHCH_2CH_3$

79

▶1247 VVS
B (0.0563)

$(CH_3)_3CCH_2CH_2C(CH_3)_3$

305

▶1247 S
CCl_4 [25%] (0.10)

$CH_3CH_2CHCH_2CH_3$
OH

196

▶1247 SSp
B (0.151)

H_3C CH_3
CH_2CH_3

215

▶1247 M
E

NO_2

20

▶1245 VS
B (0.0563)

$(CH_3)_3CCH_2CHCH(CH_3)_2$
CH_3

303

▶1245 VS
B

n-H_9C_4-O-CF_2CHFCl

264

▶1245 SB
B (film)

F CF_3 CF_3 F
F_2 F_2 F_2 F_2

231

▶1245 M
B (0.136)

$CH_2(CH_2)_2CH_3$
$CH_3(CH_2)_2H_2C$ $CH_2(CH_2)_2CH_3$

279

▶1245 MS
B (0.015)

$H_2C{=}CHCH_2C(CH_3)_3$

102, see also 273

▶1245
E or N

[CH_2] NNO
N
2

41

▶1244 W
B (0.0104)

CF_3

219

▶1244 S
B (film)

F CF_3
F_2 F_2
F_2 F_2
F CF_3

230

▶1244 MS
B (film)

n-H_7C_3-O-CF_2CHFCl

263

▶1244 VS
B (0.157)

$(CH_3)_3CCHCH(CH_3)_2$
CH_3

293

▶1244 M
B

O
H_5C_2 N-C_3H_7-n
O_2N

26

▶1244 MW
B (0.036)

$CH_3CH_2CH{=}CH(CH_2)_3CH_3$

99

▶1244 W
P

$[(C_2H_5)_2P \cdot BCl_2]_3$

[CH_2 wag]

30

▶1242 S
B

CH_3
$(CH_3)_3CCHCH_2CH_3$

340

▶1242 M
B (0.10)

$CH_2CH_2CH_2CH_3$

175

▶1242 MB
CS_2 [19 g/l] (0.51)

HO CH_3
$CH_2CH{=}CHCH_3$
O (trans)

430

▶1242 M

$[(C_2H_5)_2P \cdot BH_2]_3$

[CH_2 wag]

30

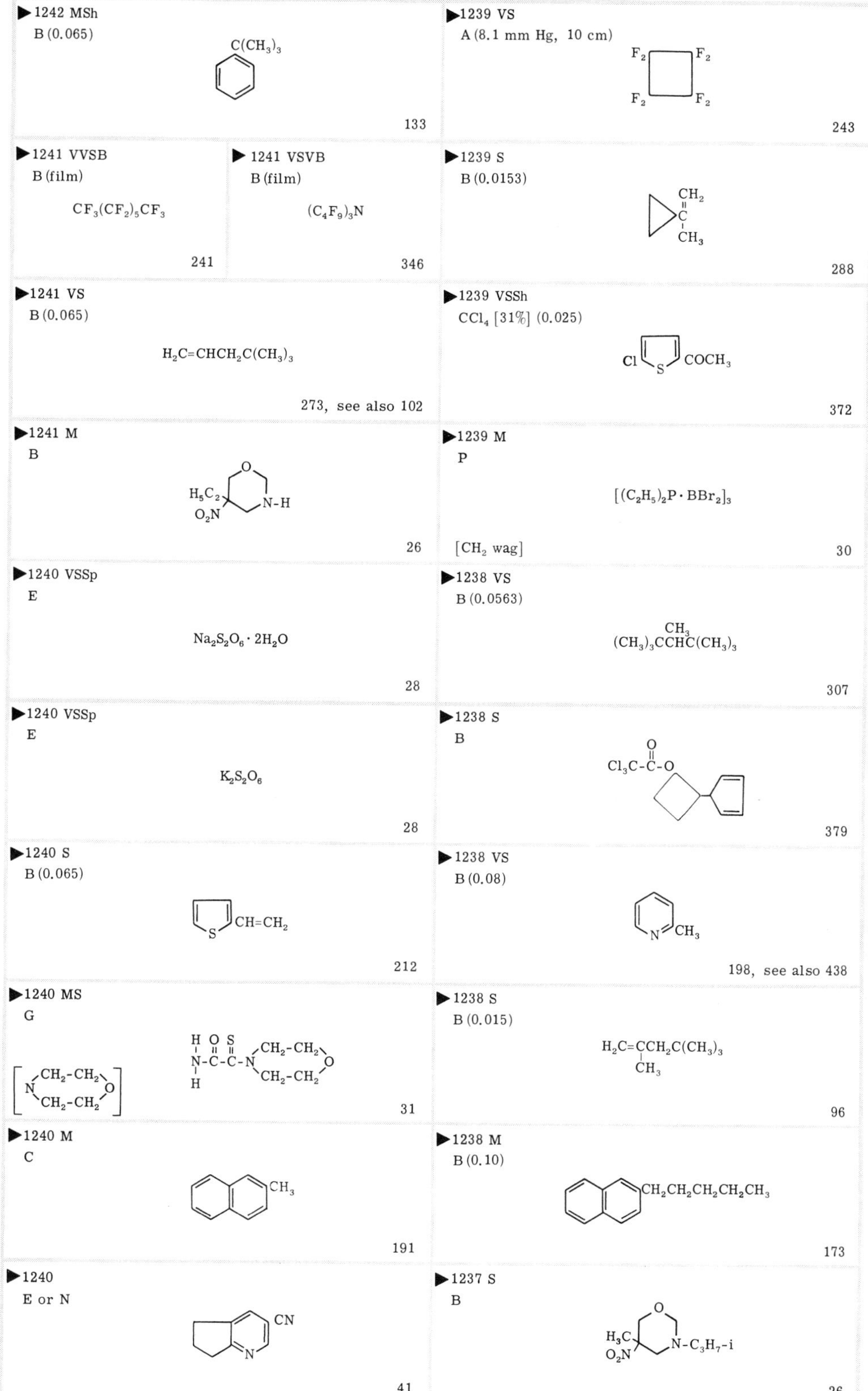
▶1242 MSh
B (0.065)
C(CH3)3
133
▶1241 VVSB
B (film)
CF3(CF2)5CF3
241
▶1241 VSVB
B (film)
(C4F9)3N
346
▶1241 VS
B (0.065)
H2C=CHCH2C(CH3)3
273, see also 102
▶1241 M
B
O
H5C2
O2N
N-H
26
▶1240 VSSp
E
Na2S2O6 · 2H2O
28
▶1240 VSSp
E
K2S2O6
28
▶1240 S
B (0.065)
S
CH=CH2
212
▶1240 MS
G
H O S
N-C-C-N
H
CH2-CH2
O
CH2-CH2
N
CH2-CH2
O
CH2-CH2
31
▶1240 M
C
CH3
191
▶1240
E or N
CN
N
41
▶1239 VS
A (8.1 mm Hg, 10 cm)
F2
F2
F2
F2
243
▶1239 S
B (0.0153)
CH2
C
CH3
288
▶1239 VSSh
CCl4 [31%] (0.025)
Cl
S
COCH3
372
▶1239 M
P
[(C2H5)2P · BBr2]3
[CH2 wag]
30
▶1238 VS
B (0.0563)
CH3
(CH3)3CCHC(CH3)3
307
▶1238 S
B
O
Cl3C-C-O
379
▶1238 VS
B (0.08)
N
CH3
198, see also 438
▶1238 S
B (0.015)
H2C=CCH2C(CH3)3
CH3
96
▶1238 M
B (0.10)
CH2CH2CH2CH2CH3
173
▶1237 S
B
O
H3C
O2N
N-C3H7-i
26

▶1237 VS
B (0.055)

H_3C, CH_2CH_3, CH_3, N–H (pyrrole)

343

▶1237 M
B

$VOBr_3$

12

▶1237 M
B

H_5C_2, O_2N, O, $N\text{-}C_6H_{13}\text{-}n$

26

▶1236 VS
B (0.10)

CH_3, CH_3 (naphthalene)

188

▶1236 VS
B (0.0157)

$(CH_3)_3CCH(CH_2CH_3)_2$

78, 300

▶1236 VS
B (0.0104)

CH_3, F (benzene)

221

▶1236 MS
CS_2 [sat.] (0.025)

Br, S, $COCH_3$ (thiophene)

348

▶1236 M
B (0.066)

$H_2C{=}CH(CH_2)_8CH_3$

352

▶1235 VSSp
E

$Na_2S_2O_6$

28

▶1235 W
B (0.163)

$H_2C{=}CH(CH_2)_7CH_3$

270

▶1235 M
B (0.169)

$(CH_3)_2CHCH(CH_3)CH_2CH_2CH_3$

139, see also 338

▶1235 M
C

CH_3, CH_3 (naphthalene)

182

▶1235

OH (benzene)

[C-O, OH]

632, see also 36

▶1234 M
B

$CH_3CH{=}CHCH{=}CHCH_3$

354

▶1233 VS
A (25 mm Hg, 10 cm)

$CF_2Cl\text{-}CF_2Cl$

257

▶1233 VS
A (12.5 mm Hg, 10 cm)

F, F, F, F (benzene)

223

▶1233 S
B (0.003)

$(CH_3)_2CHCH_2NO_2$

326

▶1233 MS
G

$H_3C\text{-}C(=O)\text{-}C(=S)\text{-}N(CH_2\text{-}CH_2)_2O$

[$N(CH_2\text{-}CH_2)_2O$]

31

▶1233 MS
G

$H_2N\text{-}C(=O)\text{-}C(=S)\text{-}N(CH_2\text{-}CH_2)_2O$

31

▶1233 M
B (0.136)

$CH_3CHCH_2CH_3$, $CH_3CH_2HC(CH_3)$, $CHCH_2CH_3(CH_3)$ (benzene)

280

▶1232 VSVB
B (film)

F_9C_4-O-C_4F_9

266

▶1232 VSB
CCl_4 [25%] (0.10)

$CH_3CH_2CH_2OH$

314

▶1232 M
B (0.018)

$CH_3CH_2CH(CH_2)_2CH_3$
CH_3

81, see also 157, 330

▶1232 MS
CCl_4 [25%] (0.10)

$CH_3(CH_2)_4OH$

393

▶1232 MS
B (film)

H_3C-O-CF_2CHFCl

356

▶1231 S
B

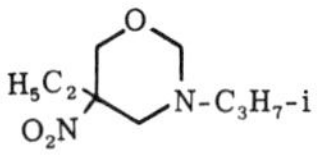

26

▶1230 VS
A (1.7 mm Hg, 10 cm)

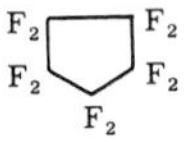

245

▶1230 S
B (0.0563)

CH_3
$(CH_3)_3CC$ - $CH_2CH_2CH_3$
CH_3

304

▶1230 S
B (0.0563)

CH_3
$(CH_3)_3CC$ - $CH(CH_3)_2$
CH_3

306

▶1230 VS
B (film)

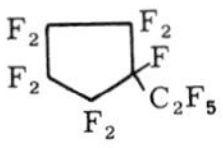

247

▶1230 S
B (0.003)

$C_3H_7NO_2$

392

▶1230 S
B (0.08)

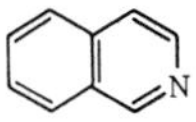

197

▶1230 S
B (0.065)

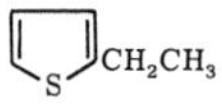

213

▶1230

CH_3CF_3

447

▶1229 VVS
B (film)

$(CF_3)_2CFCF_2CF_3$

240

▶1229 MS
B (film)

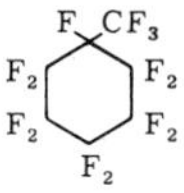

232

▶1229 VS
B (0.0563)

$(CH_3)_2C{=}CHC(CH_3)_3$

204

▶1229 S
A (25 mm Hg, 10 cm)

CF_2Cl-CF_2Cl

257

▶1227 VS
A (6.2 mm Hg, 10 cm)

CF_3-CCl_3

253

▶1227 MW
B (0.0563)

$(CH_3CH_2)_2CHCH_2CH_2CH_3$

140

▶1227 MS CCl_4 [25%] (0.10) $CH_3CH(CH_2)_2CH_3$ (OH) 318	▶1224 S B $[(CH_3)_2N\text{-}BO]_3$ $[CH_3]$ 32
▶1227 E or N $(CH_2)_3$, CH_2OH, N 41	▶1224 VS B (0.169) $(CH_3)_2CHCH_2CH_2CH(CH_3)_2$ 137, see also 339
▶1226.2 S B (0.03) $CH_2\text{-}CH$ = CH 9	▶1224 MS B (0.051) $CH(CH_3)_2$, $CH(CH_3)_2$ 296
▶1225 VS A (12.5 mm Hg, 10 cm) F, F, F, F 222	▶1224 MSh B (0.0563) $CH_3CH_2C(CH_3)(CH_3)\text{-}CH(CH_3)CH_2CH_3$ 299
▶1225 VVSB B (film) $CF_3(CF_2)_5CF_3$ 241	▶1224 O=C-OH, C-OH=O 670
▶1225 MS B (0.0563) $(CH_3)_2CHC(CH_3)_2\text{-}CH_2CH_2CH_3$ 134	▶1224 SB CCl_4 [25%] (0.10) $CH_3CH_2CH(CH_3)CH_2OH$ 321
▶1225 MW B (film) F, CF_3, F_2, F_2, F, CF_3, F_3C, F, F_2 228	▶1222 VS B (film) F, F, F, F 222
▶1224 VS B (film) CH_3, F 220	▶1222 MW B (0.036) $H_2C{=}C(CH_3)(CH_2)_4CH_3$ 98
▶1224 MB B (film) F, CF_3, CF_3, F, F_2, F_2, F_2, F_2 231	▶1222 S B O, H_5C_2, O_2N, $N\text{-}C_2H_5$ 26
▶1224 VS B (0.10) 200	▶1222 MW B (0.238) H_3C C_3H_7-n C=C H H (cis) 276

▶1222 MS
B (0.097)

$(CH_3)_2CH(CH_2)_3CH_3$

80, see also 158, 331

▶1222 M
B

H_3C, O_2N, O, N-CH_3

26

▶1221 S
B

$(CH_3)_3CCHCH_2CH_3$ (CH_3)

340

▶1220 VVS
A (1.5 mm Hg, 10 cm)

$CF_3CF_2CF_2CF_2CF_3$

237

▶1220 VS
B (0.0104)

F

226

▶1220 M
B (0.10)

$CH_2CH_2CH_2CH_3$

176

▶1220 S
B

CH_3, CH_3, CH_3

126

▶1220 SSh
B (0.300)

$C_{26}H_{54}$
(5, 14-di-n-butyloctadecane)

283

▶1220 S
B (0.157)

$(CH_3)CCHCH(CH_3)_2$ (CH_3)

293

▶1220 S
E

NH_2, N

327

▶1220

$H_2C(C{\equiv}N)_2$

[CH_2] 466

▶1219 M
B

H_3C, O_2N, O, N-C_2H_5

26

▶1219 M
E

Pt $(NH_2CH_2CH_2NH_2)_2$ $PtCl_4$

[NH_2] 8

▶1218 VS
B (0.157)

$(CH_3)_3CCH(CH_2CH_3)_2$

78, 300

▶1218 SB
B (film)

CH_3CH_2-O-CF_2CHFCl

261

▶1218 S
B

H_3C, O_2N, O, N-C_3H_7-i

26

▶1218 S
CCl_4 [25%] (0.10)

$CH_3(CH_2)_5OH$

320

▶1218 S
B (0.003)

$CH_3(CH_2)_2CH_2NO_2$

310

▶1218 M
$CHCl_3$

N, O

15

▶1218 MW
A (15.9 mm Hg, 40 cm)

$CH_3CH_2CH(CH_2)_3CH_3$ (CH_3)

106, see also 119

▶1218

$H_3C\text{-}\overset{O}{\overset{\|}{C}}\text{-}OCH_3$

644

▶1217 VS
B (0.10)

CH_3
CH_3

186

▶1217 VSVB
B (film)

$(C_4F_9)_3N$

346

▶1217 VS
B (0.0563)

$(CH_3)_3CCH_2CH_3$

145

▶1217 SSh
B (0.105)

$CH_3CH_2\overset{CH_3}{\underset{CH_3}{C}}\text{-}CH_2CH_3$

149, see also 148, 336

▶1217 S
B

O
H_3C
O_2N
$N\text{-}C_3H_7\text{-}n$

26

▶1217 VS
B (0.10)

CH_3
CH_3

187

▶1217 MS
B (0.10)

$CH_2CH_2CH_2CH_2CH_3$

174

▶1217
E or N

CN
OCH_3
N

41

▶1216 S
B (0.10)

CH_3

192, see also 42

▶1215 VS
B (0.063)

CH_3
CN

312

▶1215 M
CS_2 [20 g/l] (0.51)

$CH_3CH{=}CH(CH_2)_2CH{=}CH(CH_2)_2CO\text{-}\underset{}{NH}\text{-}CH_2CH(CH_3)_2$

357

▶1215 SSh
CCl_4 [31%] (0.025)

Cl S $COCH_3$

372

▶1215 M
B

O
H_3C
O_2N
$N\text{-}C_6H_{13}\text{-}n$

26

▶1215 WB
B (0.163)

$H_2C{=}CH(CH_2)_6CH_3$

271

▶1215 MW
CCl_4

O
C-NH
CH_3

2

▶1214 M
B (film)

F CF_3
F_2 F_2
F_2 F_2
F_2

232

▶1214 VS
B (0.0104)

CF_3
F

218

▶1214 M
CS_2 [sat.] (0.025)

Br S $COCH_3$

348

▶1214 S
CCl_4 [25%] (0.10)

$(CH_3)_2CHCH_2CH_2OH$

195

▶1214 M
B (0.10) 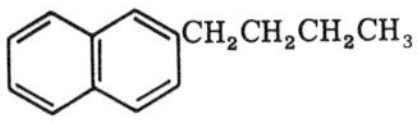
$CH_2CH_2CH_2CH_3$
175

▶1214 MW
CCl_4 [0.04 g/ml] 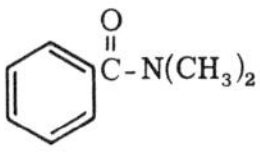
O
C-$N(CH_3)_2$
3

▶1214 S
B (0.0563)
CH_3
$(CH_3)_3CCHC(CH_3)_3$
307

▶1213 S
B (0.10) 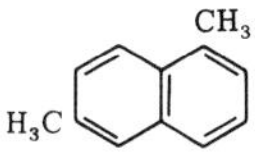
CH_3
H_3C
184

▶1213 S
B (0.10) 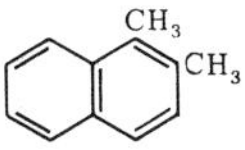
CH_3
CH_3
188

▶1213 S
B 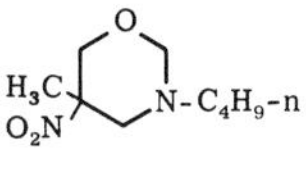
O
H_3C
N-C_4H_9-n
O_2N
26

▶1212 VSB
B (film)
n-H_7C_3-O-CF_2CHFCl
263

▶1212 VSB
B (0.0576) 
F
F
225

▶1212 MSp
E
$K_2S_2O_6$
28

▶1212 S
B 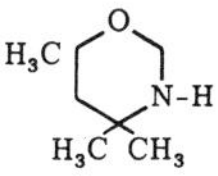
O
H_3C
N-H
H_3C CH_3
26

▶1212 M
B 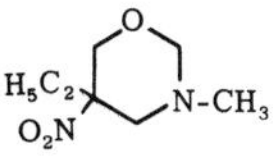
O
H_5C_2
N-CH_3
O_2N
26

▶1211 S
B (0.064)
$(CH_3)_2C$=$CHCH_2CH_3$
166

▶1211 S
B (0.030)
$(CH_3)_3CCH(CH_3)_2$
147, see also 146

▶1211 M
B (0.003)
CH_3NO_2
311

▶1211 S
B 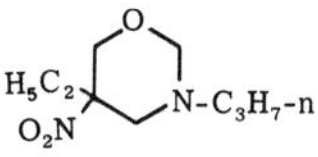
O
H_5C_2
N-C_3H_7-n
O_2N
26

▶1211 VS
B (0.08) 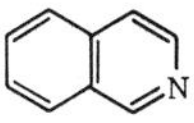
N
197

▶1211 M
B (0.064)
H CH_2CH_3
C=C
H_3C CH_3
(trans)
165

▶1210 S
CCl_4 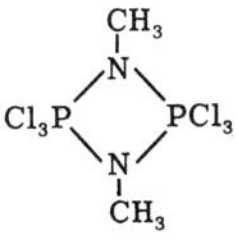
CH_3
N
Cl_3P PCl_3
N
CH_3
[CH_3]
384

▶1210 M
B (0.10) 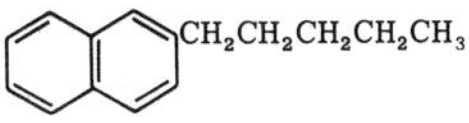
$CH_2CH_2CH_2CH_2CH_3$
173

▶1210 M
B 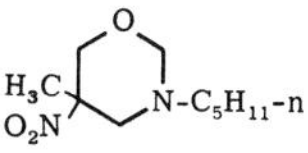
O
H_3C
N-C_5H_{11}-n
O_2N
26

▶1210 M
B

H_5C_2 O_2N (morpholine ring: O, $N-C_3H_7$-i)

26

▶1210

CF_3Cl

[CF] 447

▶1210
E or N

CN, OCH_3, N (tetrahydroquinoline ring)

41

▶1209 S
A (1.5 mm Hg, 10 cm)

$CF_3CF_2CF_3$

235

▶1209 VS
B (0.0104)

CF_2Cl-$CFCl_2$

255

▶1209 S
A

CF_3Br

11

▶1209 S
B (0.015)

H_2C=$CHC(CH_3)_3$

129

▶1209 S
$CHCl_3$

NH_2, N (pyridine ring)

327

▶1209 M
B (0.015)

H_2C=$CHCH_2C(CH_3)_3$

102, see also 273

▶1209 M
B (0.065)

CH_3, $CH(CH_3)_2$ (benzene ring)

132

▶1208 S
B

H_5C_2 O_2N (morpholine ring: O, $N-C_6H_{13}$-n)

26

▶1208 M
A (16.6 mm Hg, 40 cm)

$(CH_3)_3CCH(CH_3)CH_2CH_3$

71, see also 110

▶1208 S
B (0.169)

$CH_3(CH_2)_2CH$=$CH(CH_2)_2CH_3$
(trans)

205

▶1208 M
C

CH_3 CH_3 (naphthalene ring)

182

▶1208 M
E

CH_2OH, P (phenyl-substituted dibenzophosphole)

34

▶1207 S
B

H_5C_2 O_2N (morpholine ring: O, $N-C_5H_{11}$-n)

26

▶1207 M
B (0.10)

H_3C, CH_3 (naphthalene ring)

183

▶1206 VS
B (0.0104)

CF_3, F (benzene ring)

218

▶1206 VS
B (0.157)

$(CH_3)_3C(CH_2)_2CH(CH_3)_2$

292

▶1206 M
B (film)

F, CF_3, CF_3, F, F_2, F_2, F_2, F_2 (cyclohexane ring)

231

▶1206 MS
B (0.068)

H_3C CH_3
C=C
H CH_2CH_3
(cis)

274

▶1206 S
B

$(CH_3)_3C(CH_2)_3CH_3$

337

▶1206 MS
B (0.0563)

CH_3
$(CH_3)_2CHC - CH_2CH_2CH_3$
CH_3

134

▶1206
E or N

$[(CH_2)_3 \ldots N \ldots CH_2 + NH]_2$

41

▶1205.2 S
B (0.03)

CH_2-CH
$\Vert$
CH

9

▶1205 VS
B (0.10)

200

▶1205 VS
B (0.157)

CH_3
$(CH_3)_3CCH_2CHCH_2CH_3$

79

▶1205 VS
B (0.169)

CH_3
$(CH_3)_3CCH_2CHCH_2CH_3$

285

▶1205 M
B (film)

$F\ CF_3$
$F_2 \quad F_2$
$F_2 \quad F_2$
F_2

232

▶1205 MS
B (0.0563)

CH_3
$CH_3CH_2C - CHCH_2CH_3$
$CH_3\ CH_3$

299

▶1205 S
B

CH_3
$(CH_3)_3CCHCH_2CH_3$

340

▶1205 M
C

CH_3
CH_3

185

▶1205 M
E

CH_2OH
P
O

34

▶1205 VS
B (0.065)

$H_2C=CHCH_2C(CH_3)_3$

273, see also 102

▶1204 S
B (0.065)

$C(CH_3)_3$

133

▶1204

O
$H_3C-\overset{\Vert}{C}-OCH_3$

644

▶1203 S
B (0.0563)

$(CH_3)_3CCH_2CHCH(CH_3)_2$
CH_3

303

▶1203 VS
B (0.065)

$H_2C=CC(CH_3)$
CH_3

272

▶1203 VS
B (0.0104)

F
F
F

224

▶1203 S
B

O
H_5C_2
$N-C_4H_9-n$
O_2N

26

▶1203 M
B (film)

$F\ CF_3$
$F_2 \quad F_2$
$F \quad CF_3$
$F_3C \quad F$
F_2

228

▶1203 M
B (0.10)
$CH_2CH_2CH_2CH_3$ (on naphthalene ring)
175

▶1202 VSB
B (film)
n-H_9C_4-O-CF_2CHFCl
264

▶1202 VS
B (0.0563)
$(CH_3)_2C$=$CHC(CH_3)_3$
204

▶1202 VSB
B (0.0104)
F, F (on benzene ring)
225

▶1202 M
B (0.015)
H_2C=$CCH_2C(CH_3)_3$, CH_3
96

▶1202
E or N
NH-CH_3, N (ring structure)
41

▶1202
E or N
$(CH_2)_3$, O, C-O-C_2H_5, N (ring structure)
41

▶1201 VS
B (0.157)
$(CH_3)_3CCHCH(CH_3)_2$, CH_3
293

▶1201 VS
B (0.065)
S, CH=CH_2 (ring structure)
212

▶1200 MS
B (film)
F, CF_3, F_2, F_2, F_2, F_2, F, CF_3 (ring structure)
230

▶1200 S
G, CCl_4
$N(CH_3)_2$, NO_2 (on benzene ring)
[CH deformation]
21

▶1200 M
C
CH_3 (on naphthalene ring)
191

▶1200 MS
CCl_4 [25%] (0.10)
$CH_3(CH_2)_4OH$
393, see also 350

▶1200
H_2C=CH-O-CH=CH_2
666

▶1199 VW
B (0.0104)
CF_3, F, F (on benzene ring)
217

▶1199 S
E, F
H, N, N, CF_3, F_3C (ring structure)
37

▶1198 S
B
CH_3CH_2C - CH_2CH_3, CH_3, CH_3
336

▶1198 S
E, F
H, N, N, CH_2OH, CF_3 (ring structure)
37

▶1198 S
CCl_4
$N(C_2H_5)_2$, NO_2 (on benzene ring)
[CH deformation]
21

▶1198 MSh
B (0.300)
$C_{26}H_{54}$
(5,14-di-n-butyloctadecane)
283

▶1198 M
B (0.064)
H CH_2CH_3
C=C
H_3C CH_3
(trans)
428

▶1198
E or N 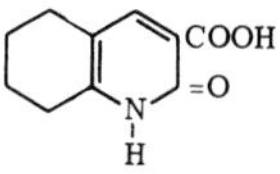
COOH
=O
N
H
41

▶1198
E or N 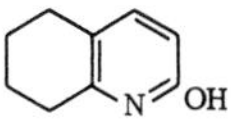
N OH
41

▶1197 VS
CCl_4 [25%]
CH_3
$H_2C=CCO_2CH_3$
193

▶1196 MSh
B (0.169)
$(CH_3)_2CHCH(CH_3)_2$
144

▶1196 MS
B (0.10) 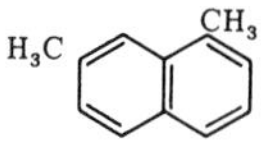
H_3C CH_3
183

▶1196 M
$CHCl_3$
O
H_3C O CH_3 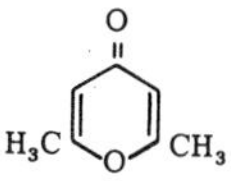
7

▶1196 M
$CHCl_3$ 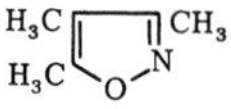
H_3C CH_3
H_3C O N
15

▶1196 MWSh
$CHCl_3$ 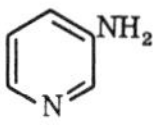
NH_2
N
433

▶1195 VS
A (6.2 mm Hg, 10 cm)
CF_2Cl-CF_2Cl
257

▶1195 VS
B (0.157)
$(CH_3)_3CCH(CH_2CH_3)_2$
78

▶1195 MS
B (0.10) 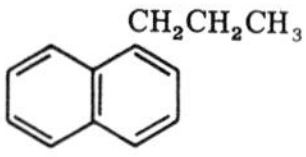
$CH_2CH_2CH_3$
178

▶1195 S
B (0.105)
CH_3
$CH_3CH_2C - CH_2CH_3$
CH_3
149

▶1195 S
E, F 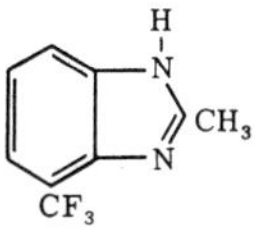
H
N
CH_3
N
CF_3
37

▶1195 S
E, F 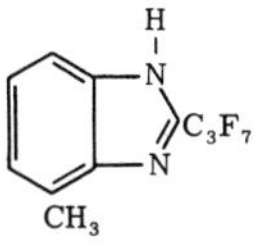
H
N
C_3F_7
N
CH_3
37

▶1195 S
E, F 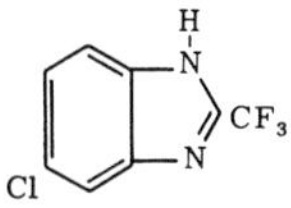
H
N
CF_3
N
Cl
37

▶1195 M
$CHCl_3$ 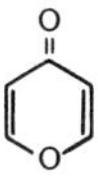
O
O
7

▶1195
E or N 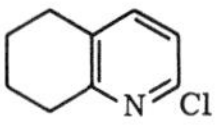
N Cl
41

▶1195
E or N 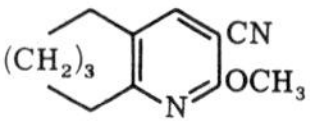
CN
$(CH_2)_3$
N OCH_3
41

▶1195
O
$H-C-O-C_2H_5$
432

▶1193 VS
$CHCl_3$
[2,6-dimethyl-4H-thiopyran-4-thione structure: S, H_3C, O, CH_3]
7

▶1193 W
B (0.036)
$H_2C{=}C(CH_3)(CH_2)_4CH_3$
98

▶1193 S
E, F
[structure: F_3C, F_3C, H, N, N]
37

▶1193 MB
B (0.136)
[structure: $CH_3(CH_2)_2H_2C$, $CH_2(CH_2)_2CH_3$, $CH_2(CH_2)_2CH_3$ on benzene]
279

▶1193 MSh
$CHCl_3$
[structure: pyridine–C(=O)–CH_3]
433

▶1193 MSh
$CHCl_3$
[structure: pyridine–C(=O)–O–C_2H_5]
433

▶1193 M
$CHCl_3$
[structure: pyridine–C(=O)–O–C_3H_7-i]
433

▶1193
E or N
[structure: $(CH_2)_3$ fused pyridine]
41

▶1192 VS
B (0.169)
$CH_3CH_2C(CH_3)_2-CH_2CH_2CH_3$
360

▶1192 VS
E
$Pt(NH_2-CH_2-CH_2-NH_2)Cl_2$
[NH_2]
8

▶1192 S
E, F
[structure: benzimidazole, H, N, N, CF_3]
37

▶1192 S
E, F
[structure: O_2N-benzimidazole, H, N, N, CF_3]
37

▶1192 S
G
[structure: benzene with $N(C_2H_5)_2$ and NO_2]
[CH deformation]
21

▶1192 S
$CHCl_3$
[structure: pyridine with NO_2]
433

▶1192 MSSh
$CHCl_3$
[structure: pyridine–C(=O)–O–C_4H_9-i]
433

▶1192
E or N
[structure: $(CH_2)_3$ fused pyridine, NH–CH_3]
41

▶1192
E or N
[structure: $(CH_2)_3$ fused pyridine, C(=O)–O–C_2H_5]
41

▶1191 S
E, F
[structure: H_3C, H_3C benzimidazole, H, N, N, CF_3]
37

▶1191 MSSh
$CHCl_3$
[structure: pyridine–C(=O)–O–C_4H_9-s]
433

▶1191 MSh
B (0.300)
$CH_3(CH_2)_5CH(C_3H_7)(CH_2)_5CH_3$
281

▶1191 MSh $CHCl_3$ 3-(methylamino)pyridine: NH-CH_3 433	▶1190 MWSh $CHCl_3$ 3-pyridylmethanol: CH_2OH 433
▶1191 MSh $CHCl_3$ pyridine-3-C(=O)-O-C_3H_7-n 433	▶1190 MWSh $CHCl_3$ 3-pyridyl-N(CH_3)-C(=O)-CH_3 433
▶1190 VS B (0.136) $(CH_3)_2HC$, CH$(CH_3)_2$, CH$(CH_3)_2$ (benzene) 278	▶1190 MWSh $CHCl_3$ 3-pyridyl-N(H)-C(=O)-phenyl 433
▶1190 MS CS_2 HO, CH_3, $CH_2CH=CHCH_3$, O (cyclopentanone) (trans) 430	▶1190 SO_2Cl, CH_3 (benzene) 389
▶1190 W B (0.0153) cyclopropyl-C(=CH_2)-CH_3 288	▶1189 S B (0.055) H_3C, CH_2CH_3, CH_3, N-H (pyrrole) 343
▶1190 S E, F H-N, N, F_3C, CF_3 (benzimidazole) 37	▶1189 S B (0.051) CH$(CH_3)_2$, CH$(CH_3)_2$ (benzene) 298
▶1190 S E, F H-N, N, CF_3, F_3C, CF_3 (benzimidazole) 37	▶1189 MS 3-pyridyl-CH$(C_2H_5)_2$ 431
▶1190 VS CCl_4 [25%] $CH_3CH_2C(CH_3)_2$OH 194	▶1189 M E Pd (NH_2-CH_2-CH_2-NH_2)$_2$ $PtCl_4$ [NH_2] 8
▶1190 M B H_3C, CH_3, CH_3 (cyclopentane) 124	▶1188 MS B (0.169) $(CH_3)_2$CHCH(CH_3)CH_2CH_3 142
▶1190 MSh $CHCl_3$ pyridine-3-C(=O)-O-C_4H_9-n 433	▶1188 S B (0.051) CH$(CH_3)_2$, CH$(CH_3)_2$ (benzene) 297

▶1188 VS
B (0.0104) 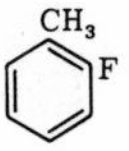
221

▶1188 S
B (film) 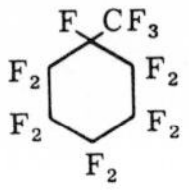
232

▶1188 S
B 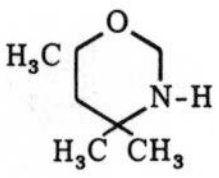
26

▶1188 S
B (0.114) 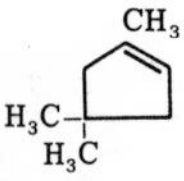
127

▶1188 S
E, F 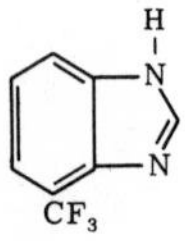
37

▶1188 S
E, F 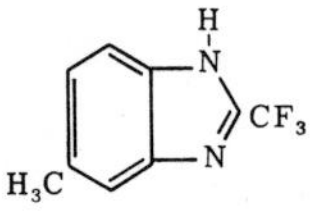
37

▶1188 S
E, F 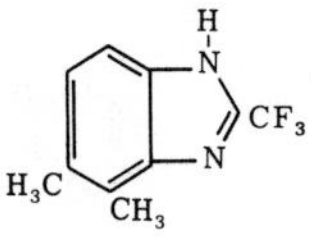
37

▶1188 VS
B (0.169)
$(CH_3)_2CHC(CH_3)_2-CH_2CH_3$
341

▶1188 M
B (0.169)
$(CH_3)_2CHCH(CH_3)CH_2CH_2CH_3$
139

▶1188 M
$CHCl_3$ 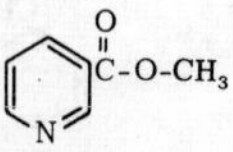
433

▶1188 M
$CHCl_3$ 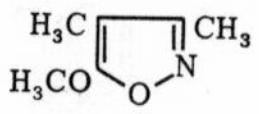
15

▶1188 M
B
$(C_6H_5)_2SiCl_2$
86

▶1188 MW
$CHCl_3$ 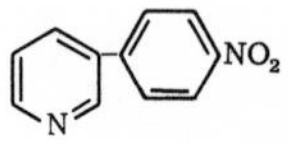
433

▶1188 W
$CHCl_3$ 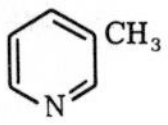
433, see also 453

▶1187 VS
B (0.151) 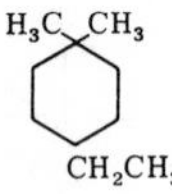
215

▶1186 VS
B (0.0563)
$(CH_3)_2CHC(CH_3)_2-CH_2CH_2CH_3$
134

▶1186 VS
A (6.2 mm Hg, 10 cm)
CF_2Cl-CF_2Cl
257

▶1186 VSSp
E
$Na_2H_2P_2O_6$
[$PO_2(OH)$ stretch]
28

▶1186 VS
$CHCl_3$ 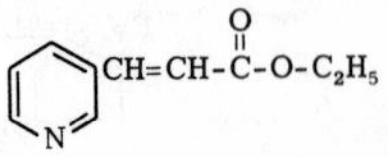
433

▶1186 VS
C
$(CH_3)_3CNO_2$
325

▶1186 S

N C_3H_7-n

431

▶1186 M
B (0°C)

$(CH_3)_2CHCHCH(CH_3)_2$
CH_3

135, 284

▶1186 VS
B (0.169)

CH_3 CH_3
$CH_3CH_2C-CH_2CCHCH_2CH_3$
CH_3

302

▶1186 MS

N $CH(C_2H_5)_2$

431

▶1185 VS
B (0.10)

CH_3
CH_3

188

▶1185 MSh
B (film)

$CF_3(CF_2)_5CF_3$

241

▶1185 VS
B (0.169)

$(CH_3)_3CCH_2CH_2C(CH_3)_3$

305

▶1185 S
E

CH_2OH
P

34

▶1185

H_3C-C-

457

▶1184 S
CCl_4

CH_3
N
Cl_3P PCl_3
N
CH_3

[CH_3] 384

▶1184 S
B (0.169)

$(CH_3)_2CHCH(CH_2CH_3)_2$

136

▶1184
E or N

N

41

▶1183 VS
B (0.0104)

F
F

225

▶1183 VSB
B (0.169)

$(CH_3)_2CHCHCH(CH_3)_2$
CH_2
CH_3

301

▶1183 VS
B (0.065)

$H_2C=C-C(CH_3)_3$
CH_3

272

▶1183 MWSh
$CHCl_3$

N -O-C_2H_5

433

▶1183 (fundamental)
A

CH_2Br_2

[CH_2] 646

▶1182 VS
B (0.136)

$CH(CH_3)_2$
H_3C CH_3

277

▶1182 M
B (film)

F CF_3
F_2 CF_3
F
F_2 F_2
F_2

231

▶1182 VS
B (0.169)

CH_3
CH_3CH_2C - $CHCH_2CH_3$
CH_3 CH_3

299

▶1182 S
C 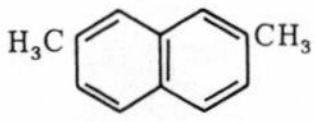
H_3C CH_3
179

▶1182 S
E, F 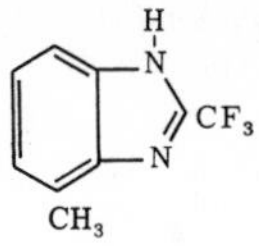
H N CF_3 N CH_3
37

▶1182 VS
A (2.5 mm Hg, 10 cm) 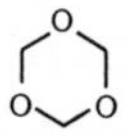
O O O
201

▶1181 S
B 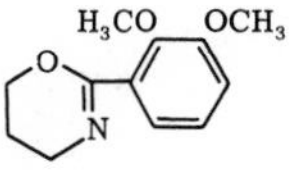
H_3CO OCH_3 O N
26

▶1181 S
G 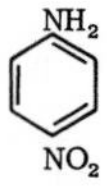
NH_2 NO_2
[CH deformation]
21

▶1181 MSh
B (0.169)
CH_3 CH_3
CH_3CH_2CH-$CHCH_2CH_3$
83

▶1181 MWB
B (0.163)
$H_2C{=}CH(CH_2)_7CH_3$
270

▶1181
E or N 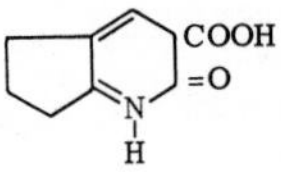
COOH =O N H
41

▶1180 VS
B (0.136) 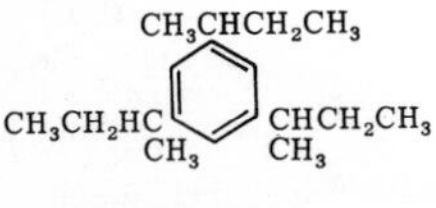
$CH_3CHCH_2CH_3$
CH_3CH_2HC $CHCH_2CH_3$
CH_3 CH_3
280

▶1180 VS
B (0.151) 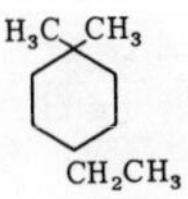
H_3C CH_3
CH_2CH_3
215

▶1180 S
B
$(C_2H_5$-NH-B-N-$C_2H_5)_3$
[CH_3]
32

▶1180 S
B
$[(CH_3)_2N$-$BO]_3$
[C_2N]
32

▶1180 S
B (0.10) 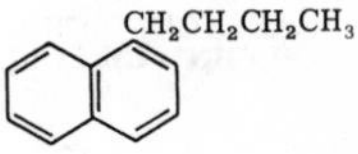
$CH_2CH_2CH_2CH_3$
176

▶1180 M
CCl_4 [0.006 m/l] 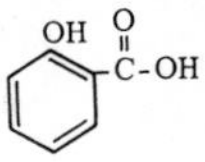
OH O C-OH
448

▶1180 M
CCl_4 [0.0128 m/l] 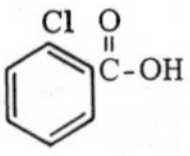
Cl O C-OH
448

▶1180
E or N 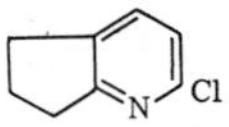
N Cl
41

▶1179 VS
B (0.063) 
CH_3 CN
214

▶1179 S
B (0.003)
$(CH_3)_2CHNO_2$
366

▶1179 MS
E, F 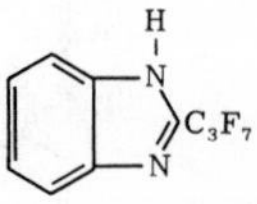
H N C_3F_7 N
37

▶1179 MW
B (0.013)
$CH_3CH{=}CHCH(CH_3)_2$
130

▶1179 MSh
B (0.157)

$(CH_3)_3CCHCH(CH_3)_2$
CH_3

293

▶1178 M
B (film)

CF_3

219

▶1178 VSSh
$CHCl_3$

CH_2-C(=O)-O-C_2H_5
N

433

▶1178 S
B (0.10)

CH_3
CH_3

187

▶1178 S
E, F

H
N
N
CF_3

37

▶1178 S
CCl_4

NH_2
NO_2

21

▶1178 MW
B (0.163)

H_2C=$CH(CH_2)_8CH_3$

352

▶1178
E or N

COOH
N
Cl

41

▶1177 M
CCl_4 [0.0082 m/l]

C(=O)-OH

25, 448, see also 38

▶1176 S
B (0.10), A (100 mm Hg, 10 cm)

$(CH_3)_2CHCH_2CH_3$

401, 424 (liquid), 423 (gas)

▶1176 S
B (film)

F CF_3
F_2 F_2
F CF_3
F_3C F
F_2

228

▶1176 MW
B (film)

F_2 F_2
F
F_2 C_2F_5
F_2

247

▶1176 VS
E, CCl_4

$[(CH_3)_2CNO_2]^-Na^+$

[NO_2 asymm. stretch]

328

▶1176 S
E, F

H
N
CH_3
N
F_3C
CF_3

37

▶1176 M
B

H_3C
CH_3
CH_3

125

▶1176 M
B (0.08)

N

197

▶1176 M
B (0.0563)

$(CH_3)_3CCH_2CHCH(CH_3)_2$
CH_3

303

▶ 1175 VSB
B (0.169)

$(CH_3)_2CHCHCH(CH_3)_2$
CH_2
CH_3

301

▶1175 W
B (0.0153)

CH_2
C
CH_3

288

▶1175 S
E

CH_2OH
P
O

34

▶1175 S
E, F 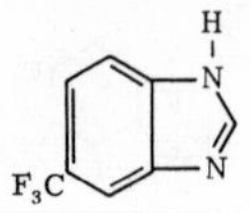
H, N, N, F_3C
37

▶1175 S
E, F 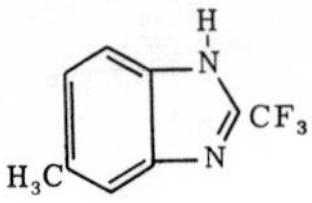
H, N, N, CF_3, H_3C
37

▶1175 S
E, F 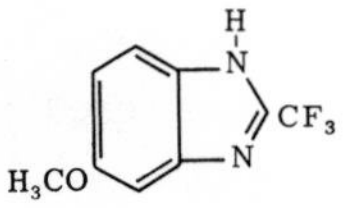
H, N, N, CF_3, H_3CO
37

▶1175 S
$CHCl_3$
S, H_3C, S, CH_3 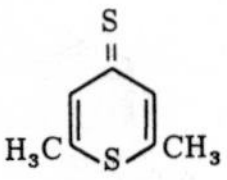
[CH_3 rock]
7

▶1175 MWSh
$CHCl_3$ 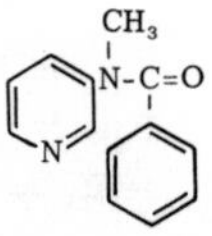
CH_3, N-C=O, N
433

▶1175 SSp
B (0.0104)
$(CF_3)_2CFCF_2CF_3$
240

▶1174 MS
B (film) 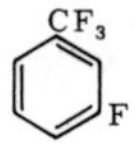
CF_3, F
218

▶1174 S
B (0.003)
$(CH_3)_2CHCH_2NO_2$
326

▶1174 VVS
B (0.0563)
$(CH_3)_3CC(CH_3)_2-CH_2CH_2CH_3$
304

▶1174 S
E or N
COOH, N 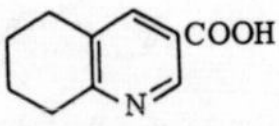
41

▶1174 M
B (0.036)
$H_2C{=}CH(CH_2)_3CH(CH_3)_2$
97

▶1174 MS
C 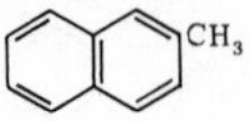
CH_3
191

▶1174 M 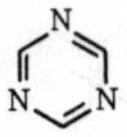
N, N, N
14

▶1173
H-C(=O)-O-CH_3
458

▶1173 VS
B (0.028) 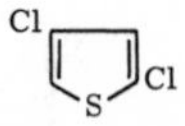
Cl, S, Cl
211

▶1173MS
B (0.10) 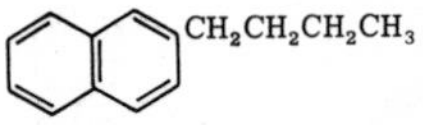
$CH_2CH_2CH_2CH_3$
175

▶1172 MS
B (0.097)
$(CH_3)_2CH(CH_2)_3CH_3$
80

▶1172 VS
A (2.5 mm Hg, 10 cm) 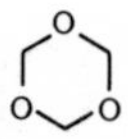
O, O, O
201

▶1172 MSh
B (0.0104) 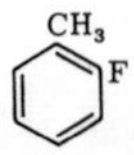
CH_3, F
221

▶1172 S
B
$(CH_3)_2CHCH_2CH(CH_3)_2$
151, 335

▶1172 S
B

$(CH_3)_2CH(CH_2)_3CH_3$

425

▶1172 S
E, F

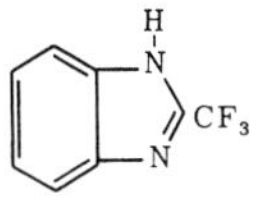

37

▶1172 S
E, F

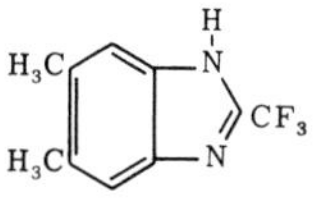

37

▶1172 S
N

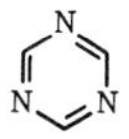

14

▶1171 MS
B (film)

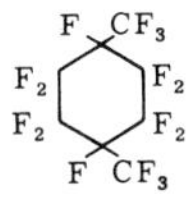

230

▶1171 VS
B (0.169), A (16.3 mm Hg, 40 cm)

$(CH_3)_2CHCH_2CH(CH_3)CH_2CH_3$

138 (liquid), 429 (gas)

▶1171 VS
B (0.169)

$(CH_3)_2CHCH_2CH_2CH(CH_3)_2$

137

▶1171 VS
C

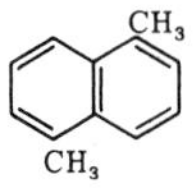

185

▶1171 MSp
A (25 mm Hg, 10 cm)

CHF_2CH_3

233

▶1171 MS
B (0.10)

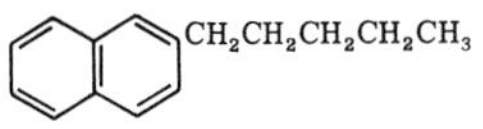

173

▶1171 M
CCl_4 [25%]

$(CH_3)_2CHCH_2CH_2OH$

195

▶1171
E or N

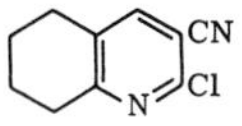

41

▶1170 M
B (film)

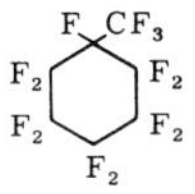

232

▶1170 VS (R branch)
A (38 mm Hg, 10 cm)

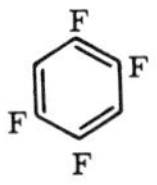

223

▶1170 MS
B (0.003)

$CH_3CH_2CH(NO_2)CH_3$

324

▶1170 SSh
B (0.10)

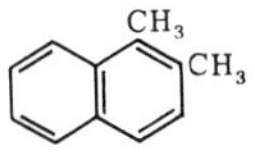

188

▶1170 S
E, F

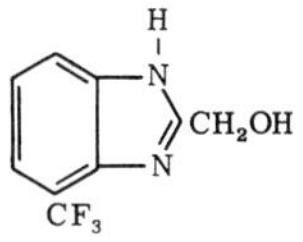

37

▶1170 S
E, F

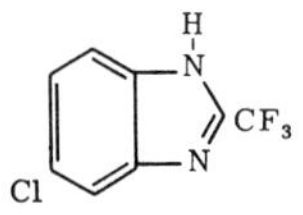

37

▶1170 S
E, F

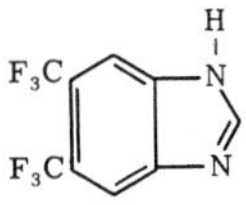

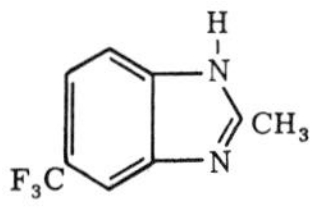

37

▶1170 S
E, F

37

▶1170 S
E, F

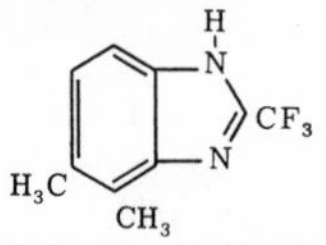

H N CF_3 N H_3C CH_3

37

▶1170 S
E, F

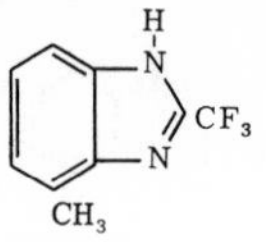

H N CF_3 N CH_3

37

▶1170 S

N C_4H_9-i

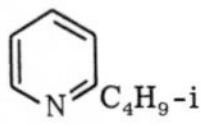

455

▶1170 MS
A (18 mm Hg, 40 cm)

$(CH_3)_2CH(CH_2)_4CH_3$

294

▶1170 M
B

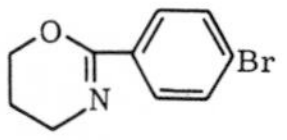

O N Br

26

▶1170

SO_2Cl CH_3

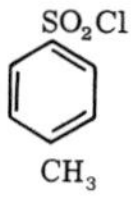

389

▶1169 S
B (0.157)

$(CH_3)_3C(CH_2)_2CH(CH_3)_2$

292

▶1169

$Pb(CH_3)_4$

402

▶1168 VS
CCl_4 [2%]

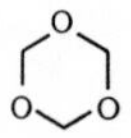

O O O

342

▶1168 VS
B (0.0563)

$(CH_3)_3CC(CH_3)_2-CH(CH_3)_2$

306

▶1168 S
B (0.028)

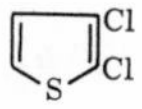

Cl Cl S

162

▶1168 S
B (0.10)

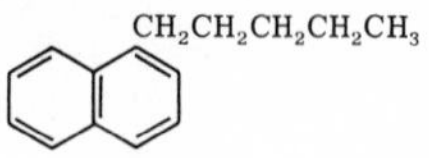

$CH_2CH_2CH_2CH_2CH_3$

174

▶1168 S
E, F

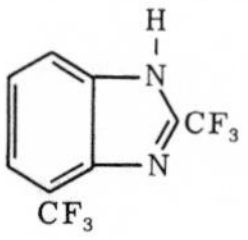

H N CF_3 N CF_3

37

▶1168 S
E, F

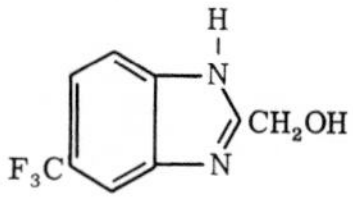

H N CH_2OH N F_3C

37

▶1168 S

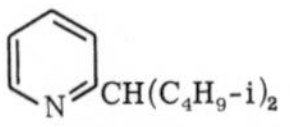

N $CH(C_4H_9\text{-}i)_2$

455

▶1168 M
B

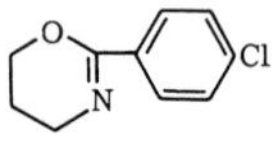

O N Cl

26

▶1168 VS
B (0.010)

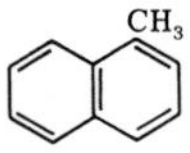

CH_3

192

▶1167 VVS
$CHCl_3$

O_2N CH_3 H_3C O N

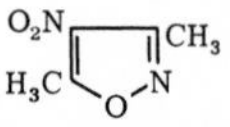

15

▶1167 VS (Q branch)
A (3.8 mm Hg, 10 cm)

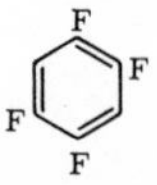

F F F F

223

▶1167 S
B

$(CH_3)_2C=C(CH_3)_2$

459

▶1167 S
C
CH_3 CH_3 (1,8-dimethylnaphthalene)
182

▶1167 S
B (0.169)
$(CH_3CH_2)_3CH$
143

▶1167 M
CCl_4 [25%]
$CH_3CH_2C(CH_3)_2$ OH
194

▶1167 M
E
N, N, N (1,3,5-triazine)
14

▶1166.2 S
B (0.03)
CH_2-CH = CH
9

▶1166 VS
E
$[(CH_3)_2CNO_2]^-Na^+$
[NO_2 stretch]
328

▶1166 S
B (0.063)
CH_3 CN
312

▶1166 S
E, F
H N CF_3 N F_3C
37

▶1166 MS
B (0.101)
CH_3
$(CH_3)_2CHCHCH_2CH_3$
142, 153

▶1166 M
CCl_4 [25%]
CH_3
$CH_3CH_2CHCH_2OH$
321

▶1166 SB
B (0.10)
CH_2CH_3
189

▶1166 MB
B (0.056)
CH_2=CHCH=CHCH=CH_2
203

▶1166 M
E
Cu (NH_2 - CH_2 - CH_2 - $NH_2$$)_2$ $PtCl_4$
[NH_2]
8

▶1165 S
E
Pd (NH_2 - CH_2 - CH_2 - NH_2) Cl_2
[NH_2]
8

▶1165 S
E, F
H N CH_3 N F_3C
37

▶1165 M
CCl_4 [0.0128 m/l]
Cl O C-OH
448

▶1165 M
$CHCl_3$
H_3C CH_3
H_3C-C(=O)-O-N(H) O N
15

▶1164 MS
C
H_3C CH_3
180

▶1164 S
B (0.10)
CH_2CH_3
190

▶1164 MS
$CHCl_3$
S
H_3C O CH_3
[CH_3 rock]
7

▶1164 MS
B (0.10) 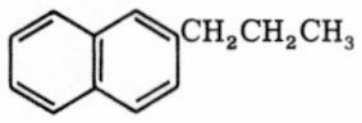
$CH_2CH_2CH_3$
177

▶1163 VS
B (0.10) 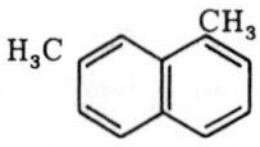
H_3C CH_3
183

▶1163 VS
A (200 mm Hg, 10 cm)
$(CH_3)_2CHCH_2CH_3$
333

▶1163 VS
B
$(CH_3)_2CHOH$
460

▶1163 VS
CCl_4
$[(CH_3)_2CNO_2]^-Na^+$
[NO_2 stretch] 328

▶1163 S
B (0.169)
$(CH_3)_2CHCHCH(CH_3)_2$
CH_3
135

▶1163 S
B (0.0563)
CH_3
$(CH_3)_3CCHC(CH_3)_3$
307

▶1163 S
E, F. 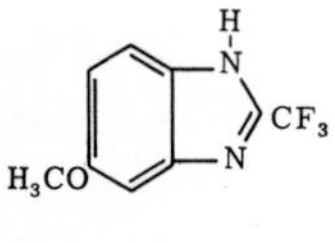
H N CF_3 N H_3CO
37

▶1163 S
E, F 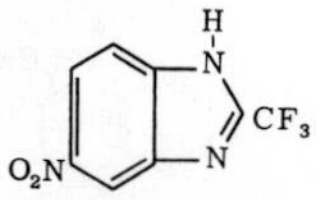
H N CF_3 N O_2N
37

▶1163 M
B (°C)
$(CH_3)_2CHCHCH(CH_3)_2$
CH_3
284, see also 135

▶1163 VS
B (0.10) 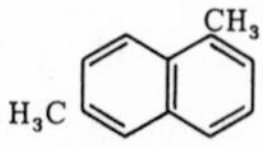
CH_3 H_3C
184

▶1162 VS
CCl_4 [25%]
CH_3
$H_2C{=}CCO_2CH_3$
193

▶1162 VS
B (0.10) 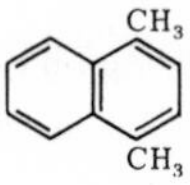
CH_3 CH_3
186

▶1162 S
E, F 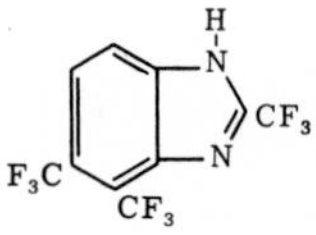
H N CF_3 N F_3C CF_3
37

▶1162 S
$CHCl_3$ 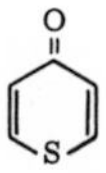
O S
[Ring] 7

▶1162 S
$CHCl_3$
O H_3C O CH_3 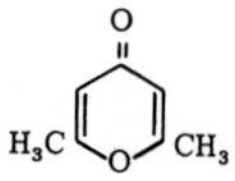
[CH_3 rock] 7

▶1162 S
CCl_4 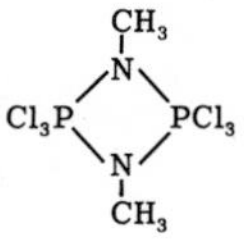
CH_3 N Cl_3P PCl_3 N CH_3
[CH_3 rock] 384

▶1162 M
CCl_4 [0.006 m/l] 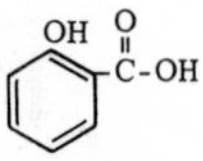
OH O C-OH
448

▶1162 VS (P branch)
B (film), A (38 mm Hg, 10 cm) 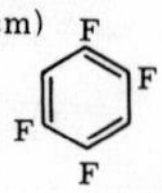
F F F F
222 (liquid), 223 (gas)

▶1161 VS
B (0.104)
$(CH_3)_3CCH(CH_3)_2$
147

▶1161 VSB
B (0.169)
$(CH_3)_2CHCHCH(CH_3)_2$
CH_2
CH_3
301

▶1161 S
B (0.169)
$(CH_3)_2CHCH(CH_2CH_3)_2$
136

▶1161 S
E 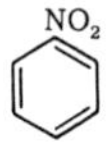
NO_2
[CH in plane]
20

▶1160 S
A
CH_3
$CH_3CH_2CHCH_2CH_3$
332

▶1160 M
B (film)
$H_3C\text{-}O\text{-}CF_2CHFCl$
356

▶1160 MW
B (0.169)
$CH_3\ CH_3$
$CH_3CH_2CH\text{-}CHCH_2CH_3$
83

▶1160 S
B (0.10) 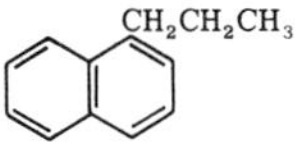
$CH_2CH_2CH_3$
178

▶1160 M
$CHCl_3$ 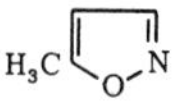
H_3C N O
15

▶1160 W
$C_6H_5SiCl_3$
86

▶1160
E or N 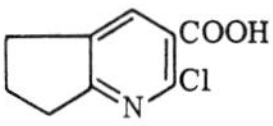
COOH
Cl
N
41

▶1159 S
B (film) 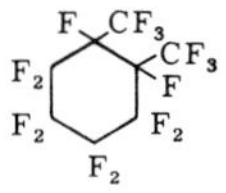
F CF_3 CF_3 F
F_2 F_2 F_2 F_2
231

▶1159 VS
B (0.0563)
$(CH_3)_2C{=}CHC(CH_3)_3$
204

▶1159 MS
B (0.10) 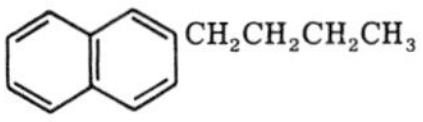
$CH_2CH_2CH_2CH_3$
175

▶1159 MS
B (0.10) 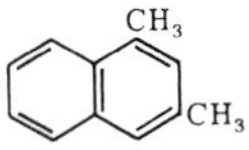
CH_3
CH_3
187

▶1159 VVS
B (0.0563)
CH_3
$(CH_3)_3CC\text{-}CH_2CH_2CH_3$
CH_3
304

▶1159 MS
A (16.6 mm Hg, 40 cm)
CH_3
$(CH_3)_3CCHCH_2CH_3$
71

▶1159 W
I
O
$H_3C\text{-}\overset{\|}{C}\text{-}NH\text{-}CH_3$
46

▶1158 MS
B (0.10) 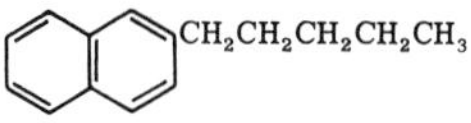
$CH_2CH_2CH_2CH_2CH_3$
173

▶1157 MS
B (film) 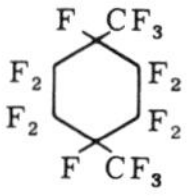
F CF_3
F_2 F_2
F_2 F_2
F CF_3
230

▶1157 VS
B (film)
$(C_4F_9)_3N$
346

▶1157 VVS
B (0.0576)

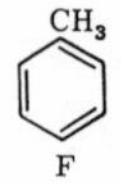

CH_3 (benzene ring) F

220

▶1157 VS
B (0.0104)

$(CF_3)_2CFCF_2CF_3$

240

▶1157 M
A (16.6 mm Hg, 40 cm), B (0.169)

$(CH_3)_2CHC(CH_3)_2CH_2CH_3$

90 (gas), 341 (liquid)

▶1157 SB
B (film)

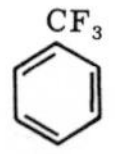

CF_3 (benzene ring)

219

▶1157
E or N

(cyclopenta[b]pyridine ring) CN, Cl, N

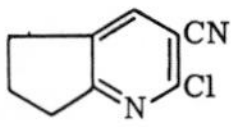

41

▶1156 VS
A (100 mm Hg, 10 cm)

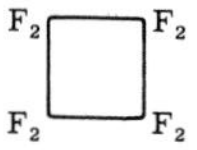

F_2, F_2, F_2, F_2 (cyclobutane ring)

243

▶1156 VS
B (0.157)

$(CH_3)_3CCH_2CH(CH_3)CH_2CH_3$

79, 285

▶1156 S
B (0.114)

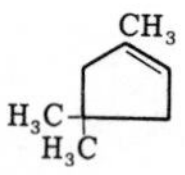

CH_3, H_3C, H_3C (cyclopentene ring)

127

▶1156 S
B (0.104)

$CH_3CH_2CH(CH_3)(CH_2)_2CH_3$

81

▶1156 S
B

$(CH_3)_2C{=}C(CH_3)_2$

459

▶1156 S
B (0.0563)

$(CH_3)_3CCH(CH_3)CH_2CH_3$

340, 426

▶1156
E or N

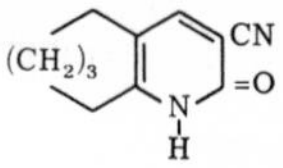

$(CH_2)_3$ (fused pyridone ring) CN, =O, N–H

41

▶1156 VS
B (0.0104)

F (benzene ring)

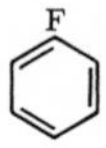

226

▶1155 VS
A (25 mm Hg, 10 cm)

$CF_3CF_2CF_3$

235

▶1155 VS
B (film)

$CF_3(CF_2)_5CF_3$

241

▶1155 SB
B (0.169)

$(CH_3CH_2)_2CHCH_2CH_2CH_3$

140

▶1155 S
P

$Pt([CH_2]_4SO)_2Cl_2$

[SO]

33

▶1155 MS
$CHCl_3$

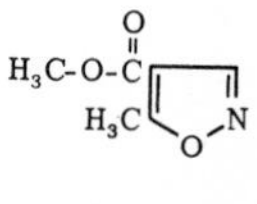

H_3C-O-C(=O) (isoxazole ring) H_3C, O, N

15

▶1155 MS
$CHCl_3$

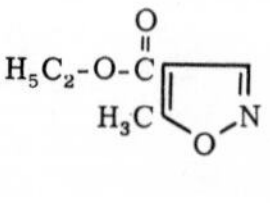

H_5C_2-O-C(=O) (isoxazole ring) H_3C, O, N

15

▶1155 MS
$CHCl_3$

n-H_7C_3-O-C(=O) (isoxazole ring) H_3C, O, N

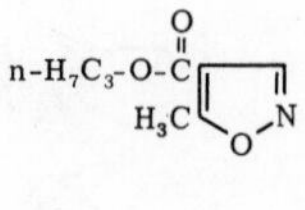

15

▶1155 VS
B (0.169)

$CH_3CH_2C(CH_3)_2-CH_2CH(CH_3)CH_2CH_3$

302

▶1155 MSh
C

(2-methylnaphthalene structure) CH_3

191

▶1155 (fundamental)
A

CH_2Cl_2

[CH_2]

376

▶1154 S
E

$Pt(NH_2-CH_2-CH_2-NH_2)_2Cl_2$

[NH_2]

8

▶1154 S
E, F

(benzimidazole, N-H) C_2F_5

37

▶1153 VS
B (0.10)

(naphthalene) CH_3, CH_3

188

▶1153 VS
B (0.169)

$(CH_3)_2CHCH_2CH(CH_3)CH_2CH_3$

138

▶1153 VS
B (0.169)

$(CH_3)_2CHCH(CH_3)_2$

144

▶1153 W
B (0.036)

$H_2C{=}C(CH_3)(CH_2)_4CH_3$

98

▶1153 S
B (0.169)

$(CH_3CH_2)_3CH$

143, 156

▶1153 S
E, F

(benzimidazole, N-H, CH_3) C_3F_7

37

▶1153 SB
B (0.10)

(naphthalene) CH_2CH_3

189

▶1152 VS
A (8.2 mm Hg, 10 cm)

$CF_3CF_2CF_2CF_2CF_3$

237

▶1152 VS
B (0.169)

$CH_3CH_2C(CH_3)(CH_3CH_3)-CHCH_2CH_3$

299

▶1152 VSSp
E

$Na_4P_2O_7$

28

▶1152 S
B (0.151)

(cyclohexane) H_3C, CH_3, CH_2CH_3

215

▶1152 S

(pyridine) $C(C_2H_5)_3$

455

▶1152 S

(pyridine) $CH(C_4H_9\text{-}i)_2$

455

▶1152 M
B (0.10)

(tetrahydronaphthalene) CH_3

269

▶1152 M
CCl_4 [0.00658 m/l]

(benzene) OH, C(=O)-OH

448

▶1151 VS
B (film)

F_9C_4-O-C_4F_9

266

▶1151 M
A (15.9 mm Hg, 40 cm)

$CH_3CH_2CH(CH_3)(CH_2)_3CH_3$

106

▶1150 S
B

[structure: 2-(p-fluorophenyl)-5,6-dihydro-4H-1,3-oxazine]

26

▶1150 S
E, F

[structure: 2-C_3F_7 benzimidazole, CH_3]

37

▶1150 S
E, F

[structure: benzimidazole, F_3C, CF_3]

37

▶1150 M
B

CH_3CH=$CHCH$=$CHCH_3$

354

▶1150 M
B

[structure: H_3C, O_2N, N-CH_3 tetrahydro-1,3-oxazine]

26

▶1150
E or N

[structure: $(CH_2)_3$ fused pyridine]

41

▶1149 VS
B (0.08)

[structure: 2-methylpyridine, CH_3]

198, see also 438

▶1149 VSB
B (film)

[structure: CF_3, F, F benzene]

217

▶1149 S
B

$CH_3CH_2CH(CH_3)(CH_2)_2CH_3$

330

▶1149 S
B

$[(CH_3)_2N$-$BO]_3$

$[CH_3]$ 32

▶1149 VS
B (0.055)

[structure: H_3C, CH_3 pyrrole, N-H]

344

▶1149 M
CCl_4

[structure: 2-aminopyridine, NH_2]

[CH in plane] 327

▶1148 VSSh
B (0.169)

$(CH_3)_2CHCH(CH_3)CH_2CH_2CH_3$

139

▶1148 S
E, F

[structure: 2-CH_3 benzimidazole, F_3C, CF_3]

37

▶1148 S
E, F

[structure: 2-CF_3 benzimidazole, F_3C, CF_3]

37

▶1148 MS
B (0.157)

$(CH_3)_3CCH(CH_2CH_3)_2$

78

▶1148 MS
B (0.10)

[structure: naphthalene, $CH_2CH_2CH_3$]

177

▶1148
E or N

[structure: tetrahydroquinoline, COOH, Cl]

41

▶1147 VS
CCl_4 [25%]
$CH_3CHCH_2CH_3$
OH
319

▶1147 MS
B (0.10)
$CH_2CH_2CH_2CH_3$
175

▶1147 VSVB
CCl_4 [25%]
$CH_3CH(CH_2)_2CH_3$
OH
318

▶1147 MW
B (0.169)
$CH_3(CH_2)_2CH{=}CH(CH_2)_2CH_3$
(trans)
205

▶1147 M
B (0.300)
$(C_2H_5)_2CH(CH_2)_{20}CH_3$
282

▶1146 S
B
H_3C O N-H H_3C CH_3
26

▶1146 S
E, F
H N CF_3 N H_3C CH_3
37

▶1146 MB
B (0.136)
$CH(CH_3)_2$
$(CH_3)_2HC$ $CH(CH_3)_2$
278

▶1146 M
B (0.104)
$CH_3CH_2CH(CH_2)_2CH_3$
CH_3
81

▶1146 MSh
B (0.10)
CH_3
192

▶1146 S
B (0.101)
$(CH_3)_2CHCH_2CH_3$
401

▶1145 S
C
CH_3
CH_3
181

▶1145 MS
B (0.127)
$CH_3(CH_2)_2CH{=}CH(CH_2)_2CH_3$
(cis)
291

▶1145 MS
B (0.10)
$CH_2CH_2CH_2CH_2CH_3$
173

▶1145 M
C
CH_3
191

▶1145 (fundamental)
A
$CHBr_3$
[CH]
454

▶1145 VS
CCl_4 [25%]
$CH_3CH_2CHCH_2CH_3$
OH
196

▶1144 SShB
B (0.169)
$(CH_3)_2CHCHCH(CH_3)_2$
CH_2
CH_3
301

▶1144 S
E, F
H N CF_3 N H_3C
37

▶1144 S
E, F
H N F_3C F_3C N
37

▶1144 MW
B (0.003)
$CH_3CH_2CH(NO_2)CH_3$
324

▶1144 MSSh
B (0.101)
$(CH_3)_2CHCH(CH_3)CH_2CH_3$
153

▶1144 M
A (40 mm Hg, 58 cm) 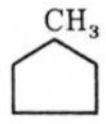
CH_3
105

▶1144 M
CCl_4 [0.0128 m/l] 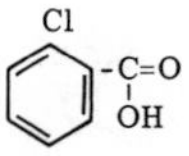
Cl
-C=O
OH
448

▶1144 VS
B (0.0104) 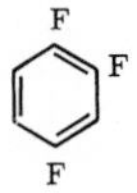
F
F
F
224

▶1143 MS
B (0.0563)
$(CH_3)_3CCH_2CH(CH_3)CH(CH_3)_2$
303

▶1143 S
E 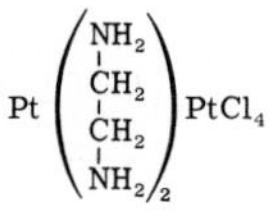
$Pt(NH_2CH_2CH_2NH_2)_2PtCl_4$
[NH_2]
8

▶1143 MS
E, F 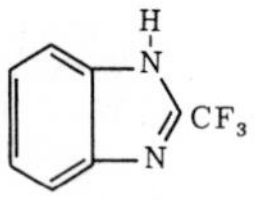
H
N
CF_3
N
37

▶1143 M
E 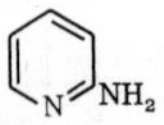
N NH_2
327

▶1143 VS
E, CCl_4
$[(CH_3)_2CNO_2]^- Na^+$
[C-C-C asymm. stretch]
328

▶1142 S
B (0.003)
$(CH_3)_2CHCH_2NO_2$
326

▶1142 S
B (0.10) 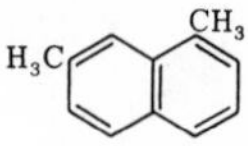
H_3C CH_3
183

▶1142 S
E, F 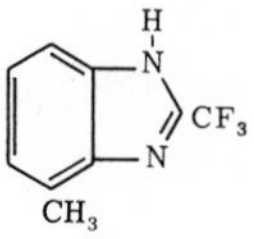
H
N
CF_3
N
CH_3
37

▶1142 VS
A (8.2 mm Hg, 10 cm)
CHF_2CH_3
233

▶1141 M
B 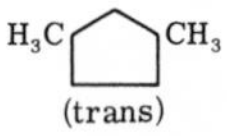
H_3C CH_3
(trans)
464

▶1141 VS
B (0.157)
$(CH_3)_3CCH(CH_3)CH(CH_3)_2$
293

▶1140 M
B (0.101)
$CH_3(CH_2)_5CH_3$
85

▶1140 S
B (0.08) 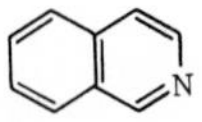
N
197

▶1140 SB
B (0.10) 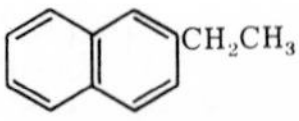
CH_2CH_3
189

▶1140 MS
B (0.300)
$CH_3(CH_2)_5CH(C_3H_7)(CH_2)_5CH_3$
281

▶1140 M
E
CH_2OH
P
34

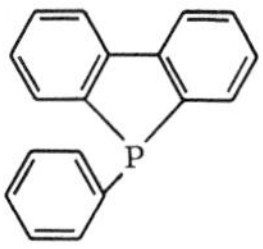

▶1140 M
E
CH_2OH
P=O
34

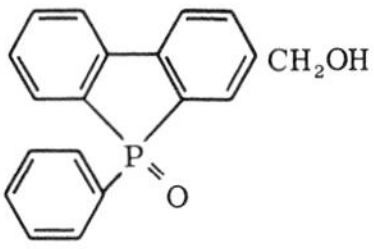

▶1140 M
CS_2 (0.728)
CH_2=CHCH=CHCH=CHCH=CH_2
202

▶1140 M
P
Pt([CH_2]$_4$SO)$_2$$Cl_2$
33

▶1140
K
La$(NO_3)_3$
45

▶1140
O
H_5C_2-C-NH_2
465

▶1140 VS
B (0.0104)
F CF_3
F_2 F_2
F_2 F_2
F_2
232

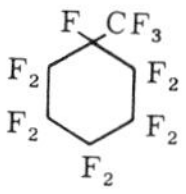

▶1139 S
B (0.10)
CH_3
CH_3
186

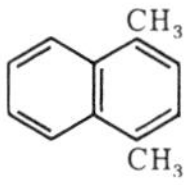

▶1139 MSh
B (film)
$CF_3(CF_2)_5CF_3$
241

▶1139 MS
B (0.300)
$C_{26}H_{54}$
(5, 14-di-n-butyloctadecane)
283

▶1139 MS
B (0.10)
$CH_2CH_2CH_3$
177

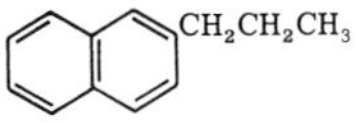

▶1139 S
B (0.224)
CH_3
104

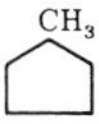

▶1138 VS
B (film)
CF_3
F
218

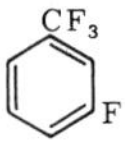

▶1138 VS
$CHCl_3$
H_3C CH_3
H_5C_2-O N
O
15

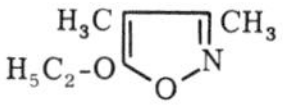

▶1138 VS
B (0.0104)
F_2 F_2
F
F_2 C_2F_5
F_2
247

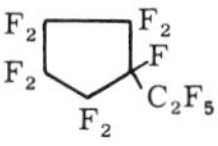

▶1138 VS
B (0.15)
128

▶1138 S
B (0.003)
$(CH_3)_2CHNO_2$
366

▶1138 S
E
Pt (NH_2-CH_2-CH_2-NH_2)$_2$ Cl_2
[NH_2]
8

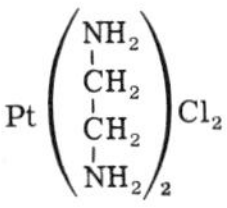

▶1138 S
E, F
H
N
H_3C CF_3
H_3C N
37

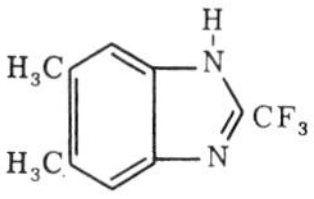

▶1137 S
E, F
H
N
CF_3
N
CF_3
37

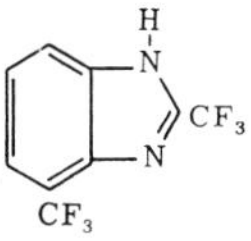

▶1136 M
B (0.0104) 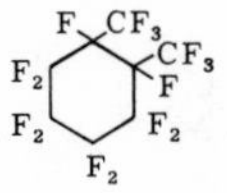
F CF$_3$ CF$_3$ F$_2$ F F$_2$ F$_2$ F$_2$
231

▶1136 VS
$CHCl_3$ 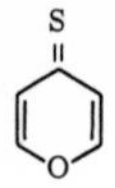
S O
7

▶1136 S
E, F 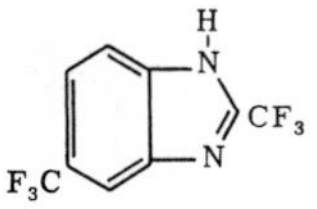
H N CF$_3$ N F$_3$C
37

▶1136 VS
A (8.2 mm Hg, 10 cm)
$CF_3CF_2CF_2CF_2CF_3$
237

▶1136 MWB
B (0.066)
$H_2C{=}CH(CH_2)_7CH_3$
270

▶1135 S
B 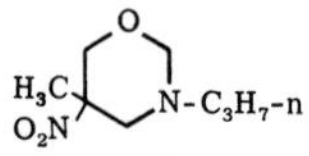
O H$_3$C N-C$_3$H$_7$-n O$_2$N
[C-O-C]
26

▶1135 S
B (0.169)
$(CH_3)_3CCH_2CH_2C(CH_3)_3$
305

▶1135 S
E, F 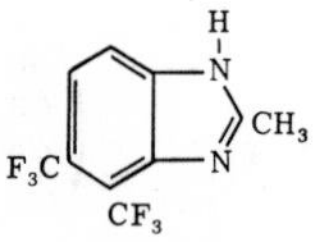
H N CH$_3$ N F$_3$C CF$_3$
37

▶1135 M
B (0.028) 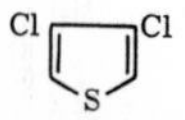
Cl Cl S
161

▶1135 M
B 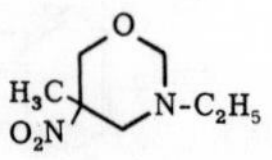
O H$_3$C N-C$_2$H$_5$ O$_2$N
[C-O-C]
26

▶1135 M
CCl_4 [0.00658 m/l] 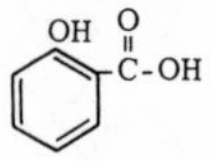
OH O C-OH
448

▶1135 W
$CHCl_3$ 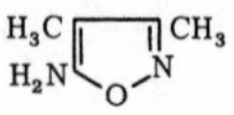
H$_3$C CH$_3$ H$_2$N O N
15

▶1135 W
(0.0563)
CHF_2CH_3
462

▶1135 VS
B (film) 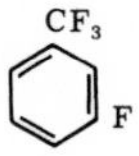
CF$_3$ F
218

▶1134 VS
$CHCl_3$ 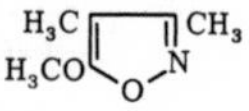
H$_3$C CH$_3$ H$_3$CO O N
15

▶1134 S
B (0.003)
$CH_3(CH_2)_2CH_2NO_2$
310

▶1134 MW
B (0.0563)
$(CH_3)_3CCH(CH_3)C(CH_3)_3$
307

▶1133 VS
B (0.0563)
$(CH_3)_3CC(CH_3)_2-CH(CH_3)_2$
306

▶1133 S
B (0.003)
$C_3H_7NO_2$
392

▶1132 SSh
B (0.101)
$(CH_3)_2CHCH(CH_3)CH_2CH_3$
153

▶1132 S
B
[C-O-C] 26

▶1132 MS
E, F
37

▶1132 M
$CHCl_3$
15

▶1132 VW
14

▶1131 VS
B
[C-O-C] 26

▶1131 M
CCl_4 [0.0128 m/l]
448

▶1131 S
B (0.10)
187

▶1131 SB
B (0.169)
$(CH_3CH_2)_2CHCH_2CH_2CH_3$
140

▶1131 S
E
[NH_2] 8

▶1131 M
E, $CHCl_3$
433

▶1131 W
$CHCl_3$
15

▶1130 S
B (0.056)
$CH_2=CHCH=CHCH=CH_2$
203

▶1130 VS
B
$(CH_3)_2CHOH$
460

▶1130 S
B (0.10)
175

▶1130 S
A (17.7 mm Hg, 40 cm)
$(CH_3)_2CHCH(CH_3)CH_2CH_2CH_3$
72

▶1130 S
B (0.10)
173

▶1130 S
E
[NH_2] 8

▶1130 S
E, F
37

▶1130 S
E, F
37

▶1130 MW
$CHCl_3$
[CH in plane] 433

▶1130 MWSh
$CHCl_3$ 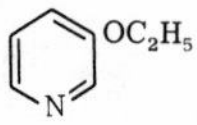
OC_2H_5 N
[CH in plane] 433

▶1128 M
B
O H_3C O_2N N-CH_3
[C-O-C] 26

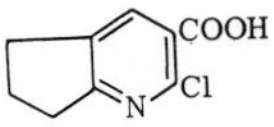

▶1130
E or N
CH_2 NNO N 2
41

▶1128 W
$CHCl_3$
NO_2 N
[CH in plane] 433

▶1130
E or N
COOH N Cl
41

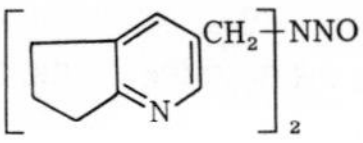

▶1127 VVS
B (0.169)
$(CH_3)_2CHCH(CH_3)CH_2CH_2CH_3$
139

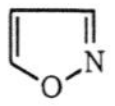

▶1129 VVS
B (0.169)
$(CH_3)_2CHCH(CH_3)_2$
144

▶1127 VSB
B (film) 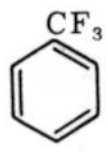
CF_3
219

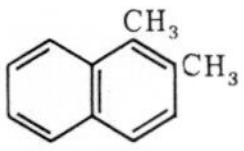

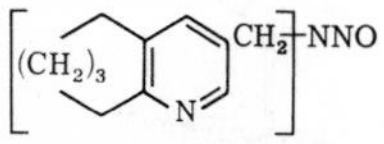

▶1129 VS
B (0.169)
$(CH_3)_2CHCH(CH_2CH_3)_2$
136

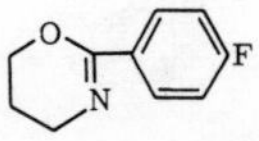

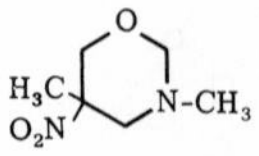

▶1127 VS
B 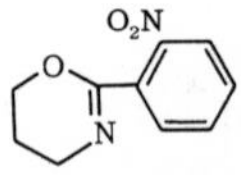
O_2N O N
[C-O-C] 26

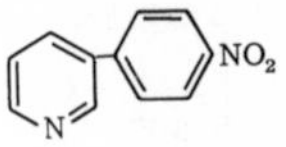

▶1129 VS
$CHCl_3$
O N
15

▶1127 S
E, F 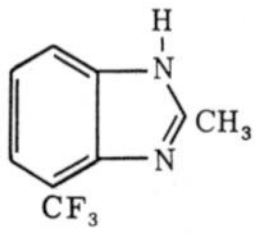
H N CH_3 N CF_3
37

▶1129 S
B (0.169)
$(CH_3CH_2)_3CH$
143

▶1127 S
P
$Pd([CH_2]_4SO)_2Cl_2$
[SO] 33

▶1129 MSh
B (0.10)
CH_3 CH_3
188

▶1127 M
B (0.101)
$(CH_3CH_2)_3CH$
156

▶1129
E or N
$(CH_2)_3$ CH_2 NNO N
41

▶1127 MWB
C
$(CH_3)_3CNO_2$
325

▶1128 VS
B
O N F
[C-O-C] 26

▶1126 S
B (0.157)
$(CH_3)_3CCH(CH_2CH_3)_2$
78

▶1126 VS
B (0.169)
$(CH_3)_2CHCHCH(CH_3)_2$
CH_2
CH_3
301

▶1126 S
E, F
37

▶1126 W
$CHCl_3$
[CH in plane] 433

▶1126 W
$CHCl_3$
15

▶1125 VVS
$CHCl_3$
15

▶1125 S
E, F
37

▶1125 MW
B (0.104)
228

▶1125 M
E, F
37

▶1125 M
CCl_4 [0.008 m/l]
25, 448, see also 38

▶1125
E or N
41

▶1124 VS
B (0°C)
$(CH_3)_2CHCHCH(CH_3)_2$
CH_3
284

▶1124 S
CCl_4 [25%]
$(CH_3)_2CHCH_2CH_2OH$
195

▶1124 VSB
CCl_4 [25%]
$CH_3CH_2CHCH_2CH_3$
OH
196

▶1124 VS
B (0.169)
$(CH_3)_2CHCHCH(CH_3)_2$
CH_3
135

▶1124 S
$CHCl_3$
7

▶1124 M
A (385 mm Hg, 10 cm)
CF_3-CCl_3
253

▶1124 MWSh
$CHCl_3$
[CH in plane] 433

▶1124 W
$CHCl_3$
[CH in plane] 433, see also 453

▶1124 W
$CHCl_3$
[CH in plane] 433

▶1123 W
$CHCl_3$
[CH in plane] 433

▶1122.7 S
B (0.03)
CH_2-CH=CH
9

▶1122 VVS
B (0.169)
$CH_3CH_2CH(CH_3)CH(CH_3)CH_2CH_3$
83

▶1122 VSSh
$CHCl_3$
(pyridine)-C(=O)-O-CH_3
[CH in plane]
433

▶1122 MW
B (0.036)
$H_2C=C(CH_3)(CH_2)_4CH_3$
98

▶1122 S
B (0.064)
$(CH_3)_2C=CHCH_2CH_3$
166

▶1122 SSp
B (0.10)
(naphthalene)-CH_2CH_3
189

▶1122 S
E, F
(benzimidazole, N-H, with CF_3)
37

▶1122 S
E, F
(benzimidazole, N-H, with CH_2OH and CF_3)
37

▶1122 MS
B (0.10)
(naphthalene)-$CH_2CH_2CH_3$
177

▶1122 MW
$CHCl_3$
(pyridine)-CH=CH-C(=O)-O-C_2H_5
[CH in plane]
433

▶1122 WSh
$CHCl_3$
(pyridine)-CH_2OH
[CH in plane]
433

▶1122 W
$CHCl_3$
(pyridine)-CN
[CH in plane]
433

▶1121 S
B (0.101)
$(CH_3)_2CHCH(CH_3)CH_2CH_3$
142, 153

▶1121 M
B (0.0104)
(cyclohexane: F, CF_3; F_2, F_2; F_2, F_2; F, CF_3)
230

▶1121 VW
B (0.0104)
(benzene with F, F, F, F)
222

▶1121 VSSp
E
$Na_4P_2O_7$
28

▶1121 VS
$C_6H_5HSiCl_2$
86

▶1121 VS
$(C_6H_5)_2SiCl_2$
86

▶1121 S
B (0.157)
$(CH_3)_3C(CH_2)_2CH(CH_3)_2$
292

▶1121 SSh
B (0.003)
$CH_3CH_2CH(NO_2)CH_3$
324

▶1121 S $CHCl_3$ [structure] 15	▶1120 WSh $CHCl_3$ [structure] [CH in plane] 433
▶1120 VS B $C_6H_5SiCl_3$ 86	▶1120 W $CHCl_3$ [structure] 15
▶1120 VS B $C_6H_5(CH_3)SiCl_2$ 86	▶1119 S B (0.0104) $CF_3(CF_2)_5CF_3$ 241
▶1120 S B (0.028) [structure] 161	▶1119 VS B $(C_6H_5)_2(CH_3)SiCl$ 86
▶1120 S B (0.063) [structure] 214	▶1119 S B [structure] [C-O-C] 26
▶1120 S E, F [structure] 37	▶1119 S CCl_4 [25%] $CH_3(CH_2)_5OH$ 320
▶1120 S E, F [structure] 37	▶1119 S CCl_4 [25%] $CH_3(CH_2)_7OH$ 317
▶1120 M E [structure] 34	▶1119 S E $[Coen_2SO_3]Cl$ 5
▶1120 M E [structure] 34	▶1119 S E $[Coen_2SO_3]I$ 5
▶1120 MWSh $CHCl_3$ [structure] [CH in plane] 433	▶1119 S E, F [structure] 37

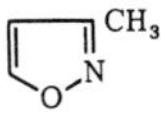

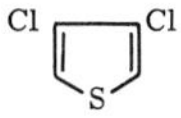

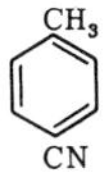

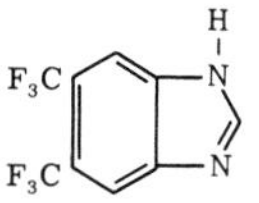

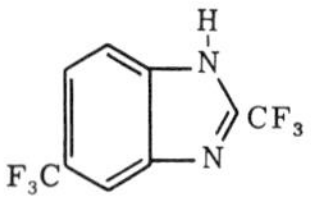

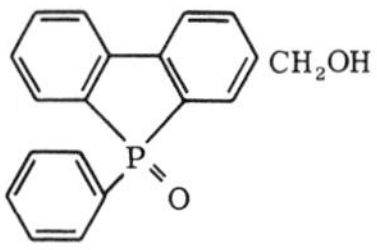

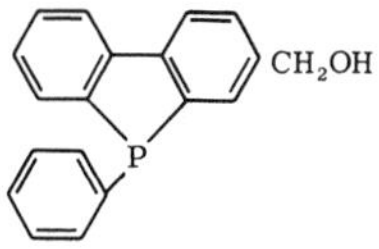

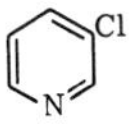

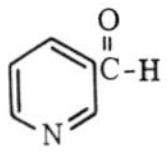

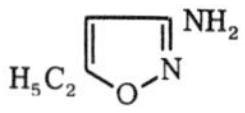

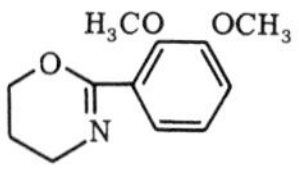

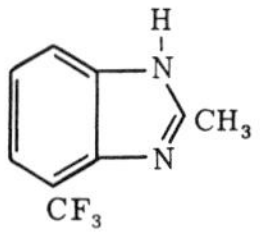

▶1119 M
B (0.169)
$CH_3CH_2C(CH_3)_2-CH(CH_3)CH_2CH_3$
299

▶1118 S
E, F
5-methoxy-2-(trifluoromethyl)benzimidazole (H_3CO, CF_3, N-H)
37

▶1118 M
B
$(C_2H_5)_2P\cdot BH_2$
[BH_2]
30

▶1118 MW
$CHCl_3$
2-pyridyl–$C(=O)-CH_3$
[CH in plane]
433

▶1117 MSh
A (25 mm Hg, 10 cm)
$CF_3CF_2CF_3$
235

▶1117 VS
B
cyclopentene with CH_3, CH_3, CH_3
126

▶1117 VW
B (0.2398)
p-difluorobenzene (F, F)
225

▶1117 S
E
$[Coen_2SO_3Cl]^0$
5

▶1117 S
E
$[Coen_2SO_3OH]^0$
5

▶1117 S
E
$[Coen_2SO_3]SCN$
5

▶1117 S
E
phthalide bearing two (HO, H_3C)-phenyl and (OH, CH_3)-phenyl groups
39

▶1117 W
B (0.0563)
$(CH_3)_3CCH(CH_3)CH_2CH_3$
340, 426

▶1117
K
$Mn(NO_3)_2\cdot xH_2O$
45

▶1117 MWSh
$CHCl_3$
pyridyl–$CH_2-C(=O)-O-C_2H_5$
[CH in plane]
433

▶1117 MWSh
$CHCl_3$
bromopyridine (Br)
[CH in plane]
433

▶1117 MS
B (0.10)
naphthyl–$CH_2CH_2CH_2CH_3$
176

▶1116 VSB
CCl_4 [25%]
$CH_3CH(OH)CH_2CH_3$
319

▶1116 MW
B (0.066)
$H_2C=CH(CH_2)_7CH_3$
270

▶1116 S
CCl_4 [25%]
CH_3OH
316

▶1116 S
A
$(i-C_4H_9)_2NH$
[CN stretch]
388

▶1116 S
B (0.064)

H CH_2CH_3
C=C
H_3C CH_3
(trans)

165

▶1116 MW
B (0.163)

$H_2C{=}CH(CH_2)_8CH_3$

352

▶1116 M
G

$\left[C_5H_4\overset{O}{\overset{\|}{C}}\text{-}CH_3\right]Ru[C_5H_5]$

[Cyclopentadienyl ring] 452

▶1116 M
G

$\left[C_5H_4\overset{O}{\overset{\|}{C}}\text{-}CH_3\right]Os[C_5H_5]$

[Cyclopentadienyl ring] 452

▶1116 W
$CHCl_3$

H_3C (isoxazole ring, O–N) NH_2

15

▶1115 MB
B (0.169)

$(CH_3)_3CCH_2\overset{CH_3}{CH}CH_2CH_3$

285

▶1115 VS
CS_2

$(C_6H_5)_3SiCl$

86

▶1115 S
B (0.065)

$C(CH_3)_3$ (benzene ring)

133

▶1115 S
B

H_3C O N-H H_3C CH_3 (morpholine ring)

26

▶1115 S
G

$N(CH_3)_2$ (benzene ring) NO_2

21

▶1115 MB
B (0.169)

$CH_3CH_2\overset{CH_3}{\underset{CH_3}{C}}\text{-}CH_2\overset{CH_3}{CH}CH_2CH_3$

302

▶1115 VW
B (0.300)

$C_{26}H_{54}$
(5, 14-di-n-butyloctadecane)

283

▶1115 M
G

$\left[C_5H_4\overset{O}{\overset{\|}{C}}\text{-}CH_3\right]Fe[C_5H_5]$

[Cyclopentadienyl ring] 452

▶1115 M
G

$\left[C_5H_4\overset{O}{\overset{\|}{C}}\text{-}CH_3\right]Fe\left[C_5H_4\overset{O}{\overset{\|}{C}}\text{-}CH_3\right]$

452

▶1115
E or N

$\left[\text{(tetrahydroquinoline ring, N)}\,CH_2\right]_2NH$

41

▶1114 VS
B (0.157)

$(CH_3)_3C\underset{CH_3}{CH}CH(CH_3)_2$

293

▶1114 VS
CCl_4 [25%]

$CH_3\underset{OH}{CH}(CH_2)_2CH_3$

318

▶1114 S
B (0.068)

H_3C CH_3
C=C
H CH_2CH_3
(cis)

274

▶1114 S
CCl_4 [25%]

$CH_3(CH_2)_4OH$

393, see also 350

▶1114 MS
B (0.0563)

$(CH_3)_3CCH_2\underset{CH_3}{CH}CH(CH_3)_2$

303

▶1114 W
B (0.127)
$H_2C{=}CH(CH_2)_5CH_3$
290

▶1114 W
$CHCl_3$
NO_2, N
[CH in plane]
433

▶1113 VS
B (0.055)
CH_3, H_3C, N, H
344

▶1113 S
B
O, H_3C, O_2N, $N\text{-}CH_3$
[C-O-C]
26

▶1113 MW
B (0.036)
$H_2C{=}C(CH_3)(CH_2)_4CH_3$
98

▶1113 S
G, CCl_4
$N(C_2H_5)_2$, NO_2
21

▶1113 MS
B (0.10)
$CH_2CH_2CH_2CH_2CH_3$
173

▶1112 MS
B (0.0563)
$(CH_3)_2CHC(CH_3)_2\text{-}CH_2CH_2CH_3$
134

▶1112 VS
B (0.063)
CH_3, CN
312

▶1112 S
CCl_4 [25%]
$CH_3(CH_2)_3OH$
353

▶1112 S
E
$[Coen_2SO_3 \cdot NH_3]Cl$
5

▶1112 S
G
$H_3C\text{-}C(=O)\text{-}C(=S)\text{-}N(CH_2\text{-}CH_2)_2O$
$[N(CH_2\text{-}CH_2)_2O]$
31

▶1112 S
G, CCl_4
NH_2, NO_2
21

▶1112 S
CCl_4
$N(CH_3)_2$, NO_2
21

▶1112-1100 SVB
E
Cl, O, O
39

▶1112 S
B (0.10)
CH_3 CH_3
182

▶1112
E or N
$[\ \ CH_2]_2NH$, N
41

▶1111 SSh
A (6.2 mm Hg, 10 cm)
$CF_2Cl\text{-}CF_2Cl$
257

▶1111 S
CCl_4 [25%]
$CH_3CH_2CH(CH_3)CH_2OH$
321

▶1111 S
B (0.165)
CH_3, $CH(CH_3)_2$
132

▶1111 SSh
A
$(CH_3)_2SO$
13

▶1111 M
A (36 mm Hg, 15 cm)
$H_2C{=}CHCH(CH_3)_2$
131

▶1111 M
B (0.10) 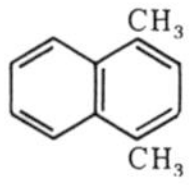
CH_3
CH_3
186

▶1111 M
P
$[(CH_3)_2P \cdot BH_2]_3$
[BH_2]
30

▶1111 W
$CHCl_3$ 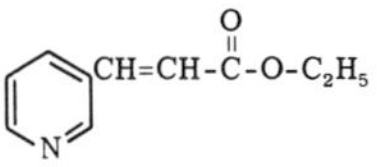
O
CH=CH-C-O-C_2H_5
N
[CH in plane]
433

▶1111 W
$CHCl_3$ 
S
O
7

▶1111 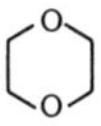
O
O
451

▶1110 M
B (0.0104) 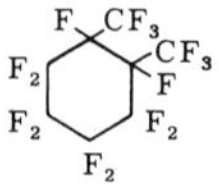
F CF_3
F_2 CF_3
F
F_2 F_2
F_2
231

▶1110 S
E, F 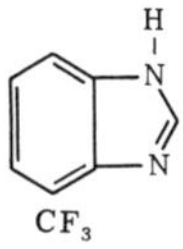
H
N
N
CF_3
37

▶1110 VS
B (0.003)
NO_2
$CH_3CH_2CHCH_3$
324

▶1110 S
E, F 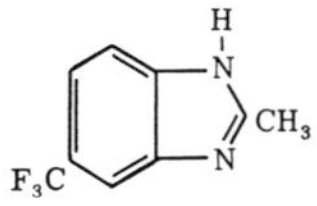
H
N
CH_3
N
F_3C
37

▶1109 W
E 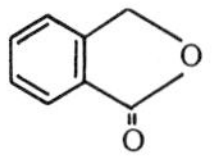
O
O
39

▶1109 MW
B (0.0104) 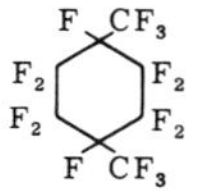
F CF_3
F_2 F_2
F_2 F_2
F CF_3
230

▶1109 VS
B (0.0104)
$CF_2Cl\text{-}CFCl_2$
255

▶1109 VSB
CCl_4 [25%]
$CH_3CH_2CHCH_2CH_3$
OH
196

▶1109 S
B 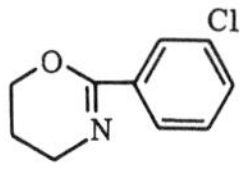
Cl
O
N
[C-O-C]
26

▶1109 S
E 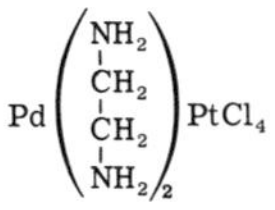
Pd (NH_2 — CH_2 — CH_2 — NH_2)$_2$ $PtCl_4$
[NH_2]
8

▶1109 M
$CHCl_3$ 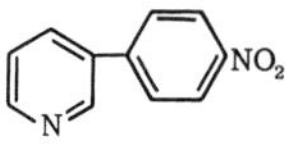
NO_2
N
[CH in plane]
433

▶1109 M
$C_6H_5(CH_3)_2SiCl$
86

▶1109 VS
B (0.169)
CH_3
$(CH_3)_2CHC - CH_2CH_3$
CH_3
341

▶1109 W
$CHCl_3$ 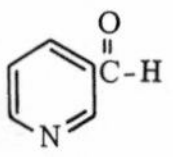
[CH in plane]
433

▶1109
CH_3OH
437

▶1108 MS
B (0.10) 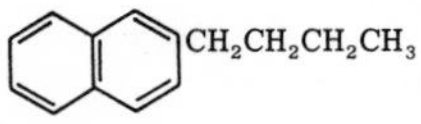
$CH_2CH_2CH_2CH_3$
175

▶1108 VS
B (0.136) 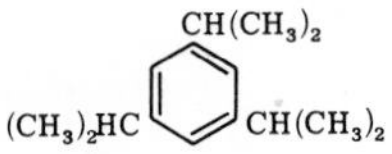
$CH(CH_3)_2$
$(CH_3)_2HC$ $CH(CH_3)_2$
278

▶1108 VS
$CHCl_3$ 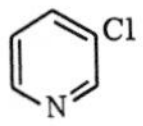
Cl
N
[CH in plane]
433

▶1107 S
E 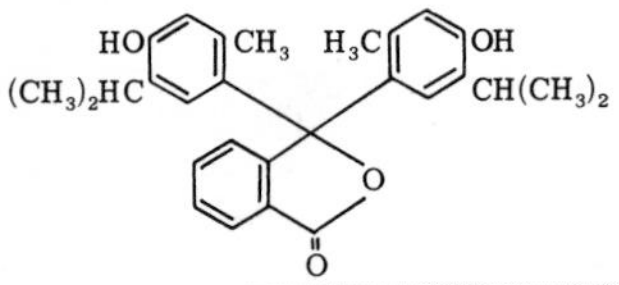
HO CH_3 H_3C OH
$(CH_3)_2HC$ $CH(CH_3)_2$
O
O
39

▶1107 S
B 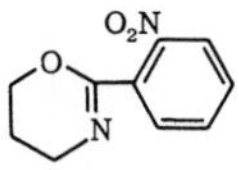
O_2N
O
N
[C-O-C]
26

▶1107 S
E 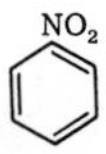
NO_2
20

▶1107 S
E, F 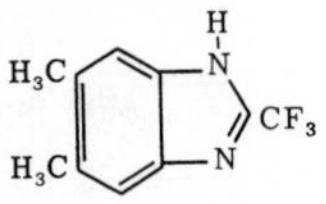
H
H_3C N
CF_3
H_3C N
37

▶1107 S
$(C_6H_5)_2SiCl_2$
86

▶1107 M
B (0.151) 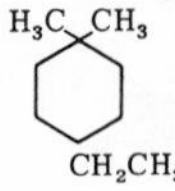
H_3C CH_3
CH_2CH_3
215

▶1107 M
B
$(CH_3)_3C(CH_2)_3CH_3$
337

▶1106.4 M
B (0.03) 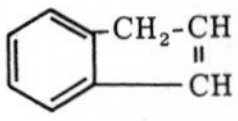
CH_2-CH
‖
CH
9

▶1106 M
E 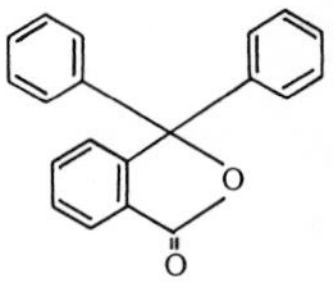
O
O
39

▶1106 VS
B (0.106)
$(CH_3)_3CCH(CH_3)_2$
147

▶1106 S
B (0.114) 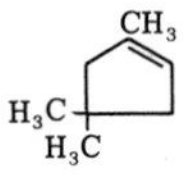
CH_3
H_3C
H_3C
127

▶1106 MS
E 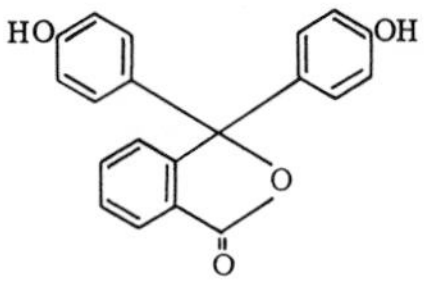
HO OH
O
O
39

▶1106 MW
B (0.029) 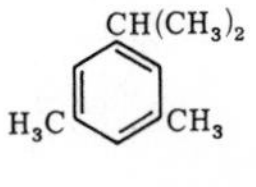
$CH(CH_3)_2$
H_3C CH_3
277

▶1105 M
G 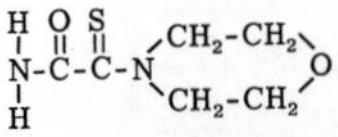
H O S
N-C-C-N CH_2-CH_2 O
H CH_2-CH_2
31

▶1105 VSB
B (0.136) 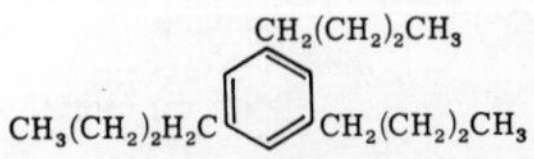
$CH_2(CH_2)_2CH_3$
$CH_3(CH_2)_2H_2C$ $CH_2(CH_2)_2CH_3$
279

▶1105 VS
B 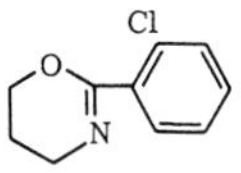
[C-O-C] 26

▶1105 S
B 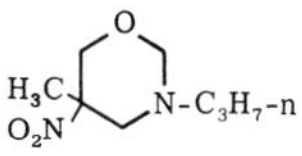
H_3C, O_2N, $N-C_3H_7$-n
[C-O-C] 26

▶1105 S
B 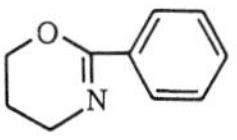
[C-O-C] 26

▶1105 W
B (0.0576) 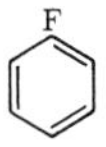
F
226

▶1105 S
B 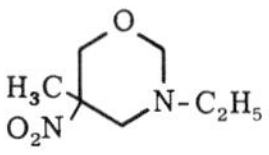
H_3C, O_2N, $N-C_2H_5$
[C-O-C] 26

▶1105 S
E 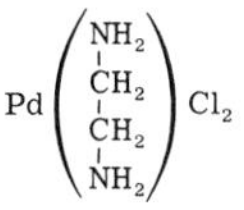
Pd $(NH_2-CH_2-CH_2-NH_2)$ Cl_2
$[NH_2]$ 8

▶1105 S
G 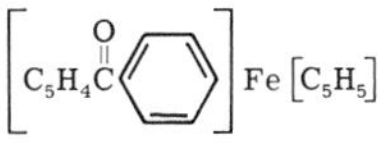
$[C_5H_4C(O)C_6H_5]Fe[C_5H_5]$
[Cyclopentadienyl ring] 450

▶1104 MW
$CHCl_3$
CH_3, N-C=O, CH_3 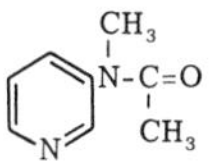
[CH in plane] 433

▶1104 M
B (0.0153) 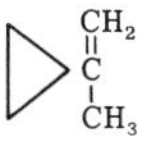
CH_2=C-CH_3
288

▶1104 M
B
$(C_2H_5-NH-B-N-C_2H_5)_3$
[C-N] 32

▶1104 MWSh
$CHCl_3$ 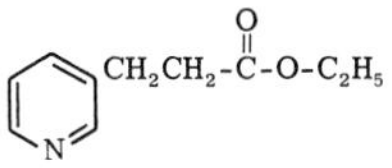
$CH_2CH_2-C(O)-O-C_2H_5$
[CH in plane] 433

▶1103 W
$CHCl_3$
H, N-C=O, CH_3 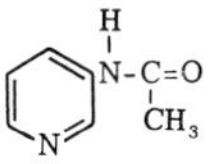
[CH in plane] 433

▶1103 VS
B (0.003)
CH_3NO_2
311

▶1103 MW
A (10 mm Hg, 58 cm), B (0.036)
$CH_3CH_2CH=CH(CH_2)_3CH_3$
99 (liquid), 421 (gas)

▶1103 S
B 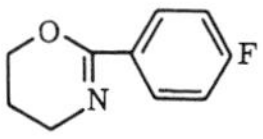
F
[C-O-C] 26

▶1103 S
B (0.003)
$(CH_3)_2CHNO_2$
366

▶1103 MWSh
B (0.003)
$(CH_3)_2CHCH_2NO_2$
326

▶1102
E or N 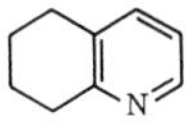
41

▶1102 VS
A
$(CH_3)_2SO$
13

▶1102 MSSh
$CHCl_3$ 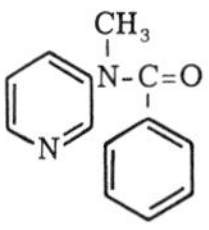
CH_3, N-C=O
[CH in plane] 433

▶1102 W
$CHCl_3$ 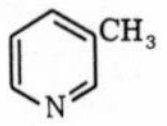
[structure: 3-methylpyridine]
[CH in plane]
433, see also 453

▶1100 M
CCl_4 [0.0128 m/l] 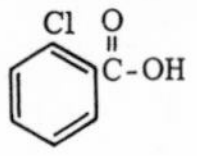
[structure: 2-chlorobenzoic acid, Cl, C(=O)-OH]
448

▶1102
Cl_3CF
[C-F]
447

▶1100 MWSh
$CHCl_3$ 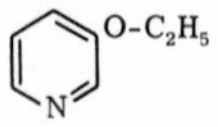
[structure: 3-ethoxypyridine, $O-C_2H_5$]
[CH in plane]
433

▶1101 M
E
$Ni\left(\begin{matrix}NH_2\\CH_2\\CH_2\\NH_2\end{matrix}\right)_3 PtCl_4$
$[NH_2]$
8

▶1100
$CaSO_4 \cdot 2H_2O$
449

▶1101 S
B (0.0104)
$(CF_3)_2CFCF_2CF_3$
240

▶1099 VS
B (0.0104) 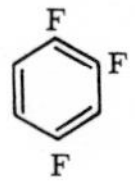
[structure: 1,2,4-trifluorobenzene, F, F, F]
224

▶1101 S
B (0.051) 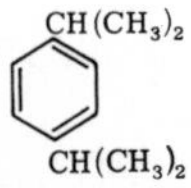
[structure: 1,4-diisopropylbenzene, $CH(CH_3)_2$, $CH(CH_3)_2$]
298

▶1099 VS
B (0.0576) 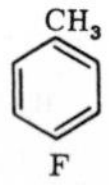
[structure: 4-fluorotoluene, CH_3, F]
220

▶1100 M
B (0.0576) 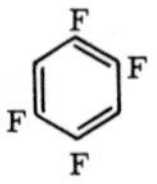
[structure: 1,2,3,5-tetrafluorobenzene, F, F, F, F]
222

▶1099 VS
CCl_4 [25%]
$CH_3CH_2CH_2OH$
314

▶1100 VS
B 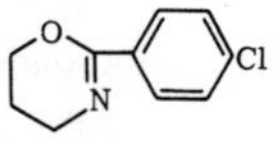
[structure: 2-(4-chlorophenyl)-5,6-dihydro-4H-1,3-oxazine, O, N, Cl]
[C-O-C]
26

▶1099 MVB
B (0.300)
$CH_3(CH_2)_5CH(C_3H_7)(CH_2)_5CH_3$
281

▶1100 S
B (0.169)
$(CH_3)_2CHCHCH(CH_3)_2$ with CH_3 on central carbon
135

▶1099 S
E
$[Coen_2SO_3NCS]^0$
5

▶1100 S
E, F 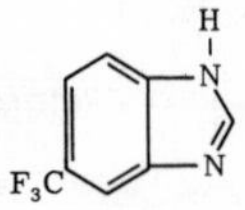
[structure: 5-trifluoromethylbenzimidazole, H, N, N, F_3C]
37

▶1099 M
B (0.157)
$(CH_3)_3C(CH_2)_2CH(CH_3)_2$
292

▶1100 S
B (0.08) 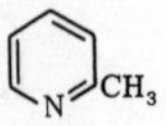
[structure: 2-methylpyridine, N, CH_3]
198

▶1099 W
$CHCl_3$ 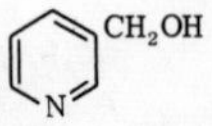
[structure: 3-pyridinemethanol, CH_2OH, N]
[CH in plane]
433

▶1098 MS
B (0.0563)
$CH_3(CH_2)_2CH{=}CH(CH_2)_2CH_3$
(trans)
205

▶1098 VS
B (0.0104)
$(C_4F_9)_3N$
346

▶1098 VVS
B (0.136)
$CH(CH_3)_2$ / $(CH_3)_2HC$ / $CH(CH_3)_2$
278

▶1098 S
B
Br
[C-O-C]
26

▶1098 M
B (0.036)
$CH_3(CH_2)_2CH{=}CH(CH_2)_2CH_3$
(cis)
291

▶1098 MW
$CHCl_3$
H / N-C=O
[CH in plane]
433

▶1097 S
B (0.0563)
$H_2C{=}C(CH_3)CH_2CH_2CH_3$
206

▶1097 S
B
H_3C / N-H / H_3C CH_3
26

▶1097
E or N
$[(CH_2)_3 \ldots CH_2]_2NH$
41

▶1096 MSh
B (0.169)
$(CH_3)_3CCH_2CH(CH_3)CH_2CH_3$
285

▶1096 M
B (0.0104)
$CF_3(CF_2)_5CF_3$
241

▶1096 VS
B (0.055)
H_3C / CH_2CH_3 / CH_3 / N-H
343

▶1096 S
$CHCl_3$
Br
[CH in plane]
433

▶1096 MB
B (0.169)
$CH_3CH_2C(CH_3)_2-CH_2CH(CH_3)CH_2CH_3$
302

▶1096 MS
A (16.6 mm Hg, 40 cm)
$(CH_3)_2CHC(CH_3)_2-CH_2CH_3$
90

▶1096 W
B
$C_6H_5(CH_3)SiCl_2$
86

▶1095 SSh
B (0.0104)
$(CF_3)_2CFCF_2CF_3$
240

▶1095 S
A (6.2 mm Hg, 10 cm)
$CH_3{-}CCl_3$
251

▶1095 S
E
$[Coen_2(SO_3)_2]Na$
(cis)
5

▶1095 M
A (36 mm Hg, 15 cm)
$H_2C{=}CHCH(CH_3)_2$
131

▶1095 W

$C_6H_5SiCl_3$

86

▶1095

Cl_2CF_2

[C-F] 646

▶1094 S
B (0.0563)

$(CH_3)_2CHC(CH_3)_2CH_2CH_2CH_3$

134

▶1094 VSB
E

$MgP_2O_6 \cdot H_2O$

28

▶1094 W
B (0.0563)

$(CH_3)_3CCH_2CH(CH_3)CH(CH_3)_2$

303

▶1094 S
E

NO_2

20

▶1094 SSh
A

$(CH_3)_2SO$

13

▶1093 MS
A (200 mm Hg, 10 cm)

F_2 F_2 F_2 F_2

243

▶1093 S
B (0.003)

$CH_3CH_2CH(NO_2)CH_3$

324

▶1093 S
E

$[Coen_2SO_3]SCN$

5

▶1093 S
E

$[Coen_2SO_3]Cl$

5

▶1093 MS
B (0.10)

$CH_2CH_2CH_2CH_3$

176

▶1093 MW
$CHCl_3$

Cl N

[CH in plane] 433

▶1092 VS
CCl_4 [5%]

CH_3CH_2OH

315

▶1092 VS
B (film)

$n\text{-}H_9C_4\text{-}O\text{-}CF_2CHFCl$

264

▶1092 S
B

$(CH_3)_2CHCH_2CH_2CH(CH_3)_2$

339

▶1092 M
B (0.064)

$H_2C{=}CHCH(CH_3)CH_2CH_3$

167

▶1092 M
$CHCl_3$

$C_4H_4SO([CH_2]_4SO)$

33

▶1091 VS
B (0.136)

$CH_3CHCH_2CH_3$
CH_3CH_2HC CH_3 $CHCH_2CH_3$ CH_3

280

▶1091 SVB
B (film)

$H_3C\text{-}O\text{-}CF_2CHFCl$

356

▶1091 S
B (0.169)

$(CH_3)_2CHCH_2CH_2CH(CH_3)_2$

137

▶1091 (fundamental)
A

CH_2Br_2

$[CH_2]$ 646

▶1090 VS
B

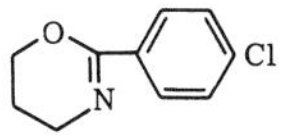

O N Cl

[C-O-C] 26

▶1090 S
B

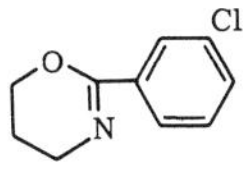

Cl O N

[C-O-C] 26

▶1090 MSB
B (0.10)

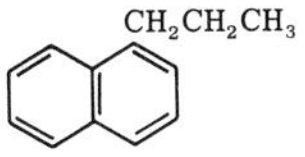

$CH_2CH_2CH_3$

178

▶1090 M
CCl_4 [0.0082 m/1]

O C-OH

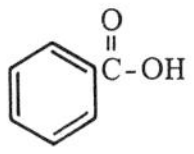

25, 448, see also 38

▶1090
E

$Hg(NO_3)_2$

45

▶1090

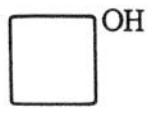

OH

445

▶1090 MSh
A (385 mm Hg, 10 cm)

CF_3-CCl_3

253

▶1089 VSB
B (film)

n-H_7C_3-O-CF_2CHFCl

263

▶1089 S
E

$[Coen_2SO_3NCS]^0$

5

▶1089 M
CS_2

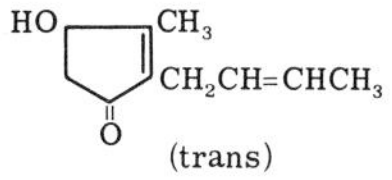

HO CH_3 CH_2CH=$CHCH_3$ O

(trans)

430

▶1089 M
B (0.10)

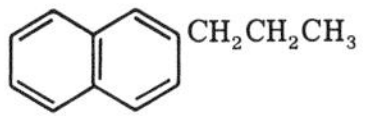

$CH_2CH_2CH_3$

177

▶1089 M
B (0.10)

CH_3 CH_3

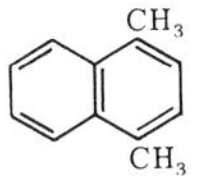

186

▶1089 W

$C_6H_5HSiCl_2$

86

▶1089 VS
B (0.028)

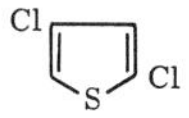

Cl S Cl

211

▶1088 S
A (6.2 mm Hg, 10 cm)

CH_3-CCl_3

251

▶1088 VS
B (0.169)

$(CH_3)_2CHC(CH_3)_2$-CH_2CH_3

341

▶1088 MW
$CHCl_3$

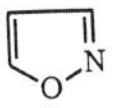

O N

15

▶1088
E or N

N

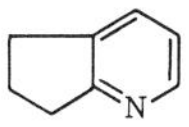

41

▶1088 E or N (structure: 2-chloro-6,7-dihydro-5H-cyclopenta[b]pyridine) 41	▶1086 S B (structure: 2-(o-nitrophenyl)-5,6-dihydro-4H-1,3-oxazine; O_2N, O, N) [C-O-C] 26
▶1087 VS B (0.0104) (CF_3, F on benzene ring) 218	▶1086 S E $S{=}C(NH_2)_2$ [NH_2 deformation] 22
▶1087 M B (0.0153) (cyclopropyl–C($=CH_2$)–CH_3) 288	▶1086 M A (385 mm Hg, 10 cm) $CF_3\text{-}CCl_3$ 253
▶1087 MS A (16.6 mm Hg, 40 cm) $(CH_3)_2CHC(CH_3)_2\text{-}CH_2CH_3$ 90	▶1086 M B (0.157) $(CH_3)_3CCH(CH_2CH_3)_2$ 78
▶1087 S B (0.064) $H_2C{=}C(CH_2CH_3)_2$ 163	▶1086 M (naphthalene with CH_3, CH_3) 181
▶1087 MS B (0.064) $H_3C(H)C{=}C(H)CH_2CH_2CH_3$ (trans) 169	▶1086 M $CHCl_3$ (pyridine with Br; N) [CH in plane] 433
▶1087 M P $Pt([CH_2]_4SO)_2Cl_2$ 33	▶1085 VS B (0.157) $(CH_3)_3CCH(CH_3)CH(CH_3)_2$ 293
▶1086 VSB B (film) $CH_3CH_2\text{-}O\text{-}CF_2CHFCl$ 261	▶1085 VS B (0.0104) (benzene with F, F para) 225
▶1086 S B (0.0563) $(CH_3)_3CC(CH_3)_2\text{-}CH(CH_3)_2$ 306	▶1085 VSB E $Na_4P_2O_6$ 28
▶1086 S B (ring with O; H_5C_2, O_2N; $N\text{-}C_4H_9\text{-}n$) 26	▶1085 S A CF_3Br 11

▶1085 S
B

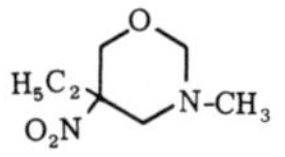

H_5C_2, O_2N, O, $N\text{-}CH_3$

[C-O-C] 26

▶1085 S
B

H_5C_2, O_2N, O, $N\text{-}C_3H_7\text{-}i$

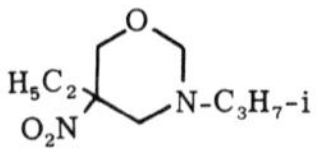

[C-O-C] 26

▶1085 VS
B (0.0563)

$(CH_3)_3CCH(CH_3)C(CH_3)_3$

307

▶1085 M
B (0.169)

$(CH_3)_2CHCH_2CH(CH_3)CH_2CH_3$

138

▶1085
E or N

$(CH_2)_3$, N

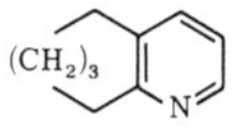

41

▶1084 VS
B (0.104)

$(CH_3)_3CCH(CH_3)_2$

147

▶1084 VS
B

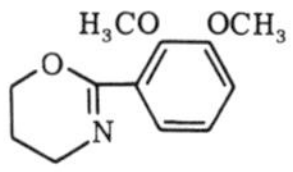

H_3CO, OCH_3, O, N

[C-O-C] 26

▶1083 VSSp
G

$Na_4P_2O_6$

28

▶1083 SB
B (0.300)

$C_{26}H_{54}$
(5, 14-di-n-butyloctadecane)

283

▶1083 S
B (0.10)

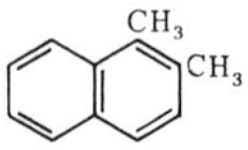

CH_3, CH_3

188

▶1083 MS
A (16.6 mm Hg, 40 cm)

$(CH_3)_3CCH(CH_3)CH_2CH_3$

71

▶1083 MS
B (0.10)

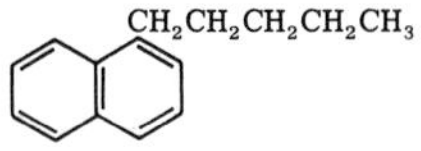

$CH_2CH_2CH_2CH_2CH_3$

174

▶1082 VW
B (0.2398)

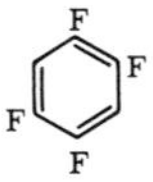

F, F, F, F

222

▶1082 VS
B (0.025)

S

199

▶1082 VS
B (0.169)

$CH_3CH_2C(CH_3)_2\text{-}CH(CH_3)CH_2CH_3$

299

▶1082 S
B

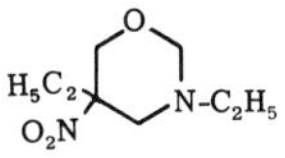

H_5C_2, O_2N, O, $N\text{-}C_2H_5$

[C-O-C] 26

▶1082 S
B

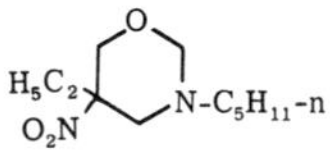

H_5C_2, O_2N, O, $N\text{-}C_5H_{11}\text{-}n$

[C-O-C] 26

▶1082 S
B (0.10)

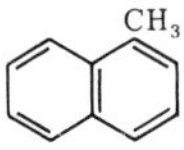

CH_3

192

▶1082 S
B

$CH_3CH_2C(CH_3)_2\text{-}CH_2CH_3$

149, 336

▶1082 MW
B (0.0563)

$(CH_3)_3CCH_2CH(CH_3)CH(CH_3)_2$

303

▶1082 MW
B (0.065)

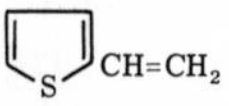

(2-thienyl)$CH=CH_2$

212

▶1081 VS
B

H_3C, O_2N – morpholine ring – $N-C_3H_7-i$

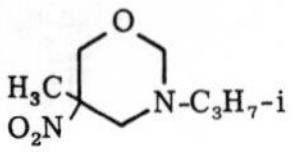

[C-O-C] 26

▶1081 VSSp
E

$Ba_2P_2O_6$

28

▶1081 SSh
E

$Pb_2P_2O_6$

28

▶1081 S
B (0.065)

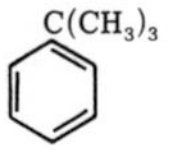

$C(CH_3)_3$ (benzene ring)

133

▶1081 S
B

$CH_3CH_2C(CH_3)_2CH_2CH_3$

336

▶1081 S
B

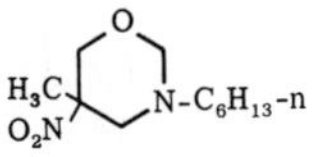

H_3C, O_2N – morpholine ring – $N-C_6H_{13}-n$

[C-O-C] 26

▶1081 MS
B (0.065)

(2-thienyl)CH_2CH_3

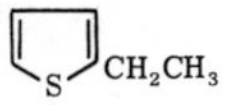

213

▶1081 W
A (12.6 mm Hg, 40 cm)

$CH_3(CH_2)_6CH_3$

89

▶1080 VS
B (0.136)

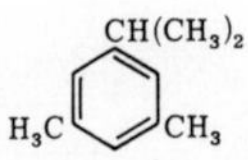

$CH(CH_3)_2$, H_3C, CH_3 (benzene ring)

277

▶1080 VS
B

H_5C_2, O_2N – morpholine ring – $N-C_3H_7-n$

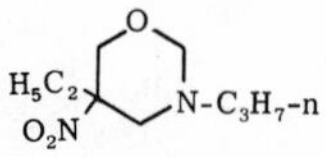

[C-O-C] 26

▶1080 VSB
E

$CeP_2O_6 \cdot H_2O$

28

▶1080 S
B (0.136)

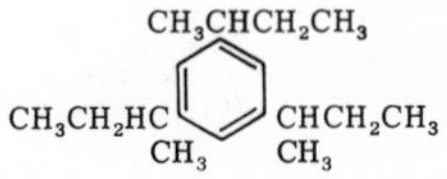

$CH_3CHCH_2CH_3$, $CH_3CH_2HC(CH_3)$, $CH(CH_3)CH_2CH_3$ (benzene ring)

280

▶1080 S
B

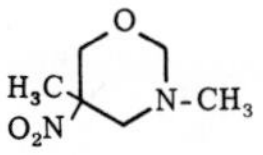

H_3C, O_2N – morpholine ring – $N-CH_3$

[C-O-C] 26

▶1080 S
B

H_3C, O_2N – morpholine ring – $N-C_4H_9-n$

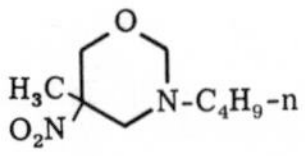

[C-O-C] 26

▶1080 MW
A (4.2 mm Hg, 40 cm)

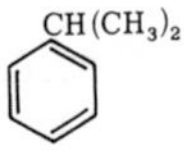

$CH(CH_3)_2$ (benzene ring)

66

▶1080 MS
B (0.169)

$(CH_3)_2CHCH(CH_2CH_3)CH(CH_3)_2$

301

▶1080 W
A (15.9 mm Hg, 40 cm)

$CH_3CH_2CH(CH_3)(CH_2)_3CH_3$

106

▶1080 W
E

$[CH_3---S=C(NH_2)_2]^+$

[NH_2 deformation] 22

▶1080
E or N

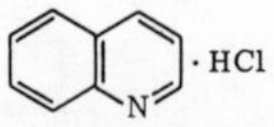

(quinoline) $\cdot HCl$

41

▶1080 VS
A (10 mm Hg, 10 cm)
(1,3,5-trioxane)
201

▶1080 VS
B (0.0563)
$(CH_3)_3CCH(CH_3)CH_2CH_3$
340, 426

▶1079 S
B
(3-methyl-3-nitro-5-n-propyl tetrahydro-1,3-oxazine: H_3C, O_2N, N-C_3H_7-n)
[C-O-C]
26

▶1079 MWSh
B (0.014)
(perfluoro-1,4-dimethylcyclohexane: F, CF_3, F_2, F_2, F_2, F_2, F, CF_3)
230

▶1079 S
E
$[Coen_2SO_3 \cdot NH_3]Cl$
5

▶1079 M
B (0.10)
(methyltetralin, CH_3)
269

▶1079 SSh
B (0.10)
(propylnaphthalene, $CH_2CH_2CH_3$)
178

▶1079 S
B (0.238)
H_3C CH_3 C=C H CH_2CH_3
(cis)
274

▶1078 VS
CCl_4 [25%]
$CH_3(CH_2)_4OH$
393, see also 350

▶1078 S
B (0.238)
H_3C CH_3 C=C H CH_2CH_3
(cis)
274

▶1078 S
B (0.10)
(dimethylnaphthalene: H_3C, CH_3)
184

▶1078 MB
B (0.300)
$(C_2H_5)_2CH(CH_2)_{20}CH_3$
282

▶1078
E or N
($(CH_2)_3$, N, Cl)
41

▶1077 S
B
(H_3C, O_2N, O, N-C_2H_5)
[C-O-C]
26

▶1077 M
P
$Pt([CH_2]_4SO)_2Cl_2$
33

▶1076 S
B
(H_3C, O_2N, O, N-C_5H_{11}-n)
[C-O-C]
26

▶1076 M
B (0.101)
$CH_3(CH_2)_5CH_3$
85, 362

▶1076 W
B (0.036), A (10 mm Hg, 58 cm)
$CH_3CH_2CH{=}CH(CH_2)_3CH_3$
99 (liquid), 421 (gas)

▶1075 VS
B (0.0563)
$(CH_3)_2C{=}CHC(CH_3)_3$
204

▶1075 VSSp
E
$K_4P_2O_6$
28

▶1075 S
B (0°C)
$(CH_3)_2CHCHCH(CH_3)_2$
CH_3
284

▶1075 M
CCl_4 [0.04 g/ml]
$C_6H_5C(=O)-N(CH_3)_2$
3

▶1075 S
B (0.114) 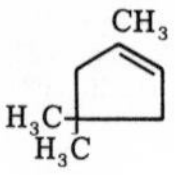
CH_3, H_3C, H_3C (cyclopentene)
127

▶1075 M
D
$[CH_2CH(CN)]_n$
[tert. CH]
23

▶1075 S
E
$[Coen_2SO_3Cl]^0$
5

▶1074 MS
B (0.025)
$CH_3CH_2-S-CH_2CH_3$
172

▶1075 M
B (0.068)
$H_3C\ C_3H_7$-n
C=C
H H
(cis)
276

▶1074 VS
B (0.169)
$(CH_3)_3CCH_2CH_3$
145

▶1075 MVB
B (0.300)
$CH_3(CH_2)_5CH(C_3H_7)(CH_2)_5CH_3$
281

▶1074 VS
B (0.051)
$C_6H_4(CH(CH_3)_2)_2$
296

▶1075 MS
B (0.064)
H_3C H
C=C
H $CH_2CH_2CH_3$
(trans)
169, 427

▶1074 S
B
$[(CH_3)_2N-BO]_3$
32

▶1075 M
B (0.097)
$(CH_3)_2CH(CH_2)_3CH_3$
80

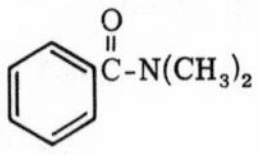

▶1074 VS
B (0.0104)
F_2, F_2, F_2, F_2, F, C_2F_5 (cyclopentane)
247

▶1075 VS
B (0.169)
$(CH_3)_2CH(CH_2)_3CH_3$
425

▶1074
$S=C(S-C_2H_5)_2$
[C=S]
434

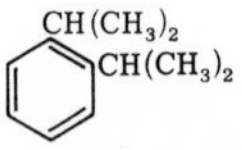

▶1075 M
B (0.163)
$H_2C=CH(CH_2)_7CH_3$
270

▶1074
E or N
$(CH_2)_3$, CN, N (pyridine)
41

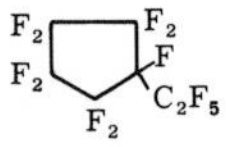

▶1075 MSB
B (0.163)
$H_2C=CH(CH_2)_8CH_3$
352

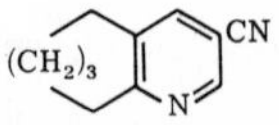

▶1074 MS
B (0.169)
$(CH_3CH_2)_2CHCH_2CH_2CH_3$
140

▶1073 S
A (200 mm Hg, 10 cm)
$(CH_3)_3CCH_2CH_3$
94

▶1073 S
B 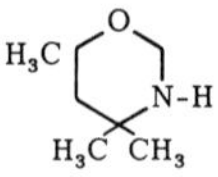
[C-O-C]
26

▶1073 M
B 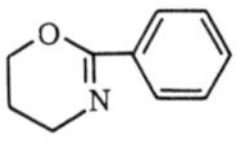
[C-O-C]
26

▶1073 W
$C_6H_5(CH_3)SiCl_2$
86

▶1073 VS
B (0.136) 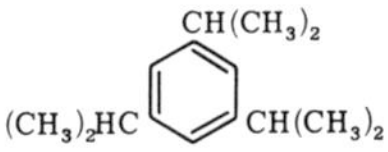
$CH(CH_3)_2$, $(CH_3)_2HC$, $CH(CH_3)_2$
278

▶1072 VS
CCl_4 [5%]
$CH_3(CH_2)_3OH$
353

▶1072 VS
B (0.10) 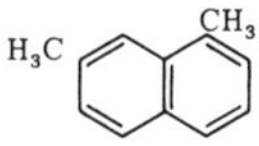
H_3C, CH_3
183

▶1072 S
B (0.064)
H CH_2CH_3
C=C
H_3C CH_3
(trans)
428

▶1072 MW
B
$CH_3CH_2CH(CH_2)_2CH_3$
CH_3
330

▶1072
$CHFCl_2$
[C-F]
442

▶1072
E or N 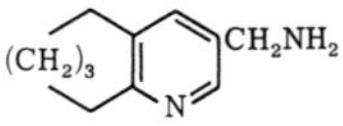
$(CH_2)_3$, CH_2NH_2, N
41

▶1072 S
CS_2 [sat.] (0.170) 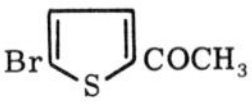
Br, S, $COCH_3$
348

▶1071 VS
B (0.0563)
$CH_3CH{=}CHCH_2CH_3$
(cis)
208

▶1071 S
B (728)
CH_3 CH_3
$CH_3CH_2CH\text{-}CHCH_2CH_3$
83

▶1071 S
B 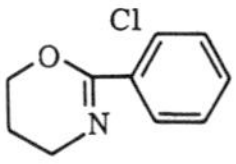
Cl
[C-O-C]
26

▶1071 S
B (0.169)
CH_3 CH_3
$CH_3CH_2CH\text{-}CHCH_2CH_3$
83

▶1071 M
CCl_4 [0.00658 m/l] 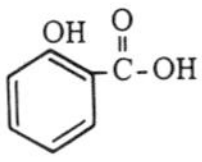
OH O
C-OH
448

▶1070 VS
B (0.0104) 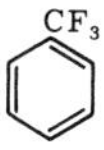
CF_3
219

▶1070 MW
B (0.036)
$CH_3CH_2CH{=}CH(CH_2)_3CH_3$
99, 421

▶1070 S
B (0.051) 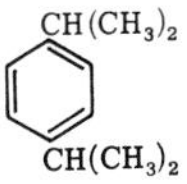
$CH(CH_3)_2$
$CH(CH_3)_2$
298

▶1070 S
E

[$Coen_2SO_3$]I

5

▶1070 S

$(C_6H_5)_3SnCH_2CH_2Si(C_6H_5)_3$

[Sn-Ph] 364

▶1070 S

$(C_6H_5)_3SnCH_2CH_2Si(C_6H_5)_2CH_2CH_2Sn(C_6H_5)_3$

[Sn-Ph] 364

▶1070 S

$(C_6H_5)_3SnCH_2CH_2Ge(C_6H_5)_3$

[Sn-Ph] 364

▶1070 S

$(C_6H_5)_3SnCH_2CH_2Ge(C_6H_5)_2CH_2CH_2Sn(C_6H_5)_3$

[Sn-Ph] 364

▶1070 M
B (0.0563)

$(CH_3)_2CHC(CH_3)_2-CH_2CH_2CH_3$

134

▶1070 W
B (0.169)

$(CH_3)_3CCH_2CH_2C(CH_3)_3$

305

▶1070

$H_2C=C=CH_2$

441

▶1069 S
E

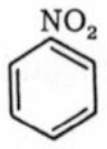

20

▶1068 VS
CCl_4 [2%]

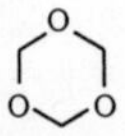

342

▶1068 VW
A (25 mm Hg, 10 cm)

$CF_3CF_2CF_2CF_2CF_3$

237

▶1068 VS
B (0.068)

$H_3CH_2C(H)C=C(H)CH_2CH_3$
(cis)

275

▶1068 VS
B (0.068)

$H_3CH_2C(H)C=C(H)CH_2CH_3$
(trans)

168

▶1068 M
B (0.0104)

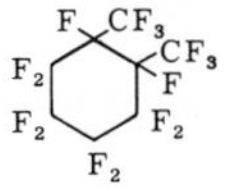

231

▶1068 VS
CCl_4 [5%]

$CH_3CH_2CH_2OH$

314

▶1068 VS
A (10 mm Hg, 10 cm)

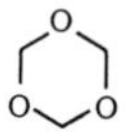

201

▶1068 MB
B (0.029)

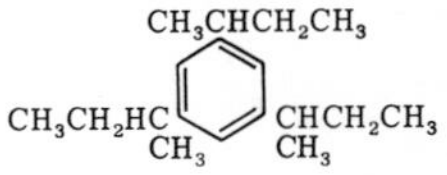

280

▶1068 S
E

[$Coen_2(SO_3)_2$]Na
(trans)

5

▶1068 S
P

$Pd([CH_2]_4SO)_2Cl_2$

33

▶1068 S

$(C_6H_5)_3SnCH_2CH_2Sn(C_6H_5)_3$

[Sn-Ph] 364

▶1068 VW
B (0.0576) 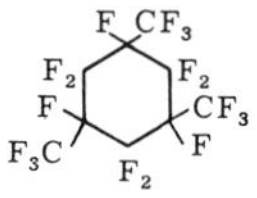
(perfluoro-1,3-dimethylcyclohexane: ring with F, CF_3; F_2; F, CF_3; F_2; F_2; F, F_3C)
228

▶1068 MW
B (0.015)
$H_2C{=}CHC(CH_3)_3$
129

▶1068 W
$C_6H_5HSiCl_2$
86

▶1067.9 S
B (0.03) 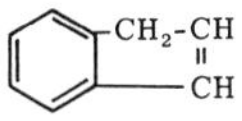
(benzene ring fused with $-CH_2-CH{=}CH$)
9

▶1067 VS
B (film)
n-H_9C_4-O-CF_2CHFCl
264

▶1067 M
B (0.0104) 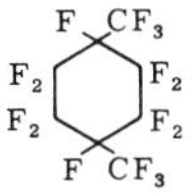
(perfluoro-1,4-dimethylcyclohexane: F, CF_3; F_2; F_2; F_2; F_2; F, CF_3)
230

▶1067 S
B 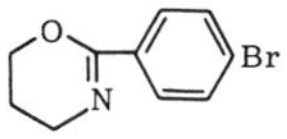
(2-(p-bromophenyl)-5,6-dihydro-4H-1,3-oxazine)
[C-O-C]
26

▶1067 W
E
$(C_6H_5)_2SiCl_2$
86

▶1066 MW
B (0.0104)
$(CF_3)_2CFCF_2CF_3$
240

▶1066 VS
B (0.0576) 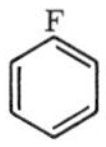
(fluorobenzene)
226

▶1066 S
B (0.169)
$(CH_3)_2CHCH(CH_3)_2$
144

▶1066 S
C 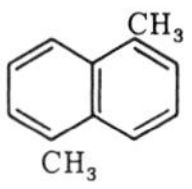
(dimethylnaphthalene, CH_3, CH_3)
185

▶1066 M
B (0.10) 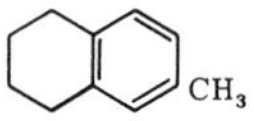
(methyltetrahydronaphthalene, CH_3)
269

▶1065 VSSh
B (film)
n-H_7C_3-O-CF_2CHFCl
263

▶1065 S
E
$Pd\left(\begin{matrix}NH_2\\CH_2\\CH_2\\NH_2\end{matrix}\right)_2 PtCl_4$
[Ring]
8

▶1065 S
E 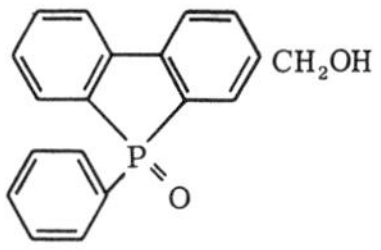
(phenyl dibenzophosphole oxide with CH_2OH; P=O)
34

▶1065 M
B 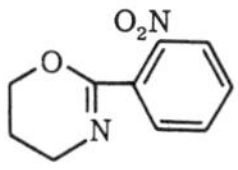
(2-(o-nitrophenyl)-5,6-dihydro-4H-1,3-oxazine, O_2N)
26

▶1065 M
CCl_4 [0.0082 m/l] 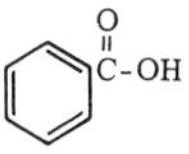
(benzoic acid, C_6H_5-C(=O)-OH)
25, 448, see also 38

▶1065 W
CS_2
$(C_6H_5)_3SiCl$
86

▶1065
E
$Pb(NO_3)_2$
45

▶1065
E or N

(tetrahydroquinoline ring) COOH, N

41

▶1064 S
B (0.0563)

$(CH_3)_3CC(CH_3)_2-CH(CH_3)_2$

306

▶1064 VSSp
E

$Li_4P_2O_6 \cdot 7H_2O$

28

▶1064 M
B (0.003)

$CH_3(CH_2)_2CH_2NO_2$

310

▶1064 S
B

H_5C_2, O_2N, O, N-H

[C-O-C]

26

▶1064 VSVB
A (420 mm Hg, 15 cm)

$H_2C{=}C{=}CHCH_3$

440

▶1063 M
B (0.0104)

F, CF_3, F_2, F_2, F_2, F_2, F_2

232

▶1063 S
B (0.025)

$CH_3CH_2SCH_3$

209

▶1063 VS
B (0.055)

H_3C, CH_2CH_3, CH_3, N-H

343

▶1063 M
B (0.728)

$CH_3CH_2CH(CH_3)-CH(CH_3)CH_2CH_3$

83

▶1063 W

$C_6H_5SiCl_3$

86

▶1063 S
B (0.0563)

$CH_3CH{=}CHCH_2CH_3$
(trans)

207

▶1062 VS
B (0.064)

$(CH_3)_2C{=}CHCH_2CH_3$

166

▶1062 S
E

$[Coen_2SO_3OH]^0$

5

▶1062 S
E

$Pd\left(NH_2-CH_2-CH_2-NH_2\right)_2Cl_2$

[Ring]

8

▶1062 S
CCl_4 [25%]

$CH_3CH(OH)(CH_2)_2CH_3$

318

▶1062 S
C_6H_6

$Zn[S-C(=S)-O-C_4H_9\text{-}n]_2$

[C=S]

434

▶1062 MS
B (0.169)

$(CH_3)_2CHCH(CH_2CH_3)CH(CH_3)_2$

301

▶1062 MS
B (0.169)

$(CH_3)_2CHCH(CH_3)CH_2CH_3$

142

▶1062 M
G

$H_3C-C(=O)-C(=S)-N(CH_2-CH_2)_2O$

[C=S?]

31

▶1062 M
G

H O S
N-C-C-N(CH₂-CH₂)₂O — $H_2N-C(=O)-C(=S)-N(CH_2-CH_2)_2O$

[C=S?] 31

▶1059 MW
B (0.0576)

(1,2,4-trifluorobenzene: F, F, F)

224

▶1061 S
Cyclohexane

$CH_3(CH_2)_{16}CH_2-S-C(=S)-O-CH_2(CH_2)_{16}CH_3$

[C=S] 434

▶1059 VS
B (0.0563)

$CH_3(CH_2)_2CH=CH(CH_2)_2CH_3$

(cis and trans)

205 (trans), 422 (cis)

▶1060 VS
CCl₄ [5%]

$(CH_3)_2CHCH_2CH_2OH$

195

▶1059 S
B (0.10)

CH_2CH_3 (naphthalene)

190

▶1060 VS
B (0.0104)

$F_9C_4-O-C_4F_9$

266

▶1059 M
B (0.169)

$(CH_3)_3CC(CH_3)_2-CH_2CH_2CH_3$

304

▶1060 SSh
CS_2

HO, CH_3, $CH_2CH=CHCH_3$, O

(trans) 430

▶1059 M
$CHCl_3$

CH_3, N, O

15

▶1060 S
E

CH_2OH, P

34

▶1058 VS
B (0.0104)

CF_3, F

218

▶1060 S
CH_3CN

$Zn[S-C(=S)-O-C_4H_9-n]_2$

[C=S] 434

▶1058 VSB
CCl_4 [25%]

$CH_3(CH_2)_5OH$

320

▶1060 S
CCl_4

$CH_3(CH_2)_{16}CH_2-S-C(=S)-O-CH_2(CH_2)_{16}CH_3$

[C=S] 434

▶1058 VS
B (0.169)

CH_3, $CH(CH_3)_2$

132

▶1060 M
B

O, N, F

26

▶1058 VSSp
B (0.0104)

$CF_3(CF_2)_5CF_3$

241

▶1059 VS
CCl_4 [25%]

$CH_3CH_2C(OH)(CH_3)_2$

194

▶1058 VSSp
E

$Ba_2P_2O_6$

28

▶1058 VSB
E
$Co(NH_3)_6NaP_2O_6 \cdot 4H_2O$
28

▶1058 M
B
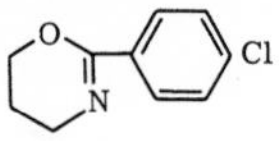

26

▶1058 M
G
$C_6H_5\text{-}NH\text{-}C(=O)\text{-}C(=S)\text{-}NH\text{-}C_6H_5$
[C=S]
31

▶1058
P
$La(NO_3)_3$
45

▶1058
P
$Co(NO_3)_2$
45

▶1057 S
B (0.169)
$CH_3CH_2C(CH_3)_2\text{-}CH_2CH(CH_3)CH_2CH_3$
302

▶1057 S
B (0.136)
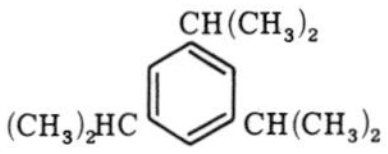

278

▶1057 S
E
$Pt(NH_2\text{-}CH_2\text{-}CH_2\text{-}NH_2)_2PtCl_4$
[Ring]
8

▶1057 S
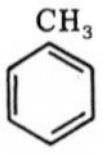

388

▶1057 M
B (0.068)
$H_3C(H)C{=}C(H)C_3H_7\text{-}n$
(cis)
276

▶1057 MS
B (0.068)
$H_3C(CH_3)C{=}C(H)CH_2CH_3$
(cis)
274

▶1057 M
B
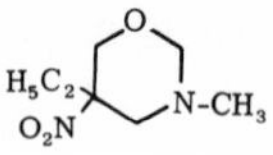

26

▶1057 S
B (0.10)
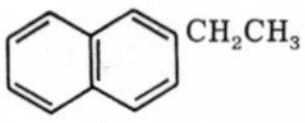

189

▶1056 VS
CCl_4 [25%]
$CH_3(CH_2)_4OH$
393, see also 350

▶1056 M
B
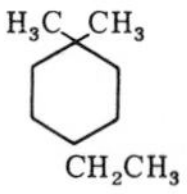

215

▶1056 VS
CCl_4 [5%]
$CH_3CH_2CH_2OH$
314

▶1056 W
A (4.2 mm Hg, 40 cm)
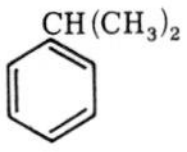

66

▶1056 VSB
CCl_4 [25%]
$CH_3(CH_2)_7OH$
317

▶1055 S
E
$Pd(NH_2\text{-}CH_2\text{-}CH_2\text{-}NH_2)Cl_2$
[Ring]
8

▶1055 MSVB
B (0.06)
$H_2C{=}CH(CH_2)_6CH_3$
271

▶1055 W
B (0.0104)

[structure: perfluorodimethylcyclohexane; labels F, CF_3, F_2, F_2, F, CF_3, F_3C, F, F_2]

228

▶1055
P

$Cr(NH_3)_5(NO_3)_3$

45

▶1054 VS
P

$[(C_2H_5)_2P \cdot BI_2]$

[CH_3 rock]
30

▶1054 S
B (0.10)

[structure: dimethylnaphthalene with CH_3 and H_3C]

184

▶1054 MS
G

H O S H
N-C-C-N
C_6H_{11} C_6H_{11}

[C=S]
31

▶1054 M
B

$CH_3CH{=}CHCH{=}CHCH_3$

354

▶1054 M
B (0.064)

$H_2C{=}C(CH_2CH_3)_2$

163

▶1053 VS
A (6.2 mm Hg, 10 cm)

$CF_2Cl\text{-}CF_2Cl$

257

▶1053 S
B

[structure: morpholine ring with O, H_5C_2, O_2N, N-C_3H_7-n]

26

▶1053 S
B

[structure: morpholine ring with O, H_5C_2, O_2N, N-C_6H_{13}-n]

26

▶1053 S
E

Pt (NH_2-CH_2-CH_2-NH_2) Cl_2

[Ring]
8

▶1053 SB
E

$Tl_4P_2O_6$

28

▶1053 S

[structure: pyridine ring with N, $CH(C_2H_5)_2$]

455

▶1053 S
B (0.169)

$(CH_3)_2CHCH(CH_3)CH_2CH_2CH_3$

139

▶1053
E

$Cu(NO_3)_2 \cdot 5H_2O$

45

▶1053

HCl_3

[HC]
439

▶1053
E or N

[structure: tetrahydroquinoline ring with N, CH_2NH_2]

41

▶1052 S
B

[structure: morpholine ring with O, H_5C_2, O_2N, N-C_5H_{11}-n]

26

▶1052 VS
B (0.08)

[structure: pyridine ring with N, CH_3]

198

▶1052
E

$Cr(NH_3)_5(NO_3)_3$

45

▶1052
E, P

$RbNO_3$

45

▶1052
P

$Zn(NO_3)_2$

45

▶1051 S
B

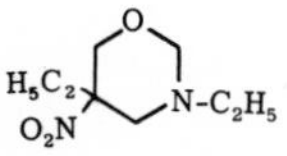

O
H_5C_2 N-C_2H_5
O_2N

26

▶1050 VS
CCl_4 [5%]

CH_3CH_2OH

315

▶1050 VS
B (0.051)

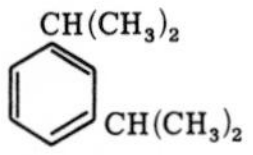

$CH(CH_3)_2$
$CH(CH_3)_2$

297

▶1050 VS
P

$[(C_2H_5)_2P \cdot BBr_2]_3$

[CH_3 rock] 30

▶1050 VSSp
E

$Na_2H_2P_2O_6$

28

▶1050 S
B (0.163)

$H_2C{=}CH(CH_2)_7CH_3$

270

▶1050 S
E

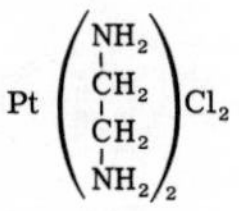

Pt (NH_2 – CH_2 – CH_2 – NH_2)$_2$ Cl_2

[Ring] 8

▶1050 S
G

NH_2
NO_2

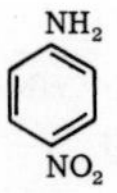

[NH_2] 21

▶1050 S
P

$[(C_2H_5)_2P \cdot BCl_2]_3$

[CH_3 rock] 30

▶1050 S
$(C_2H_5)_2O$

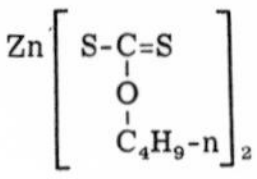

Zn [S-C=S, O, C_4H_9-n]$_2$

[C=S] 434

▶1050 M (R branch)
A (24.1 mm Hg, 40 cm)

309

▶1050 S
B (0.051)

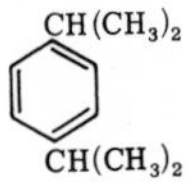

$CH(CH_3)_2$
$CH(CH_3)_2$

298

▶1050 M
B

$(CH_3)_2CHCH_2CHCH_2CH_3$
CH_3

334

▶1050 M
B

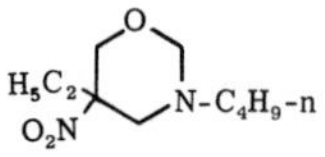

O
H_5C_2 N-C_4H_9-n
O_2N

26

▶1050 M
CCl_4 [0.0128 m/l]

Cl
C-OH (=O)

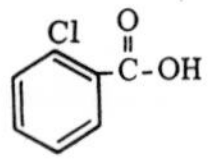

448

▶1050
P

$Ca(NO_3)_2$

45

▶1050
E or N

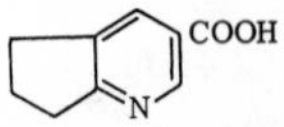

COOH
N

41

▶1050 MS
B (0.169)

CH_3
$CH_3CH_2C - CHCH_2CH_3$
CH_3CH_3

299

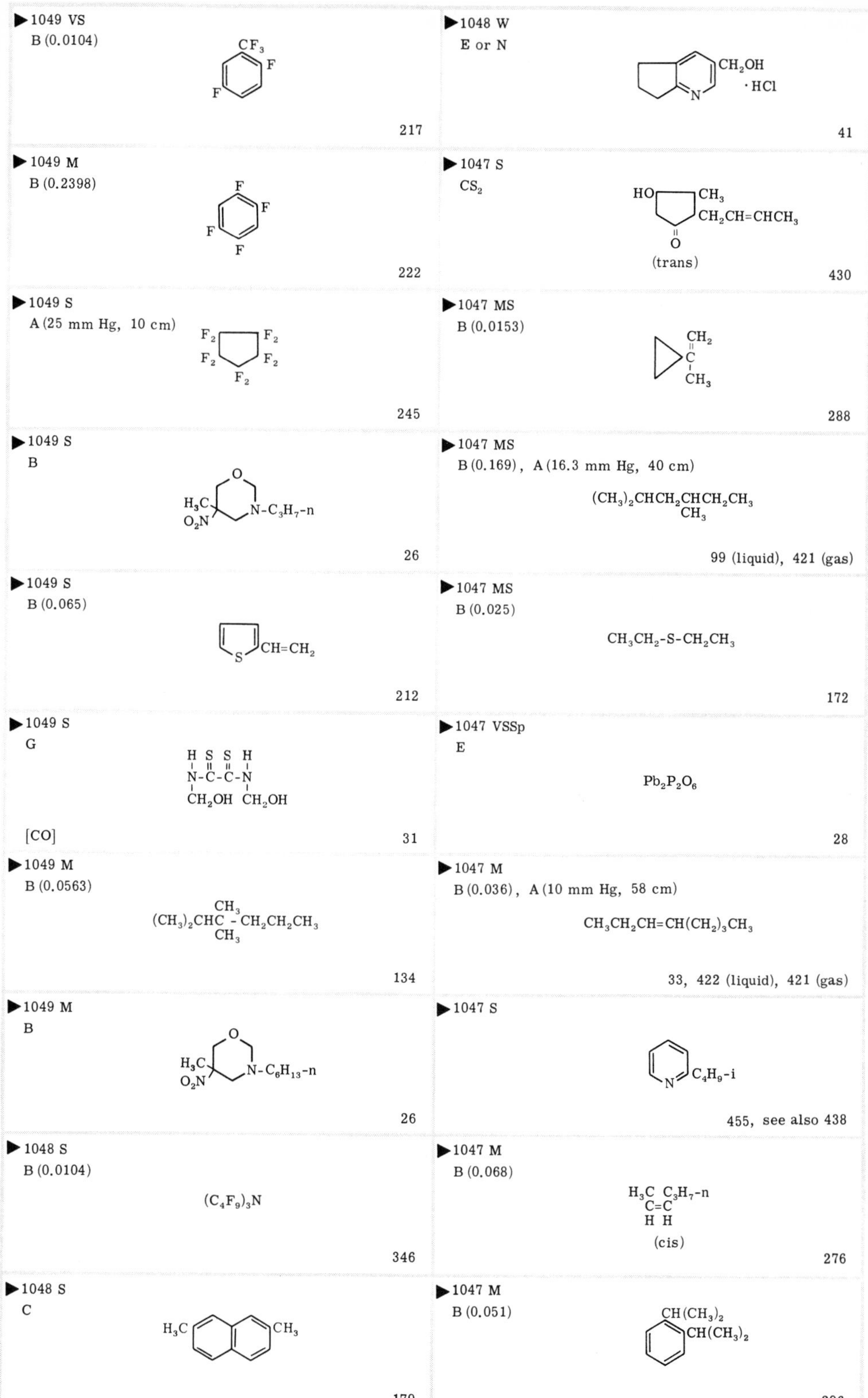

▶1049 VS
B (0.0104)
CF_3, F, F (difluorobenzotrifluoride)
217

▶1049 M
B (0.2398)
F, F, F, F (tetrafluorobenzene)
222

▶1049 S
A (25 mm Hg, 10 cm)
F_2, F_2, F_2, F_2, F_2 (perfluorocyclopentane)
245

▶1049 S
B
H_3C, O_2N, O, $N-C_3H_7-n$
26

▶1049 S
B (0.065)
S, $CH=CH_2$
212

▶1049 S
G
H S S H
N-C-C-N
CH_2OH CH_2OH
[CO]
31

▶1049 M
B (0.0563)
$(CH_3)_2CHC(CH_3)_2-CH_2CH_2CH_3$
134

▶1049 M
B
H_3C, O_2N, O, $N-C_6H_{13}-n$
26

▶1048 S
B (0.0104)
$(C_4F_9)_3N$
346

▶1048 S
C
H_3C, CH_3 (dimethylnaphthalene)
179

▶1048 W
E or N
CH_2OH $\cdot HCl$, N
41

▶1047 S
CS_2
HO, CH_3, $CH_2CH=CHCH_3$, O
(trans)
430

▶1047 MS
B (0.0153)
CH_2=C–CH_3 (cyclopropyl)
288

▶1047 MS
B (0.169), A (16.3 mm Hg, 40 cm)
$(CH_3)_2CHCH_2CH(CH_3)CH_2CH_3$
99 (liquid), 421 (gas)

▶1047 MS
B (0.025)
$CH_3CH_2-S-CH_2CH_3$
172

▶1047 VSSp
E
$Pb_2P_2O_6$
28

▶1047 M
B (0.036), A (10 mm Hg, 58 cm)
$CH_3CH_2CH=CH(CH_2)_3CH_3$
33, 422 (liquid), 421 (gas)

▶1047 S
N, C_4H_9-i
455, see also 438

▶1047 M
B (0.068)
H_3C C_3H_7-n
C=C
H H
(cis)
276

▶1047 M
B (0.051)
$CH(CH_3)_2$, $CH(CH_3)_2$
296

▶1047 M
E, $CHCl_3$
[structure: 3-aminopyridine, NH_2]
[CH in plane] 327, see also 433

▶1047 W
$CHCl_3$
[structure: 3-acetamidopyridine, N(H)-C(=O)-CH_3]
[CH in plane] 433

▶1046 MSh
A (25 mm Hg, 10 cm)
$CF_3CF_2CF_2CF_2CF_3$
237

▶1046 S
B
[structure: 5-methyl-5-nitro-tetrahydro-1,3-oxazine, H_3C, O_2N, N-C_5H_{11}]
26

▶1046 S
Cyclohexane
Zn [S-C=S, O, C_4H_9-n]$_2$
[C=S] 434

▶1046 M
B
[structure: H_3C, O_2N, N-C_2H_5 oxazine]
26

▶1046 M
CCl_4
[structure: 2-aminopyridine, NH_2]
[CH in plane] 327, see also 438

▶1045 M
B (0.0576)
[structure: CH_3 / F benzene]
220

▶1045 M
B (film)
CF_2Cl-$CFCl_2$
255

▶1045 S
B
[structure: H_3C, O_2N, N-C_4H_9-n oxazine]
26

▶1045 MW
B (0.003)
$C_3H_7NO_2$
392

▶1045 MS
B (0.0576)
[structure: CH_3 / F benzene]
220

▶1045 M
B
[structure: H_5C_2, O_2N, N-CH_3 oxazine]
26

▶1045 W
$CHCl_3$
[structure: H_3C, CH_3, H_2N isoxazole, N, O]
15

▶1045
P
$Ba(NO_3)_2$
45

▶1045
P
$La(NO_3)_3$
45

▶1045
P
$Mg(NO_3)_2$
45

▶1045
P
$Sr(NO_3)_2$
45

▶1044 VS
B (0.063)
[structure: CH_3, CN benzene]
312

▶1044 S
E
Cu (NH_2-CH_2-CH_2-NH_2)$_2$ $PtCl_4$
[Ring] 8

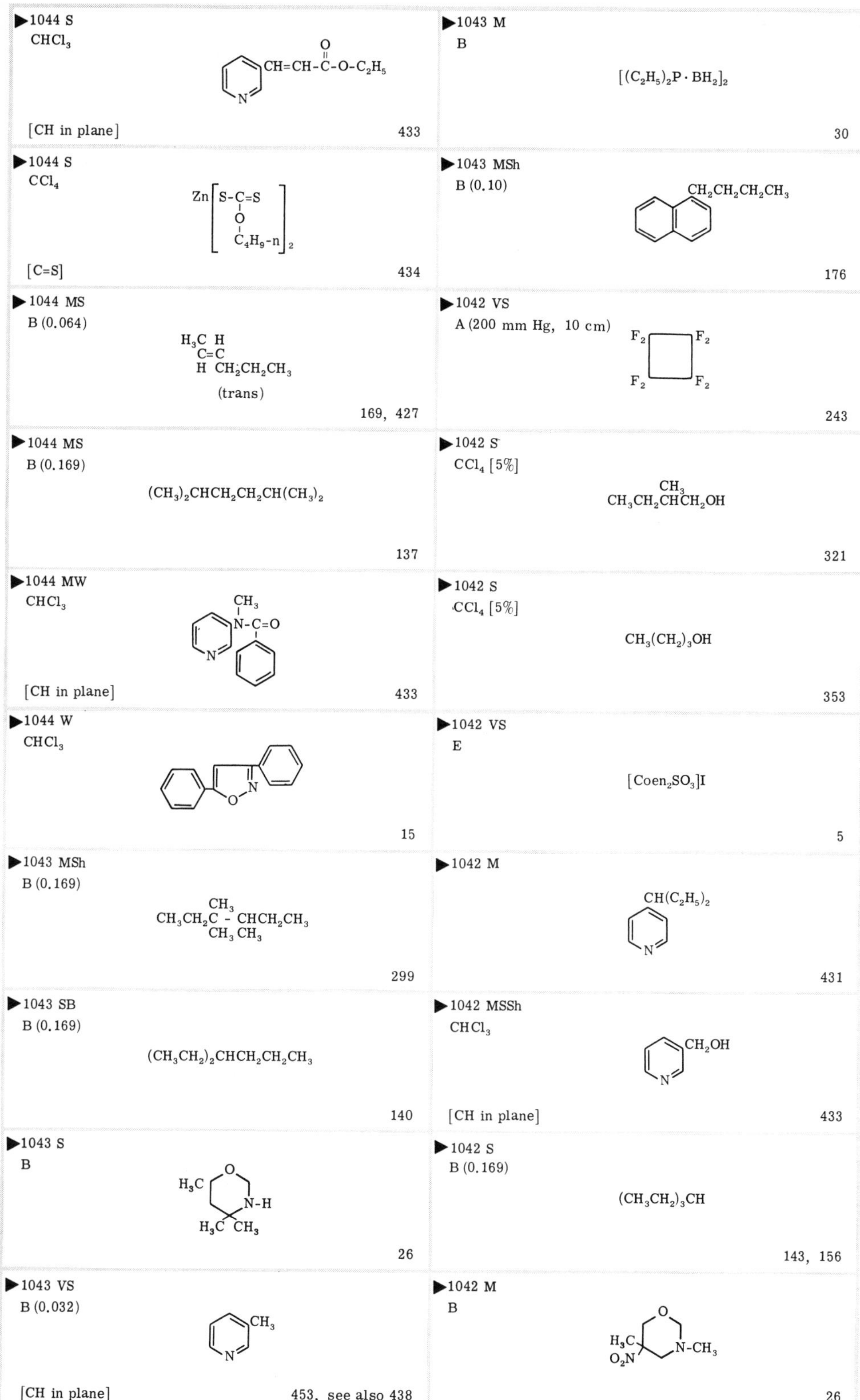

▶1044 S
$CHCl_3$

[CH in plane] 433

▶1044 S
CCl_4

[C=S] 434

▶1044 MS
B (0.064)

169, 427

▶1044 MS
B (0.169)
$(CH_3)_2CHCH_2CH_2CH(CH_3)_2$
137

▶1044 MW
$CHCl_3$

[CH in plane] 433

▶1044 W
$CHCl_3$

15

▶1043 MSh
B (0.169)

299

▶1043 SB
B (0.169)
$(CH_3CH_2)_2CHCH_2CH_2CH_3$
140

▶1043 S
B

26

▶1043 VS
B (0.032)

[CH in plane] 453, see also 438

▶1043 M
B
$[(C_2H_5)_2P \cdot BH_2]_2$
30

▶1043 MSh
B (0.10)

176

▶1042 VS
A (200 mm Hg, 10 cm)

243

▶1042 S
CCl_4 [5%]

321

▶1042 S
CCl_4 [5%]
$CH_3(CH_2)_3OH$
353

▶1042 VS
E
$[Coen_2SO_3]I$
5

▶1042 M

431

▶1042 MSSh
$CHCl_3$

[CH in plane] 433

▶1042 S
B (0.169)
$(CH_3CH_2)_3CH$
143, 156

▶1042 M
B

26

▶1042 M
$CHCl_3$
H_3CO O N
15

▶1042 M
CCl_4 [0.0128 m/l]
Cl O C-OH
448

▶1042 W
$Gd(NO_3)_3$
45

▶1042
P
$La(NO_3)_3 \cdot 3NH_4NO_3$
45

▶1041 VS
B (0.169)
$(CH_3)_3CCH_2CHCH(CH_3)_2$
CH_3
303

▶1041 VS
CCl_4 [25%]
$CH_3CH_2CHCH_2CH_3$
OH
196

▶1041 VS
B (0.0563)
$(CH_3)_2CHCHCH(CH_3)_2$
CH_3
135

▶1041 SVB
B (0.169)
$(CH_3)_2CHCH(CH_2CH_3)_2$
136

▶1041 S
B (0.169)
CH_3
$(CH_3)_2CHC - CH_2CH_2CH_3$
CH_3
134

▶1041 S
B (0.063)
CH_3
CN
214

▶1041 S
C
CH_3
191

▶1041 M
CS_2 [40%] (0.025)
Cl S $COCH_3$
372

▶1041 M
G
H O S H
N-C-C-N
C_6H_{11} C_6H_{11}
[C=S]
31

▶1041
P
$Sm(NO_3)_3$
45

▶1040 VS
B (0°C)
$(CH_3)_2CHCHCH(CH_3)_2$
CH_3
284

▶1040 VS
C
CH_3 CH_3
182

▶1040 S
B (0.169)
CH_3 CH_3
$CH_3CH_2C - CH_2CHCH_2CH_3$
CH_3
302

▶1040 S
B
$(CH_3)_2CHCH_2CH_2CH(CH_3)_2$
339

▶1040 S
$CHCl_3$
Zn[S-C=S / O / C_4H_9-n$]_2$
[C=S]
434

▶1040 MWSh
$CHCl_3$
N CH_2-C(=O)-O-C_2H_5
[CH in plane]
433

▶1040 MWSh
$CHCl_3$ 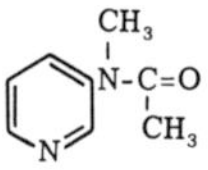
CH_3 N-C=O CH_3 N
[CH in plane]
433

▶1040 W
$CHCl_3$ 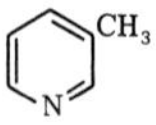
CH_3 N
[CH in plane]
433

▶1040 W
$CHCl_3$ 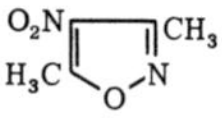
O_2N CH_3 H_3C O N
15

▶1040
K
$Ca(NO_3)_2$
45

▶1040
P
$Ce(NO_3)_3$
45

▶1040
P
$Nd(NO_3)_3$
45

▶1040
E or N 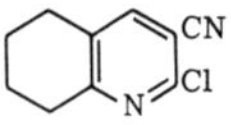
CN Cl N
41

▶1040
E or N
CH_2NH_2 N 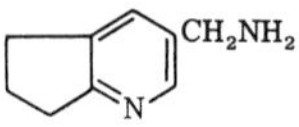
41

▶1039 S
CH_2Cl_2
Zn [S-C=S O C_4H_9-n]$_2$ 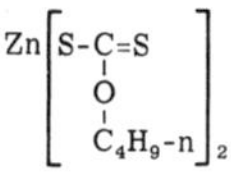
[C=S]
434

▶1039 MS
$CHCl_3$ 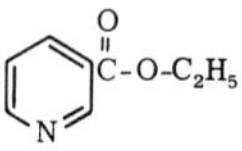
O C-O-C_2H_5 N
[CH in plane]
433

▶1039 MSB
B (0.055) 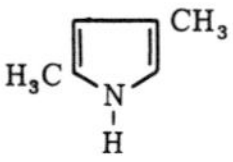
CH_3 H_3C N H
344

▶1039 M
B 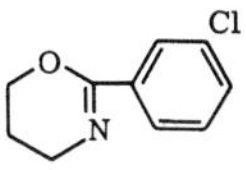
Cl O N
26

▶1039 MB
B (0.300)
$(C_2H_5)_2CH(CH_2)_{20}CH_3$
282

▶1039
E
$Al(NO_3)_3$
45

▶1039 W
$CHCl_3$ 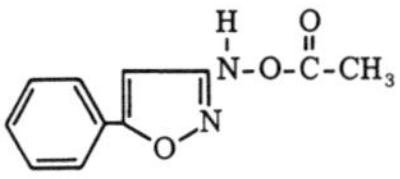
H O N-O-C-CH_3 O N
15

▶1038 VS
B (0.10) 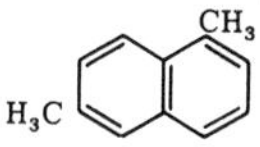
CH_3 H_3C
184

▶1038 VS
B (film) 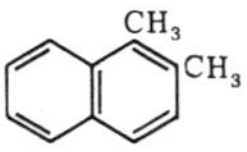
CH_3 CH_3
188

▶1038 VVS
B (0.169)
$(CH_3)_2CHCH(CH_3)_2$
144

▶1038 VS
B (0.0576) 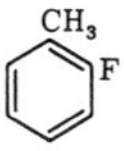
CH_3 F
221

▶1038 S (Q branch)
A (24.1 mm Hg, 40 cm) 
309

▶1038 S B (0.10) [structure: methyltetralin, CH_3] 269	▶1038 P $Y(NO_3)_3$ 45
▶1038 MS C $(CH_3)_3CNO_2$ 325	▶1038 S B (0.08) [structure: isoquinoline, N] 197
▶1038 S $CHBr_3$ $Zn[S-C=S, O, C_4H_9-n]_2$ [C=S] 434	▶1037 VS B (0.136) $CH(CH_3)_2$, H_3C, CH_3 277
▶1038 MS G H O S H; N-C-C-N; C_6H_5 C_6H_{11} [C=S] 31	▶1037 SSh $K_4P_2O_6$ 28
▶1038 S B (0.127) [structure: cyclohexane] 103	▶1037 S A (200 mm Hg, 10 cm) F_2 F_2 F_2 F_2 243
▶1038 M B (0.10) [structure: cyclooctatetraene] 200	▶1037 VS B (0.15) [structure: cyclohexene] 128
▶1038 MSh $CHCl_3$ H_3C, CH_3, H_5C_2-O, O, N 15	▶1037 S CH_3CN $Zn[S-C=S, O, C_4H_9-n]_2$ [C=S] 434
▶1038 E $Ce(NO_3)_3$ 45	▶1037 S $CHBr_2-CHBr_2$ $Zn[S-C=S, O, C_4H_9-n]_2$ [C=S] 434
▶1038 P $Ce(NH_4)_2(NO_3)_6$ 45	▶1037 MS B (0.169) $(CH_3)_2CHCH_2CH_2CH(CH_3)_2$ 137
▶1038 P $Gd(NO_3)_3$ 45	▶1037 M B (0.0576) F CF_3, F_2, F_2, F, CF_3, F_3C, F, F_2 228

▶1037 MW
$CHCl_3$
3-pyridyl-C(=O)-O-C_3H_7-n
[CH in plane]
433

▶1037 MW
$CHCl_3$
3-pyridyl-C(=O)-O-C_3H_7-i
[CH in plane]
433

▶1037 MW
$CHCl_3$
3-pyridyl-C(=O)-O-C_4H_9-i
[CH in plane]
433

▶1037 W
$C_6H_5HSiCl_2$
86

▶1037 S
B (0.10)
CH_3 (methylnaphthalene)
192

▶1037 VS
B (0.028)
Cl, Cl (dichlorothiophene)
162

▶1036 VS
B (0.10)
H_3C, CH_3 (dimethylnaphthalene)
183

▶1036 VS
E
$[Coen_2SO_3]SCN$
5

▶1036 VS
E
$[Coen_2SO_3]Cl$
5

▶1036 S
C
H_3C, CH_3 (dimethylnaphthalene)
180

▶1036 MW
$CHCl_3$
3-pyridyl-C(=O)-O-CH_3
[CH in plane]
433

▶1036 VS
A (200 mm Hg, 10 cm)
$(CH_3)_2CHCH_2CH_3$
333

▶1036 MW
$CHCl_3$
H_3C, H_3CO, N, O (isoxazoline with phenyl)
15

▶1036
E or N
CH_2OH, N (cyclopenta-pyridine)
41

▶1036 VW
$CHCl_3$
$(CH_2)_3$, CN, Cl, N
41

▶1036 VW
$CHCl_3$
CH_3, N, O (isoxazole)
15

▶1035 S
C
CH_3, CH_3 (dimethylnaphthalene)
185

▶1035 MW
B (0.0576)
F CF_3, F_2, F_2, F_2, F_2, F CF_3 (perfluorodimethylcyclohexane)
230

▶1035 MS
B (0.169)
$(CH_3)_2CHC(CH_3)_2-CH_2CH_3$
341

▶1035 MSh
B (0.068)
H_3C C_3H_7-n, C=C, H H
(cis)
276

▶1035 M
$CHCl_3$ 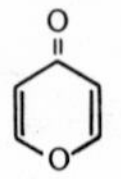
O
O
[CH] 7

▶1035 MW
$CHCl_3$ 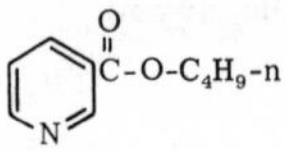
O
$C-O-C_4H_9-n$
N
[CH in plane] 433

▶1035 MW
$CHCl_3$ 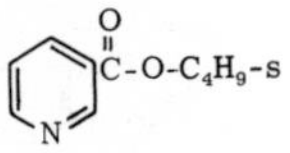
O
$C-O-C_4H_9-s$
N
[CH in plane] 433

▶1035 MW
$CHCl_3$ 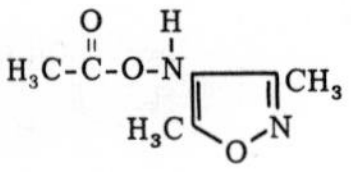
O H
$H_3C-C-O-N$ CH_3
H_3C O N
15

▶1035 W
$CHCl_3$ 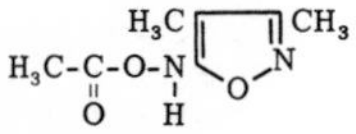
H_3C CH_3
$H_3C-C-O-N$ N
O H O
15

▶1035 VW
$CHCl_3$ 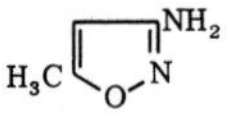
NH_2
H_3C O N
15

▶1035
E
$Hg(NO_3)_2$
45

▶1035
E
$Sm(NO_3)_3$
45

▶1035
K
$Mn(NO_3)_2 \cdot xH_2O$
45

▶1035
P
$La(NO_3)_3$
45

▶1034 MS
A (25 mm Hg, 10 cm)
$CF_3CF_2CF_3$
235

▶1034 VS
B (0.238)
$H_2C=CC(CH_3)_3$
CH_3
272

▶1034 VS
B (0.10) 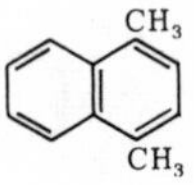
CH_3
CH_3
186

▶1034 S
B 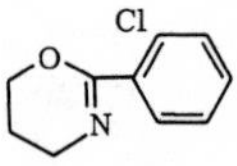
Cl
O
N
26

▶1034 S
B (0.169)
$(CH_3)_2CHCHCH(CH_3)_2$
CH_2
CH_3
301

▶1034 S
B (0.169)
CH_3
$(CH_3)_2CHCHCH_2CH_2CH_3$
139

▶1034 MS
B (0.065)
$H_2C=CHCH_2C(CH_3)_3$
273

▶1034 WSh
$CHCl_3$ 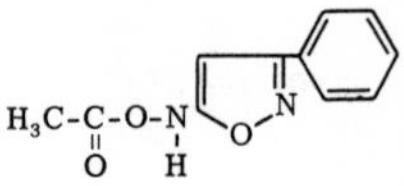
$H_3C-C-O-N$ N
O H O
15

▶1034
CH_3OH
437

▶1033 VS
B (0.0104)
$CF_2Cl-CFCl_2$
255

▶1033 VS
D, E

$[H_2CNO_2]^-Na^+$

$[NO_2]$ 328

▶1033 S
CS_2 [sat.] (0.170)

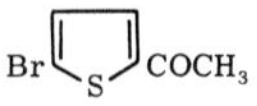

Br (thiophene) $COCH_3$

348

▶1033 S
B

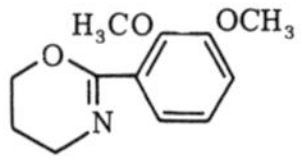

H_3CO OCH_3 O N

26

▶1033 M
E

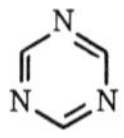

N N N

14

▶1033 M
CCl_4 [0.00658 m/l]

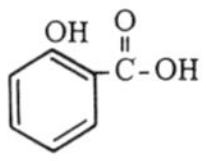

OH O C-OH

448

▶1033
E or N

COOH =O N H

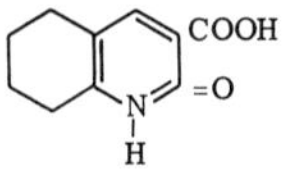

41

▶1033
E

$Ni(NO_3)_2$

45

▶1032
E or N

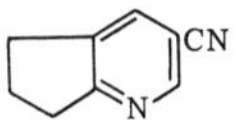

CN N

41

▶1032 S
B (0.0104)

F CF_3 CF_3 F F_2 F_2 F_2 F_2

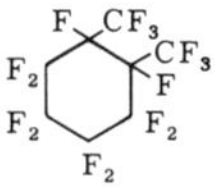

231

▶1032 MS
A (385 mm Hg, 10 cm)

CF_3-CCl_3

253

▶1032 MW
$CHCl_3$

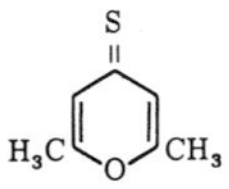

S H_3C O CH_3

[CH] 7

▶1032
E

$Be(NO_3)_2$

45

▶1032
E

$Hg_2(NO_3)_2$

45

▶1032
K

$La(NO_3)_3$

45

▶1032
E or N

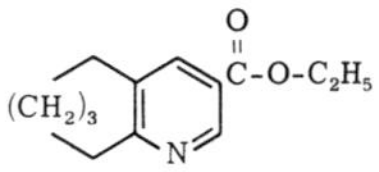

O C-O-C_2H_5 $(CH_2)_3$ N

41

▶1031 VS
CCl_4 [25%]

$CH_3CHCH_2CH_3$ OH

319

▶1031 VS
B (0.065)

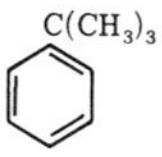

$C(CH_3)_3$

133

▶1031 VS
B (0.051)

$CH(CH_3)_2$ $CH(CH_3)_2$

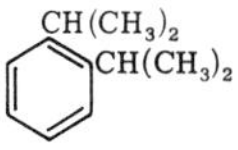

296

▶1031 VS
CCl_4 [5%]

CH_3OH

316

▶1031 VS
CCl_4

$[H_2C \cdot NO_2]^-Na^+$

$[NO_2]$ 328

▶1031 VSB
E

$Ag_4P_2O_6$

28

▶1031 M
B

(morpholine ring: O; H_3C, O_2N; $N-C_3H_7$-i)

26

▶1031 M
B

(morpholine ring: O; H_5C_2, O_2N; $N-C_3H_7$-i)

26

▶1031 S
CCl_4 [5%]

$CH_3(CH_2)_3OH$

353

▶1030 SSh
B (0.169)

$(CH_3)_3CCH_2CHCH(CH_3)_2$
CH_3

303

▶1030 MSh
B (0.064)

H CH_2CH_3
C=C
H_3C CH_3
(trans)

165

▶1030 SSp
E

$Na_4P_2O_7$

28

▶1030 S
C_6H_6

Zn[S-C(=S)-O-C_4H_9-n]$_2$

[C=S]

434

▶1030 S
$(C_2H_5)_2O$

n-C_4H_9-O-C(=S)-S-S-C(=S)-O-C_4H_9-n

[C=S]

434

▶1030 M
B

(morpholine ring: O; H_5C_2, O_2N; $N-C_3H_7$-n)

26

▶1030 MSSp
B (0.064)

H_3CH_2C H
C=C
H CH_2CH_3
(trans)

168

▶1030 M
G

H O S H
N-C-C-N
C_6H_5 C_6H_{11}

[C=S]

31

▶1030 W
CS_2

$(C_6H_5)_3SiCl$

86

▶1030 W
$CHCl_3$

(thiopyran-4-thione ring: S=; H_3C, S, CH_3)

[CH]

7

▶1030
P

$Th(NO_3)_4$

45

▶1030 VS
B (0.0104)

$CF_3(CF_2)_5CF_3$

241

▶1029 VS
B (0.0563)

$(CH_3)_2C=CHC(CH_3)_3$

204

▶1029 M (band center)
A (4.2 mm Hg, 40 cm)

(benzene ring) $CH(CH_3)_2$

66

▶1029 W

$(C_6H_5)_2SiCl_2$

86

▶1029 W

$C_6H_5SiCl_3$

86

▶ 1029 MS
B (0.169)

CH_3
$(CH_3)_2CHC - CH_2CH_2CH_3$
CH_3

134

▶ 1029 MS
B (0.065)

S CH_2CH_3

213

▶ 1029 M
$CHCl_3$

N CH_3

[Ring] 433

▶ 1029 MWVB
B (0.300)

$CH_3(CH_2)_5CH(C_3H_7)(CH_2)_5CH_3$

281

▶ 1028 MWSh
B (0.157)

$(CH_3)_3CCHCH(CH_3)_2$
CH_3

293

▶ 1028 S
B (0.10)

$CH_2CH_2CH_2CH_3$

176

▶ 1028 VS
CCl_4 [25%]

$CH_3CH(CH_2)_2CH_3$
OH

318

▶ 1028 SB
B (0.169)

$(CH_3CH_2)_2CHCH_2CH_2CH_3$

140

▶ 1028 S
Cyclohexane

S
‖
$n-C_4H_9-O-C$
|
S
|
S
|
$n-C_4H_9-O-C=S$

[C=S] 434

▶ 1028 MS
G

H O S
| ‖ ‖
N-C-C-N (CH_2-CH_2) O (CH_2-CH_2)
|
H

31

▶ 1028 MS
$CHCl_3$

N $CH_2CH_2-C(=O)-O-C_2H_5$

[Ring] 433

▶ 1028 M
B

O
H_5C_2 $N-C_5H_{11}-n$
O_2N

26

▶ 1028 M
B (0.10)

CH_2CH_3

190

▶ 1028 MW
$CHCl_3$

O N

15

▶ 1028 VS
B (0.10)

CH_3
CH_3

187

▶ 1027 VS
B (0.0104)

F CF_3
F_2 F_2
F_2 F_2
F_2

232

▶ 1027 VS
B (0.0104)

CF_3

219

▶ 1027 VS
CCl_4 [25%]

$CH_3CH(CH_2)_2CH_3$
OH

318

▶ 1027 VS
B (0.0104)

$CH_3CH_2-O-CF_2CHFCl$

261

▶ 1027 VS
$CHCl_3$

N $CH_2-C(=O)-O-C_2H_5$

[Ring] 433

▶1027 VS
$CHCl_3$ 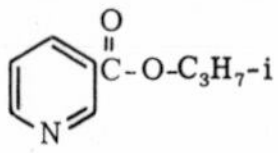
O
‖
C-O-C_3H_7-i
N
[Ring] 433

▶1027 VS
$CHCl_3$ 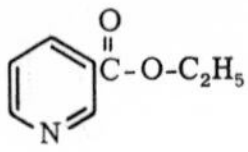
O
‖
C-O-C_2H_5
N
[Ring] 433

▶1027 S
CCl_4 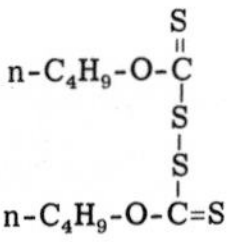
S
‖
n-C_4H_9-O-C
S
S
n-C_4H_9-O-C=S
[C=S] 434

▶1027 M
B (0.068) 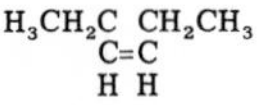
H_3CH_2C CH_2CH_3
C=C
H H
(cis)
275

▶1027 MS
B (0.0563) 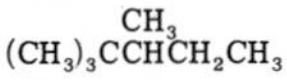
CH_3
$(CH_3)_3CCHCH_2CH_3$
340, 426

▶1027 M
B (0.157) 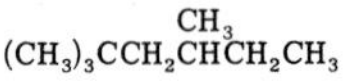
CH_3
$(CH_3)_3CCH_2CHCH_2CH_3$
79

▶1027
E or N
O
‖
C-O-C_2H_5
N 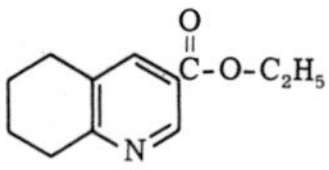
41

▶1026 VS
B (0.10) 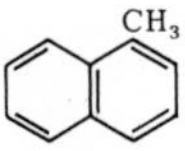
CH_3
192

▶1026 VS
$CHCl_3$ 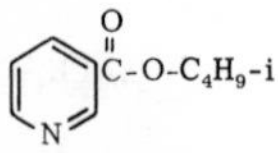
O
‖
C-O-C_4H_9-i
N
[Ring] 433

▶1026 S
$CHCl_3$ 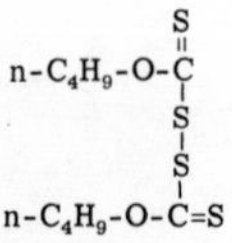
S
‖
n-C_4H_9-O-C
S
S
n-C_4H_9-O-C=S
[C=S] 434

▶1026 SSp
E
$Na_2H_2P_2O_6$
[PO_2(OH) stretch] 28

▶1026 MS
B (0.10) 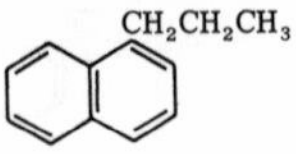
$CH_2CH_2CH_3$
178

▶1026 M
G 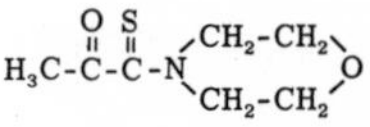
O S
‖ ‖
H_3C-C-C-N (CH_2-CH_2) O (CH_2-CH_2)
31

▶1026 M
$CHCl_3$ 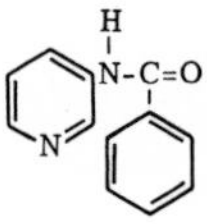
H
N-C=O
N
[Ring] 433

▶1026 W
$CHCl_3$ 
O
O
[CH] 7

▶1026
E or N
O
‖
C-O-C_2H_5
N 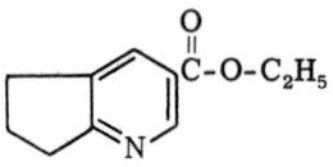
41

▶1025 VVS
B
$VOBr_3$
[V=O] 12

▶1025 SVB
B (0.169)
$(CH_3)_2CHCH(CH_2CH_3)_2$
136

▶1025 VS
B 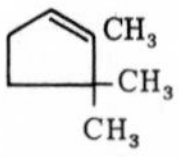
CH_3
CH_3
CH_3
126

▶1025 VS
$CHCl_3$ 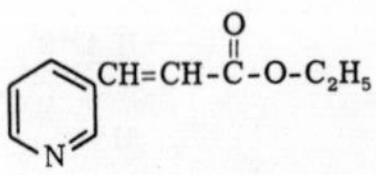
O
‖
CH=CH-C-O-C_2H_5
N
[Ring] 433

▶1025 VS $CHCl_3$ 3-pyridyl–$C(=O)-O-C_4H_9$-n [Ring] 433	▶1025 M CCl_4 [0.0082 m/l] $C_6H_5-C(=O)-OH$ 448
▶1025 VS $CHCl_3$ 3-pyridyl–$C(=O)-O-C_4H_9$-s [Ring] 433	▶1025 MW $CHCl_3$ 3-pyridyl–C_6H_4–NO_2 [CH in plane] 433
▶1025 VS 3-pyridyl–$C(=O)-O-CH_3$ [Ring] 433	▶1025 WSh $CHCl_3$ 3-bromopyridine [CH in plane] 433
▶1025 S B (0.169) $(CH_3)_2CHCH(CH_3)CH_2CH_3$ 142, 153	▶1025 E $Gd(NO_3)_3$ 45
▶1025 VS B (0.0104) $H_3C-O-CF_2CHFCl$ 356	▶1025 E or N $(CH_2)_3$-fused pyridine–CH_2OH 41
▶1025 S 3-pyridyl–C_3H_7-n 431	▶1024 VS $CHCl_3$ 3-pyridyl–$C(=O)-O-C_3H_7$-n [Ring] 433
▶1025 MS G $H_2N-C(=O)-C(=S)-NH(C_6H_{11})$ [C=S] 31	▶1024 S B (0.0563) $CH_3CH=CHCH_2CH_3$ (cis) 208
▶1025 MS $CHCl_3$ 3-pyridyl–CN [Ring] 433	▶1024 S CH_2Cl_2 $n-C_4H_9-O-C(=S)-S-S-C(=S)-O-C_4H_9-n$ [C=S] 434
▶1025 MS $CHCl_3$ 3-pyridyl–$C(=O)-H$ [Ring] 433	▶1024 M (P branch) A (24.1 mm Hg, 40 cm) benzene 309
▶1025 M B morpholine ring, H_5C_2, O_2N, $N-CH_3$ 26	▶1024 M B 2-phenyl-5,6-dihydro-4H-1,3-oxazine 26

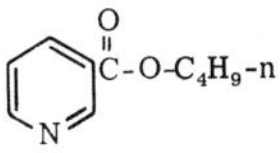

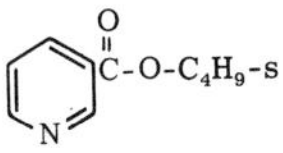

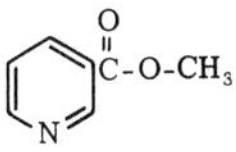

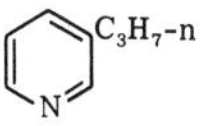

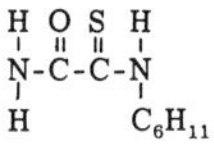

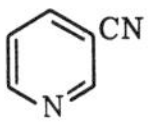

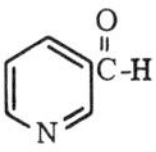

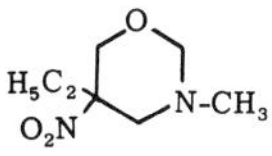

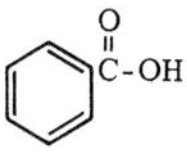

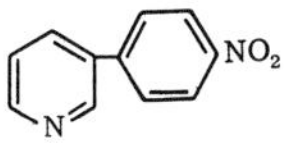

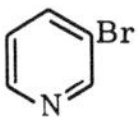

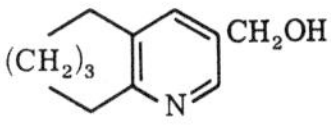

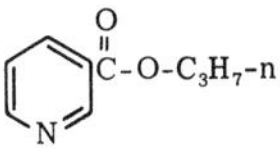

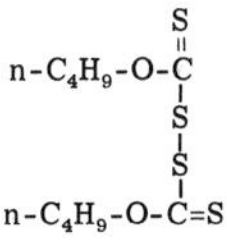

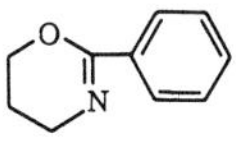

▶1024 MW
$CHCl_3$
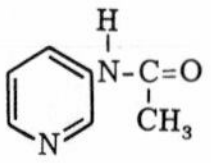
[Ring] 433

▶1024
E
$Pr(NO_3)_3$
45

▶1023 VS
B
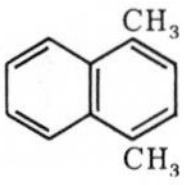
186

▶1023 VS
C
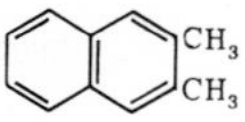
181

▶1023 VS
E
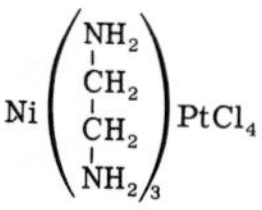
[Ring] 8

▶1023 S
$CHCl_3$
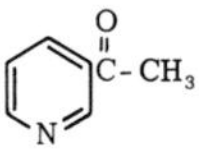
[Ring] 433

▶1023 S
C_6H_6
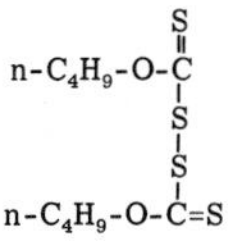
[C=S] 434

▶1023 S
CH_3NO_2
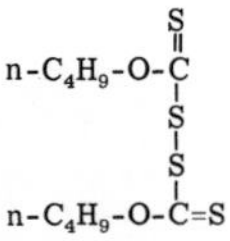
[C=S] 434

▶1023 M
$CHCl_3$
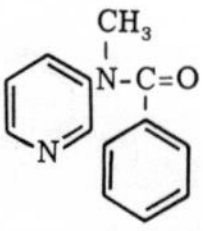
[Ring] 433

▶1023 M
$CHCl_3$
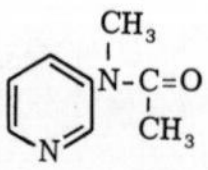
[Ring] 433

▶1023 MW
$CHCl_3$
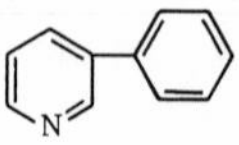
[CH in plane] 433

▶1023 W
$CHCl_3$
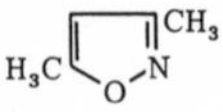
15

▶1023 Sh
B
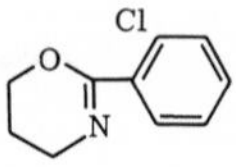
26

▶1022 VS
A (25 mm Hg, 10 cm)
$CF_3CF_2CF_2CF_2CF_3$
237

▶1022 W
B (0.0104)
$n-H_9C_4-O-CF_2CHFCl$
264

▶1022 VS
CCl_4 [25%]
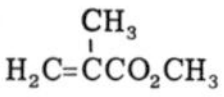
193

▶1022 S
B (0.10)
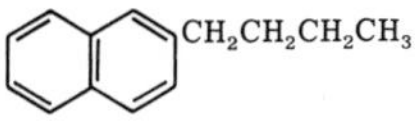
175

▶1022 S
B (0.157)
$(CH_3)_3CCH(CH_2CH_3)_2$
78

▶1022 S
B (0.063)
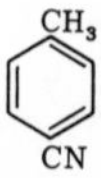
214

▶1022 S
$OCH_2OCH_2CH_2CH_2$
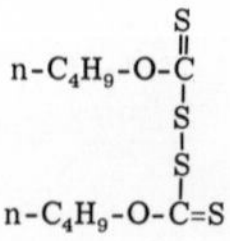
[C=S] 434

▶1022 S
$CHCl_3$
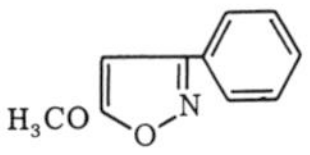
15

▶1022 MW
B (0.157)
$(CH_3)_3CCHCH(CH_3)_2$
CH_3
293

▶1022 MSh
B (0.003)
NO_2
$CH_3CH_2CHCH_3$
324

▶1022 W
$CHCl_3$
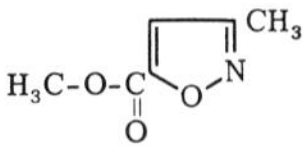
15

▶1022
P
$Pr(NO_3)_3$
45

▶1021 VS
B (0.169)
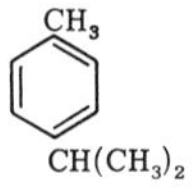
132

▶1021 MSSh
CCl_4 [25%]
$CH_3CH_2C(CH_3)_2$
OH
194

▶1021 MS
B (0.169)
CH_3
$(CH_3)_3CC - CH_2CH_2CH_3$
CH_3
304

▶1021 VS
B (0.068)
H_3C CH_3
C=C
H CH_2CH_3
(cis)
274

▶1021 S
B (0.10)
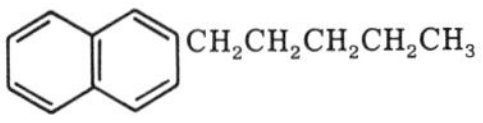
173

▶1021 S
$CHCl_3$
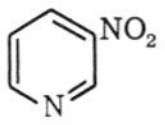
[Ring]
433

▶1021 S
CH_3CN
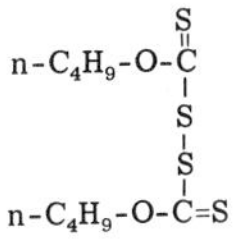
[C=S]
434

▶1021 M
$CHCl_3$
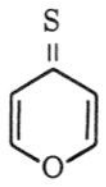
[CH]
7

▶1021 MW
$CHCl_3$
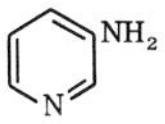
[Ring]
433

▶1021 MB
B (0.104)
$CH_3CH_2CH(CH_2)_2CH_3$
CH_3
81

▶1020 VS
B (0.0576)
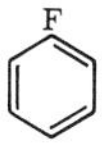
226

▶1020 VS
B (0.169)
$(CH_3)_3CCH(CH_2CH_3)_2$
300

▶1020
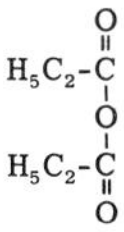
[C-O-C]
436

▶1020 S
E
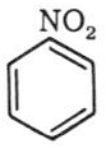
[CH in plane]
20

▶1020 S
$CHBr_3$
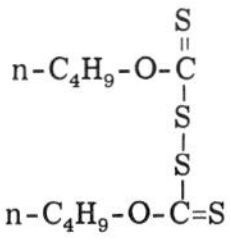
[C=S]
434

▶1020 S
CHBr₂-CHBr₂

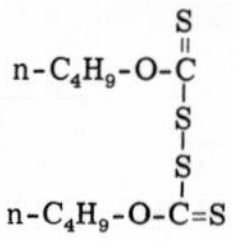

[C=S] 434

▶1020 M
B (0.064)

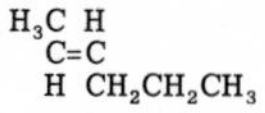

(trans)

169, 427

▶1020 W
$CHCl_3$

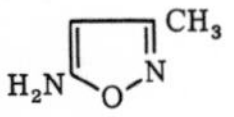

15

▶1020
E

$Cr(NH_3)_5(NO_3)_3$

45

▶1019 W
$CHCl_3$

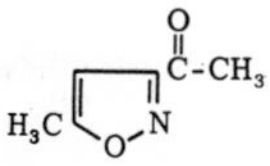

15

▶1019 S
B (0.0563)

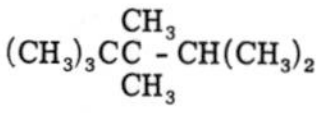

306

▶1019 MW
$CHCl_3$

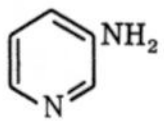

[Ring] 433

▶1018.6 VS
B (0.03)

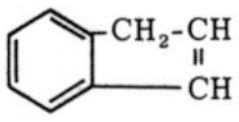

9

▶1018 VS
B (0.0104)

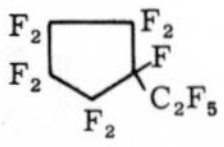

247

▶1018 VS
B (0.169)

$(CH_3)_3CCH_2CH_3$

145

▶1018 VS
D

$[H_2CNO_2]^-Na^+$

328

▶1018 VS
$CHCl_3$

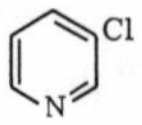

[Ring] 433

▶1018 S
A (200 mm Hg, 10 cm)

$(CH_3)_3CCH_2CH_3$

94

▶1018 VSB
CCl_4 [25%]

$CH_3(CH_2)_5OH$

320

▶1018 S
B (0.10)

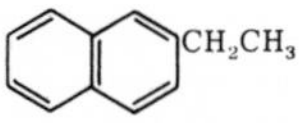

189

▶1018 S
B (0.10)

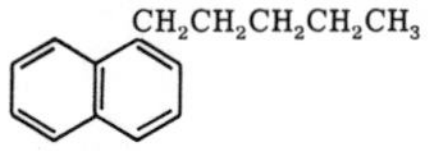

174

▶1018 MSh
B (0.169)

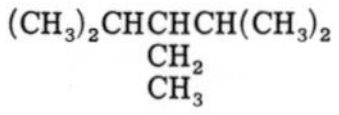

301

▶1018 M
B (0.15)

128

▶1018 W

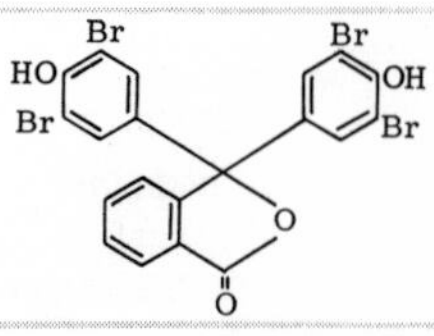

39

▶1018
E or N

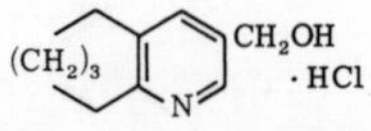

41

▶1018
E or N
$(CH_2)_3$ [fused pyridine]-C(=O)-OH · HCl
41

▶1017 VSSh
B (0.136)
$CH_3CHCH_2CH_3$ / $CH_3CH_2HC(CH_3)$ / $CHCH_2CH_3$ CH_3 [benzene]
280

▶1017 S
B (0.0563)
$CH_3CH{=}CHCH_2CH_3$
(trans)
207

▶1017 S
B (0.0576)
CH_3 [benzene] F
220

▶1017 VS
B (0.238)
$H_2C{=}C(CH_3)C(CH_3)_3$
272

▶1017 VS
E
$[H_2CNO_2]^-Na^+$
[NO_2 asymm. stretch]
328

▶1017 S
B (0.0563)
$CH_3CH_2CH(CH_3)-CH(CH_3)CH_2CH_3$
83

▶1017 S
B (0.08)
[isoquinoline] N
197

▶1017 S
CS_2 [sat.] (0.170)
Br [thiophene, S] $COCH_3$
348

▶1017 S
B
$[(CH_3)_2N\text{-}BO]_3$
32

▶1017 MS
B (0.10)
[naphthalene] $CH_2CH_2CH_3$
177

▶1017 M
[isochromanone] O, O
39

▶1017 VS
CCl_4 [25%]
$CH_3CH_2CH_2OH$
314

▶1016 VS
CCl_4
$[H_2CNO_2]^-Na^+$
[NO_2 asymm. stretch]
328

▶1016 M
B (0.955)
F, F, F, F [benzene]
222

▶1016 S
B (0.051)
$CH(CH_3)_2$ [benzene] $CH(CH_3)_2$
298

▶1016 SSh
$CHCl_3$
H_3C [isoxazole, O, N] C(=O)-O-CH_3
[Ring]
15

▶1016 M
A
$(CH_3)_2SO$
13

▶1016 MVB
B (0.300)
$C_{26}H_{54}$
(5, 14-di-n-butyloctadecane)
283

▶1016
P
$Cr(NH_3)_5(NO_3)_3$
45

▶ 1015 S
A (100 mm Hg, 10 cm)

$CH_3\text{-}CCl_3$

251

▶ 1015 VS
B

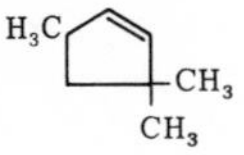

125

▶ 1015 S
B (0.0563)

CH_3
$CH_3CH_2C\text{ - }CH_2CH_2CH_3$
CH_3

360

▶ 1015 MS
E

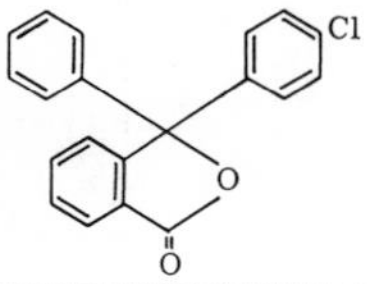

39

▶ 1015 M
B (0.0563)

CH_3
$(CH_3)_3CCHC(CH_3)_3$

307

▶ 1015 M
B

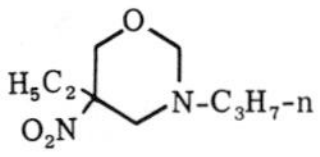

26

▶ 1015
E

$Cr(NO_3)_3$

45

▶ 1015 (fundamental)
A

CH_3Cl

[CH_3]

376

▶ 1014 VS
CCl_4 [25%]

CH_3
$H_2C{=}CCO_2CH_3$

193

▶ 1014 SVB
B (0.169)

$(CH_3)_2CHCH(CH_2CH_3)_2$

136

▶ 1014 S
CCl_4 [5%]

CH_3
$CH_3CH_2CHCH_2OH$

321

▶ 1014 S
B

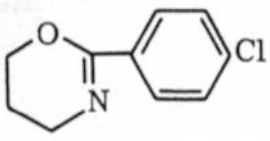

26

▶ 1014 VS
B (0.169)

CH_3
$(CH_3)_2CHCHCH_2CH_3$

142, 153

▶ 1014 M
B (0.157)

CH_3
$(CH_3)_3CCH_2CHCH_2CH_3$

79, 285

▶ 1014 M
B

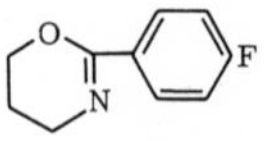

26

▶ 1014 MS
C

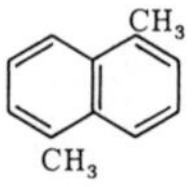

185

▶ 1014 MS
B (0.127)

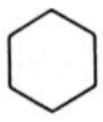

103

▶ 1014 W
E

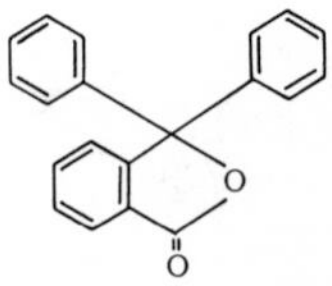

39

▶ 1013 S
B (0.169)

$(CH_3)_2CHCH_2CHCH_2CH_3$
CH_3

138, 334

▶ 1013 MS
B (0.169)

$(CH_3)_3CCH_2CH_2C(CH_3)_3$

305

▶1013 MW
$CHCl_3$ 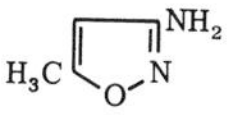
NH_2, H_3C, O, N
[Ring] 15

▶1013 MW
$CHCl_3$ 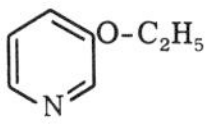
$O-C_2H_5$, N
[Ring] 433

▶1013 W
E 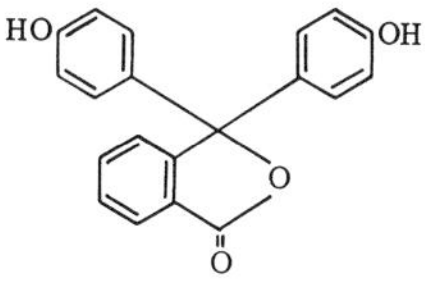
HO, OH, O, O
39

▶1013
P
$Pb(NO_3)_2$
45

▶1012 VS
B (0.0576) 
F, F
225

▶1012 VS
$CHCl_3$ 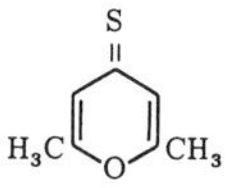
S, H_3C, O, CH_3
7

▶1012 S
B 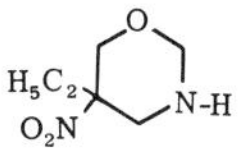
O, H_5C_2, O_2N, N-H
26

▶1012 M
B
$CH_3CH_2CH(CH_2)_2CH_3$ / CH_3
330

▶1012 MB
B (0.104)
$CH_3CH_2CH(CH_2)_2CH_3$ / CH_3
81

▶1012
E
$Sr(NO_3)_2$
45

▶1012 M
H_2O
$[H_2CNO_2]^-Na^+$
[NO_2 asymm. stretch] 328

▶1012 VS
B (0.101)
$(CH_3)_2CHCH_2CH_3$
401

▶1011 VVSB
B (0.169)
$CH_3CH_2C(CH_3)_2-CH_2CH(CH_3)CH_2CH_3$
302

▶1011 S
CS_2 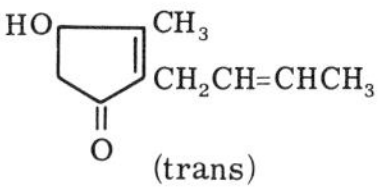
HO, CH_3, $CH_2CH=CHCH_3$, O
(trans)
430

▶1011 SB
B (0.169)
$(CH_3CH_2)_2CHCH_2CH_2CH_3$
140

▶1011 M
E 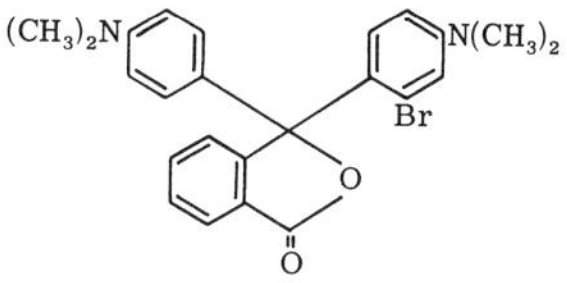
$(CH_3)_2N$, $N(CH_3)_2$, Br, O, O
39

▶1011 VS
B (0.009)
$CH_2=CHCH=CHCH=CH_2$
203

▶1010 S
A (100 mm Hg, 10 cm)
CH_3-CCl_3
251

▶1010 S
CCl_4 [5%]
$CH_3(CH_2)_3OH$
353

▶1010 VS
CCl_4 [5%]
$(CH_3)_2CHCH_2CH_2OH$
195

▶1010 S
B (0.0104)

F CF$_3$ / F$_2$ F$_2$ / F CF$_3$ / F$_3$C F / F$_2$

228

▶1010 VS
A (400 mm Hg, 15 cm)

$H_2C{=}CHCH(CH_3)_2$

131

▶1010 S
$CHCl_3$

$C_4H_4SO([CH_2]_4SO)$

[S-O] 33

▶1010 MB
B (0.136)

$CH_2(CH_2)_2CH_3$ / $CH_3(CH_2)_2H_2C$ / $CH_2(CH_2)_2CH_3$

279

▶1010 W
$CHCl_3$

NH_2 / H_5C_2 / O N

15

▶1009 VS
$CHCl_3$

Br / N

[Ring] 433

▶1009 S
B (0.169)

CH_3
$(CH_3)_2CHCHCH_2CH_2CH_3$

139

▶1009 S
B

CH_2CH_3

190

▶1009 M
CCl_4 [25%]

$CH_3CH_2CHCH_2CH_3$
OH

196

▶1009 M
$CHCl_3$

CH_3 / H_3C / O N

[Ring] 15

▶1008 VS
CS_2 [40%] (0.025)

Cl S $COCH_3$

372

▶1008 VS
CS_2 (0.169)

$CH_2{=}CHCH{=}CHCH{=}CHCH{=}CH_2$

202

▶1008 VVS
B (0.169)

CH_3
$(CH_3)_2CHC - CH_2CH_3$
CH_3

341

▶1008 S
B

CH_3
$(CH_3)_3CC - CH_2CH_2CH_3$
CH_3

304

▶1008 S
B

O N Br

26

▶1008 MB
B (0.300)

$(C_2H_5)_2CH(CH_2)_{20}CH_3$

282

▶1008 MW
$CHCl_3$

NH-CH / N

[Ring] 433

▶1008
E

$Pb(NO_3)_2$

45

▶1007 VS
A (25 mm Hg, 10 cm)

$CF_3CF_2CF_3$

235

▶1007 MW
B (0.0104)

n-H_7C_3-O-CF_2CHFCl

263

▶ 1007 S
B (0.003)

$CH_3(CH_2)_2CH_2NO_2$

310

▶ 1007 M
$CHCl_3$

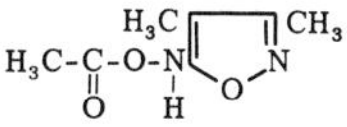
H3C
CH3
H3C-C-O-N
N
O
O
H

[Ring] 15

▶ 1007 MW
$CHCl_3$

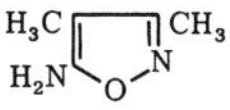
H3C
CH3
H2N
N
O

[Ring] 15

▶ 1007
E or N

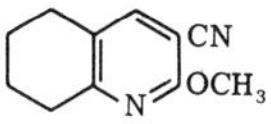
CN
OCH3
N

41

▶ 1007
E or N

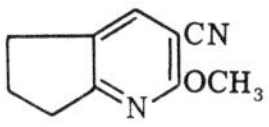
CN
OCH3
N

41

▶ 1006 VS
B

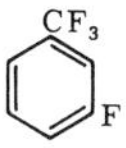
CF3
F

218

▶ 1006 S
B (0.0576), CCl_4 [25%]

$CH_3(CH_2)_4OH$

393, see also 350

▶ 1006 VS
B (0.169)

$CH_3CH_2C(CH_3)(CH_3) - CH(CH_3)CH_2CH_3$

299

▶ 1006 SVB
B (0.169)

$(CH_3)_2CHCH(CH_2CH_3)_2$

136

▶ 1006 S
B (0.003)

$CH_3CH_2CH(NO_2)CH_3$

324

▶ 1006 M
A

$(CH_3)_2SO$

13

▶ 1006 MS
B (0.10)

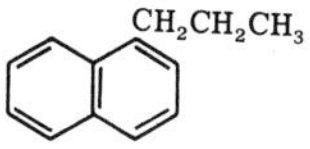
CH2CH2CH3

178

▶ 1006 S
B (0.0576)

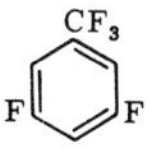
CF3
F
F

218

▶ 1006 M
$CHCl_3$

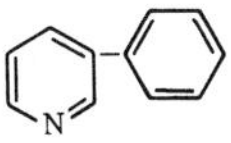
N

[Ring] 433

▶ 1006 VW
A (769.6 mm Hg, 43 cm)

$(CH_3)_3CH$

373

▶ 1006 W
E

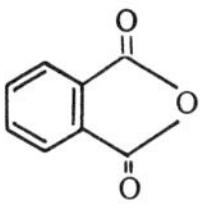
O
O
O

39

▶ 1006
E or N

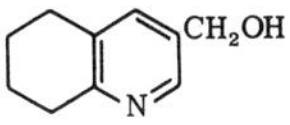
CH2OH
N

41

▶ 1006
E or N

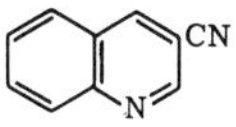
CN
N

41

▶ 1005 VW
B (0.068)

$H_3CH_2C(H)C=C(H)CH_2CH_3$
(cis)

275

▶ 1005 VS
$CHCl_3$

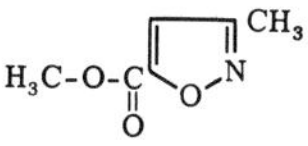
CH3
H3C-O-C
O
N
O

15

▶1005 S
B
HO (ring structure)
379

▶1005 S
B (0.068)
H_3C CH_3
C=C
H CH_2CH_3
(cis)
274

▶1005 S
B
H_3AsO_3
[AsO stretch]
692

▶1005 M
B
O
H_3C N-C_3H_7-n
O_2N
26

▶1005 WVB
B (0.300)
$CH_3(CH_2)_5CH(C_3H_7)(CH_2)_5CH_3$
281

▶1005 MS
B (0.101)
$(CH_3CH_2)_3CH$
156

▶1005 M
B (0.10)
H_3C CH_3
183

▶1005 MW
B (0.10)
CH_3
269

▶1005 M
$CHCl_3$
CH_3
H_2N O N
[Ring]
15

▶1004 VS
B (0.169)
$(CH_3CH_2)_3CH$
143

▶1004 VS
CH_3
CH_3CH_2C - CH_2CH_3
CH_3
336

▶1004 WB
B (0.0104)
n-H_9C_4-O-CF_2CHFCl
264

▶1004 VS
B (0.136)
$CH_3CHCH_2CH_3$
CH_3CH_2HC $CHCH_2CH_3$
CH_3 CH_3
280

▶1004 S
B (0.169)
$(CH_3)_3CCH_2CHCH(CH_3)_2$
CH_3
303

▶1004 S
B
H_3CO OCH_3
O N
26

▶1004 MS
$CHCl_3$
O
C-O-C_2H_5
H_3C O N
[Ring]
15

▶1004 M
B (0.10)
CH_3
H_3C
184

▶1004 W
$CHCl_3$
H_2N O N
[Ring]
15

▶1003 MW
A (200 mm Hg, 10 cm)
CF_2Cl-CF_2Cl
257

▶1003 S
CCl_4 [25%]
$CH_3CH_2C(CH_3)_2$
OH
194

▶1003 VS
$CHCl_3$

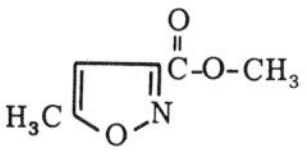

[Ring] 15

▶1003 WSh
B (0.0104)

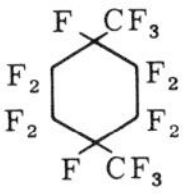

230

▶1003 MS
G

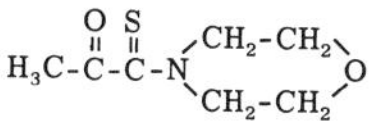

31

▶1003 M
B

$[(C_2H_5)_2P \cdot BH_2]_3$

30

▶1003 M
$CHCl_3$

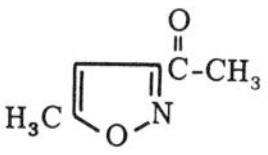

[Ring] 15

▶1002 VS
B

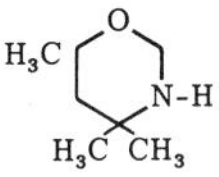

26

▶1002 VS
B (0.0153)

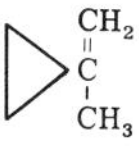

288

▶1002 MSh
CCl_4 [25%]

CH_3
$H_2C{=}CCO_2CH_3$

193

▶1002 S
B (0.0563)

CH_3
$(CH_3)_3CCHCH_2CH_3$

340, 426

▶1002 M
E

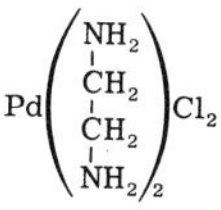

$[NH_2]$ 8

▶1002 M
E

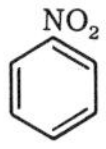

[Ring] 20

▶1002 W
B (0.136)

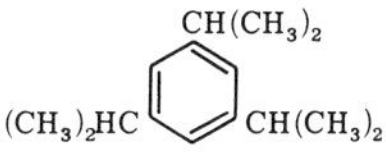

278

▶1002
E or N

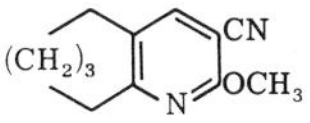

41

▶1001 VS
B (0.08)

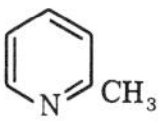

198, see also 438

▶1001 M
$CHCl_3$

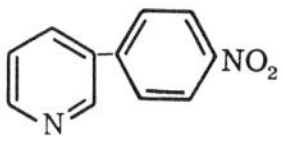

[Ring] 433

▶1001 W
$CHCl_3$

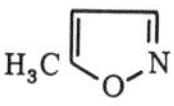

[Ring] 15

▶1001 W
$CHCl_3$

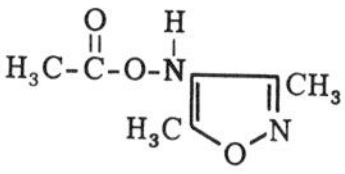

[Ring] 15

▶1000 VS
B (0.105)

CH_3
$CH_3CH_2C - CH_2CH_3$
CH_3

149

▶1000 VS
B (0.015)

$H_2C{=}CHC(CH_3)_3$

129

▶1000 VVSB
B (0.169)

CH_3 CH_3
$CH_3CH_2C - CH_2CHCH_2CH_3$
CH_3

302

▶1000 VSSp
E

$Na_2S_2O_6$

28

▶1000 M
CCl_4

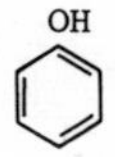

OH (benzene ring)

36

▶1000 S
A

$(CH_3)_3CCH_2CH_3$

94

▶1000 S
B (0.136)

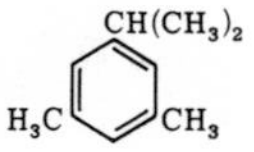

$CH(CH_3)_2$, H_3C, CH_3 (benzene ring)

227

▶1000 S
B

$(CH_3)_2CHCH_2CHCH_2CH_3$
CH_3

334

▶1000 S
B (0.114)

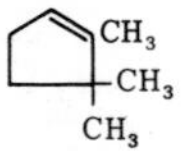

CH_3, CH_3, CH_3 (cyclopentene ring)

126

▶1000 SSh
B (0.2398)

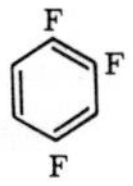

F, F, F (benzene ring)

224

▶1000 S
CS_2

$N(CH_3)_2$, NO_2 (benzene ring)

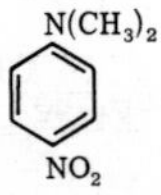

21

▶1000 MB
A (15.9 mm Hg, 40 cm)

$CH_3CH_2CH(CH_2)_3CH_3$
CH_3

106

▶1000 M
P

$[(CH_3)_2P \cdot BH_2]_3$

[BH_2 wag]

30

▶1000 M
$CHCl_3$

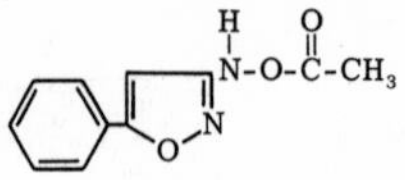

H, O, N-O-C-CH_3, O, N (phenyl-substituted isoxazole ring)

[Ring]

15

▶999.1 M

$C_6H_5HSiCl_2$

86

▶999 VS
A (36 mm Hg, 15 cm)

$H_2C{=}CHCH(CH_3)_2$

131

▶999 VS
B

$(CH_3)_2CHCH(CH_3)_2$

345

▶999 VS
B (0.064)

$H_2C{=}CHCHCH_2CH_3$
CH_3

167

▶999 VS
$CHCl_3$

H_3C, H_3CO, N, O (phenyl-substituted isoxazole ring)

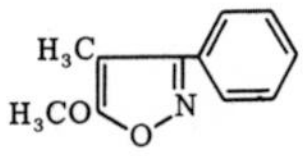

[Ring]

15

▶999 S
B (0.064)

H CH_2CH_3
C=C
H_3C CH_3
(trans)

428

▶999 MS
B (0.104)

$(CH_3)_3CCH(CH_3)_2$

147

▶999 M
B (0°C)

$(CH_3)_2CHCHCH(CH_3)_2$
CH_3

284

▶999 M
E

$Pt\left(\begin{matrix}NH_2\\ CH_2\\ CH_2\\ NH_2\end{matrix}\right)_2 Cl_2$

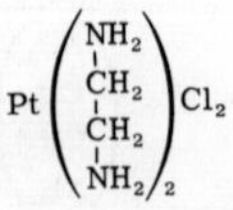

[NH_2]

8

▶998.2 S

$C_6H_5SiCl_3$

86

▶998 MS
B (capillary)

CH_3 CH_3

188, see also 42

▶998 S
CCl_4 [25%]

$CH_3CH(CH_2)_2CH_3$ OH

318

▶998 S
G

NH_2 NO_2

[CH] 21

▶998 S
CS_2

$N(C_2H_5)_2$ NO_2

21

▶998 VS
B (0.0104)

CH_3OCF_2CHFCl

356

▶997.3 S
B

$(C_6H_5)_2SiCl_2$

86

▶997 VS
B (0.169)

$(CH_3)_3CCH_2CH_3$

145

▶997 VS
B (0.169)

$(CH_3)_2CHCHCH(CH_3)_2$ CH_3

135

▶997 SVB
B (0.169)

CH_3 CH_3
$CH_3CH_2CH-CHCH_2CH_3$

83

▶996 VS
A

CH_3
$CH_3CH_2CHCH_2CH_3$

332

▶996 VS
B (0.169)

$(CH_3)_3CCH_2CH_3$

145

▶996 W
B (0.053)

CH_2 = C − CH_3

288

▶996 VS
B (0.169)

$(CH_3)_2CHCH_2CHCH_2CH_3$ CH_3

138

▶996 VSB
B (0.169)

CH_3 CH_3
$CH_3CH_2CH-CHCH_2CH_3$

83

▶996 VS
B (0.065)

$H_2C=CHCH_2C(CH_3)_3$

102, 273

▶996 VS
B (0.238)

$H_2C=CC(CH_3)_3$ CH_3

272

▶996 S
B (0.169)

CH_3
$(CH_3)_2CHCHCH_2CH_2CH_3$

139

▶996 S
B (0.157)

CH_3
$(CH_3)_3CCH_2CHCH_2CH_3$

79, 285

▶996 SSp
E

$K_2S_2O_6$

28

▶996 M
B (0.114)

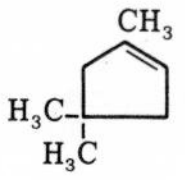

CH_3 / H_3C / H_3C

127

▶996 M
B (0.102)

H_3C CH_3 / CH_3

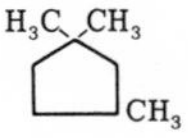

124

▶996
E

$Hg_2(NO_3)_2$

45

▶995 VS
B (0.036)

CH_2=$CHCH_2CH_2CH$=CH_2

374

▶995 VS
B (0.10)

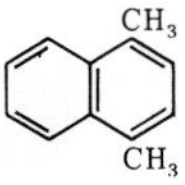

CH_3 / CH_3

186

▶995 VS
B (0.0184)

H_2C=$CHCH_2CH_2CH$=CH_2

374

▶995 S
B (0.169)

$(CH_3)_2CHCH(CH_3)CH_2CH_3$

142, 153

▶995 S
G

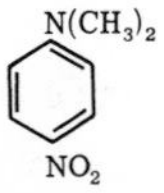

$N(CH_3)_2$ / NO_2

[CH]

21

▶995 M
B (0.169)

$(CH_3)_2CHC(CH_3)_2$ - $CH_2CH_2CH_3$

134

▶994 SSh
B (0.0104)

$(C_4F_9)_3N$

346

▶994 VS
E, $CHCl_3$

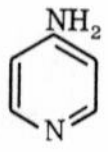

NH_2 / N

327

▶994 S
A (10 mm Hg, 58 cm)

H_2C=$CH(CH_2)_5CH_3$

289

▶994 VW
B (0.2398)

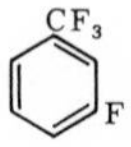

CF_3 / F

218

▶994 MW
B (0.169)

$(CH_3)_3CC(CH_3)_2$ - $CH_2CH_2CH_3$

304

▶994
P

$Zn(NO_3)_2$

45

▶993 VS
B (0.0064)

$C_{17}H_{34}$
(1-heptadecene)

159

▶993 VW
B (0.2398)

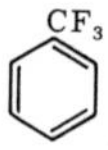

CF_3

219

▶993 M
B (0.169)

$(CH_3)_3CC(CH_3)_2$ - $CH(CH_3)_2$

306

▶993 VS
B (0.036)

H_2C=$CH(CH_2)_5CH_3$

290

▶992 VS
CCl_4 [25%]

$CH_3CH(OH)CH_2CH_3$

319

▶992 M
A (385 mm Hg, 10 cm)

CF_3-CCl_3

253

▶992 S
B (0.008)

H_2C=$CH(CH_2)_8CH_3$

352

▶992 M
B (0.10)

200

▶991 VS
B (0.0104)

F_9C_4-O-C_4F_9

266

▶991 S
G

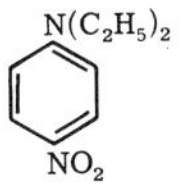
N(C2H5)2
NO2

21

▶991 M
B (0.068)

H_3C C_3H_7-n
C=C
H H

276

▶991 W
C

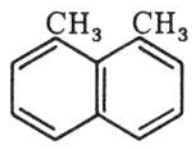
CH3 CH3

182

▶991 M
E

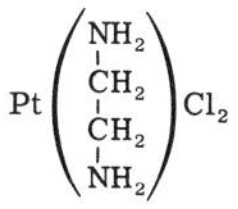
Pt
NH2
CH2
CH2
NH2
Cl2

[NH_2]

8

▶991 M
$CHCl_3$

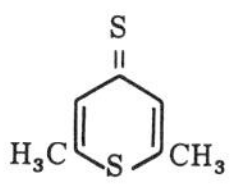
S
H3C
S
CH3

7

▶990 VS
A (36 mm Hg, 15 cm)

H_2C=$CHCH(CH_3)_2$

131

▶990 SSh
B (0.157)

$(CH_3)_3CCHCH(CH_3)_2$
CH_3

293

▶990 S
CCl_4 [25%]

$CH_3(CH_2)_3OH$

353

▶990 VW
B (0.2398)

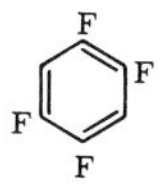
F
F
F
F

222

▶990 M
B

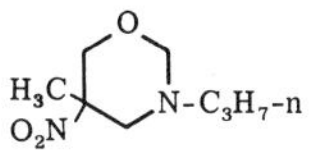
O
H3C
O2N
N-C3H7-n

26

▶990 M
E, CCl_4

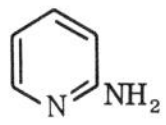
N
NH2

327

▶989 VS
E

[$Coen_2SO_3$]Cl

5

▶989 S (R branch)
A (10 mm Hg, 10 cm)

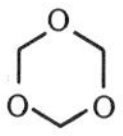
O
O
O

201

▶989 W

$CHCl_3$

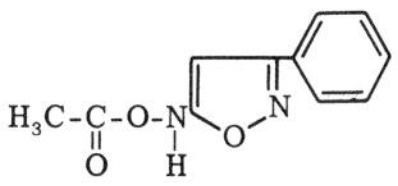
H3C-C-O-N
O
H
N
O

[Ring]

15

▶989 VS
B (0.169)

$(CH_3)_2CHCH(CH_3)_2$

144

▶988.4 W

$C_6H_5SiCl_3$

86

▶988 VS
B (film)
$(CF_2)_2CFCF_2CF_3$
240

▶988 VSSh
B (0.013)
$CH_3CH{=}CHCH(CH_3)_2$
130

▶988 VS
E
$[Coen_2SO_3]SCN$
5

▶988 M
A (40 mm Hg, 58 cm) 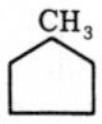
CH_3
105

▶988 M
CCl_4 [25%]
$CH_3(CH_2)_7OH$
317

▶987.4 W
$C_6H_5(CH_3)SiCl_2$
86

▶987 MS
B (0.0576) 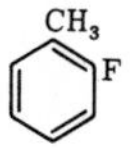
CH_3
F
221

▶987 VW
B (0.101)
$CH_3(CH_2)_5CH_3$
85

▶987 S
B (0.10) 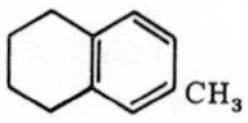
CH_3
269

▶987 SSp
E
$Na_4P_2O_7$
28

▶986 VS
A
$(CH_3)_2CHCH_2CH_3$
333

▶986 S
B
$CH_3CH{=}CHCH{=}CHCH_3$
354

▶986 S
B (0.055) 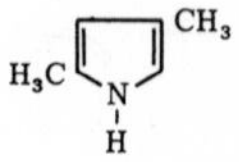
CH_3
H_3C N
H
344

▶986 M
B
$(CH_3)_2CHCH_2CH(CH_3)_2$
335

▶986 M
B (0.064)
$(CH_3)_2C{=}CHCH_2CH_3$
166

▶986 M
E
Grating
$[H_2CNO_2]^-Na^+$
[CH_2 wag]
328

▶985.6 W
$(C_6H_5)_2SiCl_2$
86

▶985 VS
B (0.169)
$(CH_3)_2CHCH_2CH(CH_3)_2$
141, see also 151

▶985 SB
B (0.0104)
$(C_4F_9)_3N$
346

▶985 VS
B (0.105)
$(CH_3)_2CHCH_2CH(CH_3)_2$
151, see also 141

▶985 MS
D
Grating
$[H_2CNO_2]^-Na^+$
[CH_2 wag]
328

▶985 S
B (0.0104)
H_3C-O-CF_2CHFCl
356

▶984 VS
A (100 mm Hg, 15 cm)
CH_3CH_2Cl
88

▶984 VS
E
$[Coen_2SO_3Cl]^0$
5

▶984 S
B (0.20) 
H_3C (cyclopentane) CH_3
(trans)
308

▶984 S
B (0.104)
$CH_3CH_2CH(CH_2)_2CH_3$
CH_3
81

▶984 S
B (0.0563)
$(CH_3)_2C$=$CHC(CH_3)_3$
204

▶984 M
CCl_4
Grating
$[H_2C\cdot NO_2]^-Na^+$
[CH_2 wag]
328

▶983 VWSh
B (0.726)
CH_3
CH_3CH_2C - $CHCH_2CH_3$
CH_3 CH_3
299

▶983 VS
E
$[Coen_2SO_3]I$
5

▶983 S
B (0.10) 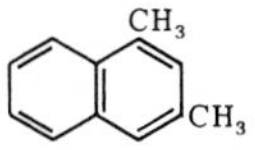
CH_3, CH_3 (naphthalene)
187

▶983 S
B (0.157)
$(CH_3)_3CCHCH(CH_3)_2$
CH_3
293

▶983 W
B (0.25)
$CH_3(CH_2)_2CH_2NO_2$
310

▶983 M
B
$CH_3CH_2CH(CH_2)_2CH_3$
CH_3
330

▶982 M
B (0.2398) 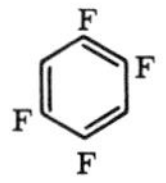
F, F, F, F (benzene)
222

▶982 MW
B
$[(C_2H_5)_2P\cdot BH_2]_3$
[BH_2 wag]
30

▶981 VW
A (775 mm Hg, 10 cm)
CF_2Cl-CF_2Cl
257

▶981 VS
CS_2 [sat.] (0.025) 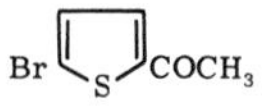
Br (thiophene, S) $COCH_3$
348

▶981 M
B (0.055) 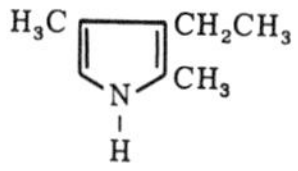
H_3C (pyrrole, N-H) CH_2CH_3, CH_3
343

▶980 VS
B (0.065) 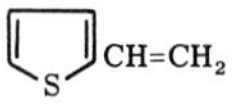
(thiophene, S) CH=CH_2
212

▶980 VS
E

$[Coen_2SO_3NCS]^0$

5

▶980 MS
B (0.015)

$CH_3(CH_2)_4OH$

350, 393

▶980 S
B (0.08)

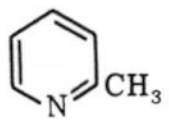

198

▶980 MS
B (0.10)

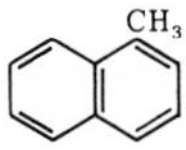

192

▶980 MW
$CHCl_3$

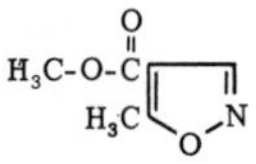

[Ring] 15

▶980

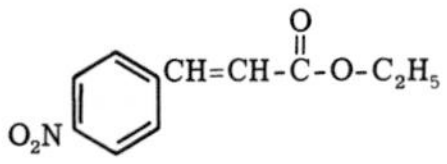

[CH] 349

▶980-974

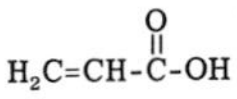

[-CH=CH-] 351

▶979 S
B (0.0104)

n-H_7C_3-O-CF_2CHFCl

363

▶979 S
B (0.064)

H CH_2CH_3
C=C
H_3C CH_3
(trans)

428

▶979 S
B (0.174)

$H_2C=CCH_2C(CH_3)_3$
CH_3

96

▶979 VS
B (0.151)

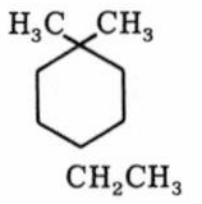

215

▶978 S
B (0.10)

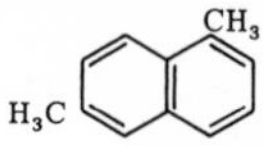

184

▶978 M
A (40 mm Hg, 58 cm)

105

▶978 VS
B (0.169)

CH_3
$(CH_3)_3CCHC(CH_3)_3$

307

▶978 M
$CHCl_3$

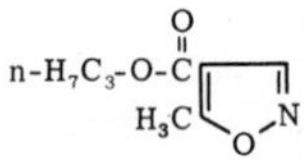

[Ring] 15

▶978 VS (Q branch)
A (10 mm Hg, 10 cm)

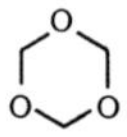

201

▶977 VS
B (0.127)

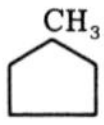

104

▶977 MS
B (0.0576)

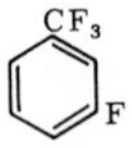

218

▶976 M
B (0.114)

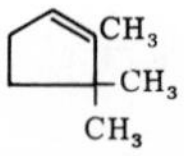

126

▶976 M
B (0.10)

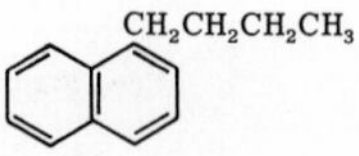

176

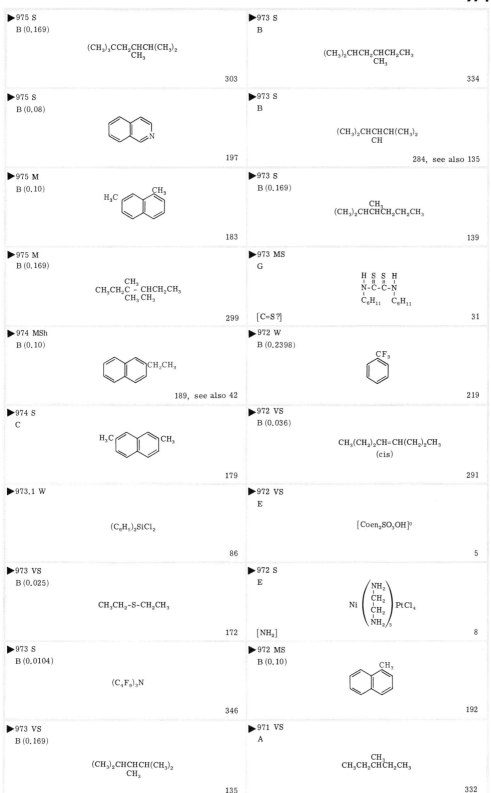

▶975 S
B (0.169)

$(CH_3)_3CCH_2CHCH(CH_3)_2$
CH_3

303

▶975 S
B (0.08)

N

197

▶975 M
B (0.10)

H_3C CH_3

183

▶975 M
B (0.169)

CH_3
$CH_3CH_2C - CHCH_2CH_3$
$CH_3\ CH_3$

299

▶974 MSh
B (0.10)

CH_2CH_3

189, see also 42

▶974 S
C

H_3C CH_3

179

▶973.1 W

$(C_6H_5)_2SiCl_2$

86

▶973 VS
B (0.025)

$CH_3CH_2\text{-}S\text{-}CH_2CH_3$

172

▶973 S
B (0.0104)

$(C_4F_9)_3N$

346

▶973 VS
B (0.169)

$(CH_3)_2CHCHCH(CH_3)_2$
CH_3

135

▶973 S
B

$(CH_3)_2CHCH_2CHCH_2CH_3$
CH_3

334

▶973 S
B

$(CH_3)_2CHCHCH(CH_3)_2$
CH

284, see also 135

▶973 S
B (0.169)

CH_3
$(CH_3)_2CHCHCH_2CH_2CH_3$

139

▶973 MS
G

H S S H
N-C-C-N
C_6H_{11} C_6H_{11}

[C=S ?]

31

▶972 W
B (0.2398)

CF_3

219

▶972 VS
B (0.036)

$CH_3(CH_2)_2CH{=}CH(CH_2)_2CH_3$
(cis)

291

▶972 VS
E

$[Coen_2SO_3OH]^0$

5

▶972 S
E

$Ni\left(NH_2\text{-}CH_2\text{-}CH_2\text{-}NH_2\right)_3PtCl_4$

$[NH_2]$

8

▶972 MS
B (0.10)

CH_3

192

▶971 VS
A

CH_3
$CH_3CH_2CHCH_2CH_3$

332

▶ 971 VS
A (100 mm Hg, 10 cm)
CH_3CH_2Cl
88

▶ 971 M
C
CH_3 CH_3
182

▶ 971 M
B (0.036)
CH_3
$H_2C{=}C(CH_2)_4CH_3$
98

▶ 971 SSh
CS_2 [sat.] (0.170)
Br S $COCH_3$
348

▶ 971 S
B (0.003)
NO_2
$CH_3CH_2CHCH_3$
324

▶ 971 S
P
$Ni([CH_2]_4SO)_6NiCl_4$
[SO] 33

▶ 970.9 W
$C_6H_5HSiCl_2$
86

▶ 970 VS
B (0.169)
$(CH_3)_2CHCH_2CHCH_2CH_3$
CH_3
138

▶ 970 VS
E
$[Coen_2SO_3 \cdot NH_3]Cl$
5

▶ 970 S
CCl_4 [2%]
O O O
342

▶ 970 S
G
$Ni([CH_2]_4SO)_4Br_2$
[SO] 33

▶ 970 M
B (0.025)
$CH_3CH_2SCH_3$
209

▶ 969 VS
A (10 mm Hg, 58 cm)
$CH_3CH{=}CH(CH_2)_4CH_3$
(cis and trans)
101

▶ 969 S
B (0.0104)
$n\text{-}H_9C_4\text{-}O\text{-}CF_2CHFCl$
264

▶ 969 VS
B (0.169)
CH_3
$(CH_3)_3CCH_2CHCH_2CH_3$
285

▶ 969 VS
B (0.10)
200

▶ 969 VS
CCl_4 [25%]
$CH_3CH_2CH_2OH$
314

▶ 969 S
B (0.114)
H_3C CH_3 CH_3
125

▶ 969 S
CCl_4 [25%]
$CH_3CHCH_2CH_3$
OH
319

▶ 969 M
B (0.300)
$CH_3(CH_2)_5CH(C_3H_7)(CH_2)_5CH_3$
281

▶969 MW
B (0.300)
$C_{26}H_{54}$
(5, 14-di-n-butyloctadecane)
283

▶968 VS
CS_2
$CH_3CH=CH(CH_2)_2CH=CH(CH_2)_2CO-NH-CH_2CH(CH_3)_2$
357

▶968 VS
CCl_4 [25%]
$(CH_3)_2CHCH_2CH_2OH$
195

▶968 VS
B (0.036)
$CH_3CH_2CH=CH(CH_2)_3CH_3$
99

▶968 S
B (0.0088)
$CH_3(CH_2)_2CH=CH(CH_2)_2CH_3$
(trans)
205

▶968 VS
B (0.157)
$(CH_3)_3CCH_2CH(CH_3)CH_2CH_3$
79, 285

▶968 MS
B (0.10)
2-naphthyl-$CH_2CH_2CH_2CH_2CH_3$
173

▶968 M
C
naphthalene with CH_3, CH_3
185

▶967 VS
B (0.013)
$CH_3CH=CHCH(CH_3)_2$
130

▶967 VS
CCl_4 [5%]
$CH_3CH_2CH(OH)CH_2CH_3$
196

▶967 VS
B (0.169)
$CH_3CH_2C(CH_3)_2-CH_2CH(CH_3)CH_2CH_3$
302

▶967 S
G
$Co([CH_2]_4SO)_6Br_2$
[SO]
33

▶967 M
B
5-ethyl-5-nitro-3-n-propyl-tetrahydro-1,3-oxazine (H_5C_2, O_2N, $N-C_3H_7-n$)
26

▶967 M
B (0.10)
naphthalene-CH_2CH_3
190

▶966 VS
B (0.036)
$CH_3CH=CH(CH_2)_4CH_3$
100

▶966 MS
B (capillary)
naphthalene with CH_3, CH_3
188

▶966 S
G
$Co([CH_2]_4SO)_6CoBr_4$
[SO]
33

▶966 M
A (15.9 mm Hg, 40 cm)
$CH_3CH_2CH(CH_3)(CH_2)_3CH_3$
106

▶965 VS
B (0.169)
$(CH_3)_2CHCH(CH_2CH_3)CH(CH_3)_2$
301

▶965 VS
B (0.064)
$CH_3CH=CH(CH_2)_4CH_3$
(cis)
160

▶965 S
B (0.10) 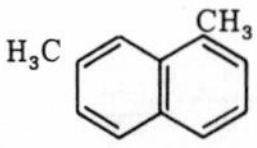
H_3C CH_3
183

▶965 S
P
$Co([CH_2]_4SO)_6CoCl_4$
[SO] 33

▶965 S
Q
$Co([CH_2]_4SO)_6CoI_4$
[SO] 33

▶965 M
B (0.10) 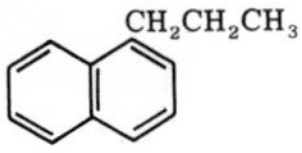
$CH_2CH_2CH_3$
178, see also 42

▶965 M
B (0.068)
H_3C C_3H_7-n
C=C
H H
(cis)
276

▶965 VS
B (0.169)
CH_3
$(CH_3)_2CHCHCH_2CH_3$
142, 153

▶965
$H_2C{=}CHNO_3$
359

▶964 VS
B (0.169)
CH_3
$CH_3CH_2C - CHCH_2CH_3$
CH_3 CH_3
299

▶964 VS
B (0.0088)
$CH_3CH{=}CHCH_2CH_3$
(trans)
207

▶964 S
C 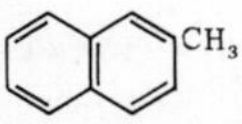
CH_3
191

▶964
E or N 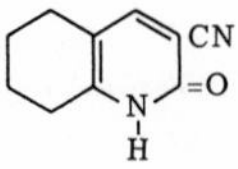
CN
=O
N
H
41

▶963 VS
A (100 mm Hg, 15 cm)
CH_3CH_2Cl
88

▶963 VS
B (0.0576) 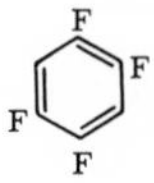
F
F
F
F
222

▶963 VS
B (0.0153) 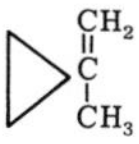
CH_2
C
CH_3
288

▶963 VS
E
$KCrO_3Cl$
358

▶963 S
B 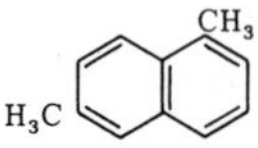
CH_3
H_3C
184

▶963 S
B (0.104)
$CH_3CH_2CH(CH_2)_2CH_3$
CH_3
81

▶963 S
P
$Co([CH_2]_4SO)_6(ClO_4)_2$
[SO] 33

▶963 S
Q
$Co([CH_2]_4SO)_6I_2$
[SO] 33

▶963 MS
B
$CH_3CH_2CH(CH_2)_2CH_3$
CH_3
330

▶963 M
$CHCl_3$

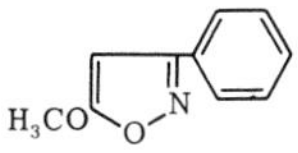

15

▶962 VS
B (0.136)

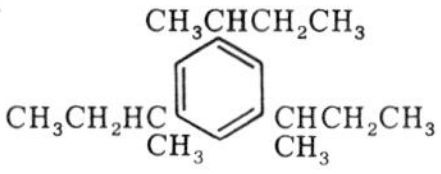

280

▶962 SB
B (0.169)

$(CH_3)_3CCH_2CHCH(CH_3)_2$
CH_3

303

▶962 MSh
CS_2 [sat.] (0.170)

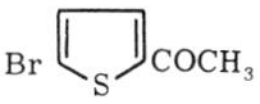

348

▶962
E

$Hg_2(NO_3)_2$

45

▶961 M
B (0.169)

CH_3
$(CH_3)_2CHC - CH_2CH_2CH_3$
CH_3

134

▶961 S
B (0.10)

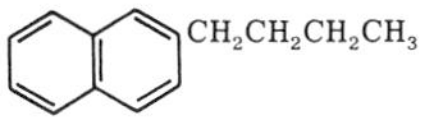

175

▶960 MS
B (0.064)

$H_2C{=}CHCHCH_2CH_3$
CH_3

167

▶960 VW
B (0.003)

CH_3NO_2

311

▶960 MW
B (0.003)

$(CH_3)_2CHCH_2NO_2$

326

▶960 S
C

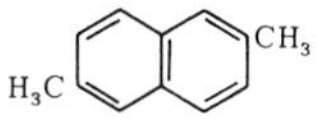

180

▶960
E or N

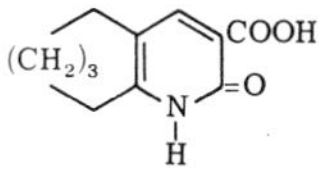

41

▶958 S
B (0.0104)

$n\text{-}H_7C_3\text{-}O\text{-}CF_2CHFCl$

263

▶958 W
B (0.169)

$(CH_3)_2CHCH(CH_3)_2$

144

▶958 M
B (0.10)

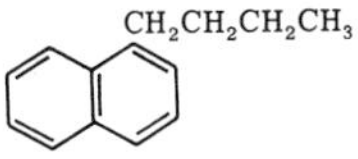

176

▶958 M
B (0.10)

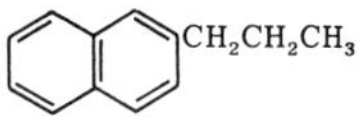

177, see also 42

▶958 MB
B (0.300)

$(C_2H_5)_2CH(CH_2)_{20}CH_3$

282

▶958 M
H(CS_2)

$CH_2{=}CHCH{=}CHCH{=}CHCH{=}CH_2$

202

▶958
E or N

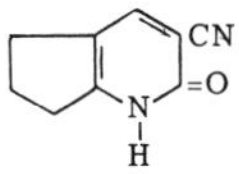

41

▶957 S
A (10 mm Hg, 10 cm)

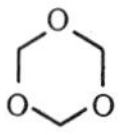

201

▶957 VS
B (0.028)

Cl Cl S

161

▶957 WSh
B (0.169)

$(CH_3)_2CHCHCH(CH_3)_2$
CH_3

284

▶957 S
B (0.055)

CH_3
H_3C N H

344

▶957 S
B (0.055)

H_3C CH_2CH_3
CH_3
N H

343

▶956 VSSh
B (0.169)

CH_3 CH_3
$CH_3CH_2C-CH_2CHCH_2CH_3$
CH_3

302

▶956 VS
B (0.0104)

F_9C_4-O-C_4F_9

266

▶956 MS
B (0.10)

$CH_2CH_2CH_2CH_3$

175

▶956 M
B (0.025)

$CH_3CH_2SCH_3$

209

▶956 M
CCl_4 [5%]

CH_3
$CH_3CH_2CHCH_2OH$

321

▶955 W
$CHCl_3$

O
H_3C O CH_3

7

▶955 M
B (0.169)

CH_3
$(CH_3)_2CHCHCH_2CH_2CH_3$

139

▶954 M
B

$(CH_3)_2CHCH_2CH_2CH(CH_3)_2$

339

▶954 VSB
B (0.169)

CH_3 CH_3
$CH_3CH_2CH-CHCH_2CH_3$

83

▶954 M
B (0.003)

$CH_3(CH_2)_2CH_2NO_2$

310

▶954 S
B (0.169)

CH_3
$(CH_3)_3CC-CH_2CH_2CH_3$
CH_3

304

▶954 M
CCl_4 [25%]

$CH_3(CH_2)_7OH$

317

▶953 S
B (0.169)

CH_3
$CH_3CH_2C-CH_2CH_2CH_3$
CH_3

360

▶953 S
B

O
H_3C N-C_3H_7-n
O_2N

26

▶953 S
C

CH_3

191

▶953 MS
B (0.10)

CH_2CH_3

190

▶953 MS
B (0.10)
1-methylnaphthalene structure (CH_3)
192

▶952 VS
B
$(CH_3)_2CHOH$
361

▶952 S
B (0.0104)
n-H_9C_4-O-CF_2CHFCl
264

▶952 S
CCl_4 [25%]
$CH_3(CH_2)_3OH$
353

▶952 MSB
B (0.169)
$(CH_3)_3CCH_2CHCH(CH_3)_2$
CH_3
303

▶952 MWSh
CCl_4 [25%]
$(CH_3)_2CHCH_2CH_2OH$
195

▶952 M
B (0.10)
$CH_2CH_2CH_3$
178

▶952
CH_3Br
[CH_3]
363

▶951 S
B (0.169)
$(CH_3)_2CHCHCH(CH_3)_2$
CH_2
CH_3
301

▶951 M
B (0.063)
CH_3
CN
214

▶950.6 W
$(C_6H_5)_2SiCl_2$
86

▶950 VS
A
O
HC≡C-C-H
[C-C]
18

▶950 S
B (0.10)
$CH_2CH_2CH_2CH_2CH_3$
173

▶950 VS
O
H_3C-C-CH_3
$KCrO_3Cl$
358

▶950 VVW
B (0.2398)
CF_3
F
218

▶950 S
$(C_6H_5)_3SnCH{=}CH_2$
[$SnCH{=}CH_2$]
364

▶950 S
KBr
$N(C_2H_5)_2$
NO_2
21

▶950 M
B (0.169)
$(CH_3)_3C(CH_2)_2CH(CH_3)_2$
365

▶950 M
B (0.0104)
F CF_3
F_2 F_2
F CF_3
F_3C F
F_2
228

▶950
E or N
O
C-O-C_2H_5
N
41

▶949 VSB
B (0.169)

$CH_3CH_2CH(CH_3)\text{-}CH(CH_3)CH_2CH_3$

83

▶949 VS
B (0.169)

$(CH_3)_2CHCH_2CH_2CH(CH_3)_2$

137

▶949 S
CCl_4 [25%]

$CH_3CH(OH)(CH_2)_2CH_3$

318

▶949 S
B (0.10)

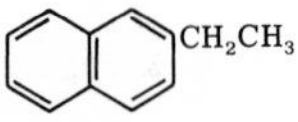

189

▶949 M
B

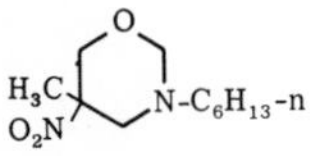

[C-O-C]

26

▶949 MS
B (0.063)

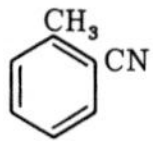

312

▶949 M
B (0.157)

$(CH_3)_3CCH_2CH(CH_3)CH_2CH_3$

79, 285

▶949 M
$CHCl_3$

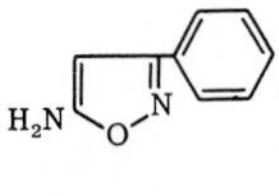

15

▶949
E or N

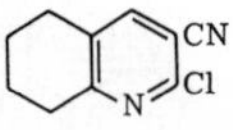

41

▶948 VS
B (0.0104)

$(C_4F_9)_3N$

346

▶948 S
A (25 mm Hg, 10 cm)

$CF_2Cl\text{-}CF_2Cl$

257

▶948 S
B (0.10)

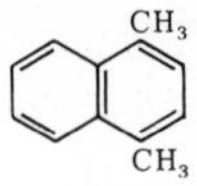

186

▶948 S
B

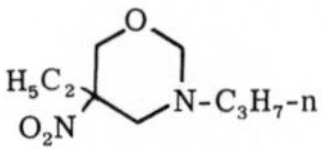

[C-O-C]

26

▶948 S
B (0.112)

$(CH_3)_3CCH_2CH_2CH_3$

155

▶948 S
B (0.10)

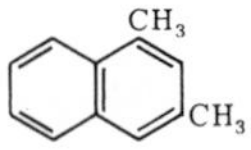

187

▶948 S
C

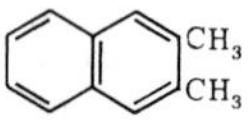

181

▶948 S
E

$KCrO_3Cl$

358

▶948 MS
$CHCl_3$

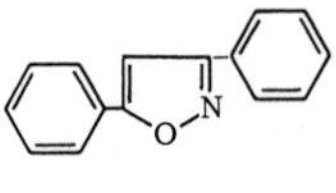

15

▶947.2 VS
B (0.03)

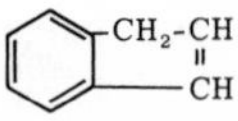

9

▶947 M
B (0.169)

$(CH_3)_2CHC(CH_3)_2\text{-}CH_2CH_2CH_3$

134

▶947 S
B

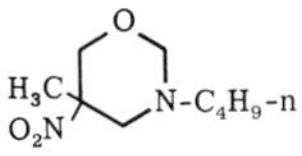

O
H_3C N-C_4H_9-n
O_2N

[C-O-C] 26

▶947 S
B

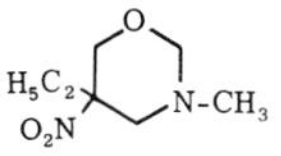

O
H_5C_2 N-CH_3
O_2N

[C-O-C] 26

▶947 S
CS_2

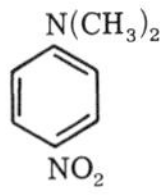

$N(CH_3)_2$

NO_2

21

▶947 M
B

O
H_5C_2 N-C_5H_{11}-n
O_2N

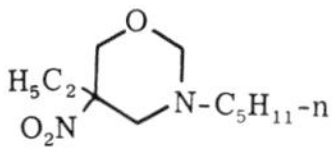

26

▶946 VS
B (0.169)

$(CH_3)_2CHCH(CH_2CH_3)_2$

136

▶946 S
B (0.136)

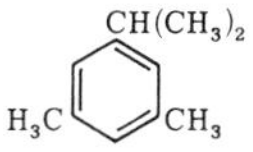

$CH(CH_3)_2$
H_3C CH_3

277

▶946 S
B

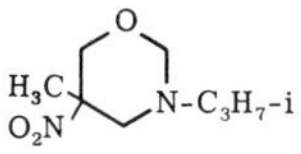

O
H_3C N-C_3H_7-i
O_2N

[C-O-C] 26

▶946 M
B

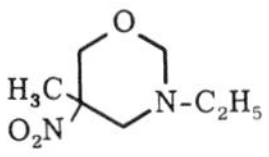

O
H_3C N-C_2H_5
O_2N

[C-O-C] 26

▶946 M
C

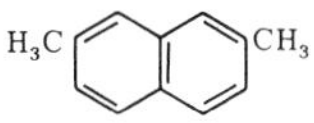

H_3C CH_3

179

▶946 M
$CHCl_3$

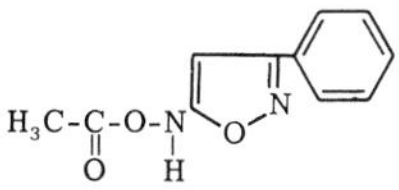

H_3C-C-O-N O N
O H

15

▶946 MW

$CHCl_3$

N
H_3CO O

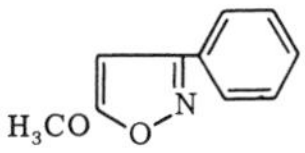

15

▶945 S
A (100 mm Hg, 10 cm)

CH_3-CCl_3

251

▶945 S
B

CH_3CH=$CHCH$=$CHCH_3$

354

▶945
E or N

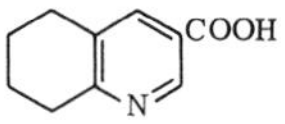

COOH
N

41

▶945
E or N

O
C-O-C_2H_5
N

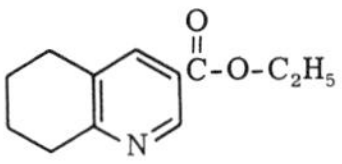

41

▶944 VS
A

CHF_2CH_3

233

▶944 VS (Q branch)
A (10 mm Hg, 10 cm)

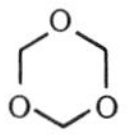

O
O O

201

▶944 VS
B (0.08)

N

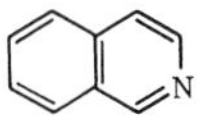

197

▶944 S
B (0.25)

$(CH_3)_2CHNO_2$

366

▶944 S
E, CCl_4

$[(CH_3)_2CNO_2]^-Na^+$

[NO_2] 328

▶ 944 M B (0.169) $(CH_3)_2CHCH(CH_3)CH_2CH_2CH_3$ 139	▶ 942.3 VS B (0.03) [structure: benzene fused ring with CH_2-CH=CH] 9
▶ 944 M B (0.0576) [structure: benzene ring with CF_3, F, F] 217	▶ 942 S E, G $Na_4P_2O_6$ 28
▶ 943 VS B (0.10) [structure: cyclooctatetraene] 200	▶ 942 S $(C_6H_5)_3PbCH{=}CH_2$ [$PbCH{=}CH_2$] 364
▶943 VS E [$Coen_2(SO_3)_2$]Na (cis) 5	▶ 941 MSh $CHCl_3$ [structure: 2,6-dimethyl-4H-pyran-4-thione; S, H_3C, O, CH_3] 7
▶ 943 S B (0.10) [structure: naphthalene-$CH_2CH_2CH_3$] 177	▶941 S C [structure: naphthalene with CH_3, CH_3] 182
▶ 943 MSh B (0.0088) $CH_3CH{=}CHCH_2CH_3$ (trans) 207	▶941 Sh B [structure: 2-(chlorophenyl)-5,6-dihydro-4H-1,3-oxazine; O, N, Cl] 26
▶ 943 S B [structure: morpholine-type ring with O, H_5C_2, O_2N, N-C_6H_{13}-n] [C-O-C] 26	▶940 VS P $[(CH_3)_2P \cdot BCl_2]_3$ [CH_3 rock] 30
▶ 943 MB B (0.0576) [structure: benzene ring with F, F] 225	▶940 M B (0.0104) CH_3CH_2-O-CF_2CHFCl 261
▶ 943 SSp E $Na_2H_2P_2O_6$ [PO_2(OH) stretch] 28	▶940 S E $Mg_2P_2O_6 \cdot H_2O$ 28
▶ 943 M B [structure: ring with O, H_5C_2, O_2N, N-C_4H_9-n] [C-O-C] 26	▶940 S G [structure: benzene ring with $N(CH_3)_2$, NO_2] 21

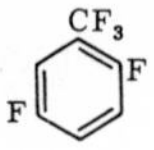

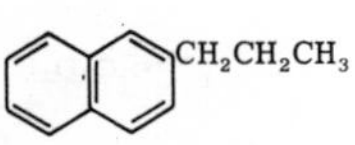

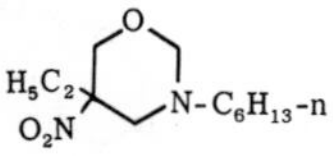

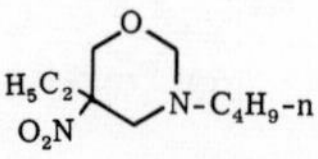

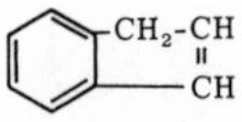

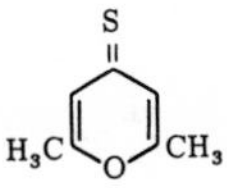

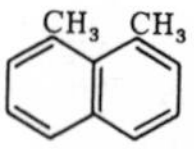

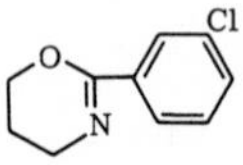

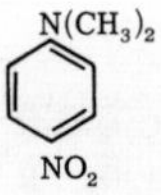

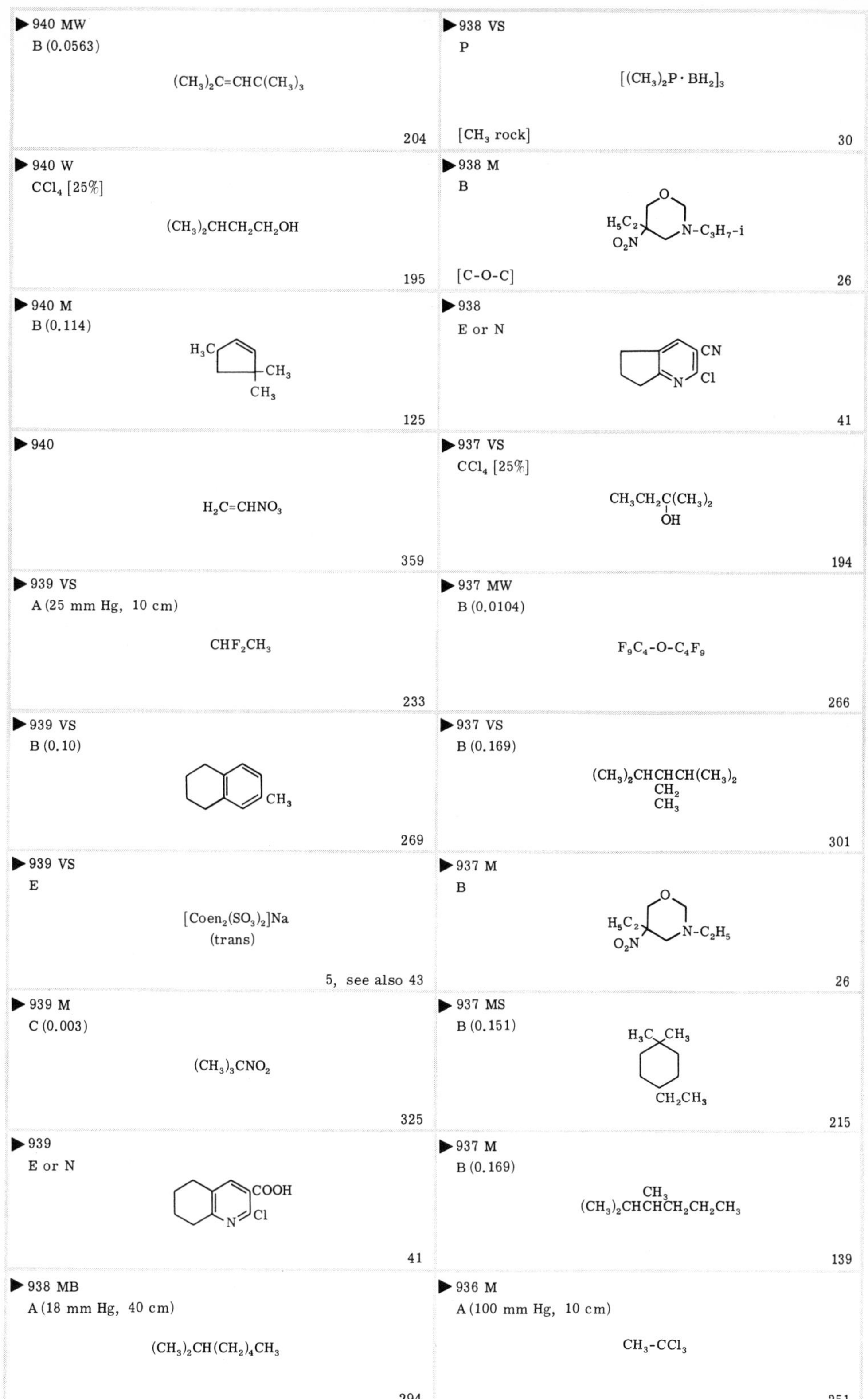

▶ 940 MW
B (0.0563)
$(CH_3)_2C{=}CHC(CH_3)_3$
204

▶ 940 W
CCl_4 [25%]
$(CH_3)_2CHCH_2CH_2OH$
195

▶ 940 M
B (0.114)
H_3C, CH_3, CH_3
125

▶ 940
$H_2C{=}CHNO_3$
359

▶ 939 VS
A (25 mm Hg, 10 cm)
CHF_2CH_3
233

▶ 939 VS
B (0.10)
CH_3
269

▶ 939 VS
E
$[Coen_2(SO_3)_2]Na$
(trans)
5, see also 43

▶ 939 M
C (0.003)
$(CH_3)_3CNO_2$
325

▶ 939
E or N
COOH, N, Cl
41

▶ 938 MB
A (18 mm Hg, 40 cm)
$(CH_3)_2CH(CH_2)_4CH_3$
294

▶ 938 VS
P
$[(CH_3)_2P \cdot BH_2]_3$
[CH_3 rock]
30

▶ 938 M
B
O, H_5C_2, O_2N, $N{-}C_3H_7{-}i$
[C-O-C]
26

▶ 938
E or N
CN, N, Cl
41

▶ 937 VS
CCl_4 [25%]
$CH_3CH_2C(OH)(CH_3)_2$
194

▶ 937 MW
B (0.0104)
$F_9C_4{-}O{-}C_4F_9$
266

▶ 937 VS
B (0.169)
$(CH_3)_2CHCH(CH_2CH_3)CH(CH_3)_2$
301

▶ 937 M
B
O, H_5C_2, O_2N, $N{-}C_2H_5$
26

▶ 937 MS
B (0.151)
H_3C, CH_3, CH_2CH_3
215

▶ 937 M
B (0.169)
$(CH_3)_2CHCH(CH_3)CH_2CH_2CH_3$
139

▶ 936 M
A (100 mm Hg, 10 cm)
$CH_3{-}CCl_3$
251

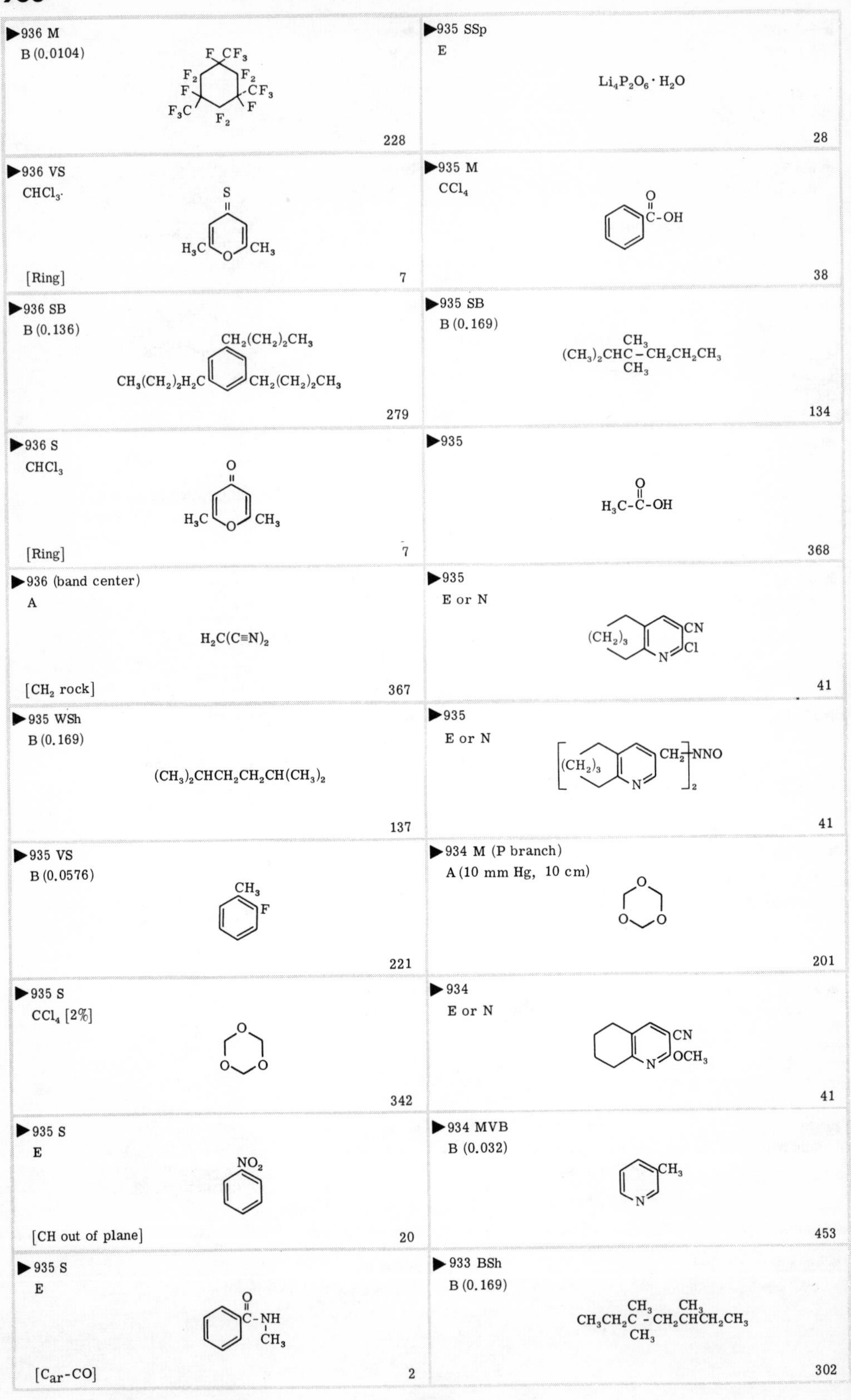

Band	Conditions	Compound	Assignment	Ref.
▶936 M	B (0.0104)	perfluorotrimethylcyclohexane (CF_3, F_2, F, F_3C substituents)		228
▶936 VS	$CHCl_3$	2,6-dimethyl-4H-thiopyran-4-thione ring (H_3C, CH_3, S, O)	[Ring]	7
▶936 SB	B (0.136)	$CH_2(CH_2)_2CH_3$ / $CH_3(CH_2)_2H_2C$ / $CH_2(CH_2)_2CH_3$ substituted benzene		279
▶936 S	$CHCl_3$	2,6-dimethyl-4-pyrone (H_3C, CH_3, O)	[Ring]	7
▶936 (band center)	A	$H_2C(C{\equiv}N)_2$	[CH_2 rock]	367
▶935 WSh	B (0.169)	$(CH_3)_2CHCH_2CH_2CH(CH_3)_2$		137
▶935 VS	B (0.0576)	fluorotoluene (CH_3, F)		221
▶935 S	CCl_4 [2%]	1,3,5-trioxane		342
▶935 S	E	nitrobenzene (NO_2)	[CH out of plane]	20
▶935 S	E	N-methylbenzamide (C(=O)-NH-CH_3)	[C_{ar}-CO]	2
▶935 SSp	E	$Li_4P_2O_6 \cdot H_2O$		28
▶935 M	CCl_4	benzoic acid (C(=O)-OH)		38
▶935 SB	B (0.169)	$(CH_3)_2CHC(CH_3)_2-CH_2CH_2CH_3$		134
▶935		$H_3C-C(=O)-OH$		368
▶935	E or N	$(CH_2)_3$-fused pyridine with CN, Cl		41
▶935	E or N	$[(CH_2)_3$-fused pyridine-$CH_2]_2$NNO		41
▶934 M (P branch)	A (10 mm Hg, 10 cm)	1,3,5-trioxane		201
▶934	E or N	tetrahydroquinoline with CN, OCH_3		41
▶934 MVB	B (0.032)	methylpyridine (CH_3, N)		453
▶933 BSh	B (0.169)	$CH_3CH_2C(CH_3)_2-CH_2CH(CH_3)CH_2CH_3$		302

▶933 VS
B (0.0563)

$CH_3CH{=}CHCH_2CH_3$
(cis)

208

▶933 W
B (0.0576)

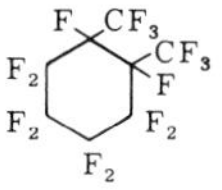
F CF3
CF3
F2
F
F2
F2
F2

231

▶933 MW
B (0.003)

$(CH_3)_2CHCH_2NO_2$

326

▶933 W
B (0.0576)

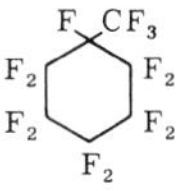
F CF3
F2
F2
F2
F2
F2

232

▶933 SB
B (0.169)

CH_3
$(CH_3)_2CHC - CH_2CH_2CH_3$
CH_3

134

▶933 M
B

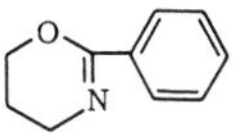
O
N

[C-O-C]

26

▶933
E or N

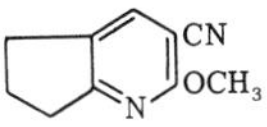
CN
OCH3
N

41

▶933
E or N

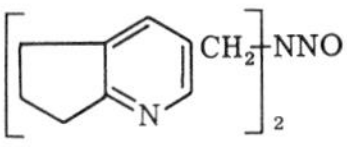
CH2
NNO
N
2

41

▶932 S
B (0.169)

CH_3
$CH_3CH_2C - CHCH_2CH_3$
$CH_3\ CH_3$

299

▶932 SB
B (0.169)

CH_3
$(CH_3)_2CHC - CH_2CH_2CH_3$
CH_3

134

▶932 W
B (0.2398)

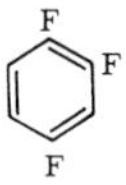
F
F
F

224

▶932 W
$CHCl_3$

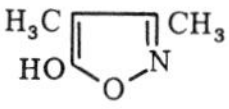
H3C
CH3
HO
N
O

[Ring]

15

▶932
E or N

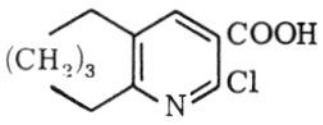
(CH2)3
COOH
Cl
N

41

▶932
E or N

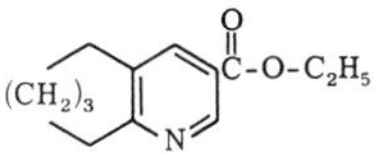
O
C-O-C2H5
(CH2)3
N

41

▶932
E or N

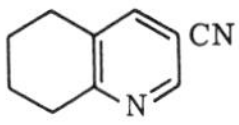
CN
N

41

▶932 M
CCl_4 [25%]

CH_3
$CH_3CH_2CHCH_2OH$

321

▶931 MS
B (0.036)

$CH_3CH_2CH{=}CH(CH_2)_3CH_3$

99

▶931 MSh
B (0.157)

$(CH_3)_3CCH(CH_2CH_3)_2$

78

▶931 M
B

$CH_3CH_2CH(CH_2)_2CH_2$
CH_3

330

▶931 M
B

CH_3
$(CH_3)_2CHC - CH_2CH_3$
CH_3

341

▶931 M CHCl₃	
▶931 M $CHCl_3$ n-H_7C_3-O-C(=O)- (5-methylisoxazole, H_3C, O, N) 15	▶929 S B (0.169) $(CH_3)_3CCH_2CH_3$ 145
▶930 VS B (0.169) $(CH_3)_3CC(CH_3)_2-CH(CH_3)_2$ 306	▶929 S B (0.0576) CH_3-benzene-F 220
▶930 SSh B (0.238) $H_2C=C(CH_3)C(CH_3)_3$ 272	▶929 S Benzene HNO_3 [N-O stretch] 642
▶930 VS B (0.169) $(CH_3)_3CCH_2CH_2C(CH_3)_3$ 305	▶929 S B (0.151) H_3C, CH_3 cyclohexane CH_2CH_3 215
▶930 M A (200 mm Hg, 10 cm) $CH_3CH_2C(CH_3)_3$ 94	▶929 S E $Li_4P_2O_6 \cdot 7H_2O$ 28
▶930 MSB B (0.169) $(CH_3)_3CCHC(CH_3)_3$ with CH_3 307	▶929 MS B (0.104) $CH_3CH_2CH(CH_3)(CH_2)_2CH_3$ 81
▶930 M $CHCl_3$ H_5C_2-O-C(=O)- (5-methylisoxazole, H_3C, O, N) 15	▶928 VS B (0.136) $CH(CH_3)_2$, H_3C, CH_3 benzene 277
▶930 E or N $(CH_2)_3$ fused pyridine, CN, OCH_3, N 41	▶928 S A (8.2 mm Hg, 10 cm) CF_2Cl-CF_2Cl 257
▶930 benzene-C(=O)-OH 370	▶928 S B (0.112) $(CH_3)_3CCH_2CH_2CH_3$ 155
▶929 VS A (400 mm Hg, 15 cm) $CH_3CH_2CH_2CH_2CH_3$ 371	▶928 S B (0.169) $(CH_3)_3CCH_2CH(CH_3)CH(CH_3)_2$ 303

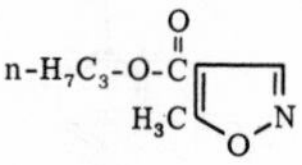

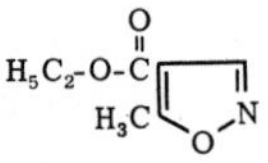

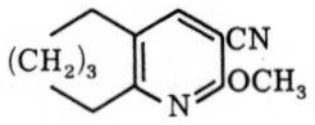

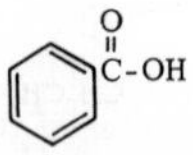

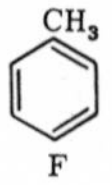

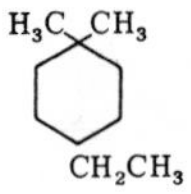

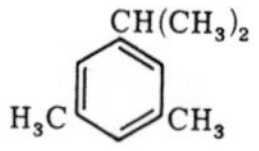

▶928 SB
B (0.2398)

225

▶928 SB
B (0.169)

CH_3
$(CH_3)_2CHC - CH_2CH_2CH_3$
CH_3

134

▶928
E or N

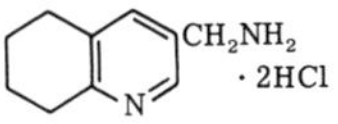

41

▶928
E or N

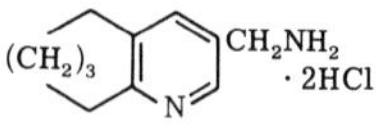

41

▶927 S
B (0.169)

CH_3
$(CH_3)_3CC - CH_2CH_2CH_3$
CH_3

304

▶927 M
B (0.157)

$(CH_3)_3CCHCH(CH_3)_2$
CH_3

293

▶927 M
CS_2 [40%]

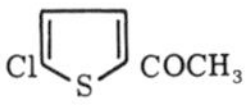

372

▶927 M
B

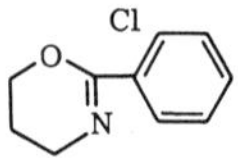

[C-O-C]

26

▶927 M
B

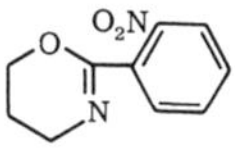

[C-O-C]

26

▶927 M
$CHCl_3$

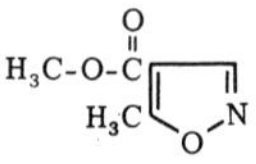

15

▶927 MW
B (0.20)

(trans)

308

▶926 M
B (0.10)

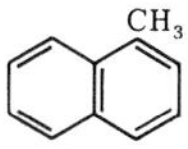

192

▶926 MS
B (0.157)

CH_3
$(CH_3)_3CCH_2CHCH_2CH_3$

79, 285

▶926 M
B

$(CH_3)_2CHCH_2CHCH_2CH_3$
CH_3

334

▶925 VS
B (0.0576)

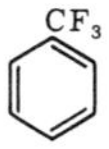

219

▶925 M
B

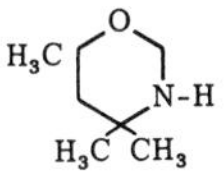

26

▶925 MW
B (0.0104)

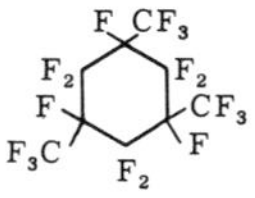

228

▶925 M
B (0.064)

$CH_3CH{=}CH(CH_2)_4CH_3$
(cis)

160

▶925 M
CCl_4

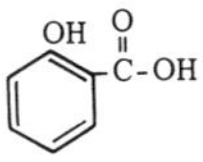

27

▶925 VW

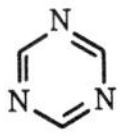

14

▶ 924 VS
B (0.051)
(structure) $CH(CH_3)_2$, $CH(CH_3)_2$
297

▶ 924 SSh
B (0.0104)
$CF_2Cl\text{-}CFCl_2$
255

▶ 924 M
B (0.10)
(structure) CH_3, H_3C
184

▶ 924 M
CCl_4 [25%]
$CH_3CH_2CH(OH)CH_2CH_3$
196

▶ 924
E or N
(structure) [CH_2]NH · HCl, N, subscript 2
41

▶ 923 VS
A (8.2 mm Hg, 10 cm)
$CF_2Cl\text{-}CF_2Cl$
257

▶ 923 MSB
B (0.104)
$(CH_3)_3CCH(CH_3)_2$
147

▶ 923 VS
CS_2 [sat.] (0.170)
(structure) Br, S, $COCH_3$
348

▶ 923 MS
B (0.0563)
$(CH_3)_2C{=}CHC(CH_3)_3$
204

▶ 923 S
B
$CH_3CH{=}CHCH{=}CHCH_3$
354

▶ 923 S
E
$K_4P_2O_6$
28

▶ 923 M
B
(structure) O, N, Cl
26

▶ 923 M
B (0.064)
H CH_2CH_3 / C=C / H_3C CH_3
(trans)
428

▶ 923
E or N
(structure) COOH, Cl, N
41

▶ 922 VS
B (0.169)
$(CH_3)_2CHCH_2CH(CH_3)CH_2CH_3$
138

▶ 922 SBSh
B (0.169)
$CH_3CH_2C(CH_3)_2\text{-}CH_2CH(CH_3)CH_2CH_3$
302

▶ 922 SB
B (0.003)
$C_3H_7NO_2$
377, 392

▶ 922 SB
B (0.169)
$(CH_3)_2CHC(CH_3)_2\text{-}CH_2CH_2CH_3$
134

▶ 922 S
E
$Ca_2P_2O_6 \cdot 2H_2O$
28

▶ 922 MW
B (0.10)
(structure) CH_3, CH_3
187, see also 42

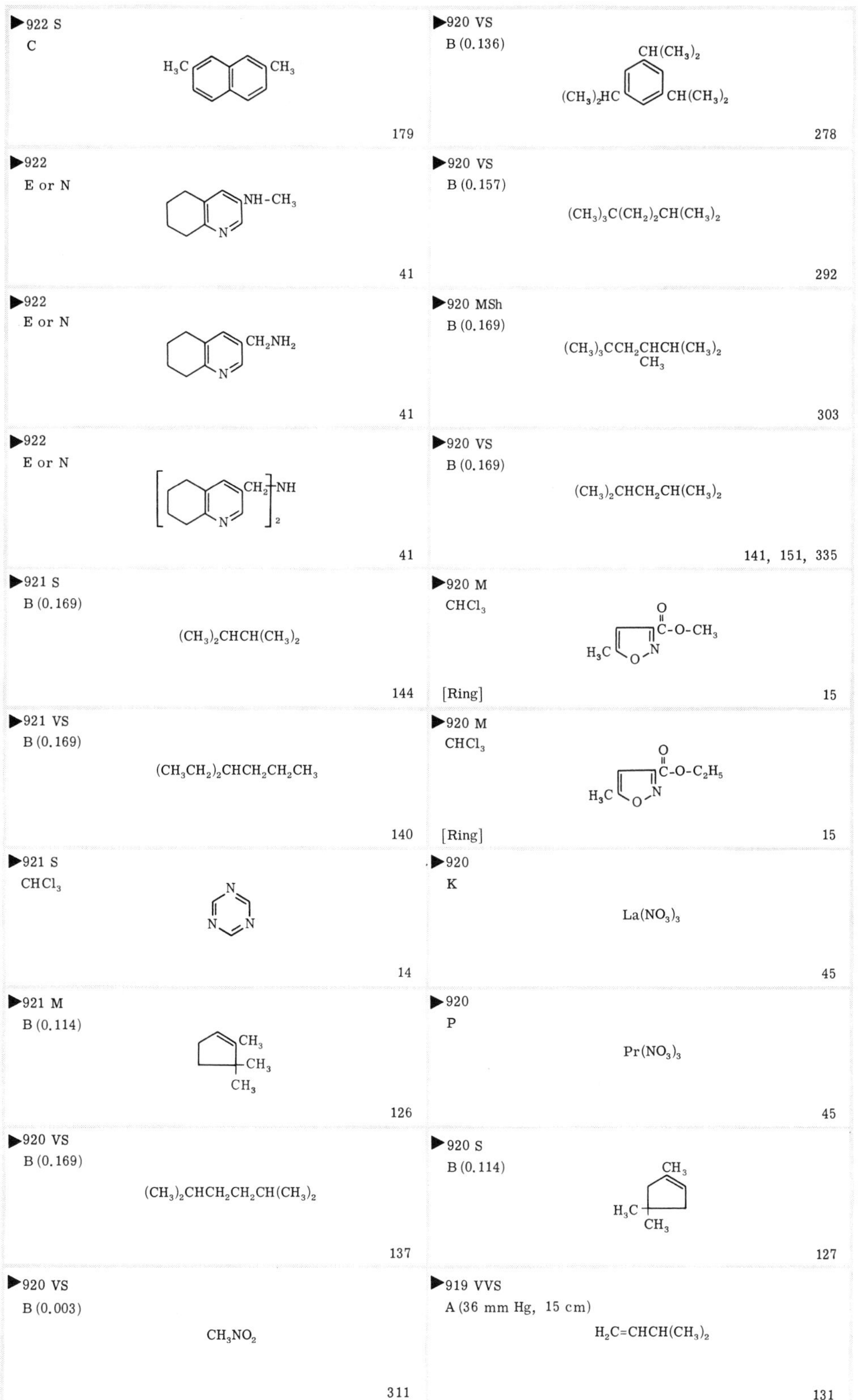

▶922 S
C
H_3C, CH_3
179

▶922
E or N
NH-CH_3, N
41

▶922
E or N
CH_2NH_2, N
41

▶922
E or N
CH_2-NH, N, 2
41

▶921 S
B (0.169)
$(CH_3)_2CHCH(CH_3)_2$
144

▶921 VS
B (0.169)
$(CH_3CH_2)_2CHCH_2CH_2CH_3$
140

▶921 S
$CHCl_3$
N, N, N
14

▶921 M
B (0.114)
CH_3, CH_3, CH_3
126

▶920 VS
B (0.169)
$(CH_3)_2CHCH_2CH_2CH(CH_3)_2$
137

▶920 VS
B (0.003)
CH_3NO_2
311

▶920 VS
B (0.136)
$CH(CH_3)_2$, $(CH_3)_2HC$, $CH(CH_3)_2$
278

▶920 VS
B (0.157)
$(CH_3)_3C(CH_2)_2CH(CH_3)_2$
292

▶920 MSh
B (0.169)
$(CH_3)_3CCH_2CHCH(CH_3)_2$
CH_3
303

▶920 VS
B (0.169)
$(CH_3)_2CHCH_2CH(CH_3)_2$
141, 151, 335

▶920 M
$CHCl_3$
O, C-O-CH_3, H_3C, O, N
[Ring]
15

▶920 M
$CHCl_3$
O, C-O-C_2H_5, H_3C, O, N
[Ring]
15

▶920
K
$La(NO_3)_3$
45

▶920
P
$Pr(NO_3)_3$
45

▶920 S
B (0.114)
CH_3, H_3C, CH_3
127

▶919 VVS
A (36 mm Hg, 15 cm)
$H_2C=CHCH(CH_3)_2$
131

▶919 VS
B (0.157)
$(CH_3)_3CCH(CH_2CH_3)_2$
78, see also 300

▶919 VS
$CHCl_3$ 
O
O
[Ring]
7

▶919 VS
B (0.169)
$(CH_3)_2CHCH_2CH(CH_3)_2$
141, 151

▶919 S
B
$(CH_3)_2CHCH_2CH_2CH(CH_3)_2$
339

▶919 S
B (0.097)
$(CH_3)_2CH(CH_2)_3CH_3$
80

▶919 S
B (0.169)
$(CH_3)_2CHCHCH(CH_3)_2$
CH_2
CH_3
301

▶919 M
$CHCl_3$ 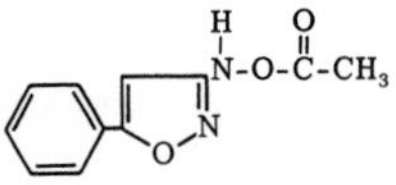
H O
N-O-C-CH_3
O N
[Ring]
15

▶919
E or N
$(CH_2)_3$
CN
N 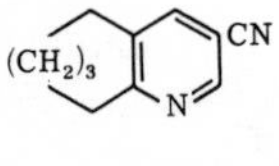
41

▶919 S
B (0.169)
$(CH_3)_2CHCH(CH_2CH_3)_2$
136

▶918 VVS
$CHCl_3$ 
S
O
[Ring]
7

▶918 S
B (0.0104)
CH_3CH_2-O-CF_2CHFCl
261

▶918 S
B (0.169)
CH_3
$(CH_3)_2CHCHCH_2CH_3$
142, 153

▶918 M
B (0.169)
CH_3
$(CH_3)_2CHCHCH_2CH_2CH_3$
139

▶918 MS
B (0.068)
H_3C CH_3
C=C
H CH_2CH_3
(cis)
274

▶918 S
CCl_4 [25%]
$CH_3(CH_2)_5OH$
320

▶917 S
B (0.169)
CH_3
CH_3CH_2C - $CHCH_2CH_3$
CH_3 CH_3
299

▶917 VS
B (0.15) 
128

▶917 S
B (0.151) 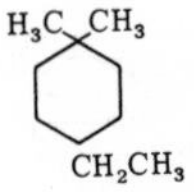
H_3C CH_3
CH_2CH_3
215

▶917 S
B (0.068)
H_3C CH_3
C=C
H CH_2CH_3
(cis)
164, 274

▶917 S
E
$Ba_2P_2O_6$
28

▶917 S
B (0.0104)

$F_9C_4\text{-}O\text{-}C_4F_9$

266

▶917 M
E

$Na_4P_2O_7$

28

▶917
E or N

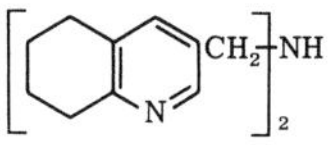

41

▶916 VS
B (0.169)

CH_3
$(CH_3)_3CC - CH_2CH_2CH_3$
CH_3

304

▶916 VS
B (0.008)

$H_2C{=}CHCH_2C(CH_3)_3$

273

▶916 S
A

$(CH_3)_2CHCH_2CH_3$

333

▶916 S
B (0.064)

$H_3C\ \ H$
$C{=}C$
$H\ \ CH_2CH_2CH_3$
(trans)

169

▶916 S
E

$Na_4P_2O_6 \cdot 10H_2O$

28

▶916 S
$CHCl_3$

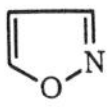

[CH ?]

15

▶916 MS
G

H S S H
N-C-C-N
CH_2OH CH_2OH

[C=S ?]

31

▶916 VS
B (0.169)

$(CH_3)_2CHCH\,CH(CH_3)_2$
CH_3

135

▶916 M
$CHCl_3$

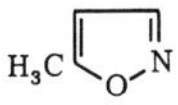

15

▶916 M
$CHCl_3$

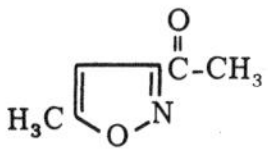

15

▶916 MW
$CHCl_3$

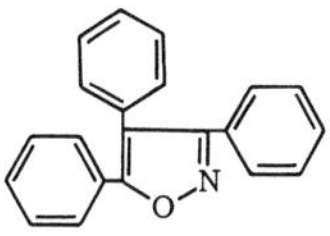

[Ring]

15

▶916
E or N

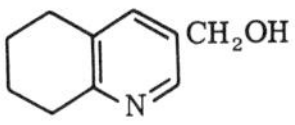

41

▶916 S
B (0.169)

$(CH_3)_2CHCH(CH_2CH_3)_2$

136

▶915.6 W
CS_2

$(C_6H_5)_3SiCl$

86

▶915 WB
B (0.0104)

$n\text{-}H_7C_3\text{-}O\text{-}CF_2CHFCl$

263

▶915 S
E

$CeP_2O_6 \cdot H_2O$

28

▶915 M
$CHCl_3$

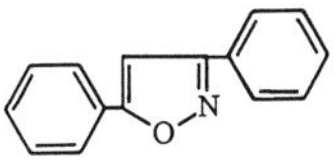

[Ring]

15

▶915 M B $CH_3CH_2C(CH_3)_2-CH_2CH_3$ 336	▶914 K $La(NO_3)_3$ 45
▶915 E $Be(NO_3)_2$ 45	▶913 VS (P branch) A (600 mm Hg, 15 cm) $(CH_3)_3CH$ 373
▶914.8 VS B (0.03) [structure: benzene ring with $CH_2-CH=CH$] 9	▶913 S B (0.169) $(CH_3)_3CC(CH_3)_2-CH(CH_3)_2$ 306
▶914 VS B (0.014) [structure: benzene ring with CF_3, F, F] 217	▶913 VS B (0.008) $H_2C=CHCH_2C(CH_3)_3$ 273
▶914 M B (0.0104) [structure: cyclohexane ring, F CF_3, F_2, F_2, F_2, F_2, F CF_3] 230	▶913 VS B (0.169) $(CH_3CH_2)_2CHCH_2CH_2CH_3$ 140
▶914 S B [structure: ring with O, H_5C_2, O_2N, $N-C_6H_{13}$-n] 26	▶913 M B (0.003) $CH_3(CH_2)_2CH_2NO_2$ 310
▶914 MS $CHCl_3$ [structure: isoxazoline ring with H_3C, CH_3, H_3CO, O, N] [Ring] 15	▶913 M $CHCl_3$ [structure: pyridine ring with NH_2, N] 327
▶914 M B [structure: ring with O, H_5C_2, O_2N, $N-C_5H_{11}$-n] 26	▶913 M B (0.157) $(CH_3)_3CCH(CH_3)CH(CH_3)_2$ 293
▶914 VS B (0.169) $(CH_3)_2CHCH(CH_2CH_3)_2$ 136	▶913 E or N [structure: cyclopenta-fused pyridine with CN, N] 41
▶914 MW $CHCl_3$ [structure: phenyl-isoxazole with NH_2, O, N] 15	▶912 VS A (10 mm Hg, 58 cm) $H_2C=CH(CH_2)_5CH_3$ 289

▶912 VS CS_2 [20%] (0.0184) $CH_2=CHCH_2CH_2CH=CH_2$ 374	▶911 S E $Co(NH_3)_6NaP_2O_6 \cdot 4H_2O$ 28
▶912 MSh B (0.0104) (structure: ring with F, CF_3, F_2, F_2, F, CF_3, F_3C, F_2) 228	▶911 S E $K_2Na_2P_2O_6 \cdot 10H_2O$ 28
▶912 MSh B (0.003) $(CH_3)_2CHCH_2NO_2$ 326	▶911 VS CCl_4 [25%] $CH_3CH(OH)CH_2CH_3$ 319
▶912 MS B $(CH_3)_3C(CH_2)_3CH_3$ 337	▶911 M B (0.068) $H_3C(H)C=C(H)C_3H_7$-n (cis) 276
▶912 M B (0.105) $CH_3CH_2C(CH_3)_2-CH_2CH_3$ 149	▶911 M B (structure: morpholine ring with H_5C_2, O_2N, N-C_2H_5) 26
▶912 MSh B (0.157) $(CH_3)_3C(CH_2)_2CH(CH_3)_2$ 292	▶910 VS B (0.036) $H_2C=CH(CH_2)_5CH_3$ 290
▶912 K $Ca(NO_3)_2$ 45	▶910 VS B (0.169) $(CH_3)_3CCH_2CH_2C(CH_3)_3$ 305
▶912 E or N (structure: $(CH_2)_3$ fused pyridine with CH_2-NH_2, N) 41	▶910 W $CHCl_3$ (structure: H_5C_2-O-C(=O)- isoxazole with CH_3, O, N) [Ring] 15
▶911 VVS B (0.036) $H_2C=CH(CH_2)_3CH(CH_3)_2$ 97	▶910 $La(NO_3)_3$ 45
▶911 VS B (0.015) $H_2C=CHC(CH_3)_3$ 129	▶910 E or N (structure: [cyclopenta-pyridine CH_2]$_2$NH, N) 41

▶910
E or N
$(CH_2)_3$ NH-CH_3 N 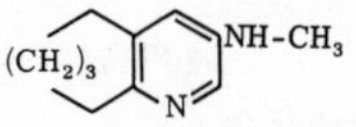
41

▶909 VS
A (82 mm Hg, 10 cm)
CF_3-CCl_3
253

▶909 VS
B (0.136) 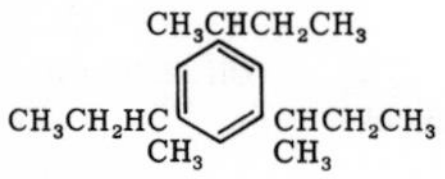
$CH_3CHCH_2CH_3$
CH_3CH_2HC $CHCH_2CH_3$
CH_3 CH_3
280

▶909 VVS
B (0.064)
H_2C=$CHCHCH_2CH_3$
CH_3
167

▶909 VS
B (0.10) 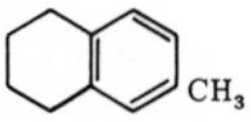
CH_3
269

▶909 VS
B (0.064)
$C_{17}H_{34}$
(1-heptadecene)
159

▶909 S
B (0.169)
$(CH_3)_3CCH_2CHCH(CH_3)_2$
CH_3
303

▶909 S
B
$(CH_3)_2CH(CH_2)_3CH_3$
331

▶909 VS
B (0.169)
$(CH_3)_2CHCH(CH_2CH_3)_2$
136

▶908 S
B (0.151) 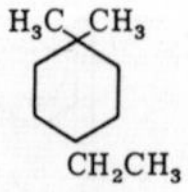
H_3C CH_3
CH_2CH_3
215

▶908 VVW
B (0.2398) 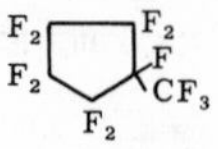
F_2 F_2 F_2 F CF_3 F_2
247

▶908 M
B (0.064)
$(CH_3)_2C$=$CHCH_2CH_3$
166

▶908 VS
B (0.008)
H_2C=$CH(CH_2)_8CH_3$
352

▶908 M
$CHCl_3$ 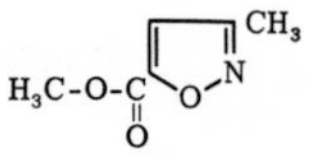
CH_3
H_3C-O-C O N
O
[Ring]
15

▶907 M
B
CH_3
$(CH_3)_2CHCHCH_2CH_2CH_3$
338

▶907 MB
A (40 mm Hg, 58 cm) 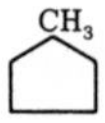
CH_3
105

▶907 W
$CHCl_3$ 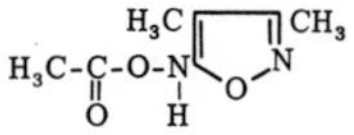
H_3C CH_3
H_3C-C-O-N O N
O H
[Ring]
15

▶906 MB
B (0.0104)
n-H_7C_3-O-CF_2CHFCl
263

▶906 M
B (0.097)
$(CH_3)_2CH(CH_2)_3CH_3$
80

▶906 M
B (0.114) 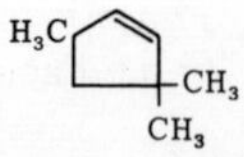
H_3C
CH_3
CH_3
125

▶906 M
C 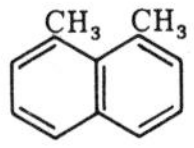
CH_3 CH_3
182

▶906 MW
$CHCl_3$ 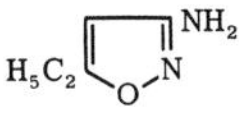
NH_2
H_5C_2 O N
[Ring]
15

▶905 VS
B (0.036) 
103

▶905 S
B (0.10) 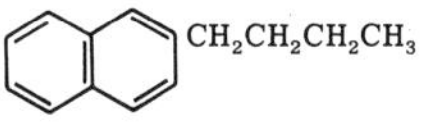
$CH_2CH_2CH_2CH_3$
175

▶905 S
B (0.068)
H_3CH_2C CH_2CH_3
C=C
H H
(cis)
275

▶905 SSh
B (0.169)
$(CH_3)_3CCH(CH_2CH_3)_2$
300

▶905 VS
B (0.169)
$(CH_3)_2CHCH(CH_2CH_3)_2$
136

▶905 S
O
$H_3C-\overset{\|}{C}-CH_3$
$KCrO_3Cl$
358

▶905 MSh
B (0.157)
$(CH_3)_3CCH(CH_2CH_3)_2$
78, see also 300

▶905 M
CS_2 [25%]
$CH_3CH_2CH_2OH$
314

▶905
E or N 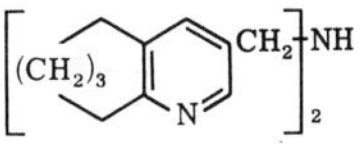
$[(CH_2)_3$ N $CH_2]_2NH$
41

▶905
O
$(CH_3)_2CH-\overset{\|}{C}-OH$
375

▶904 VSSh
B (0.169)
$(CH_3)_2CHCH(CH_2CH_3)_2$
136

▶904 S
B (0.169)
CH_3
$(CH_3)_2CHCHCH_2CH_2CH_3$
139

▶904 S
B
CF_4
10

▶904 S
B (0.15) 
128

▶904 M
B (0.10) 
S
199

▶904
K
$Mn(NO_3)_2 \cdot xH_2O$
45

▶903 VS
A (8 mm Hg, 15 cm)
CH_3
$H_2C=\overset{|}{C}CH_2CH_3$
92

▶903 S
CS_2 [25%]
$CH_3CH(CH_2)_2CH_3$
OH
318

▶903 MSh
B (0.169)

CH_3
$(CH_3)_2CHC - CH_2CH_2CH_3$
CH_3

134

▶903 M
CCl_4

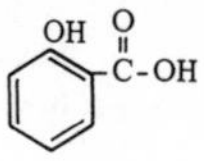

27

▶903
E or N

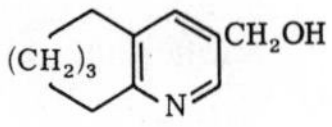

41

▶903
E or N

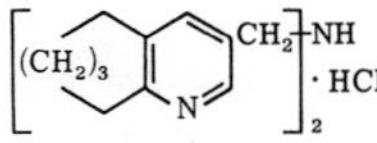

41

▶902 VVS
B (0.104)

$CF_2Cl-CFCl_2$

255

▶902 S
B (0.25)

$(CH_3)_2CHNO_2$

366

▶902
E or N

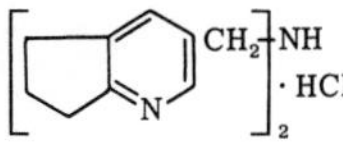

41

▶901 VS
B (0.10)

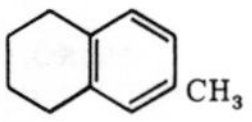

269*

▶901 MS
G

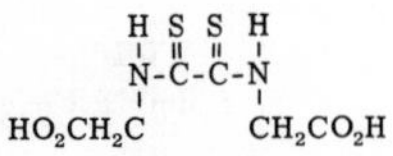

31

▶901 M
B

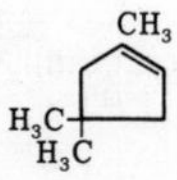

127

▶901 M
B (0.20)

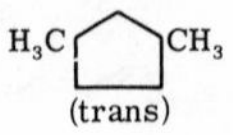

308

▶900 S
B (0.003)

$(CH_3)_2CHCH_2NO_2$

326

▶900 SSh
B (0.136)

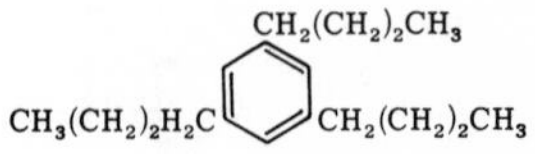

279

▶900 S
B (0.300)

$(C_2H_5)_2CH(CH_2)_{20}CH_3$

282

▶900 VS
B (0.065)

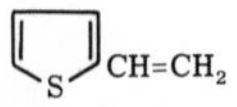

212

▶900 M
CS_2 [25%]

CH_3
$CH_3CH_2CHCH_2OH$

321

▶900 M
B (0.127)

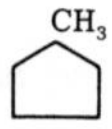

104

▶900 M
E

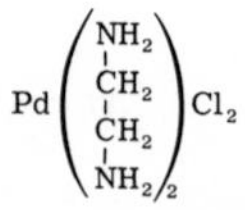

[CH_2]

8

▶899 VVS
B (0.169)

$(CH_3CH_2)_3CH$

143, 156

▶899 S
B (0.0104)

$CH_3CH_2-O-CF_2CHFCl$

261

▶899 VS
B (0.009)

$CH_2{=}CHCH{=}CHCH{=}CH_2$

203

▶899 S
B (0.169)

CH_3
$(CH_3)_2CHCHCH_2CH_2CH_3$

139

▶899
E or N

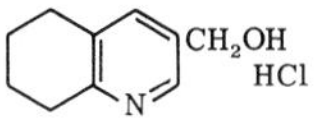

41

▶899

CH_2Cl_2

[CH_2] 376

▶899 VSSh
A (400 mm Hg, 15 cm)

$CH_3CH_2CH_2CH_2CH_3$

371

▶898 VS
H (CS_2)

$CH_2{=}CHCH{=}CHCH{=}CHCH{=}CH_2$

202

▶898 S
B

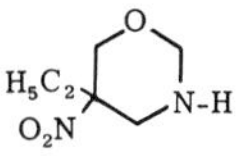

26

▶898 S

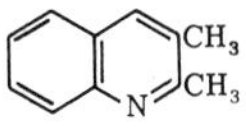

[H def. in pyridine] 378

▶897 S
B (0.10)

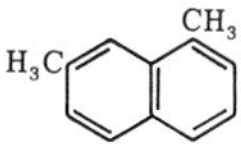

183

▶897 S
B

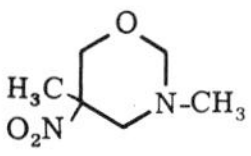

26

▶897 S
B

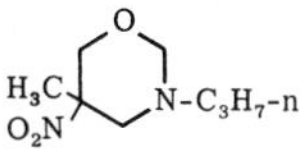

[COC] 26

▶897 S
B (0.10)

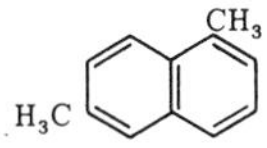

184

▶897 S
B (0.003)

$C_3H_7NO_2$

377, 392

▶897 M
B (0.114)

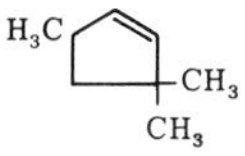

125

▶897 M
E

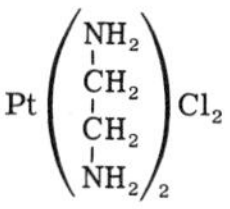

[CH_2] 8

▶897 M
$CHCl_3$

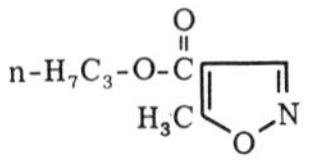

[Ring] 15

▶897 M
$CHCl_3$

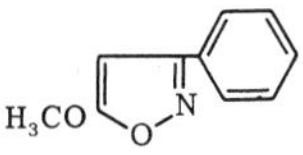

[Ring] 15

▶897 MW
$CHCl_3$

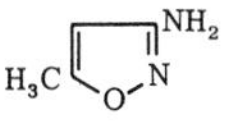

[Ring] 15

▶897 MW
$CHCl_3$

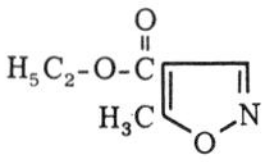

[Ring] 15

▶897 W
$CHCl_3$

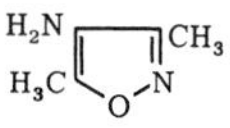

[Ring] 15

▶897
E or N

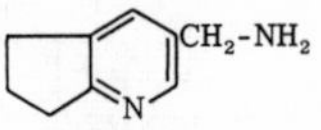

CH_2-NH_2
N

41

▶896 VS
B (0.015)

$H_2C=CCH_2C(CH_3)_3$
CH_3

96

▶896 S
E

$Pb_2P_2O_6$

28

▶896 M
B

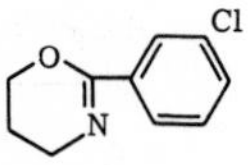

Cl
O
N

26

▶896 M
B

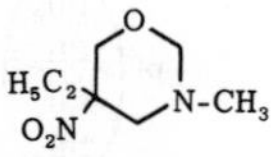

O
H_5C_2
O_2N
N-CH_3

26

▶896 M
B

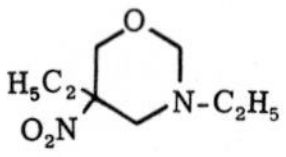

O
H_5C_2
O_2N
N-C_2H_5

26

▶896 M
$CHCl_3$

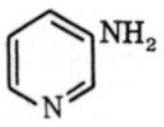

NH_2
N

327

▶896 MW
$CHCl_3$

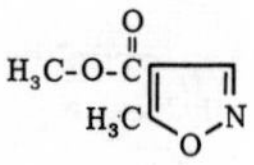

O
H_3C-O-C
H_3C
N
O

[Ring]

15

▶896 MB
A (40 mm Hg, 58 cm)

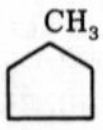

CH_3

105

▶896
E or N

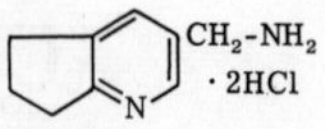

CH_2-NH_2
· 2HCl
N

41

▶895 VS (Q branch)
A (7.6 mm Hg, 40 cm)

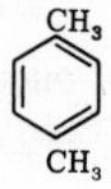

CH_3
CH_3

68

▶895 S
B (0.300)

$CH_3(CH_2)_5CH(C_3H_7)(CH_2)_5CH_3$

281

▶895 MB
B (0.036)

$CH_3(CH_2)_2CH=CH(CH_2)_2CH_3$
(cis)

291

▶895 M
B

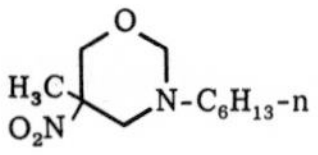

O
H_3C
O_2N
N-C_6H_{13}-n

26

▶895 M
B (0.097)

$(CH_3)_2CH(CH_2)_3CH_3$

80

▶894 MS
B (0.169)

CH_3
$(CH_3)_3CC - CH(CH_3)_2$
CH_3

306

▶894 VS
B (0.0104)

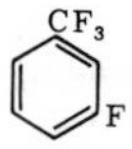

CF_3
F

218

▶894 S
B (0.300)

$C_{26}H_{54}$
(5, 14-di-n-butyloctadecane)

283

▶894 S
B

$(CH_3)_3C(CH_2)_3CH_3$

337

▶894 M
B

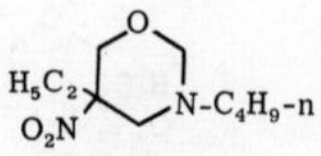

O
H_5C_2
O_2N
N-C_4H_9-n

26

▶894 M
B

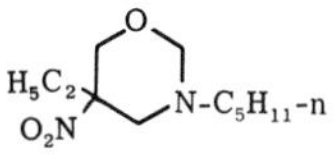
O
H_5C_2
$N-C_5H_{11}$-n
O_2N

26

▶893 VS
B (0.10)

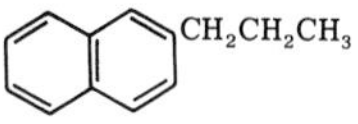
$CH_2CH_2CH_3$

177

▶893 S
B (0.169)

$(CH_3)_2CHC(CH_3)_2-CH_2CH_2CH_3$

134

▶893 S
B

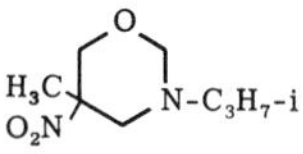
O
H_3C
$N-C_3H_7$-i
O_2N

26

▶893 S
B

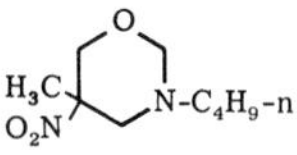
O
H_3C
$N-C_4H_9$-n
O_2N

26

▶893 S
B (0.051)

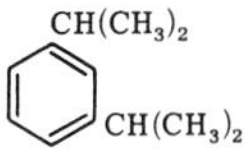
$CH(CH_3)_2$
$CH(CH_3)_2$

297

▶893
E or N

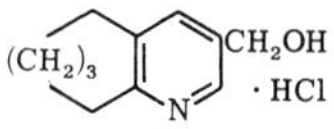
$(CH_2)_3$
CH_2OH
· HCl
N

41

▶892 MW
CS_2 [25%]

$CH_3(CH_2)_5OH$

320

▶892 S
B

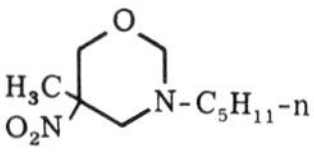
O
H_3C
$N-C_5H_{11}$-n
O_2N

26

▶892 S
C

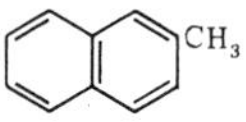
CH_3

191

▶892 M
B

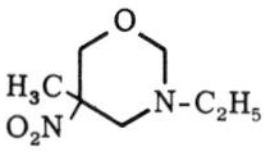
O
H_3C
$N-C_2H_5$
O_2N

26

▶892 M
$CHCl_3$

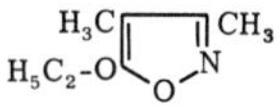
H_3C
CH_3
H_5C_2-O
O
N

[Ring] 15

▶892 W
$CHCl_3$

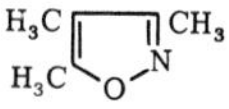
H_3C
CH_3
H_3C
O
N

[Ring] 15

▶891 SVB
A (200 mm Hg, 10 cm)

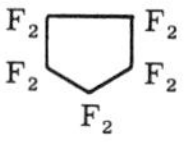
F_2
F_2
F_2
F_2
F_2

245

▶891 S
CS_2 [25%]

$CH_3CH(OH)(CH_2)_2CH_3$

318

▶891 W
B (0.036)

$CH_3CH_2CH=CH(CH_2)_3CH_3$

99

▶891 M
C

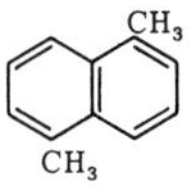
CH_3
CH_3

185

▶891 W
B (0°C)

$(CH_3)_2CHCH(CH_3)CH(CH_3)_2$

284

▶890 VS
A (8 mm Hg, 15 cm)

$H_2C=C(CH_3)CH_2CH_3$

92

▶890 VVS
B (0.064)

$H_2C=C(CH_2CH_3)_2$

163

▶890 VS B (0.10) [structure: naphthalene-$CH_2CH_2CH_2CH_2CH_3$] 173	▶888 VS B (0.0088) $H_2C{=}C(CH_3)CH_2CH_2CH_3$ 206
▶890 S B (0.10) [structure: naphthalene-CH_2CH_3] 189	▶888 VS B (0.169) $(CH_3CH_2)_2CHCH_2CH_2CH_3$ 140
▶890 M B (0.015) $CH_3(CH_2)_4OH$ 350, 393	▶888 S B (0.136) [structure: benzene with $CH(CH_3)_2$, H_3C, CH_3] 277
▶890 W $CHCl_3$ [structure: $H_3C-C(=O)-O-N(H)$-isoxazole with phenyl] [Ring] 15	▶888 S B (0.10) [structure: naphthalene-$CH_2CH_2CH_2CH_3$] 175
▶890 E or N [structure: cyclopentapyridine with CH_2OH, N] 41	▶887 VS B (0.036) $H_2C{=}C(CH_3)(CH_2)_4CH_3$ 98
▶889 S B [structure: oxazine ring with H_5C_2, O_2N, $N-C_6H_{13}$-n] 26	▶887 M CS_2 [25%] $CH_3CH_2CH_2OH$ 314
▶889 S B (0.10) [structure: naphthalene with CH_3, CH_3] 187	▶887 W B (0.0576) [structure: benzene with CH_3, F] 221
▶889 M B (0.169) $(CH_3)_3CCH(CH_3)C(CH_3)_3$ 307	▶886 VS C [structure: naphthalene with H_3C, CH_3] 180
▶889 MSB B (0.127) [structure: cyclopentane-CH_3] 104	▶886 MB B (0.036) $CH_3(CH_2)_2CH{=}CH(CH_2)_2CH_3$ (cis) 291
▶888 VS A (775 mm Hg, 10 cm) CF_2Cl-CF_2Cl 257	▶886 S B (0.0104) $CF_2Cl-CFCl_2$ 255

▶886 M
CHCl$_3$

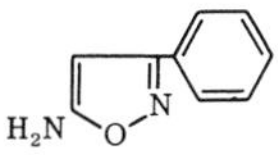

[Ring] 15

▶886 MB
A (40 mm Hg, 58 cm)

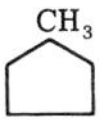

105

▶885 VS
B (0.0104)

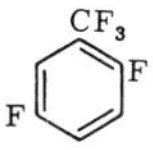

217

▶885 MWB
B (0.169)

$CH_3CH_2C(CH_3)_2-CH_2CH(CH_3)CH_2CH_3$

302

▶885 S

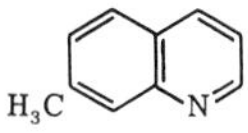

[H def. in benzene] 378

▶885 M
CHCl$_3$

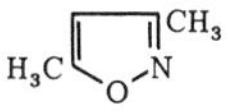

[Ring] 15

▶884 VS
B (0.028)

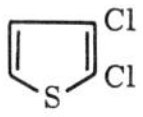

162

▶884 S
B (0.0576)

225

▶884 S
B

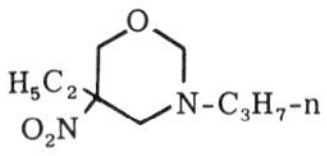

26

▶883 SB
A (100 mm Hg, 10 cm)

CHF_2CH_3

233

▶883 VS
A (775 mm Hg, 10 cm)

CF_2Cl-CF_2Cl

257

▶883 S
B (0.0104)

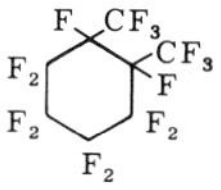

231

▶883 MS
G

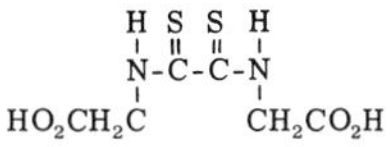

[C=S?] 31

▶882 VS
CS$_2$ [25%]

$CH_3CH_2C(CH_3)_2OH$

194

▶882 W
B (film)

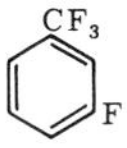

218

▶882 M
B (0.0104)

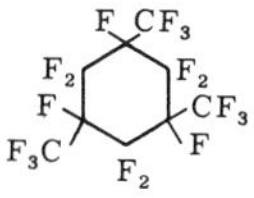

228

▶882 VS
A (25 mm Hg, 10 cm)

$CF_3CF_2CF_2CF_2CF_3$

237

▶882 M
B

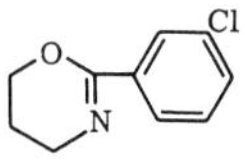

26

▶881 VS
CS$_2$ [25%]

$CH_3CH_2C(CH_3)_2OH$

194

▶881 M
B

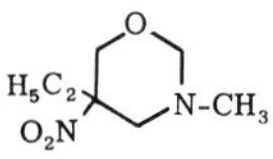

26

▶880.5 W B $C_6H_5(CH_3)SiCl_2$ 86	▶879 Sh E $K_4P_2O_6$ 28
▶880 VS A (8 mm Hg, 15 cm) $H_2C{=}C(CH_3)CH_2CH_3$ 92	▶878 W B (0.169) $CH_3CH_2C(CH_3)_2-CH(CH_3)CH_2CH_3$ 299
▶880 M B (0.169) $(CH_3)_2CHCH_2CH(CH_3)CH_2CH_3$ 138	▶878 S G (indoline)-CH=N-NH-C(=S)-NH$_2$ 380
▶880 VS CS_2 [25%] CH_3CH_2OH 315	▶877 M B (0.169) $(CH_3)_3CCH_2CH(CH_3)CH(CH_3)_2$ 303
▶880 S B (0.169) $(CH_3)_2CHCH(CH_2CH_3)CH(CH_3)_2$ 301	▶877 S B $Cl_3C-C(=O)-O$-(cyclopentyl)-(cyclopentenyl) 379
▶880 M B (0.104) $CH_3CH_2CH(CH_3)(CH_2)_2CH_3$ 81	▶877 S E $Ag_4P_2O_6$ 28
▶880 MW $CHCl_3$ 4-Cl-3-CH_3-5-H_3C-isoxazole [Ring] 15	▶877 S E $Li_4P_2O_6 \cdot 7H_2O$ 28
▶879 S B (0.003) $CH_3CH_2CH(NO_2)CH_3$ 324	▶877 M B (0.114) H_3C-cyclopentene-$(CH_3)_2$ 125
▶879 MW B $CH_3CH_2CH(CH_3)(CH_2)_2CH_3$ 330	▶877 W $CHCl_3$ $H_3C-C(=O)-O-N(H)$-(3-CH_3-5-H_3C-isoxazol-4-yl) [Ring] 15
▶879 W $CHCl_3$ 4-H_3C-3-CH_3-5-H_2N-isoxazole [Ring] 15	▶876 VS B cyclohexene 128

▶876 VVSB
B (0.0153)

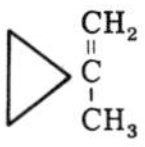

288

▶876 VS
B (0.028)

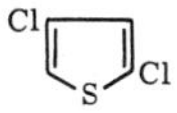

211

▶876 S
G

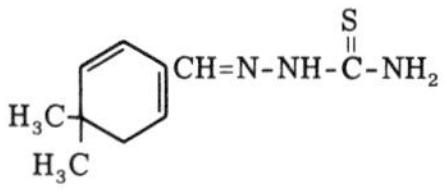

380

▶876 M
B (0.169)

$CH_3CH{=}CHCH_2CH_3$
(trans)

207

▶876 M
B (0.114)

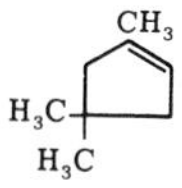

127

▶875 MVB
A (200 mm Hg, 10 cm)

CHF_2CH_3

233

▶875 W
B (0.0576)

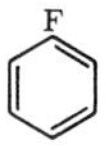

226

▶874 S

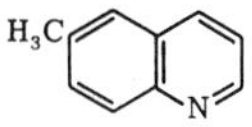

[H def. in benzene] 378

▶873 MVB
A (200 mm Hg, 10 cm)

CHF_2CH_3

233

▶873 VS
A (12.5 mm Hg, 10 cm)

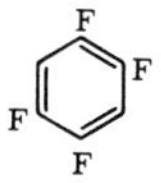

223

▶873 W
B (0.0104)

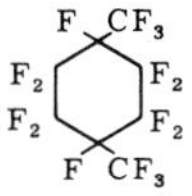

230

▶873 S
B (0.0104)

$CF_2Cl\text{-}CFCl_2$

255

▶873 S
E

$Tl_4P_2O_6$

28

▶873 MW
B (0.10)

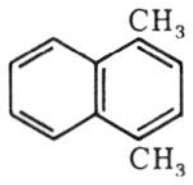

186

▶873 MSh
B (0.300)

$C_{26}H_{54}$
(5, 14-di-n-butyloctadecane)

283

▶873 M
E

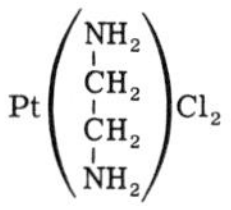

[CH_2] 8

▶873 M
$CHCl_3$

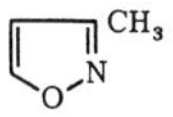

[Ring] 15

▶872 VS
B (0.029)

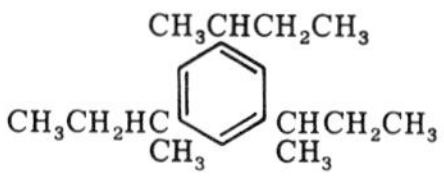

280

▶872 VS
C

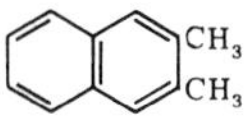

181

▶871 S
B (0.003)

$C_3H_7NO_2$

377, 392

▶871 MW
B (0.169)

$(CH_3)_2CHC(CH_3)_2-CH_2CH_2CH_3$

134

▶871 S
B (0.10)

199

▶871 W
B (0.169)

$(CH_3)_2CHCH_2CH_2CH(CH_3)_2$

137

▶871 VVW
A (100 mm Hg, 10 cm)

CH_3-CCl_3

251

▶871 MW
C

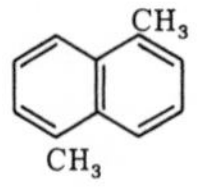

185

▶870 VS
B (0.169)

$(CH_3)_2CHCH_2CH(CH_3)_2$

141

▶870 VS
B (0.029)

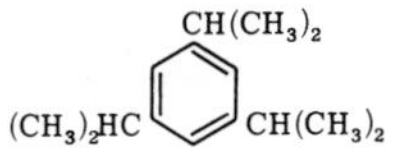

278

▶870 VS
B (0.10)

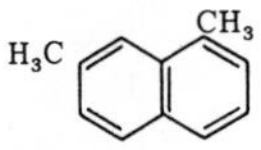

183

▶870 MW
B (0.169)

$(CH_3)_3CCH_2CH_3$

145

▶870 MSh
B (0.163)

$H_2C=CH(CH_2)_8CH_3$

352

▶870 MW
$CHCl_3$

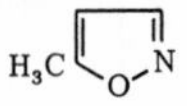

[Ring] 15

▶870 S
B (0.169)

$(CH_3)_2CHCH(CH_3)CH_2CH_2CH_3$

139, 338

▶870 W
E

$Li_4P_2O_6 \cdot H_2O$

28

▶869 VS
B (0.0104)

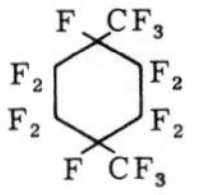

230

▶869 SB
B (film)

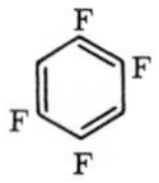

222

▶869 VS
B (0.169)

$(CH_3)_2CHCH(CH_3)_2$

144

▶869 M
B

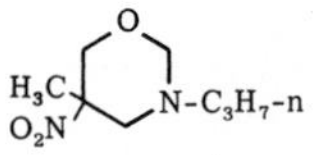

26

▶869 MW
$CHCl_3$

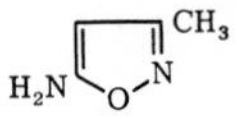

[Ring] 15

▶868 VS
A (100 mm Hg, 10 cm)

CHF_2CH_3

233

▶868 VVS
B (0.10)

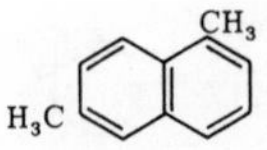

184

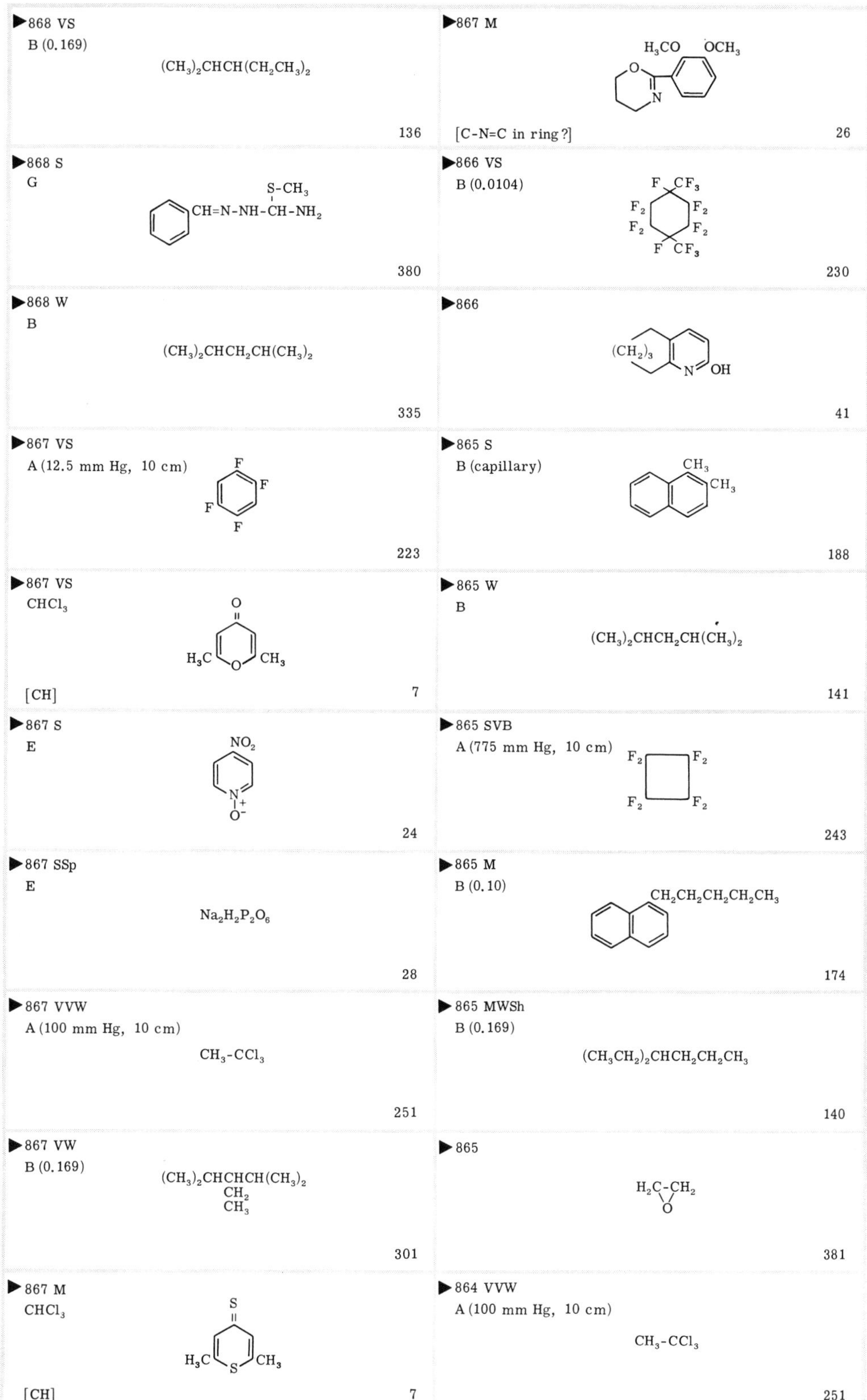

▶868 VS
B (0.169)
$(CH_3)_2CHCH(CH_2CH_3)_2$
136

▶868 S
G
S-CH_3
CH=N-NH-CH-NH_2
380

▶868 W
B
$(CH_3)_2CHCH_2CH(CH_3)_2$
335

▶867 VS
A (12.5 mm Hg, 10 cm)
F F F F
223

▶867 VS
$CHCl_3$
O
H_3C O CH_3
[CH]
7

▶867 S
E
NO_2
N+
O-
24

▶867 SSp
E
$Na_2H_2P_2O_6$
28

▶867 VVW
A (100 mm Hg, 10 cm)
CH_3-CCl_3
251

▶867 VW
B (0.169)
$(CH_3)_2CHCHCH(CH_3)_2$
CH_2
CH_3
301

▶867 M
$CHCl_3$
S
H_3C S CH_3
[CH]
7

▶867 M
H_3CO OCH_3
O N
[C-N=C in ring?]
26

▶866 VS
B (0.0104)
F CF_3
F_2 F_2
F_2 F_2
F CF_3
230

▶866
$(CH_2)_3$ N OH
41

▶865 S
B (capillary)
CH_3
CH_3
188

▶865 W
B
$(CH_3)_2CHCH_2CH(CH_3)_2$
141

▶865 SVB
A (775 mm Hg, 10 cm)
F_2 F_2
F_2 F_2
243

▶865 M
B (0.10)
$CH_2CH_2CH_2CH_2CH_3$
174

▶865 MWSh
B (0.169)
$(CH_3CH_2)_2CHCH_2CH_2CH_3$
140

▶865
H_2C-CH_2
O
381

▶864 VVW
A (100 mm Hg, 10 cm)
CH_3-CCl_3
251

▶864 S
B (0.0104)

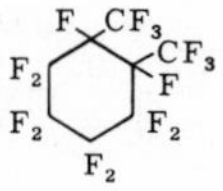

231

▶864 VS
B (0.08)

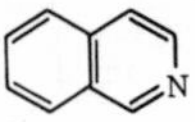

197

▶864 VVW
A (200 mm Hg, 10 cm)

$CF_2Cl\text{-}CF_2Cl$

257

▶864 W
CS_2 [25%]

$CH_3CH(CH_2)_2CH_3$
OH

318

▶864 W
B

$(CH_3)_2CHCH_2CHCH_2CH_3$
CH_3

334

▶863 S
B (0.169)

CH_3
$(CH_3)_3CC - CH_2CH_2CH_3$
CH_3

304

▶862 VS
B (0.036)

103

▶862 S
B

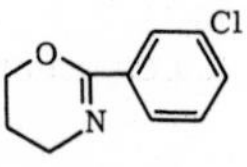

26

▶861.3 S
B (0.03)

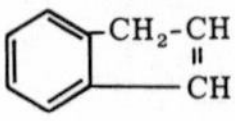

9

▶861 S
B (0.169)

$(CH_3)_2CHCH_2CHCH_2CH_3$
CH_3

138

▶861 S
B (0.10)

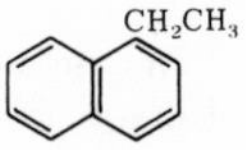

190

▶861 MW
B (0.036)

$CH_3(CH_2)_2CH{=}CH(CH_2)_2CH_3$
(cis)

291

▶860 VVS
$CHCl_3$

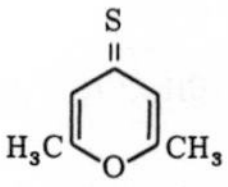

[CH] 7

▶860 S
B (0.0104)

$n\text{-}H_9C_4\text{-}O\text{-}CF_2CHFCl$

264

▶860 M
B (0.169)

$CH_3CH{=}CHCH_2CH_3$
(cis)

208

▶860 M
B

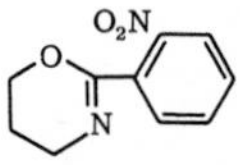

26

▶860 MS
B (0.10)

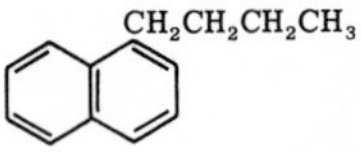

176

▶860

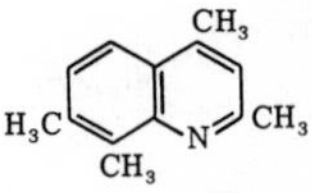

378

▶859 VS
A (8.2 mm Hg, 10 cm)

$CF_3\text{-}CCl_3$

253

▶859 VS
B (capillary)

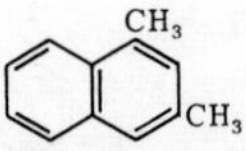

187

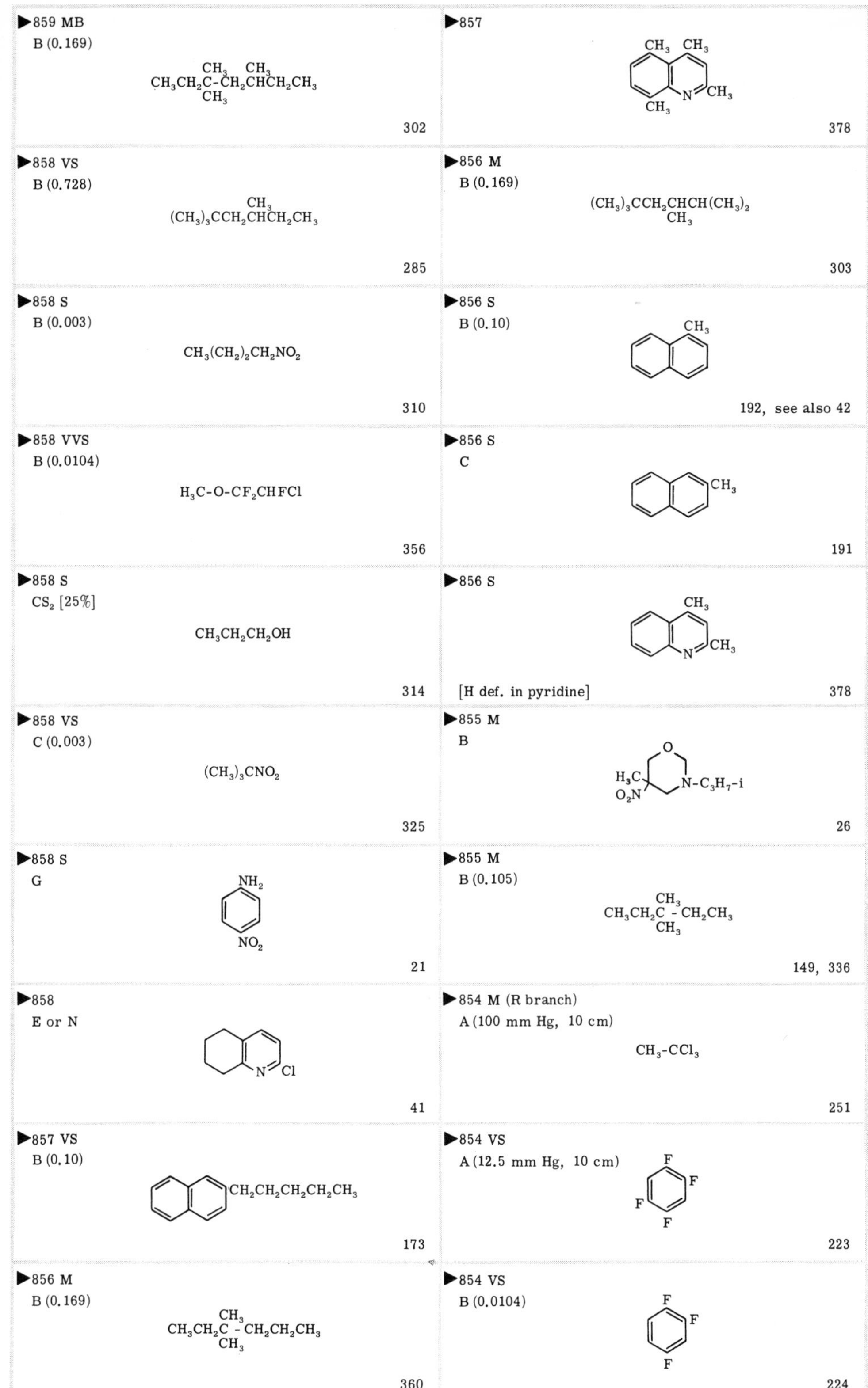

▶859 MB
B (0.169)
$CH_3CH_2C(CH_3)_2\text{-}CH_2CH(CH_3)CH_2CH_3$
302

▶858 VS
B (0.728)
$(CH_3)_3CCH_2CH(CH_3)CH_2CH_3$
285

▶858 S
B (0.003)
$CH_3(CH_2)_2CH_2NO_2$
310

▶858 VVS
B (0.0104)
$H_3C\text{-}O\text{-}CF_2CHFCl$
356

▶858 S
CS_2 [25%]
$CH_3CH_2CH_2OH$
314

▶858 VS
C (0.003)
$(CH_3)_3CNO_2$
325

▶858 S
G

21

▶858
E or N

41

▶857 VS
B (0.10)

173

▶856 M
B (0.169)
$CH_3CH_2C(CH_3)_2\text{-}CH_2CH_2CH_3$
360

▶857

378

▶856 M
B (0.169)
$(CH_3)_3CCH_2CH(CH_3)CH(CH_3)_2$
303

▶856 S
B (0.10)

192, see also 42

▶856 S
C

191

▶856 S

[H def. in pyridine]
378

▶855 M
B

26

▶855 M
B (0.105)
$CH_3CH_2C(CH_3)_2\text{-}CH_2CH_3$
149, 336

▶854 M (R branch)
A (100 mm Hg, 10 cm)
$CH_3\text{-}CCl_3$
251

▶854 VS
A (12.5 mm Hg, 10 cm)

223

▶854 VS
B (0.0104)

224

▶854 VS
B (0.10) 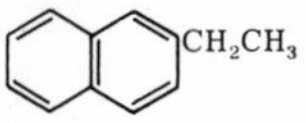
CH_2CH_3
189

▶854 S
B (0.10) 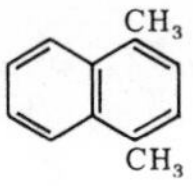
CH_3
CH_3
186

▶854 MS
CS_2 [25%]
$CH_3CH_2CHCH_2CH_3$
OH
196

▶854
E or N 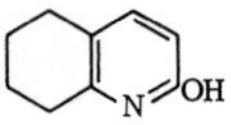
N OH
41

▶853 VS
B (0.065) 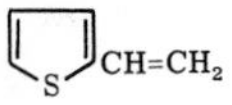
S $CH=CH_2$
212

▶853 VS
B (0.028) 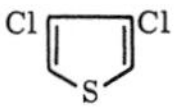
Cl Cl
S
161

▶853 S
B (0.10) 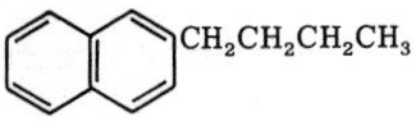
$CH_2CH_2CH_2CH_3$
175

▶853 SB
B (0.029) 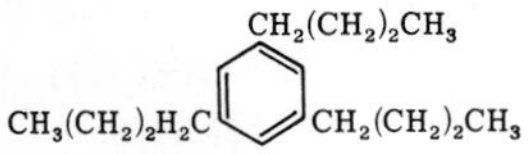
$CH_2(CH_2)_2CH_3$
$CH_3(CH_2)_2H_2C$ $CH_2(CH_2)_2CH_3$
279

▶853 M
B (0.169)
CH_3
$(CH_3)_2CHCHCH_2CH_2CH_3$
139

▶852 VW
B (0.0104)
$CF_2Cl-CFCl_2$
255

▶852 S
E 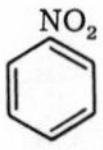
NO_2
20

▶852 M
B 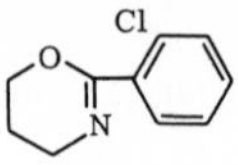
Cl
O N
26

▶852 S
B (0.20) 
H_3C CH_3
(trans)
308

▶851 VS
B (0.0104) 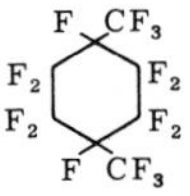
F CF_3
F_2 F_2
F_2 F_2
F CF_3
230

▶851 VS
B (0.065) 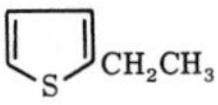
S CH_2CH_3
213

▶851 M
B 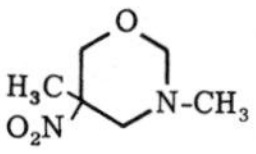
O
H_3C $N-CH_3$
O_2N
26

▶851 VWVB
B (0.066)
$H_2C=CH(CH_2)_6CH_3$
271

▶850.2 W
$(C_6H_5)_2SiCl_2$
86

▶850 VS
A (8.2 mm Hg, 10 cm)
CF_2Cl-CF_2Cl
257

▶850 S
B 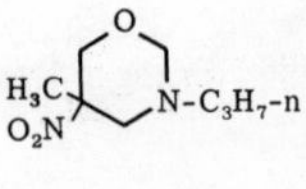
O
H_3C $N-C_3H_7$-n
O_2N
[C-O-C]
26

▶850 VSB
B (0.25)

$(CH_3)_2CHNO_2$

366

▶850 M
B (0.157)

$(CH_3)_3CCH(CH_2CH_3)_2$

78, see also 300

▶850

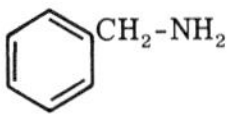

[NH_2]
383

▶849 VS
A (8.2 mm Hg, 10 cm)

CF_2Cl-CF_2Cl

257

▶849 VS
B (0.0104)

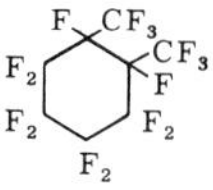

231

▶849 VS
$CHCl_3$

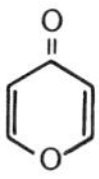

[CH]
7

▶849 S
B

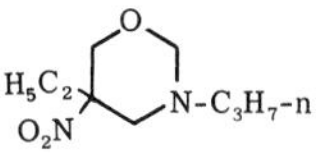

[C-O-C]
26

▶849 S
B

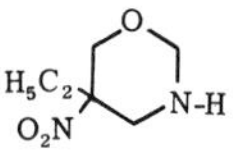

26

▶848 VS
B (0.0104)

$CH_3CH_2-O-CF_2CHFCl$

261

▶848 VS
B (0.10)

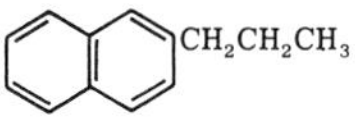

177

▶848 S
B

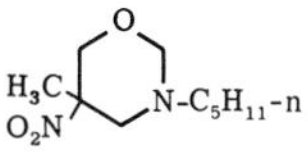

[C-O-C]
26

▶848 M
B

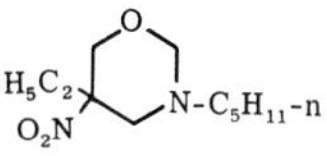

26

▶848
E or N

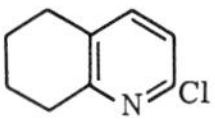

41

▶847 MW
B (0.0576)

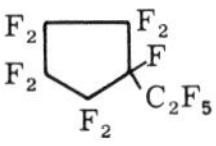

247

▶847 VS
CS_2

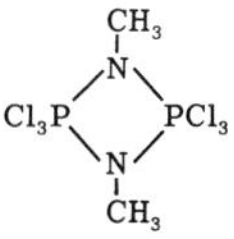

384

▶846.7 W

$C_6H_5SiCl_3$

86

▶846 S
B (0.169)

$(CH_3CH_2)_3CH$

143, 156

▶846 S
CS_2 [25%]

$CH_3(CH_2)_3OH$

353

▶846 W
B (0.036)

CH_3
$H_2C=C(CH_2)_4CH_3$

98

▶846 VS
B (0.029)

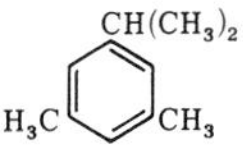

277

▶846 S $CHCl_3$ (4H-thiopyran-4-one) 7	▶844 MS B (0.0104) $CF_3(CF_2)_5CF_3$ 240
▶845 VS B (0.065) $C(CH_3)_3$ (benzene ring) 133	▶844 S E (N^+, O^-; 2-phenylpyridine N-oxide) 24
▶845 MB B (0.0104) $n\text{-}H_9C_4\text{-}O\text{-}CF_2CHFCl$ 264	▶844 M B (0.114) H_3C, CH_3, CH_3 (cyclopentene ring) 125
▶845 VS B (capillary) CH_3, CH_3 (naphthalene) 187	▶843 S B O, N, Cl 26
▶845 S B (0.003) NO_2 $CH_3CH_2CHCH_3$ 324	▶843 S E, $CHCl_3$ NH_2, N (pyridine) 327
▶845 MVB B (0.300) $CH_3(CH_2)_5CH(C_3H_7)(CH_2)_5CH_3$ 281	▶843 W B (0.0576) $F_9C_4\text{-}O\text{-}C_4F_9$ 266
▶845 M $CHCl_3$ O, N (isoxazole) [Ring] 15	▶843 M B O, H_3C, O_2N, $N\text{-}C_2H_5$ [C-O-C] 26
▶845 E $Cr(NO_3)_3$ 45	▶842 S B (0.169) $(CH_3)_2CHCHCH(CH_3)_2$ CH_2 CH_3 301
▶845 O_2N, $CH{=}CH\text{-}\overset{O}{\overset{\|\|}{C}}\text{-}O\text{-}C_2H_5$ [CH] 349	▶842 MS B (0.0104) CH_3, F (benzene ring) 220
▶844 S B O, N, F 26	▶842 S G NH_2, NO_2 (benzene ring) [CH out of plane] 21

▶842 M
B

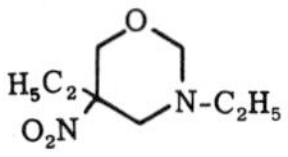

[C-O-C] 26

▶842 M
B (0.136)

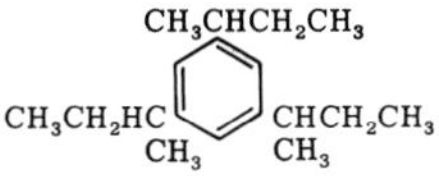

280

▶842
E or N

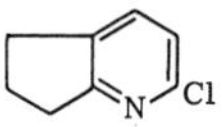

41

▶841 VS
B (0.0104)

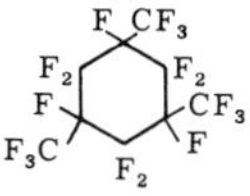

228

▶841 M
B (0.169)

$(CH_3)_3C\overset{CH_3}{\underset{CH_3}{C}}-CH(CH_3)_2$

306

▶841 M
B

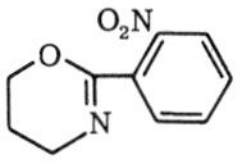

26

▶841

$Si(CH_3)_4$

386

▶840 VVS
C

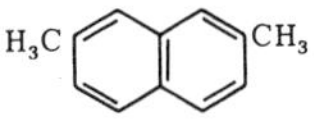

179

▶840 S

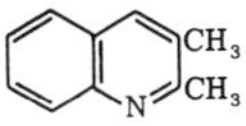

[H def. in benzene] 378

▶840 M
B (0.114)

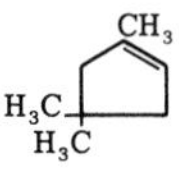

127

▶840

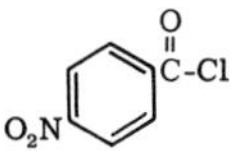

[CH] 385

▶839 VS
B (0.0576)

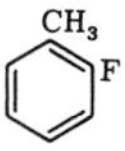

221

▶839 W
B (0.169)

$(CH_3)_2CHCH_2CH_2CH(CH_3)_2$

137

▶839 S
CCl_4

21

▶838 W
B (0.0104)

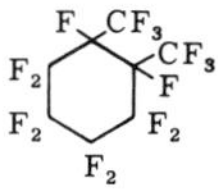

231

▶838 W
B (0.013)

$CH_3CH=CHCH(CH_3)_2$

130

▶838 M
B (0.0104)

$H_3C-O-CF_2CHFCl$

356

▶838
E

$Zn(NO_3)_2$

45

▶838
E, P

$LiNO_3$

45

▶837 MW
B (0.169)

$(CH_3)_2CH\overset{CH_3}{\underset{CH_3}{C}}-CH_2CH_2CH_3$

134

▶ 837 VWB
B (0.163)

$H_2C{=}CH(CH_2)_8CH_3$

352

▶ 836 VS
B (0.028)

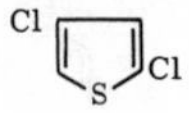
Cl
Cl
S

211

▶ 836 S
B (0.10)

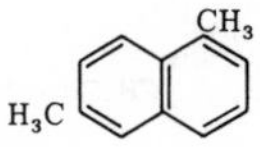
CH3
H3C

184

▶ 836 S
B (0.003)

$(CH_3)_2CHCH_2NO_2$

326

▶ 836 S
E

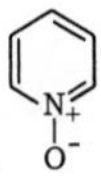
N+
O-

24

▶ 836 S

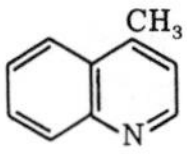
CH3
N

[H def. in pyridine]

378

▶ 836
E

$NaNO_3$

45

▶ 836
E

$RbNO_3$

45

▶ 835 VS
B (0.025)

S

199

▶ 835 S
B

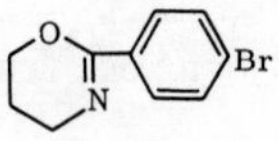
O
N
Br

26

▶ 835 S
CS_2

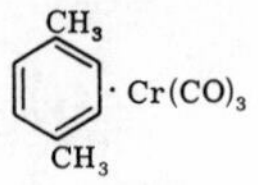
CH3
· Cr(CO)3
CH3

17

▶ 835 W
B (0.0576)

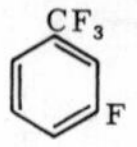
CF3
F

218

▶ 835
E

$NaNO_3$

45

▶ 835
E

$Ba(NO_3)_2$

45

▶ 835
E

$Bi(NO_3)_3$

45

▶ 835
E

$Cu(NO_3)_2 \cdot 5H_2O$

45

▶ 835
E

$Fe(NO_3)_3$

45

▶ 835
E

$Hg_2(NO_3)_2$

45

▶ 835
E

$Hg(NO_3)_2$

45

▶ 835
E

$Y(NO_3)_3$

45

▶834 MS
B (0.104)

$(CH_3)_3CCH(CH_3)_2$

147

▶834 VS
A (25 mm Hg, 10 cm)

$CF_3\dot{C}F_2CF_2CF_2CF_3$

237

▶833 MSp
B (0.0576)

$(CF_3)_2CFCF_2CF_3$

240

▶833 VS
B (0.0104)

F
F

225

▶833 VVS
B (0.064)

$(CH_3)_2C{=}CHCH_2CH_3$

166

▶833 VW
B (0.2398)

$(C_4F_9)_3N$

346

▶833 VW
B (0.169)

$(CH_3)_3CCH(CH_2CH_3)_2$

300

▶833 S
B

$Cl_3C\text{-}\overset{O}{\overset{\|}{C}}\text{-}O$

379

▶833 VW
B (0.25)

$CH_3(CH_2)_2CH_2NO_2$

310

▶833
P

$Co(NO_3)_2$

45

▶832 VS
B (0.2398)

$(CF_3)_2CFCF_2CF_3$

240

▶832 S
B (0.101)

$(CH_3CH_2)_3CH$

156, see also 143

▶832 S
CS_2 [25%]

$CH_3\underset{OH}{CH}(CH_2)_2CH_3$

318

▶832 VS
B (0.169)

$(CH_3CH_2)_3CH$

143

▶832
E

$Cr(NH_3)_5(NO_3)_3$

45

▶832
P

$Sm(NO_3)_3$

45

▶832
P

$Y(NO_3)_3$

45

▶831 W
B (0.0104)

F

226

▶831 SSh
CS_2 [25%]

$H_2C{=}\overset{CH_3}{\overset{|}{C}}CO_2CH_3$

193

▶831 M
E

$Pt\left(\begin{matrix}NH_2\\ |\\ CH_2\\ |\\ CH_2\\ |\\ NH_2\end{matrix}\right)_2 Cl_2$

[NH_2]

8

▶831 H_3C, CH_3 (quinoline) 378	▶830 M N (1,3,5-triazine) 14
▶830.7 W $(C_6H_5)_2SiCl_2$ 86	▶829 VS B (0.0104) CF_3, F, F 217
▶830.5 M B (0.03) CH_2-CH = CH 9	▶829 VS B (0.065) S, CH=CH_2 212
▶830 VS B (0.051) CH$(CH_3)_2$, CH$(CH_3)_2$ 298	▶829 VS B (0.04) N (isoquinoline) 197
▶830 VS B (0.136) CH$(CH_3)_2$, $(CH_3)_2$HC, CH$(CH_3)_2$ 278	▶829 MB A (682 mm Hg, 10 cm) F_2, F_2, F_2, F_2, F_2 245
▶830 W B (0.036) H_2C=C$(CH_3)(CH_2)_4CH_3$ 98	▶829 WB B (0.066) H_2C=CH$(CH_2)_6CH_3$ 271
▶830 S E C(=O)-NH-CH_3 [Car-H] 2	▶828 S H_3C, N [H def. in benzene] 378
▶830 S H_3C, N [H def. in pyridine] 378	▶828 S H_3C, N [H def. in pyridine] 378
▶830 M E CH_2OH, P=O 34	▶828 M B (0.169) $CH_3CH_2C(CH_3)(CH_2CH_3)$-$CHCH_2CH_3$ 299
▶830 M E CH_2OH, P 34	▶828 P Cr$(NH_3)_5(NO_3)_3$ 45

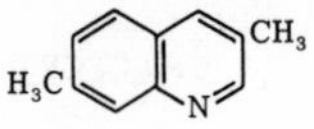

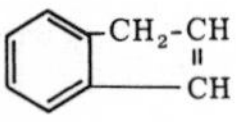

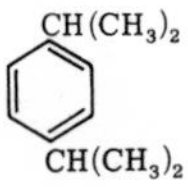

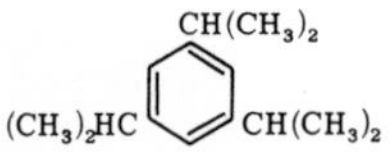

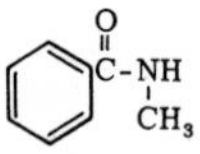

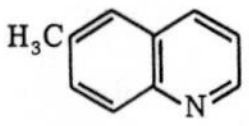

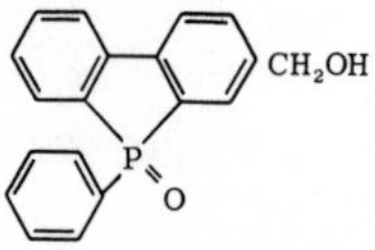

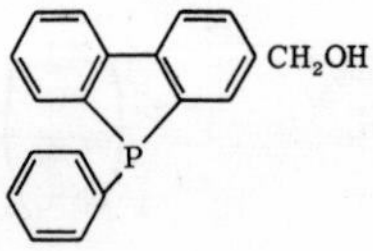

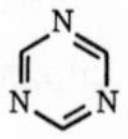

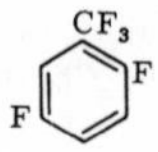

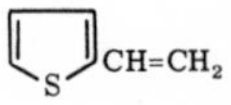

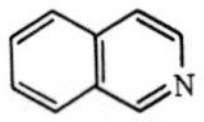

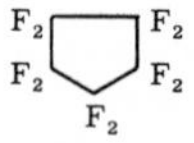

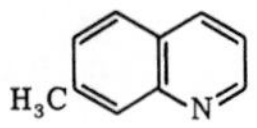

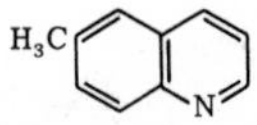

▶827 VS
B (0.0104)

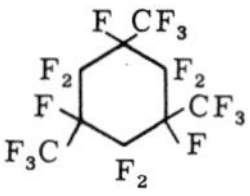

F, CF_3, F_2, F, F_2, CF_3, F_3C, F, F_2

228

▶827
P

$Co(NO_3)_2$

45

▶826 S
B (0.169)

$(CH_3)_2CHCH_2CHCH_2CH_3$
CH_3

138

▶826 S
B

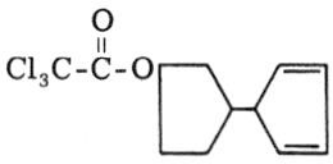

$Cl_3C\text{-}\overset{O}{\overset{\|}{C}}\text{-}O$

379

▶826 MW
B (0.015)

$H_2C{=}CCH_2C(CH_3)_3$
CH_3

96

▶826 M
B (0.169)

CH_3
$H_2C{=}CCH_2CH_2CH_3$

206

▶826
E

$LiNO_3$

45

▶826
P

$Cr(NO_3)_3$

45

▶826
P

$Gd(NO_3)_3$

45

▶826
P

KNO_3

45

▶825 VS
B (0.065)

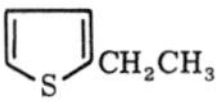

S CH_2CH_3

213

▶825 VS
E

NH_2
N

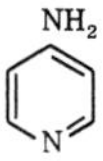

327

▶825
E

KNO_3

45

▶825
E, P

$Al(NO_3)_3$

45

▶825
P

$Be(NO_3)_2$

45

▶825
P

$Fe(NO_3)_3$

45

▶825
P

$Hg_2(NO_3)_2$

45

▶825
P

$Hg(NO_3)_2$

45

▶825
P

$Mg(NO_3)_2$

45

▶825
P

$Zn(NO_3)_2$

45

▶825

$CsNO_3$

45

▶824 VS
B (capillary)

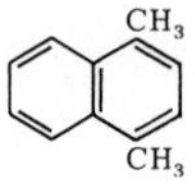

186

▶824 M
B (0.0576)

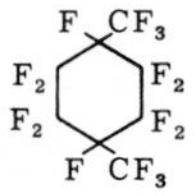

230

▶824 VS
B (0.0563)

$(CH_3)_2C{=}CHC(CH_3)_3$

204

▶824 S
CS_2

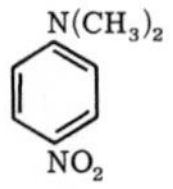

$N(CH_3)_2$ / NO_2

[CH out of plane]

21

▶824
P

$Ca(NO_3)_2$

45

▶823 VVS
B (0.064)

H CH_2CH_3
C=C
H_3C CH_3
(trans)

165

▶823 VS
B (0.114)

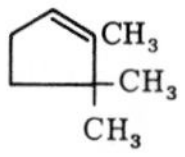

126

▶823 VVS
B (0.10)

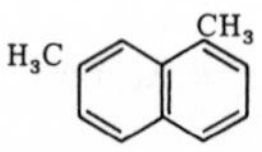

183, see also 42

▶823 VS
B (0.025)

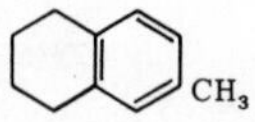

269

▶823 S
B

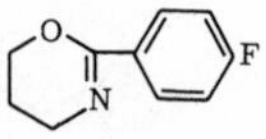

[C-O-C]

26

▶823
P

$Sr(NO_3)_2$

45

▶822 WSh
B (0.0576)

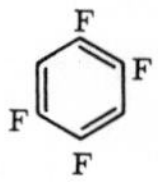

222

▶822 VS
B (0.169)

$(CH_3)_2CHCH(CH_2CH_3)_2$

136

▶822
P

$AgNO_3$

45

▶822
P

$Cu(NO_3)_2 \cdot 5H_2O$

45

▶821 VS
B (0.169)

$(CH_3CH_2)_2CHCH_2CH_2CH_3$

140

▶821 M
B

$CH_3CH_2CH(CH_2)_2CH_3$
CH_3

330

▶821 VW
B (0.036)

$CH_3(CH_2)_2CH{=}CH(CH_2)_2CH_3$
(cis)

291

▶821
P

$Ba(NO_3)_2$

45

▶820 SSh
B (0.10)

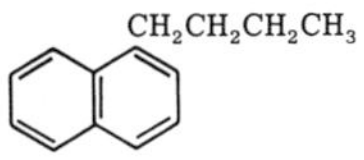

176

▶820 S
G

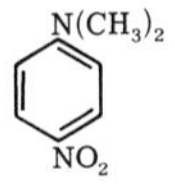

[CH out of plane]
21

▶820 MS
CS_2 [25%]

$CH_3CHCH_2CH_3$
OH

319

▶820 VVSB
B (0.065)

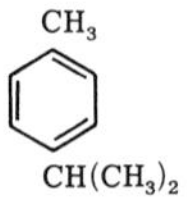

132

▶820
$CHCl_3$

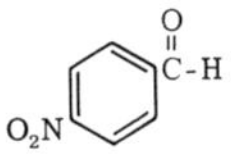

387

▶819 VS
B (0.028)

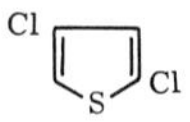

211

▶819 S
B (0.0576)

F_9C_4-O-C_4F_9

266

▶819 VS
B

$(CH_3)_2CHOH$

361

▶819 M
B (0.169)

CH_3
H_2C=$CCH_2CH_2CH_3$

206

▶818 S
B (0.0153)

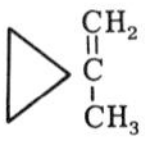

288

▶818 VS
B (0.10)

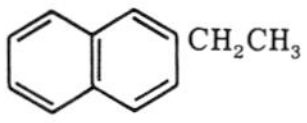

189

▶818 M
B (0.169)

CH_3
$(CH_3)_2CHC$ - $CH_2CH_2CH_3$
CH_3

134

▶818 S
C

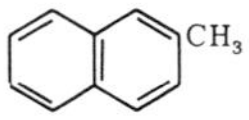

191

▶818
E

$Be(NO_3)_2$

45

▶818
K

$Ca(NO_3)_2$

45

▶818

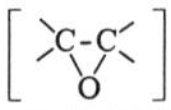

388

▶818

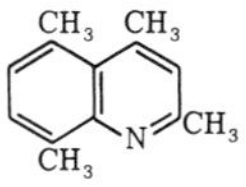

378

▶817 S
B (0.15)

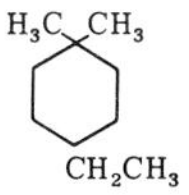

215

▶817 S

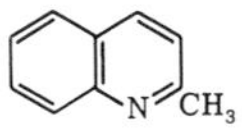

[H def. in pyridine]
378

▶817 MSh
B (0.136)

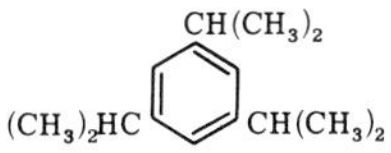

278

▶816 VS
C 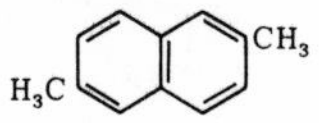
H_3C CH_3
180

▶816 VS
B (0.10) 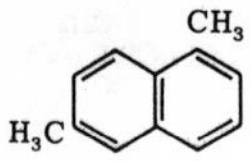
CH_3 H_3C
184

▶816 M
B 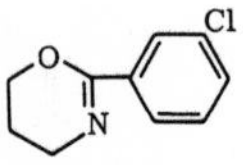
O N Cl
[C-O-C]
26

▶816 MW
B (0.003)
NO_2
$CH_3CH_2CHCH_3$
324

▶816 VVS
B (0.10) 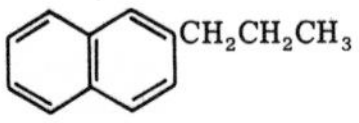
$CH_2CH_2CH_3$
177

▶816
E
$Ba(NO_3)_2$
45

▶816
P
$Ba(NO_3)_2$
45

▶816
P
$La(NO_3)_2$
45

▶816
P
$La(NO_3)_3 \cdot 3NH_4NO_3$
45

▶816
P
$Ni(NO_3)_2$
45

▶815 S
CS_2 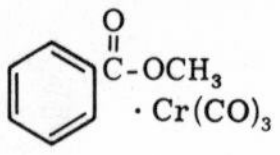
O
C-OCH_3
$\cdot Cr(CO)_3$
17

▶815 S
CS_2 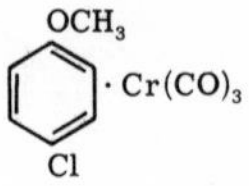
OCH_3
$\cdot Cr(CO)_3$
Cl
17

▶815 M
A
CH_3
$CH_3CH_2CHCH_2CH_3$
332

▶815 M
B (0.136) 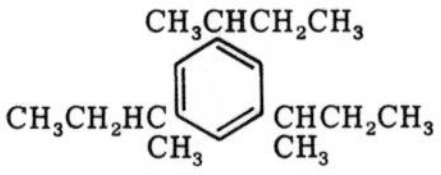
$CH_3CHCH_2CH_3$
CH_3CH_2HC $CHCH_2CH_3$
CH_3 CH_3
280

▶815 MS VVS
CS_2 [5%] B (0.063) 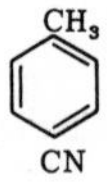
CH_3
CN
214

▶815
E
$Sm(NO_3)_3$
45

▶815
E
$Sr(NO_3)_2$
45

▶815
P
$Cd(NO_3)_2$
45

▶815
P
$Ce(NH_4)_2(NO_3)_6$
45

▶815
P
$Nd(NO_3)_3$
45

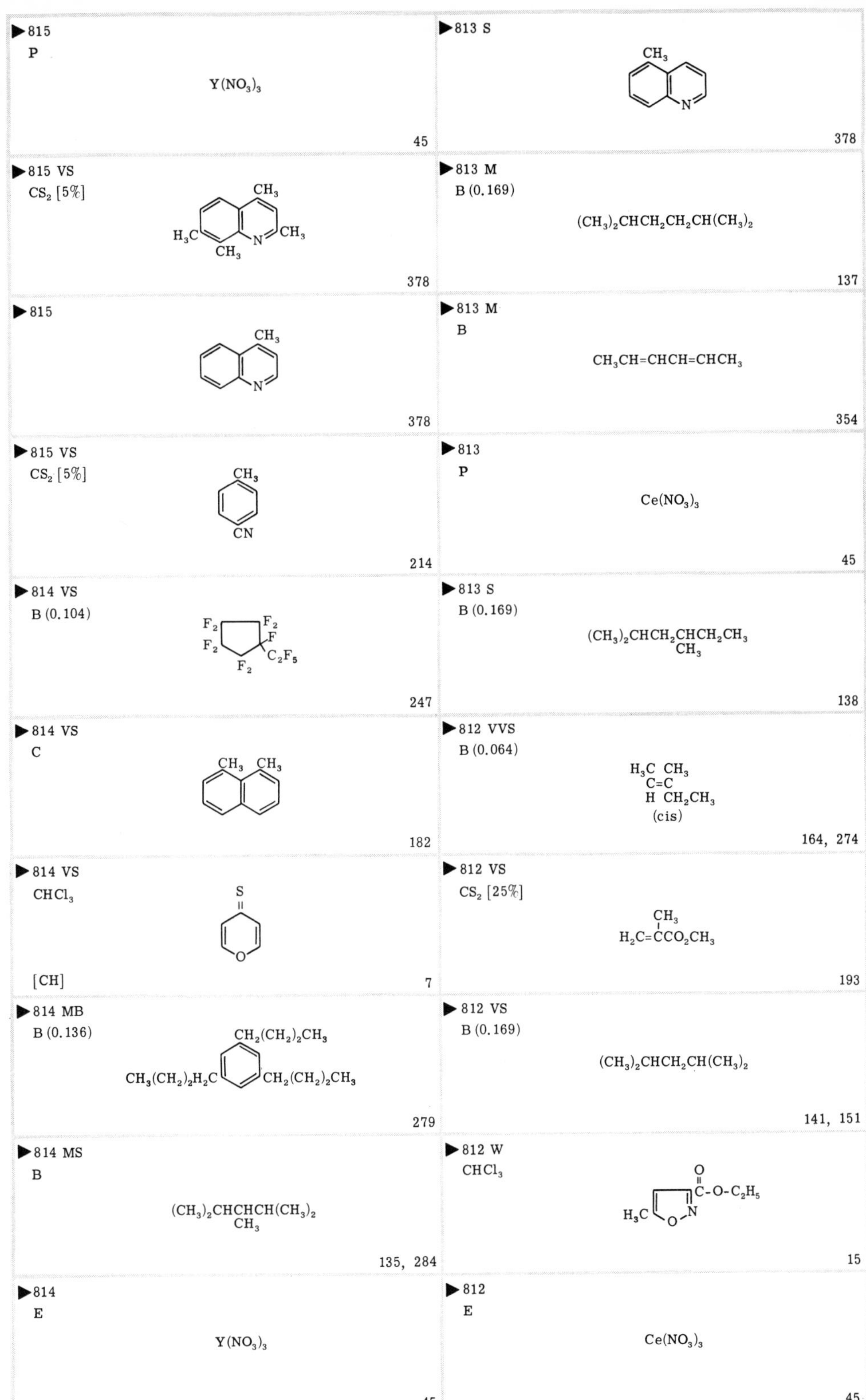
▶815
P
Y(NO3)3
45
▶815 VS
CS2 [5%]
CH3
H3C
CH3
N
CH3
378
▶815
CH3
N
378
▶815 VS
CS2 [5%]
CH3
CN
214
▶814 VS
B (0.104)
F2
F2
F2
F
F2
C2F5
247
▶814 VS
C
CH3 CH3
182
▶814 VS
CHCl3
S
O
[CH]
7
▶814 MB
B (0.136)
CH2(CH2)2CH3
CH3(CH2)2H2C
CH2(CH2)2CH3
279
▶814 MS
B
(CH3)2CHCHCH(CH3)2
CH3
135, 284
▶814
E
Y(NO3)3
45
▶813 S
CH3
N
378
▶813 M
B (0.169)
(CH3)2CHCH2CH2CH(CH3)2
137
▶813 M
B
CH3CH=CHCH=CHCH3
354
▶813
P
Ce(NO3)3
45
▶813 S
B (0.169)
(CH3)2CHCH2CHCH2CH3
CH3
138
▶812 VVS
B (0.064)
H3C CH3
C=C
H CH2CH3
(cis)
164, 274
▶812 VS
CS2 [25%]
CH3
H2C=CCO2CH3
193
▶812 VS
B (0.169)
(CH3)2CHCH2CH(CH3)2
141, 151
▶812 W
CHCl3
O
C-O-C2H5
H3C
O
N
15
▶812
E
Ce(NO3)3
45

▶812
E

$Gd(NO_3)_3$

45

▶812
P

$Gd(NO_3)_3$

45

▶812
P

$Sm(NO_3)_3$

45

▶812

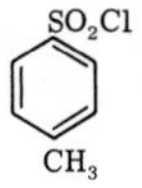
SO2Cl
CH3

389

▶811 M
B

$(CH_3)_2CHCH_2CH(CH_3)_2$

335

▶810.4 VS

$C_6H_5HSiCl_2$

86

▶810 MW
B (0.0576)

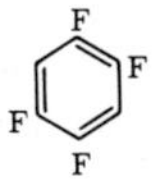
F
F
F
F

222

▶810 MW
B

$H_2C{=}CH(CH_2)_3CH(CH_3)_2$

97

▶810 M
B

128

▶809 VS
B (0.0104)

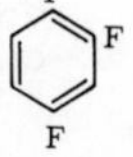
F
F
F

224

▶809 VS
B (0.169)

$(CH_3)_2CHCH_2CH(CH_3)_2$

141

▶809 MS
$CHCl_3$

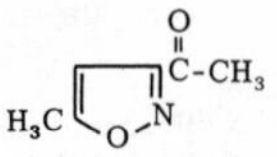
O
C-CH3
N
H3C
O

[CH ?]

15

▶809
E

$Pr(NO_3)_3$

45

▶808 VS
B (capillary)

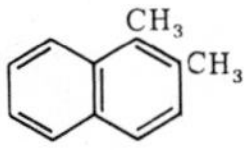
CH3
CH3

188

▶808 W
B (0.136)

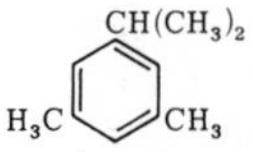
CH(CH3)2
H3C
CH3

277

▶808
E

$Hg_2(NO_3)_2$

45

▶808
P

$Th(NO_3)_4$

45

▶807 S

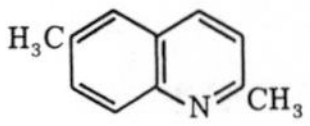
H3C
N
CH3

[H def. in benzene]

378

▶807 MW
B (0.0576)

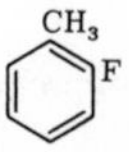
CH3
F

221

▶807 M
B

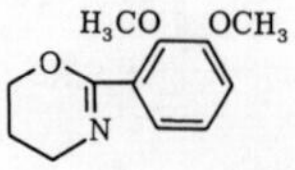
H3CO
OCH3
O
N

[C-O-C]

26

▶807

CH_2Br_2

[CH_2] 646

▶805 VS
B (0.0104)

$n\text{-}H_9C_4\text{-}O\text{-}CF_2CHFCl$

264

▶806 M
B (0.169)

$(CH_3)_2CHCHCH(CH_3)_2$
CH_2
CH_3

301

▶805 SB
B (0.068)

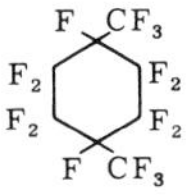

$H_3CH_2C\ CH_2CH_3$
C=C
H H
(cis)

275

▶806 W
B (0.2398)

F CF₃ / F₂ F₂ / F₂ F₂ / F CF₃ (perfluoro-1,4-dimethylcyclohexane ring)

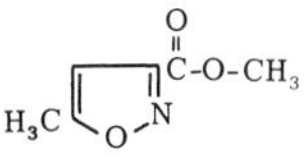

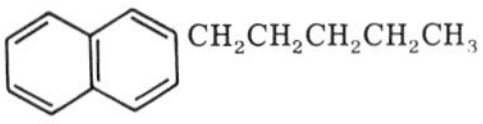

230

▶805 S
CS_2

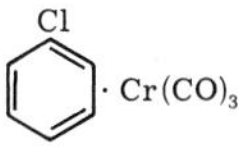

Cl
· $Cr(CO)_3$

17

▶806 VSSh
B (0.170)

Br S $COCH_3$

348

▶805 S
CS_2

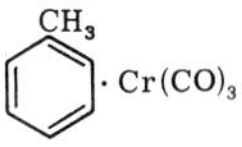

CH_3
· $Cr(CO)_3$

17

▶806 S
B

F
F

225

▶805 M
B

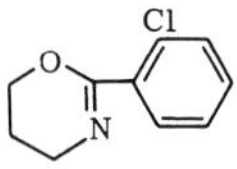

Cl
O N

[C-O-C] 26

▶806 S
$CHCl_3$

O
C-O-CH_3
H_3C O N

15

▶805
E

$Pb(NO_3)_2$

45

▶806
E

$Hg(NO_3)_2$

45

▶805
P

$Pr(NO_3)_3$

45

▶805 VVS
B (0.10)

$CH_2CH_2CH_2CH_2CH_3$

173

▶804 M
E

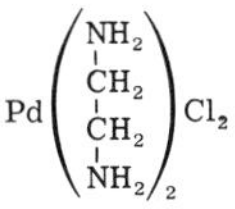

$Pd(NH_2\text{-}CH_2\text{-}CH_2\text{-}NH_2)_2Cl_2$

[NH_2] 8

▶805 VS
B (0.0104)

$CH_3CH_2\text{-}O\text{-}CF_2CHFCl$

261

▶804
P

$Ce(NH_4)_2(NO_3)_6$

45

▶805 S
B (0.169)

$(CH_3)_2CHCH_2CHCH_2CH_3$
CH_3

138

▶803 S
B (0.04)

N

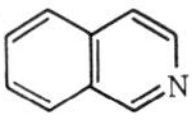

197

▶ 803 MW B (0.0576) F_9C_4-O-C_4F_9 266	▶ 802 E or N · HCl 41
▶ 803 VS B (0.0104) H_3C-O-CF_2CHFCl 356	▶ 801 MB CS_2 [25%] $CH_3CHCH_2CH_3$ OH 319
▶ 803 S $CHCl_3$ H O N-O-C-CH_3 O N 15	▶ 801 S C (0.003) $(CH_3)_3CNO_2$ 325
▶ 803 MS B (0.064) H_2C=$C(CH_2CH_3)_2$ 163	▶ 801 S E Pt (NH_2 CH_2 CH_2 NH_2)$_2$ $PtCl_4$ [NH_2] 8
▶ 803 E $Pr(NO_3)_3$ 45	▶ 801 S $CHCl_3$ O H_3C-O-C H_3C O N 15
▶ 803 E or N $(CH_2)_3$ N 41	▶ 801 MS B (0.169) CH_3 $(CH_3)_2CHCHCH_2CH_3$ 142
▶ 802 VSB C CH_3 CH_3 182	▶ 800 SB A (400 mm Hg, 15 cm) CH_3 H_2C=CCH_2CH_3 92
▶ 802 VSB CS_2 [25%] CH_3CH_2OH 315	▶ 800 VS B (0.114) H_3C CH_3 CH_3 125
▶ 802 S E O C-NH CH_3 [C_{ar}-H] 2	▶ 800 SB B (0.136) $CH_2(CH_2)_2CH_3$ $CH_3(CH_2)_2H_2C$ $CH_2(CH_2)_2CH_3$ 279
▶ 802 M $CHCl_3$ CH_3 H_3C-O-C O N O 15	▶ 800 S $CHBr_3$ O C-NH_2 29

▶800 M
B (0.08)

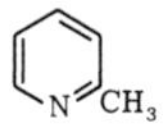

198

▶800
E

$AgNO_3$

45

▶800
P

$Pb(NO_3)_2$

45

▶800 M
B (0.08)

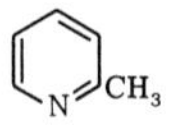

198

▶799.4 VS

$C_6H_5(CH_3)SiCl_2$

86

▶799 VS
B (0.025)

200

▶799 S (Q branch)
A (213.5 mm Hg, 43 cm)

$(CH_3)_3CH$

373

▶799 SSh
B (0.169)

$CH_3CH_2CH(CH_3)-CH(CH_3)CH_2CH_3$

83

▶799 S

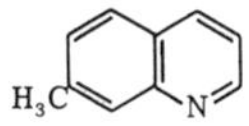

[H def. in pyridine]

378

▶799 VW
B (0.003)

$(CH_3)_2CHCH_2NO_2$

326

▶798 VW
A (100 mm Hg, 10 cm)

CH_3-CCl_3

251

▶798 M
B (0.169)

$CH_3CH=CHCH_2CH_3$
(trans)

207

▶797 VS
B (0.0104)

$(C_4F_9)_3N$

346

▶797 VS
CS_2 [40%] (0.025)

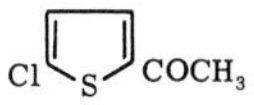

372

▶797
E

$Hg_2(NO_3)_2$

45

▶797 W
B (film)

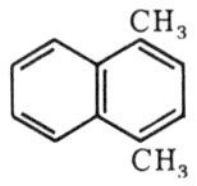

186

▶797 S
$CHBr_3$

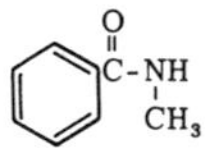

2

▶797 MVB
B (0.003)

$C_3H_7NO_2$

392

▶796 M
CS_2 [25%]

$CH_3CH(OH)CH_2CH_3$

319

▶796 VS
B (0.170)

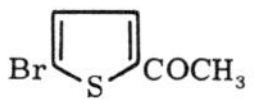

348

▶796 S
B 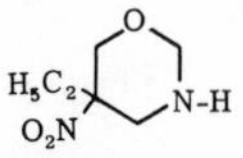
H_5C_2 O_2N O N-H
26

▶796 VS
B (0.003)
NO_2
$CH_3CH_2CHCH_3$
324

▶796 M
CS_2 [20%]
CH_3
$CH_3CH_2CHCH_2OH$
321

▶795 VS
B (0.10) 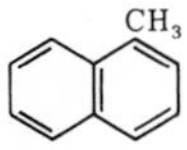
CH_3
192, see also 42

▶795 S
B 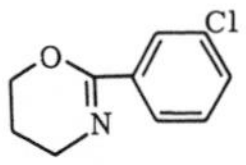
Cl O N
26

▶795 S
CH_3CN 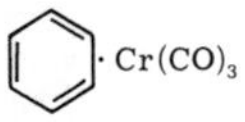
$\cdot$ Cr(CO)$_3$
17

▶795 VW
B (0.003)
$CH_3(CH_2)_2CH_2NO_2$
310

▶795
E or N 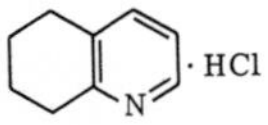
N $\cdot$ HCl
41

▶795
O
$(CH_3)_2CH-\overset{\|}{C}-H$
394

▶794.2 VS
$C_6H_5HSiCl_2$
86

▶794 VS
A (385 mm Hg, 10 cm)
CF_3-CCl_3
253

▶794 VS
B (0.114) 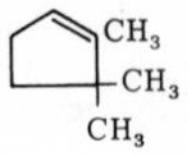
CH_3 CH_3 CH_3
126

▶794 S
E 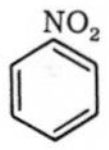
NO_2
[CH out of plane]
20

▶793 VS
B (0.051) 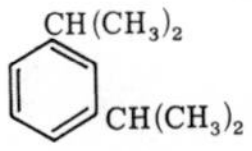
$CH(CH_3)_2$
$CH(CH_3)_2$
297

▶793 VS
B (0.0104) 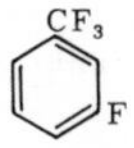
CF_3
F
218

▶793 VS
B (0.10) 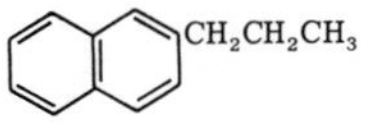
$CH_2CH_2CH_3$
177, see also 42

▶793 M
B 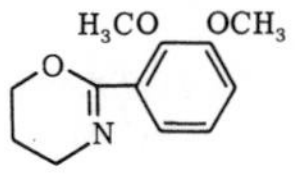
H_3CO OCH_3 O N
26

▶793
E or N
$(CH_2)_3$ N 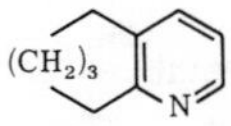
41

▶792 MW
$CHCl_3$ 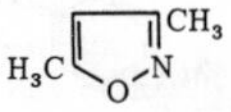
CH_3 H_3C O N
[CH ?]
15

▶791 S
B (0.0563)
$CH_3CH=CHCH_2CH_3$
(cis)
208

▶791 S
E

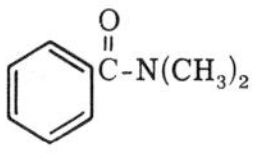
O
C-N(CH3)2

3

▶791 MB
B (0.169)

$(CH_3CH_2)_3CH$

143

▶791
E

$Hg_2(NO_3)_2$

45

▶790 VS
B (0.10)

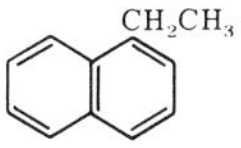
CH2CH3

190, see also 42

▶790 VVS
B (0.055)

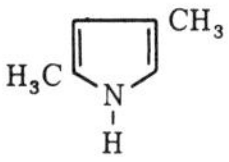
CH3
H3C
N
H

344

▶790 VS
C

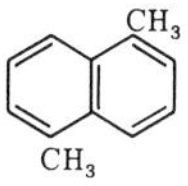
CH3
CH3

185, see also 42

▶790 MB
E

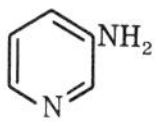
NH2
N

327

▶790
E or N

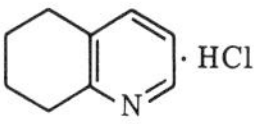
N
· HCl

41

▶789.2 S

$C_6H_5(CH_3)SiCl_2$

86

▶789 S
B (0.169)

$(CH_3)_3C\overset{CH_3}{\underset{CH_3}{C}}-CH_2CH_2CH_3$

304

▶789 S
C

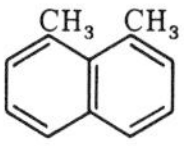
CH3 CH3

182, see also 42

▶789 M
B (0.169)

$(CH_3CH_2)_3CH$

143

▶789 M
B

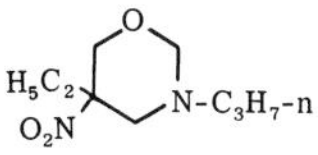
O
H5C2
N-C3H7-n
O2N

26

▶788 S
B (0.068)

H_3C CH_3
C=C
H CH_2CH_3
(cis)

164, 274

▶788 MS
$CHCl_3$

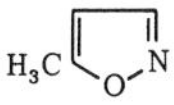
H3C
N
O

15

▶788 M
B (0.0104)

$H_3C-O-CF_2CHFCl$

356

▶788 MB
B (0.068)

H_3CH_2C CH_2CH_3
C=C
H H
(cis)

275

▶788 M
B

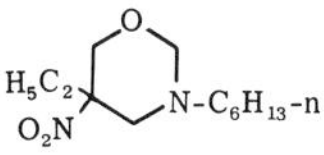
O
H5C2
N-C6H13-n
O2N

26

▶787 VSSh
B (0.0104)

$(C_4F_9)_3N$

346

▶787 S
E

$Pt\left(\begin{matrix}NH_2\\CH_2\\CH_2\\NH_2\end{matrix}\right)_2PtCl_4$

[NH_2] 8

▶787 MB
B (0.169)

$(CH_3CH_2)_3CH$

143

▶787 M
C

CH_3

191, see also 42

▶787

$CH_2ClCH{=}CHCH_2Cl$
(cis)

396

▶786 VS
B (film)

CH_3
CH_3

188, see also 42

▶786 S
B (0.169)

$(CH_3)_2CHCH(CH_3)CH_2CH_3$

142

▶786 M
B (0.157)

$(CH_3)_3CCH(CH_3)CH(CH_3)_2$

293

▶786 M
CS_2 [25%]

$CH_3CH_2C(OH)(CH_3)_2$

194

▶786 M
B (0.055)

H_3C, CH_2CH_3, CH_3, N, H

343

▶786
E or N

N

41

▶785 S
A

$CH_3CH_2CH(CH_3)CH_2CH_3$

332

▶785 S
B

$CH_3CH_2C(CH_3)_2-CH_2CH_3$

336

▶785 S
B, C, CS_2

CCl_4

395

▶785 MS
B (0.157)

$(CH_3)_3CCH(CH_2CH_3)_2$

78, 300

▶785 M
B (0.10)

$(CH_3)_2CHCH(CH_3)CH_2CH_3$

153

▶785 M
B (0.064)

$H(H_3C)C{=}C(CH_2CH_3)CH_3$
(trans)

165

▶785
P

$Pr(NO_3)_3$

45

▶784 S
B (0.10)

$CH_3CH_2SCH_3$

209

▶784 S
B (0.105)

$CH_3CH_2C(CH_3)_2-CH_2CH_3$

149

▶784 S
B (0.169)

$CH_3CH_2C(CH_3)_2-CH_2CH_2CH_3$

360

▶784 M (P branch)
A (213.5 mm Hg, 43 cm)

$(CH_3)_3CH$

373

▶784 W
B
$(CH_3)_3C(CH_2)_3CH_3$
337

▶783 VSSh
B (0.10)
$CH_2CH_2CH_2CH_2CH_3$ (naphthalene)
173, see also 42

▶783 S
B (0.300)
$CH_3(CH_2)_5CH(C_3H_7)(CH_2)_5CH_3$
281

▶783 S
B
O_2N (structure)
26

▶783 MB
B (0.127)
$CH_3(CH_2)_2CH{=}CH(CH_2)_2CH_3$
(cis)
291

▶783 MB
CS_2 [25%]
$CH_3(CH_2)_4OH$
393

▶782 W
B (0.169)
$(CH_3)_2CHCHCH(CH_3)_2$
CH_2
CH_3
301

▶782 VS
B (0.169)
$(CH_3)_3CCH_2CH_3$
145

▶782 VS
B (0.10)
CH_2CH_3 (naphthalene)
189, see also 42

▶782 VVS
B (0.10)
$CH_2CH_2CH_3$ (naphthalene)
178, see also 42

▶782 M
$CH(C_2H_5)_2$ (pyridine, N)
431

▶782
E or N
N (structure)
41

▶781 VS (R branch)
A (202.3 mm Hg, 43 cm)
$CH_3CH_2CH_3$
397

▶781 VS
B (0.025)
$CH_3CH_2\text{-}S\text{-}CH_2CH_3$
172

▶781 VS
B (0.08)
N (isoquinoline)
197

▶781 VS
B (0.0104)
F, F, F (benzene)
224

▶781 S
B (0.0576)
F, F, F, F (benzene)
222

▶781
E
$Cr(NH_3)_5(NO_3)_3$
45

▶780 VS
B (0.169)
CH_3
$(CH_3)_3CCH_2CHCH_2CH_3$
285

▶780 S
B
CH_3
$(CH_3)_3CCHCH_2CH_3$
340

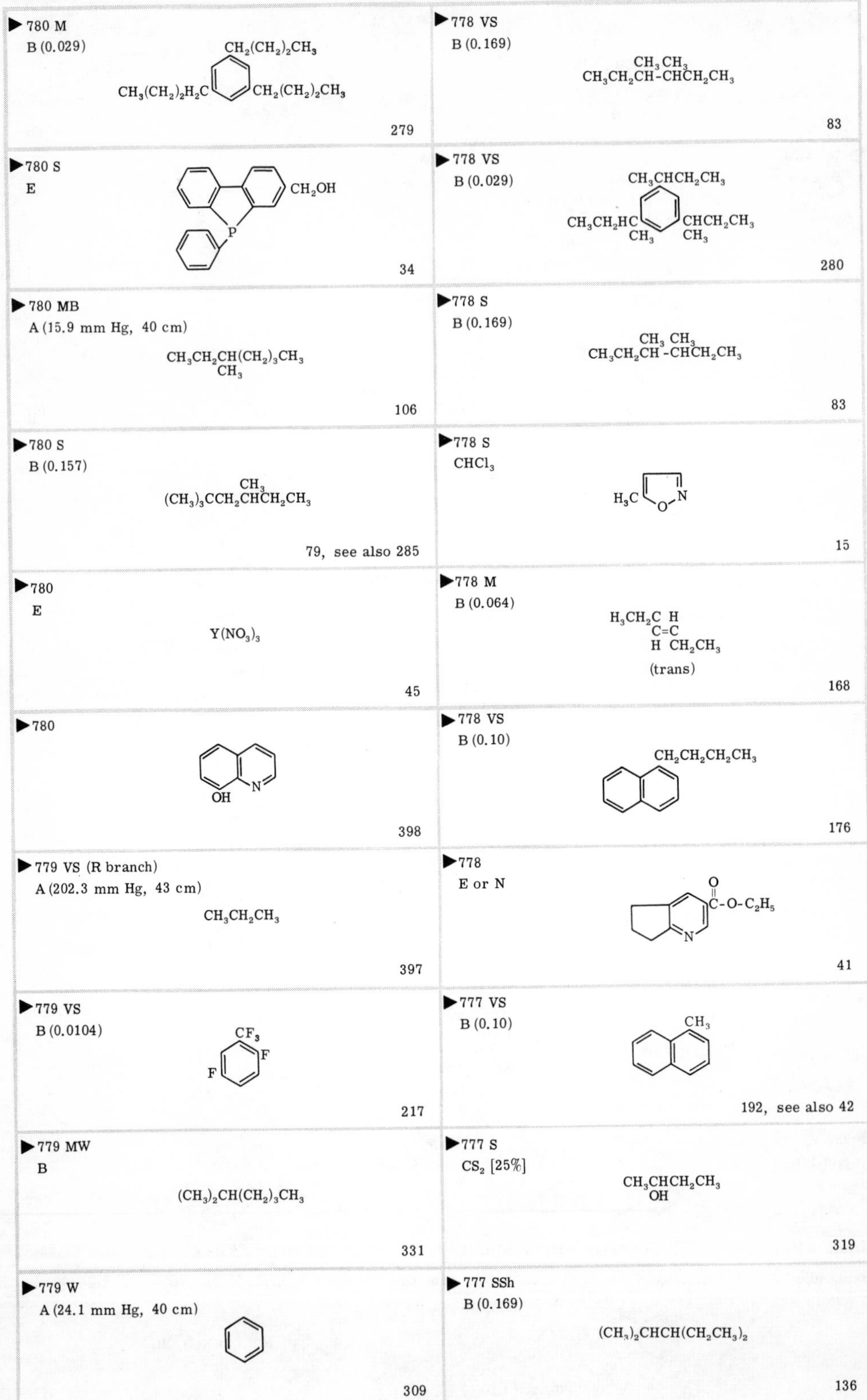

▶780 M B (0.029) $CH_3(CH_2)_2H_2C$–, $CH_2(CH_2)_2CH_3$, $CH_2(CH_2)_2CH_3$ (benzene) 279	▶778 VS B (0.169) $CH_3CH_2CH(CH_3)$-$CH(CH_3)CH_2CH_3$ 83
▶780 S E CH_2OH, P 34	▶778 VS B (0.029) $CH_3CHCH_2CH_3$, $CH_3CH_2HC(CH_3)$, $CHCH_2CH_3$ CH_3 280
▶780 MB A (15.9 mm Hg, 40 cm) $CH_3CH_2CH(CH_3)(CH_2)_3CH_3$ 106	▶778 S B (0.169) $CH_3CH_2CH(CH_3)$-$CH(CH_3)CH_2CH_3$ 83
▶780 S B (0.157) $(CH_3)_3CCH_2CH(CH_3)CH_2CH_3$ 79, see also 285	▶778 S $CHCl_3$ H_3C, O, N 15
▶780 E $Y(NO_3)_3$ 45	▶778 M B (0.064) H_3CH_2C H C=C H CH_2CH_3 (trans) 168
▶780 OH, N 398	▶778 VS B (0.10) $CH_2CH_2CH_2CH_3$ 176
▶779 VS (R branch) A (202.3 mm Hg, 43 cm) $CH_3CH_2CH_3$ 397	▶778 E or N O ‖ C-O-C_2H_5, N 41
▶779 VS B (0.0104) CF_3, F, F 217	▶777 VS B (0.10) CH_3 192, see also 42
▶779 MW B $(CH_3)_2CH(CH_2)_3CH_3$ 331	▶777 S CS_2 [25%] $CH_3CH(OH)CH_2CH_3$ 319
▶779 W A (24.1 mm Hg, 40 cm) (benzene) 309	▶777 SSh B (0.169) $(CH_3)_2CHCH(CH_2CH_3)_2$ 136

▶777 MSh
B (0.300)
$C_{26}H_{54}$
(5, 14-di-n-butyloctadecane)
283

▶777 M
B
$(CH_3)_2CHC(CH_3)_2-CH_2CH_3$
341

▶777 M
E
Grating
$[(CH_3)_2CNO_2]^-Na^+$
[C-C-C]
328

▶777
E or N
[structure: $(CH_2)_3$-fused pyridine, C(=O)-O-C_2H_5]
41

▶776 VS
B (0.169)
$(CH_3CH_2)_2CHCH_2CH_2CH_3$
140

▶776 S
B (0.10)
[structure: naphthalene with H_3C and CH_3]
183, see also 42

▶776 S
$CHCl_3$
[structure: isoxazole with H_5C_2-O-C(=O) and H_3C]
[CH ?]
15

▶776 S
$CHCl_3$
[structure: isoxazole with H_3C and C(=O)-O-C_2H_5]
15

▶776 WSh
B (0.169)
$(CH_3)_2CHCH_2CH_2CH(CH_3)_2$
137

▶776 MB
B (0.127)
$CH_3CH_2CH{=}CH(CH_2)_3CH_3$
99

▶776 M
A (16.6 mm Hg, 40 cm)
$(CH_3)_2CHC(CH_3)_2-CH_2CH_3$
90

▶776
E or N
[structure: $(CH_2)_3$-fused pyridine with CN and OCH_3]
41

▶776
$Sn(CH_3)_4$
399

▶776 VS
A
[structure: naphthalene with $CH_2CH_2CH_2CH_2CH_3$]
174

▶775 VS
B (0.169)
$CH_3CH_2C(CH_3)_2-CH(CH_3)CH_2CH_3$
299

▶775 VS
E
[structure: pyridine with NH_2]
327

▶775 MB
B (0.20)
[structure: cyclopentane with H_3C and CH_3]
(trans)
308

▶774 VS
B (film)
[structure: naphthalene with CH_3 and CH_3]
187, see also 42

▶774 S
B (0.101)
$CH_3(CH_2)_5CH_3$
85

▶774 S
$CHCl_3$
[structure: isoxazole]
15

▶774 S
$CHCl_3$
$n\text{-}H_7C_3\text{-}O\text{-}C(=O)$, H_3C, N, O
[CH ?]
15

▶774 S
$CHCl_3$
$H_3C\text{-}O\text{-}C(=O)$, H_3C, N, O
15

▶774 M
B (0.101)
$CH_3(CH_2)_5CH_3$
85

▶774
E or N
COOH, N
41

▶774
E or N
$(CH_2)_3$, CN, =O, N, H
41

▶773 VS
B (0.169)
$(CH_3)_3CCH_2\overset{CH_3}{CH}CH_2CH_3$
285

▶773 VS
C
CH_3 CH_3
182, see also 42

▶773 SSh
B (0.300)
$(C_2H_5)_2CH(CH_2)_{20}CH_3$
282

▶773 S
B (0.157)
$(CH_3)_3CCH_2\overset{CH_3}{CH}CH_2CH_3$
79, see also 285

▶773 S (Q branch)
A (8.3 mm Hg, 40 cm)
CH_2CH_3
70

▶773
E or N
CN, =O, N, H
41

▶772 VS
B (0.169)
$(CH_3CH_2)_3CH$
143

▶772 VS
B (0.157)
$(CH_3)_3CCH(CH_2CH_3)_2$
78, see also 300

▶772 M
B (0.169)
$(CH_3)_3CCH_2\underset{CH_3}{CH}CH(CH_3)_2$
303

▶772 VS
B (0.169)
$(CH_3)_3CCH(CH_2CH_3)_2$
300

▶772 S
B (0.104)
$CH_3CH_2\underset{CH_3}{CH}(CH_2)_2CH_3$
81

▶772 M
E
Grating
$[(CH_3)_2CNO_2]^-Na^+$
[C-C-C]
328

▶772
P
$Cr(NH_3)_5(NO_3)_3$
45

▶771 S
E
N^+, O^-
24

▶771 M
C
H_3C, CH_3
179, see also 42

▶770 VS
B (0.0104)

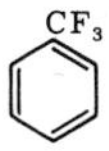

CF_3

219

▶770 S
A

CH_3
$CH_3CH_2CHCH_2CH_3$

332

▶770 MS
B

$CH_3CH_2CH(CH_2)_2CH_3$
CH_3

330

▶770 MB
B (0.127)

CH_3
$H_2C{=}C(CH_2)_4CH_3$

98

▶770 M
E

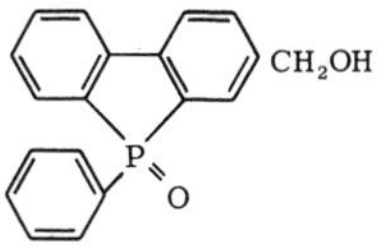

CH_2OH
P
O

34

▶770 M
E

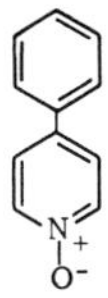

N^+
O^-

24

▶770 Sh
E

$Na_4P_2O_6$

28

▶770

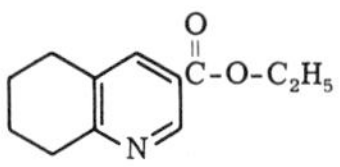

O
$C{-}O{-}C_2H_5$
N

41

▶770
E or N

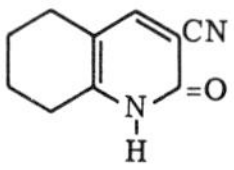

CN
=O
N
H

211

▶769 VS (Q branch)
A (6.3 mm Hg, 40 cm)

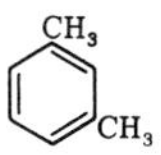

CH_3
CH_3

69

▶769 VVW
B (0.0104)

$F_9C_4{-}O{-}C_4F_9$

266

▶769 S
B (0.169)

$(CH_3)_2CHCH(CH_2CH_3)_2$

136

▶769 S
E

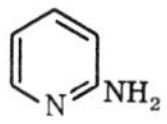

N
NH_2

327

▶769 M
B

$(CH_3)_2CHCH_2CHCH_2CH_3$
CH_3

334

▶769

$CH_2ClCH{=}CHCH_3$

400

▶768 S
C

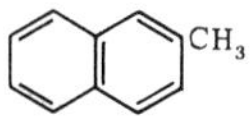

CH_3

191, see also 42

▶768 S
$CHCl_3$

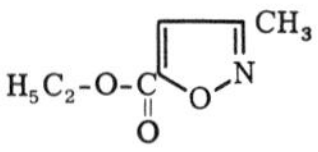

CH_3
$H_5C_2{-}O{-}C$
O
N
O

15

▶768 VW
A (200 mm Hg, 10 cm)

$CF_2Cl{-}CF_2Cl$

257

▶767 VW
A (200 mm Hg, 10 cm)

$CH_3{-}CCl_3$

251

▶767
E or N

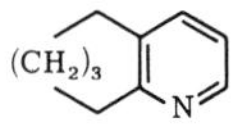

$(CH_2)_3$
N

41

▶767 VS
B (0.169)

$(CH_3)_2CHCH_2CHCH_2CH_3$
CH_3

138

▶767 M
B (0.151)

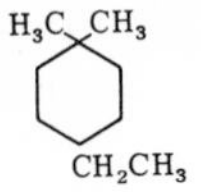

215

▶767 M
B (0.105)

CH_3
$CH_3CH_2C - CH_2CH_3$
CH_3

149

▶767

$Pb(CH_3)_4$

402

▶766 VS
B (0.169)

$(CH_3)_2CHCH_2CH_3$

401

▶766 S
A

$(CH_3)_2CHCH_2CH_3$

333

▶766 S
CS_2 [20%]

CH_3
$CH_3CH_2CHCH_2OH$

321

▶766 S
E

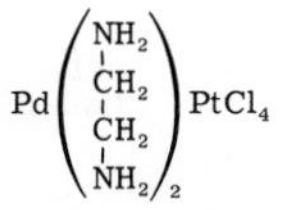

[NH_2]

8

▶766 M
B

CH_3
$CH_3CH_2C - CH_2CH_3$
CH_3

336

▶766 MVB
CS_2 [25%]

$CH_3CH_2CHCH_2CH_3$
OH

196

▶765.4 S
B

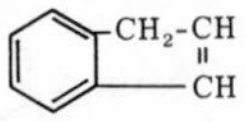

9

▶765 VS
B (0.169)

$(CH_3CH_2)_3CH$

143

▶765 VVW
B (0.0576)

218

▶765 S
B (0.101)

$(CH_3CH_2)_3CH$

156

▶765 S
B (0.13)

$H_2C=CCH_2C(CH_3)_3$
CH_3

96

▶765 W
A (12.6 mm Hg, 40 cm)

$CH_3(CH_2)_6CH_3$

89

▶764 MSh
B (0.025)

$CH_3CH_2\text{-}S\text{-}CH_2CH_3$

172

▶764 VSSh
B

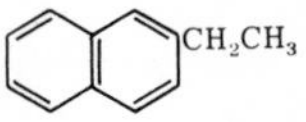

189, see also 42

▶764 S
B (0.10)

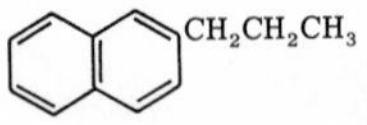

177, see also 42

▶764 S
B

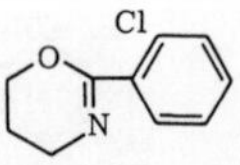

26

▶764 M
B

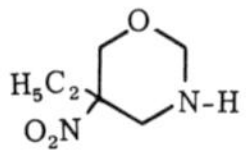

26

▶764
E or N

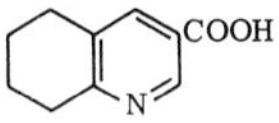

41

▶763 VVSB
B (0.065)

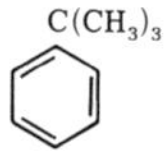

133

▶763 VS
B (0.114)

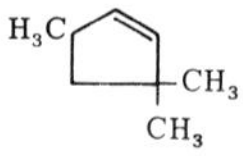

125

▶763 MW
B (0.169)

$$\begin{array}{c}\quad CH_3\ CH_3 \\ CH_3CH_2CH\text{-}CHCH_2CH_3\end{array}$$

83

▶762 S
CS_2

CCl_4

395

▶762 VSB
B (0.065)

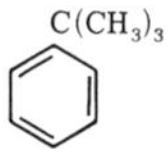

133

▶762 M
B (0.064)

$$\begin{array}{c}H_2C{=}CHCHCH_2CH_3 \\ CH_3\end{array}$$

167

▶762 M
B

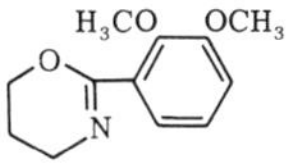

26

▶762

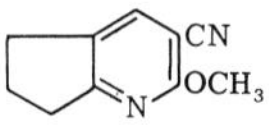

41

▶762
E or N

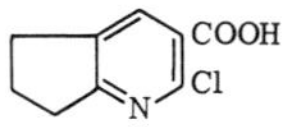

41

▶761 VS
B (0.114)

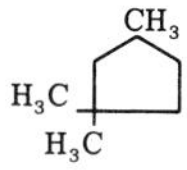

127

▶761 S
A

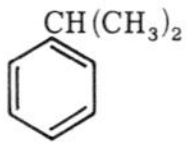

66

▶761 S
A

CF_3Br

11

▶760 W
B (0.0104)

n-H_7C_3-O-CF_2CHFCl

263

▶760 M
B (0.127)

$CH_3(CH_2)_2CH{=}CH(CH_2)_2CH_3$
(cis and trans)

205 (trans), 291 (cis)

▶760 S
B (0.003)

$CH_3(CH_2)_2CH_2NO_2$

310

▶760 S
E

$Na_2S_2O_6 \cdot 2H_2O$

28

▶760 M
B (0.065)

$H_2C{=}CHCH_2C(CH_3)_3$

273

▶760 M
B (0.157)

$(CH_3)_3C(CH_2)_2CH(CH_3)_2$

292

▶760

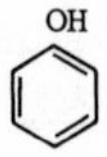

OH

36

▶759 WSp
A (100 mm Hg, 10 cm)

$CH_3\text{-}CCl_3$

251

▶759 VSB
B (0.169)

$CH_3(CH_2)_2CH{=}CH(CH_2)_2CH_3$
(trans)

205

▶759 S
B, C

CCl_4

395

▶759 M
E

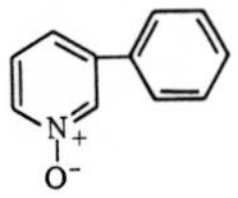

N^+
O^-

24

▶758 VS
B (0.10)

$CH_3CH_2SCH_3$

209

▶758 S
B (0.169)

$(CH_3)_2CHCHCH(CH_3)_2$
CH_2
CH_3

301

▶758 S
CS_2 (0.063)

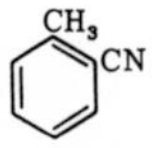

CH_3
CN

312

▶758 M
B (0.003)

$(CH_3)_2CHCH_2NO_2$

326

▶758 M
B

CH_3
$(CH_3)_2CHCHCH_2CH_2CH_3$

338

▶758
E

$Ni(NO_3)_2$

45

▶757.6 S

$C_6H_5(CH_3)SiCl_2$

86

▶757 VVS
B (0.10)

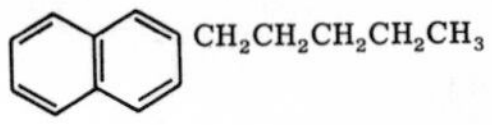

$CH_2CH_2CH_2CH_2CH_3$

173, see also 42

▶757 M
B

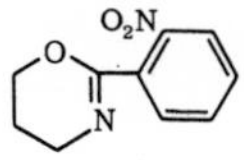

O_2N
O
N

26

▶757
E or N

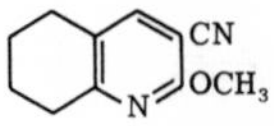

CN
OCH_3
N

41

▶756 VS
E

Pt $\left(\begin{array}{c}NH_2\\ CH_2\\ CH_2\\ NH_2\end{array}\right)$ Cl_2

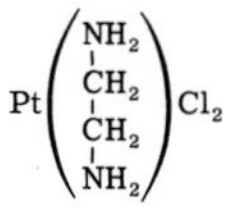

$[NH_2]$

8

▶756 MVB
CS_2 [25%]

$CH_3CH_2CH_2OH$

314

▶756 M
B (0.169)

CH_3
$(CH_3)_2CHCHCH_2CH_2CH_3$

139

▶755 MB
B (0.169)

$(CH_3)_3CCH_2CH_2C(CH_3)_3$

305

▶755 VVS
B (0.02)

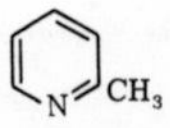

N
CH_3

198

▶755 VS
B (film) 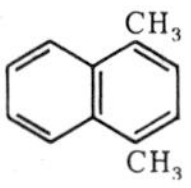
CH_3
CH_3
186, see also 42

▶755 W
B (0.169)
$(CH_3CH_2)_2CHCH_2CH_2CH_3$
140

▶755 M
B
$(CH_3)_2CHCH_2CH_2CH(CH_3)_2$
339

▶755 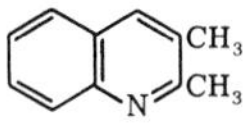
CH_3
CH_3
N
378

▶754 VS
B (0.051) 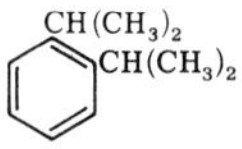
$CH(CH_3)_2$
$CH(CH_3)_2$
296

▶754 VS
B (0.728)
$(CH_3)_2CHCHCH(CH_3)_2$
CH_3
135

▶754 VS
B 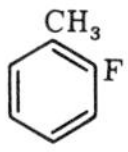
CH_3
F
221

▶753 VS
B (0.169)
$(CH_3)_2CHCH_2CH_2CH(CH_3)_2$
137

▶753 S 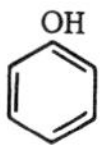
OH
36

▶753
K
$Mn(NO_3)_2 \cdot xH_2O$
45

▶752 S
B
$PSCl_3$
403

▶752 VS
B (0.10) 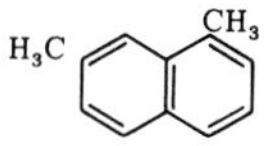
H_3C
CH_3
183, see also 42

▶752 VSB
B (0.10) 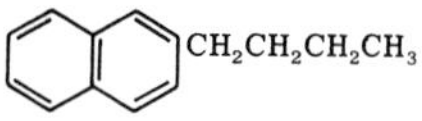
$CH_2CH_2CH_2CH_3$
175, see also 42

▶752 S
B (0.169)
$(CH_3CH_2)_3CH$
143

▶752 S
G 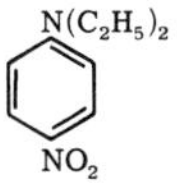
$N(C_2H_5)_2$
NO_2
[NO_2 deformation]
21

▶752 M
A (8.3 mm Hg, 40 cm) 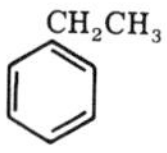
CH_2CH_3
70

▶752
E
$Pr(NO_3)_3$
45

▶752
P
$La(NO_3)_3$
45

▶752
E
$Y(NO_3)_3$
45

▶751 VSSh
B (0.0104)
$(C_4F_9)_3N$
346

▶751 S
CS_2 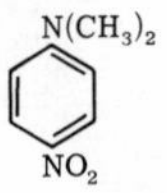
$N(CH_3)_2$ / NO_2 (benzene ring)
[NO_2 deformation] 21

▶751 S
CS_2 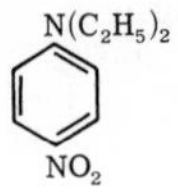
$N(C_2H_5)_2$ / NO_2 (benzene ring)
[NO_2 deformation] 21

▶751 M
B (0.169)
$(CH_3)_2CHCH(CH_3)CH_2CH_3$
142

▶751 WB
B (0.0576)
$H_3C\text{-}O\text{-}CF_2CHFCl$
356

▶750 VVS
B (0.10) 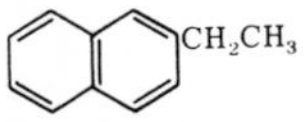
CH_2CH_3 (naphthalene)
189, see also 42

▶750 VS
B (0.10) 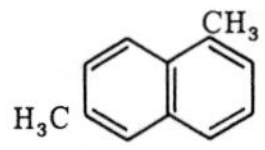
CH_3, H_3C (naphthalene)
184, see also 42

▶750 S
B 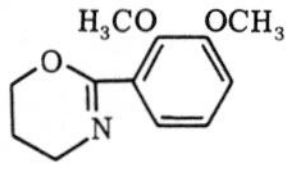
H_3CO OCH_3 (O, N ring)
26

▶750 SShB
B (0.163)
$H_2C{=}CH(CH_2)_8CH_3$
352

▶750 S
E 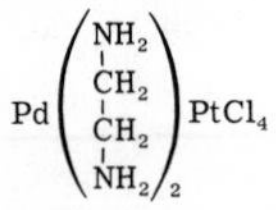
$Pd\left(\begin{matrix}NH_2\\CH_2\\CH_2\\NH_2\end{matrix}\right)_2PtCl_4$
[NH_2] 8, see also 43

▶750 S
E 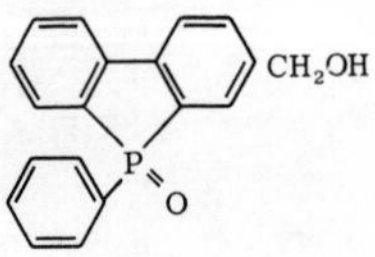
CH_2OH, P, O
34

▶750 S
CS_2 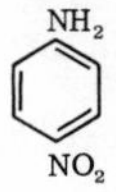
NH_2 / NO_2 (benzene ring)
[NO_2 deformation] 21

▶750
P
$Y(NO_3)_3$
45

▶750 VS
B (0.034) 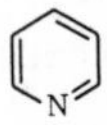
N (pyridine)
405

▶750
NO_2^+
404

▶749 M
B (0.0104) 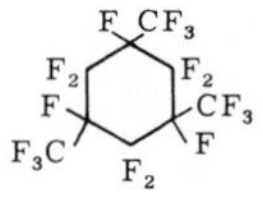
F CF_3, F_2, F, F_2, CF_3, F_3C, F, F_2
228

▶749 W
B (0.169)
$(CH_3CH_2)_2CHCH_2CH_2CH_3$
140

▶749 S
B (0.10) 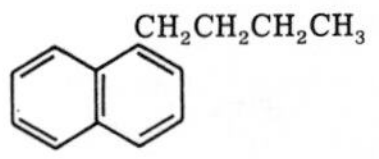
$CH_2CH_2CH_2CH_3$ (naphthalene)
176, see also 42

▶749 M
A
$CH_3CH_2CH(CH_3)CH_2CH_3$
332

▶749 M
B (0.064)
H CH_2CH_3
C=C
H_3C CH_3
(trans)
165

▶749 M
B (0.20) 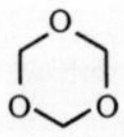
O, O, O (ring)
342

▶749 M
E

NO_2
N^+
O^-

24

▶748 S
B (0.169)

CH_2=CHCH=CHCH=CH_2

203, see also 286

▶748 M
B (0.169)

CH_3
$(CH_3)_3CCHC(CH_3)_3$

307

▶748 M
C

CH_3
CH_3

181, see also 42

▶748
E, P

$Gd(NO_3)_3$

45

▶748
P

$Ca(NO_3)_2$

45

▶747 W
B (0.0104)

F CF_3
F_2 F_2
F_2 F_2
F CF_3

230

▶747 VS
B (0.0104)

F_9C_4-O-C_4F_9

266

▶747 VS
B (film)

CH_3
CH_3

187, see also 42

▶746 VS
B (0.0104)

CF_3
F

218

▶746 VW
B (0.2398)

CF_2Cl-$CFCl_2$

255

▶746
E

$Ce(NO_3)_3$

45

▶745 VS
B (0.04)

N

[CH]

197

▶745 VS
B (0.0104)

CF_3
F
F

217

▶745 VS
B (0.136)

$CH_2(CH_2)_2CH_3$
$CH_3(CH_2)_2H_2C$ $CH_2(CH_2)_2CH_3$

279

▶745 VS
B (0.114)

CH_3
H_3C
H_3C

127

▶745
P

$Th(NO_3)_4$

45

▶745
P

$Sm(NO_3)_3$

45

▶745 S
A

CH_3

[CH out of plane of 4H ring]

452

▶745 VS
B (0.064)

$(CH_3)_2C$=$CHCH_2CH_3$

166

▶744 VS
B (0.10)
CH$_2$CH$_2$CH$_3$
177, see also 42

▶744 VS
B (0.728)
(CH$_3$)$_3$CCH$_2$CH(CH$_3$)CH$_2$CH$_3$
285

▶744
E
Ni(NO$_3$)$_2$
45

▶743 VS
B (film)
CH$_3$ CH$_3$
188, see also 42

▶743 VSVB
B (0.174)
CH$_3$(CH$_2$)$_5$CH$_3$
362

▶743 VS
B (0.169)
(CH$_3$)$_3$CC(CH$_3$)$_2$ - CH$_2$CH$_2$CH$_3$
304

▶743 S
CS$_2$ [25%]
CH$_3$CH(OH)(CH$_2$)$_2$CH$_3$
318

▶743 S
CH$_3$ N CH$_3$
[H def. in benzene]
378

▶743 MW
B (0.169)
(CH$_3$)$_3$CCH$_2$CH(CH$_3$)CH(CH$_3$)$_2$
303

▶742 S
N CH$_3$
[H def. in benzene]
378

▶742 MS
A (3.2 mm Hg, 40 cm)
CH$_2$CH$_2$CH$_3$
67

▶742 VS
B (0.169)
H$_2$C=C(CH$_3$)CH$_2$CH$_2$CH$_3$
206

▶742 MB
CS$_2$ [25%]
CH$_3$(CH$_2$)$_4$OH
393

▶742
P
La(NO$_3$)$_3$
45

▶742
P
Nd(NO$_3$)$_3$
45

▶742
E or N
COOH N Cl
41

▶742
E or N
(CH$_2$)$_3$ COOH N Cl
41

▶741 VS (Q branch)
A (25 mm Hg, 10 cm)
CF$_3$CF$_2$CF$_2$CF$_2$CF$_3$
237

▶741 S
B (0.112)
(CH$_3$)$_3$CCH$_2$CH$_2$CH$_3$
155

▶741 S
B (0.127)
CH$_3$(CH$_2$)$_2$CH=CH(CH$_2$)$_2$CH$_3$
(cis)
291

▶741 S
B (0.169)
$CH_3CH_2C(CH_3)_2-CH_2CH_2CH_3$
360

▶741 M
CS_2 [sat.] (0.170) 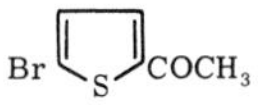
Br-(thiophene)-$COCH_3$
348

▶741
E
$Hg_2(NO_3)_2$
45

▶740 VS
B (0.169)
$CH_3(CH_2)_2CH{=}CH(CH_2)_2CH_3$
(trans)
205

▶740 VS
B (0.169)
$(CH_3)_2CHCH(CH_3)CH_2CH_2CH_3$
139, see also 72

▶740 SVB
B (0.0104)
$n\text{-}H_9C_4\text{-}O\text{-}CF_2CHFCl$
264

▶740 S (R branch)
A (200 mm Hg, 10 cm)
$CF_2Cl\text{-}CF_2Cl$
257

▶740 S
B 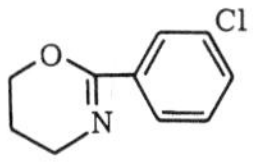
Cl
O
N
26

▶740 S
E 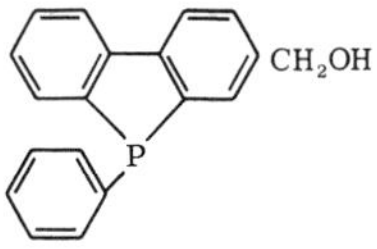
CH_2OH
P
34

▶740 M
E 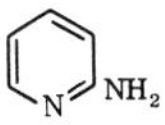
N
NH_2
327

▶740 W
B (0.003)
$CH_3(CH_2)_2CH_2NO_2$
310

▶740
P
$La(NO_3)_3$
45

▶740 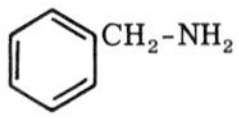
$C_6H_5CH_2\text{-}NH_2$
383

▶740
P
$Ce(NH_4)_2(NO_3)_6$
45

▶739.6 VS
$(C_6H_5)_2SiCl_2$
86

▶739.2 VS
$C_6H_5SiCl_3$
86

▶739 W
B (0.068)
$H_3C(H)C{=}C(CH_3)CH_2CH_3$
(cis)
274

▶739 MSh
B (0.10)
$CH_3(CH_2)_5CH_3$
85

▶739 VS
B (0.169)
$H_2C{=}C(CH_3)CH_2CH_2CH_3$
206

▶738 VSShB
A (400 mm Hg, 15 cm)
$CH_3CH_2CH_2CH_2CH_3$
371

▶738 M
B (0.025)
$CH_3CH_2\text{-}S\text{-}CH_2CH_3$
172

▶738 S
CS_2 [25%]
$CH_3(CH_2)_3OH$
353

▶738 S
B (0.104)
$CH_3CH_2\underset{CH_3}{CH}(CH_2)_2CH_3$
81

▶738 S
P
$[(CH_3)_2P \cdot BCl_2]_3$
[B-Cl]
30

▶738 M
B (0.169)
$CH_3CH_2\overset{CH_3}{CH}\text{-}\overset{CH_3}{CH}CH_2CH_3$
83

▶738
E
$Sm(NO_3)_3$
45

▶738
E
$Sr(NO_3)_2$
45

▶738
K
$Ca(NO_3)_2$
45

▶737.4 S
CS_2
$(C_6H_5)_3SiCl$
86

▶737 VS
B (0.0104)

225

▶737 M
B (0.0104)
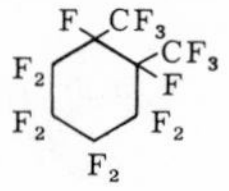
231

▶737 SB
B (0.127)
$CH_3CH_2CH{=}CH(CH_2)_3CH_3$
99

▶737 M
B
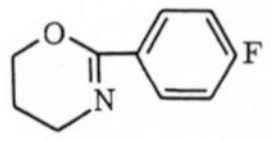
26

▶737 MS
B
$CH_3CH_2\underset{CH_3}{CH}(CH_2)_2CH_3$
330

▶737
E
$Pr(NO_3)_3$
45

▶737 S
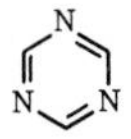
14

▶736 VS (P branch)
A (202.3 mm Hg, 43 cm)
$CH_3CH_2CH_3$
397

▶736 VS
D, E
Grating
$[H_2C \cdot NO_2]^- Na^+$
[$C{=}NO_2$ wag]
328

▶736 VS
B (0.0104)
$(CF_3)_2CFCF_2CF_3$
240

▶736
E, P
$LiNO_3$
45

▶735-743
P
$Pr(NO_3)_3$
45

▶735 VS (Q branch)
A (200 mm Hg, 10 cm)
$CF_2Cl\text{-}CF_2Cl$
257

▶735 M
B (0.0104)
F CF$_3$ F$_2$ F$_2$ F CF$_3$ F$_3$C F F$_2$
288

▶735 VS
B (0.0104)
$F_9C_4\text{-}O\text{-}C_4F_9$
266

▶735 VSSp
E
$Na_4P_2O_7$
28

▶735 S
E
CH$_2$OH P O
34

▶735
E or N
CH$_2$ NNO N 2
41

▶734.2 VS
B (0.10)
$C_6H_5(CH_3)SiCl_2$
86

▶734 VS
B (0.169)
$(CH_3CH_2)_2CHCH_2CH_2CH_3$
140

▶733 VS
A (400 mm Hg, 15 cm)
$CH_3CH_2CH_2CH_2CH_3$
371

▶734 MB
B (0.2398)
$CF_2Cl\text{-}CFCl_2$
255

▶734 VS
B (0.0104)
F CF$_3$ F$_2$ F$_2$ F$_2$ F$_2$ F CF$_3$
230

▶734 VS
D, E
Grating
$[H_2C \cdot NO_2]^{-}Na^{+}$
[C=NO$_2$ wag]
328

▶734 S
B (0.0104)
$CH_3CH_2\text{-}O\text{-}CF_2CHFCl$
261

▶734 VS
B (0.003)
$(CH_3)_2CHCH_2NO_2$
326

▶733 VS
B (0.0104)
F$_2$ F$_2$ F F$_2$ C$_2$F$_5$ F$_2$
247

▶733 VS
B (0.0104)
$CF_3(CF_2)_5CF_3$
241

▶733 S
B (0.10)
CH$_3$
192, see also 42

▶733 S
B
Cl O N
26

▶733 S
E
O C- N(CH$_3$)$_2$
3

▶733 M
B (0.101)
$(CH_3CH_2)_3CH$
143, 156

▶733 M
B (0.169)
$(CH_3)_2CHCH(CH_2CH_3)_2$
136

▶732 VS (Q branch)
A
$CF_3CF_2CF_3$
235

▶732 M
B (0.0104)
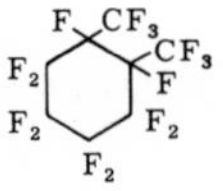
231

▶732 VVS
B (0.0104)
$(C_4F_9)_3N$
346

▶732 VS
B (0.08)
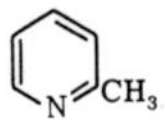
198

▶732 VS
B (0.0104)
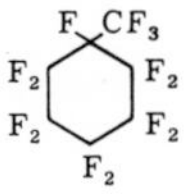
231

▶732 VS
C
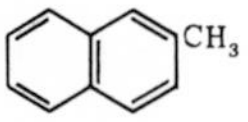
191

▶732 S
B (0.10)
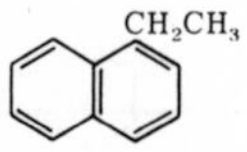
190

▶732 S
B (0.10)
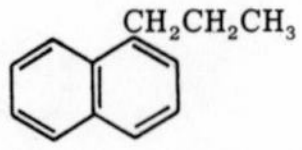
178, see also 42

▶732 S
B (0.10)
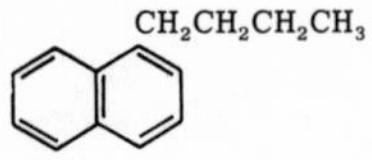
176

▶732 S
B (0.10)
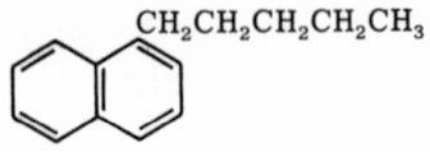
174

▶732 S
B (0.055)
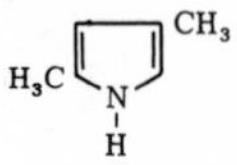
344

▶731 VS (P branch)
A (202.3 mm Hg, 43 cm)
$CH_3CH_2CH_3$
397

▶731 S (P branch)
A (200 mm Hg, 10 cm)
$CF_2Cl\text{-}CF_2Cl$
257

▶731
E
$AgNO_3$
45

▶731
E
$Cr(NH_3)_5(NO_3)_3$
45

▶731
E or N
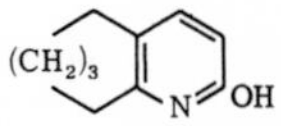
41

▶730.4 VS
$C_6H_5HSiCl_2$
86

▶730.1
B
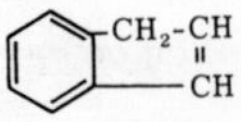
9

▶730 S
B

$(CH_3)_3C(CH_2)_3CH_3$

337

▶730 S
B (0.064)

$CH_3CH{=}CH(CH_2)_4CH_3$
(cis)

160

▶730 M
C (0.003)

$(CH_3)_3CNO_2$

325

▶730 S
E

$S{=}C(NH_2)_2$

22

▶730 MB
A (15.9 mm Hg, 40 cm)

$CH_3CH_2CH(CH_3)(CH_2)_3CH_3$

106

▶730 MSh
B (0.0576)

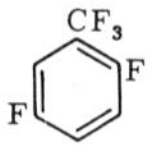
CF3
F
F

217

▶730
E

$Ba(NO_3)_2$

45

▶729 S
B (0.169)

$(CH_3)_3CCH(CH_2CH_3)_2$

300

▶729 VS
B (0.0104)

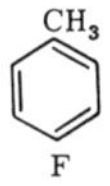
CH3
F

220

▶729 VS
E

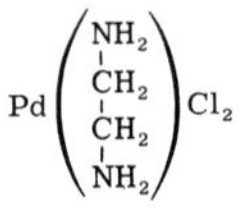
Pd
NH2
CH2
CH2
NH2
Cl2

$[NH_2]$

8, see also 43

▶729 MB
CS_2 [25%]

$CH_3(CH_2)_4OH$

393

▶728 M
B (0.136)

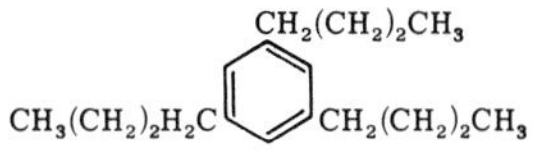
CH2(CH2)2CH3
CH3(CH2)2H2C
CH2(CH2)2CH3

279

▶728 VS
B (0.0104)

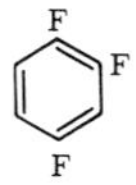
F
F
F

224

▶728 MS
B (0.0576)

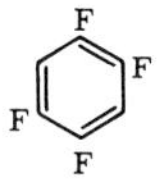
F
F
F
F

222

▶728 VS
B (0.728)

$(CH_3)_2CHCH(CH_3)_2$

144

▶728 S
A

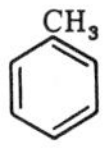
CH3

388

▶728 M
B

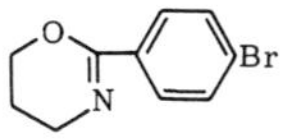
O
N
Br

26

▶728 S
B (0.097)

$(CH_3)_2CH(CH_2)_3CH_3$

80, 331

▶728 M
B (0.157)

$(CH_3)_3CCH(CH_2CH_3)_2$

78, 300

▶728
P

$Ba(NO_3)_2$

45

▶728
E or N

N

41

▶727 VVS
B (0.300)

$C_{26}H_{54}$
(5, 14-di-n-butyloctadecane)

283

▶727 S
A (18 mm Hg, 40 cm)

$(CH_3)_2CH(CH_2)_4CH_3$

294

▶727 S
B (0.127)

$H_2C{=}C(CH_3)(CH_2)_4CH_3$

98

▶727 M
B (0.127)

$CH_3CH{=}CH(CH_2)_4CH_3$
(cis)

100

▶726 VS
B (0.174)

$CH_3(CH_2)_4CH_3$

355

▶726 VS (Q branch)
A (8.2 mm Hg, 10 cm)

$CF_3CF_2CF_2CF_2CF_3$

237

▶726 VS
A (8.1 mm Hg, 10 cm)

$CH_3\text{-}CCl_3$

251

▶726 S
B (0.10)

$CH_3CH_2SCH_3$

209

▶726 S
E

$[H\text{-}S\text{-}C(NH_2)_2]ClO_4$

22

▶725 VS
B (0.300)

$CH_3(CH_2)_5CH(C_3H_7)(CH_2)_5CH_3$

281

▶725 M
B (0.0104)

F, CF_3, CF_3, F, F_2, F_2, F_2, F_2

231

▶725 VS
CS_2 [25%]

$CH_3(CH_2)_5OH$

320

▶725 VSB
B (0.028)

Cl, Cl, S

211

▶725 S
B (0.127)

$H_2C{=}CH(CH_2)_5CH_3$

290

▶725 S
B

HO

379

▶725 M
E

$[H_3C\text{-}S\text{-}C(NH_2)_2]I$

22

▶725
E

$Bi(NO_3)_3$

45

▶725
P

$Zn(NO_3)_2$

45

▶725
E or N

N, OH

41

▶725

N N N N (tetrazine ring)

406

▶724 W
B (film)

CH_3 CH_3

188

▶724 VS
B (0.055)

H_3C CH_2CH_3 CH_3 N H

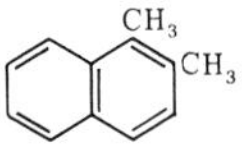

343

▶724 S
CS_2 [25%]

$CH_3(CH_2)_7OH$

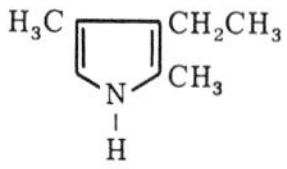

317

▶724 S
B (0.077)

$H_2C{=}CH(CH_2)_6CH_3$

271

▶724 MB
B (0.003)

NO_2
$CH_3CH_2CHCH_3$

324

▶724 S
$CHBr_3$

O C-NH_2

29

▶724
E or N

COOH =O N H

41

▶724
E or N

CH_2 NNO N 2

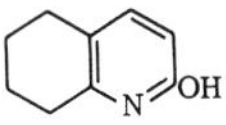

41

▶723 VS
B (0.101)

$CH_3(CH_2)_5CH_3$

85

▶723 M
A (12.6 mm Hg, 40 cm)

$CH_3(CH_2)_6CH_3$

89

▶723
E

$Ce(NO_3)_3$

45

▶723 VS
B (0.077)

$H_2C{=}CH(CH_2)_7CH_3$

270

▶722 VS
B (0.065)

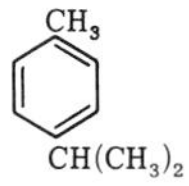

CH_3
$CH(CH_3)_2$

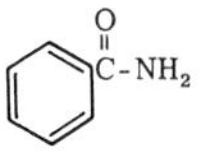

132

▶722 S
B (0.06)

$C_{17}H_{34}$
(1-heptadecene)

159

▶722 VS
B (0.066)

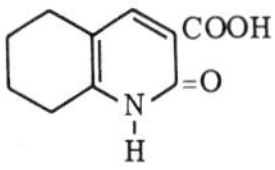

$H_2C{=}CH(CH_2)_8CH_3$

352

▶722 VW
B (0.003)

$(CH_3)_2CHNO_2$

366

▶722
E, P

$Pb(NO_3)_2$

45

▶722
E or N

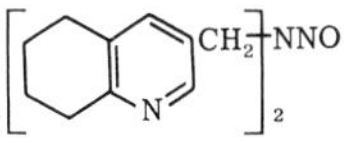

N OH

41

▶722
E or N

N

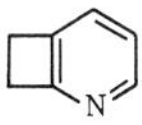

41

▶721 VVS
B (0.300)
$(C_2H_5)_2CH(CH_2)_{20}CH_3$
282

▶721 VS
B (0.0104)
$(C_4F_9)_3N$
346

▶721 M
B (0.169)
CH_3 CH_3
$CH_3CH_2C - CH_2CHCH_2CH_3$
CH_3
302

▶721 S
B 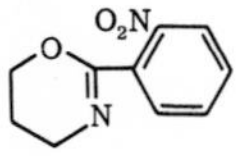
O_2N
O
N
26

▶721
E
$Sm(NO_3)_3$

▶721
E or N 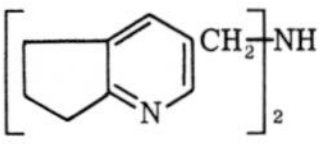
CH_2 NH
N
2
41

▶720 VS
B 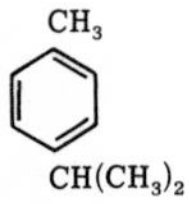
CH_3
$CH(CH_3)_2$
132

▶720 M
B (0.0576) 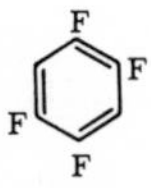
F
F
F
F
222

▶720 VS
B (0.0104)
H_3C-O-CF_2CHFCl
356

▶720 M
B (0.10) 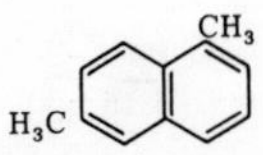
CH_3
H_3C
184

▶720 M
B (0.13)
$CH_3CH{=}CHCH(CH_3)_2$
130

▶720
E or N 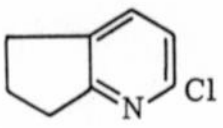
N Cl
41

▶720
E or N 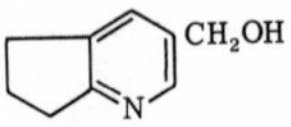
CH_2OH
N
41

▶720
E or N
CH_2-NH_2
N 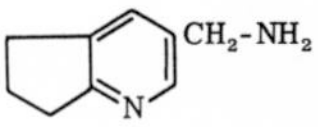
41

▶719 VS (R branch)
A (25 mm Hg, 10 cm)
CF_3-CCl_3
253

▶719 VS
B (0.06) 
128

▶719 S
B 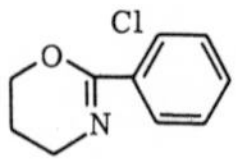
Cl
O
N
26

▶719 VS
B (0.0104)
$CF_3(CF_2)_5CF_3$
241

▶719 M
B (0.169)
CH_3 CH_3
$CH_3CH_2C - CH_2CHCH_2CH_3$
CH_3
302

▶718.3 VS
B
$(C_6H_5)_2SiCl_2$
86

▶718.2
B 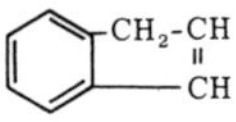
CH_2-CH = CH
9

▶718 MW
B (0.169)
$(CH_3)_2CHCH(CH_3)CH_2CH_2CH_3$
139

▶718 M
B (0.169)
$(CH_3)_2CHCH(CH_2CH_3)_2$
136

▶718 M
E
$[H\text{-}S\text{-}C(NH_2)_2]NO_3$
22

▶718
$PSBr_3$
407

▶717 VS
B (0.479)
$H_2C{=}CC(CH_3)_3$ / CH_3
272

▶717 MSh
B (0.0576) 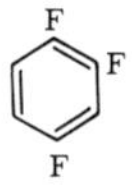
F, F, F
224

▶717 MSh
B (?)
$(C_2H_5\text{-}NH\text{-}B\text{-}N\text{-}C_2H_5)_3$
[Ring]
32

▶717 W
B
$(CH_3)_3CCH(CH_3)CH_2CH_3$
340

▶717
E
$Co(NO_3)_2$
45

▶716
E or N 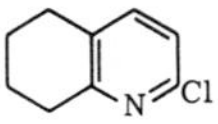
N, Cl
41

▶715.6 VS
$C_6H_5SiCl_3$
86

▶715 VW
B (0.728)
$(CH_3)_2CHC(CH_3)_2\text{-}CH_2CH_2CH_3$
134

▶715
E
$Cr(NH_3)_5(NO_3)_3$
45

▶715
E or N
$(CH_2)_3$, CH_2OH, N 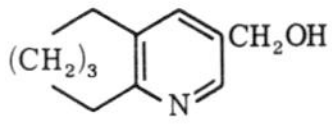
41

▶714 VS (Q branch)
A (25 mm Hg, 10 cm)
$CF_3\text{-}CCl_3$
253

▶714 VSVB
B (0.0197)
H_3CH_2C CH_2CH_3
C=C
H H
(cis)
275

▶714 S
B (?)
$[(CH_3)_2N\text{-}BO]_3$
[Ring]
32

▶714 VS
B (0.029) 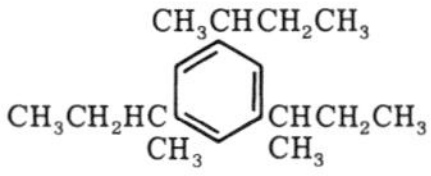
$CH_3CHCH_2CH_3$
$CH_3CH_2HC(CH_3)$ $CHCH_2CH_3$ / CH_3
280

▶713 VS
CS_2 [5%] 
S
199

▶713 WSh
B (0.055)

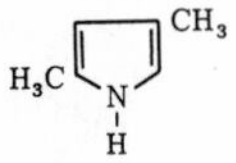

344

▶713
E or N

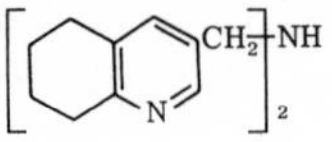

41

▶713
E or N

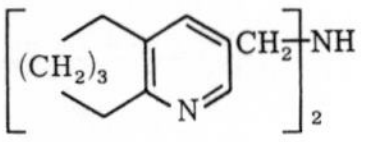

41

▶712.4 VS
CS_2

$(C_6H_5)_3SiCl$

86

▶712 MW
B (0.2398)

$CF_2Cl\text{-}CFCl_2$

255

▶712 VS
B (0.0104)

$F_9C_4\text{-}O\text{-}C_4F_9$

267

▶712 M
B (0.151)

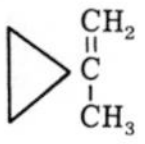

288

▶712
P

$Cr(NH_3)_5(NO_3)_3$

45

▶712
E or N

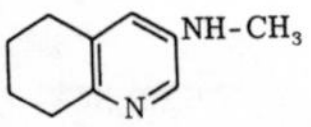

41

▶712
E or N

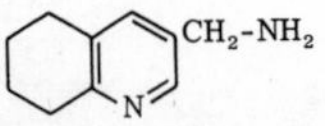

41

▶712

$CHCl_3$

408

▶711 VS
B (0.025), CS_2 (0.063)

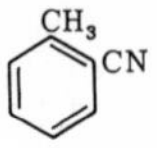

312

▶711 MB
A (200 mm Hg, 10 cm)

$CF_3CF{=}CF_2$

409

▶711 M
B (0.0576)

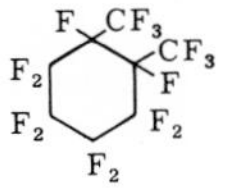

231

▶711 M
B (0.0576)

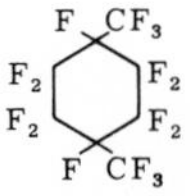

229, 230

▶711 VS
B (0.0104)

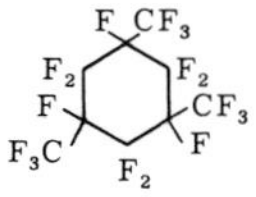

228

▶711 S
B (0.728)

$(CH_3)_3CCH_2CH_3$

145

▶711 M
B

CH_3
$(CH_3)_3CCH_2CHCH_2CH_3$

410

▶711 W
B (0.728)

CH_3
$(CH_3)_2CHC - CH_2CH_2CH_3$
CH_3

134

▶711 S
B

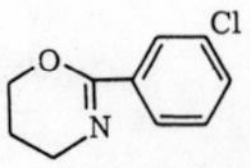

26

▶711 MW
A (775 mm Hg, 10 cm)

$CF_2{=}CFCl$

260

▶711 MW
B (0.10)

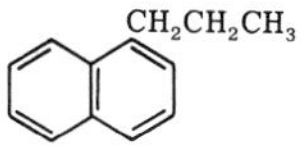
CH2CH2CH3

178, see also 42

▶711 MW
B (0.10)

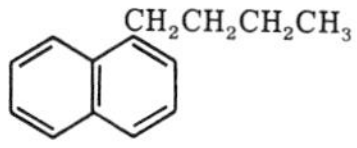
CH2CH2CH2CH3

176

▶711 M
B (0.10)

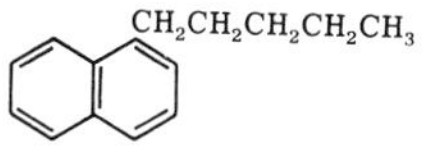
CH2CH2CH2CH2CH3

174

▶710
E or N

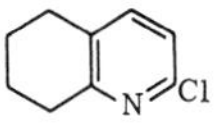
N
Cl

41

▶710 WB
B (0.0104)

$(C_4F_9)_3N$

268

▶710 S
B (?)

$(C_2H_5\text{-}NH\text{-}B\text{-}N\text{-}C_2H_5)_3$

[Ring]

32

▶710 M
A

$(CH_3)_3CCH_2CH_3$

329

▶710 M
G

$Na_4P_2O_6$

28

▶710 M
M

$K_2S_2O_6$

28

▶709
E or N

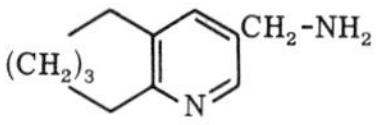
(CH2)3
CH2-NH2
N

41

▶709 VS (P branch)
A (25 mm Hg, 10 cm)

$CF_3\text{-}CCl_3$

253

▶709 Sh
E

$[H\text{-}S\text{-}C(NH_2)_2]Cl$

22

▶709
E or N

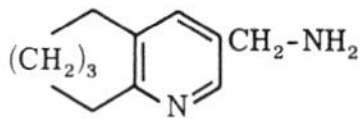
(CH2)3
CH2-NH2
N

41

▶708
E or N

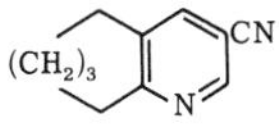
(CH2)3
CN
N

41

▶708 S
B (0.029)

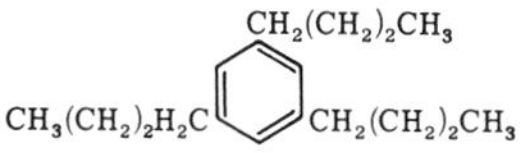
CH2(CH2)2CH3
CH3(CH2)2H2C
CH2(CH2)2CH3

279

▶707 S
G

$Na_2S_2O_6$

28

▶707 VS
B (0.0104)

$CF_3(CF_2)_5CF_3$

240

▶706
E or N

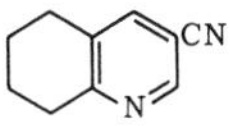
CN
N

41

▶706 MB
A (200 mm Hg, 10 cm)

$CF_3CF{=}CF_2$

409

▶705-720

$C_6H_5COO^-$

[Ring out of plane] 683

▶705 VSSh
E

$Cu\left(\begin{matrix}NH_2\\CH_2\\CH_2\\NH_2\end{matrix}\right)_2PtCl_4$

$[NH_2]$ 8

▶705 S
CS_2

CH_3, CH_3, CH_3

411

▶705 MW
B (0.10)

H_3C, CH_3

183

▶705

CH_3SH

[CS] 412

▶704 S
B (0.063)

CH_3, CN

214

▶704 VS
A (12.5 mm Hg, 10 cm)

F, F, F, F

223

▶704 M
B (0.169)

$CH_3CH_2\overset{CH_3}{\underset{CH_3}{C}}-\underset{CH_3}{C}HCH_2CH_3$

299

▶704 VS
B (0.0576)

CH_3, F

221

▶704 S
E

NO_2

[CC] 20

▶703.2 W

$C_6H_5HSiCl_2$

86

▶703 VS
B (0.051)

$CH(CH_3)_2$, $CH(CH_3)_2$

297

▶703 VS
B (0.029)

$CH(CH_3)_2$, H_3C, CH_3

277

▶703 S
B (0.063)

CH_3, CN

214

▶703 S
B

$[(CH_3)_2N\text{-}BO]_3$

[Ring] 32

▶703 MW
B (0.169)

$CH_3CH{=}CHCH{=}CHCH_3$

354

▶703 M
E

$[H\text{-}S\text{-}C(NH_2)_2]NO_3$

22

▶703 W
B (0.169)

$H_2C{=}\overset{CH_3}{C}CH_2CH_2CH_3$

206

▶702 M
B (0.114)

CH_3, CH_3, CH_3

126

▶702 MWB
E

$[H\text{-}S\text{-}C(NH_2)_2]ClO_4$

22

▶701 VS
B (0.104)

$H_3C\text{-}O\text{-}CF_2CHFCl$

323

▶701 VS
B (0.0576)

$(C_4F_9)_3N$

268

▶701 S
B (0.055)

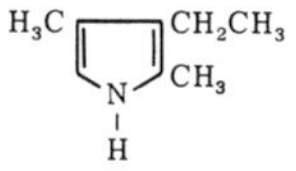
H3C
CH2CH3
CH3
N
H

343

▶701 S
B (0.0104)

$H_3C\text{-}O\text{-}CF_2CHFCl$

356

▶701 MB
B (0.127)

$CH_3CH{=}CH(CH_2)_4CH_3$
(cis)

100

▶700 VS
B (0.0104)

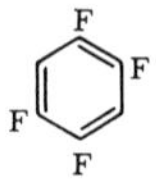
F
F
F
F

222

▶700 M
A

$(CH_3)_3CCH_2CH_3$

94

▶700 M
CS_2

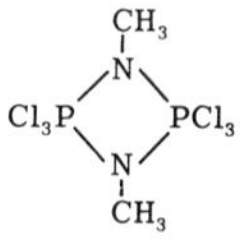
CH3
N
Cl3P
PCl3
N
CH3

384

▶700

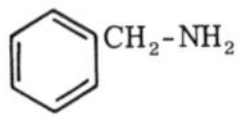
CH2-NH2

383

▶699 S
B

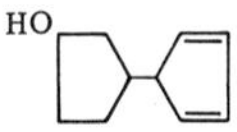
HO

379

▶699
E or N

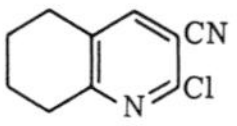
CN
N
Cl

41

▶698 VS
A (4.2 mm Hg, 40 cm)

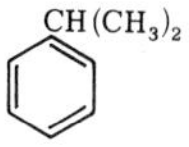
CH(CH3)2

66

▶698 S
B (0.10)

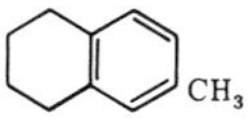
CH3

269

▶698 VS
E

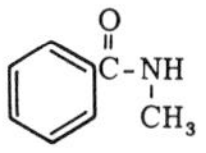
O
C-NH
CH3

3

▶698 MS
A (3.2 mm Hg, 40 cm)

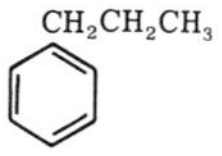
CH2CH2CH3

67

▶698 VS
B (0.0563)

$CH_3CH{=}CHCH_2CH_3$
(cis)

208

▶697 VS
A (12.5 mm Hg, 10 cm)

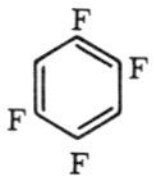
F
F
F
F

223

▶697 VSB
B (0.028)

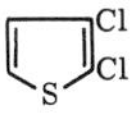
Cl
Cl
S

162

▶697 S
A (8.3 mm Hg, 10 cm)

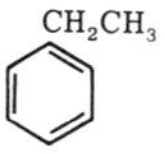
CH2CH3

70

▶696 VS*
E

* A broad band from 716-706 also appears

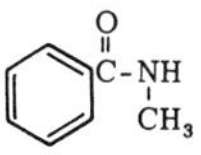
O
C-NH
CH3

$[C_{ar}\text{-}H]$

2

▶696 S
E

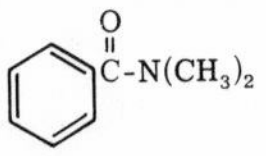

3

▶696 S
G

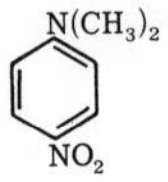

21

▶696 MS
E

$[H\text{-}S\text{-}C(NH_2)_2]_2^+\ SnCl_6^=$

22

▶696 M
B (0.169)

$(CH_3)_3CCH(CH_3)C(CH_3)_3$

307

▶696 M
B (0.10)

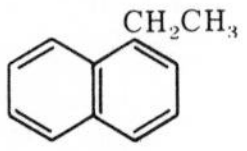

190

▶695 S
B (0.2398)

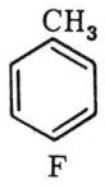

220

▶695 S
G

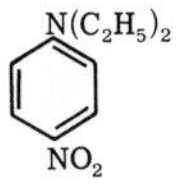

21

▶695 VSB
B (0.065)

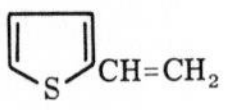

212

▶695

406

▶694 VS
B (0.0197)

$H_3C\ C_3H_7\text{-}n$
$C=C$
$H\ H$
(cis)

276

▶694 VS
B (0.0104)

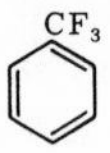

219

▶694 VS
E

$Cu(NH_2CH_2CH_2NH_2)_2PtCl_4$

$[NH_2]$

8

▶694 S
B

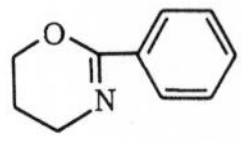

26

▶694
E or N

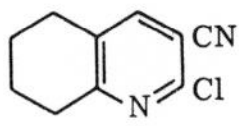

41

▶692.8 M
B

$(C_6H_5)_2SiCl_2$

9

▶692 VSB
B (0.064)

$CH_3CH=CH(CH_2)_4CH_3$
(cis)

160

▶693 VS
B (0.40)

$CH_3CH_2\text{-}S\text{-}CH_2CH_3$

172

▶692.8 M
B

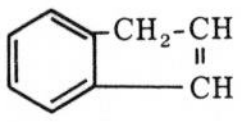

9

▶692 M
A (100 mm Hg, 10 cm)

$CH_3CH=CHCH_2CH_3$
(cis)

287

▶692 VS
B

$C_6H_5HSiCl_2$

86

▶691.1 VS
B
$C_6H_5SiCl_3$
86

▶691 VS
P
$[(C_2H_5)_2P \cdot BCl_2]_3$
[B-Cl]
30

▶691 SSh
A
$HC{\equiv}C{-}\overset{O}{\overset{\|}{C}}{-}H$
[HC≡C bend]
18

▶691 M
B (0.151) 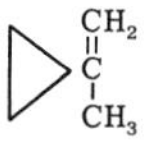
CH_2 = C(–CH_3) (cyclopropyl)
288

▶690 VS (R branch)
A (10 mm Hg, 40 cm) 
309

▶690 S
E
Grating
$[H_2C \cdot NO_2]^- Na^+$
[NO_2 deformation]
328

▶690 S
CS_2 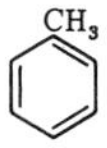
CH_3
413

▶690 MW
E
$[H_3C{-}S{-}C(NH_2)_2]I$
[CS]
22

▶690 S (Q branch)
A (6.3 mm Hg, 40 cm) 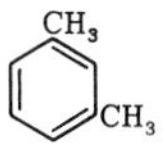
CH_3
CH_3
69

▶690
$CH_3CH{=}CH\overset{O}{\overset{\|}{C}}Cl$
(trans)
672

▶689 SSh
A (100 mm Hg, 10 cm)
$CH_3{-}CCl_3$
251

▶689 MW
B (0.169)
$(CH_3)_2CHC(CH_3)_2{-}CH_2CH_2CH_3$
134

▶689 S
D
Grating
$[H_2C \cdot NO_2]^- Na^+$
[NO_2 deformation]
328

▶689 M
A
$(CH_3)_2SO$
13

▶688 VS
B (0.2398) 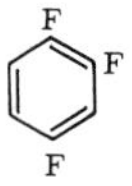
F
F
F
224

▶688 S
B (0.0576) 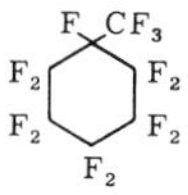
F CF₃
F_2 F_2
F_2 F_2
F_2
232

▶687 S
B (0.0104) 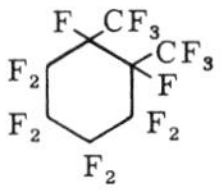
F CF_3
CF_3
F
F_2 F_2
F_2 F_2
F_2
231

▶687 VS
B (0.056)
$CH_2{=}CHCH{=}CHCH{=}CH_2$
286

▶687 VS
B (0.0576) 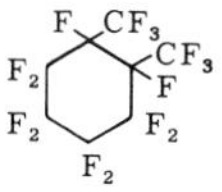
F CF_3
CF_3
F
F_2 F_2
F_2 F_2
F_2
230

▶687 S
CCl_4 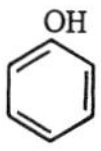
OH
36

▶686 VS
A (100 mm Hg, 15 cm)
CH_3CH_2Cl
88

▶686 S
B (0.0576)
$n\text{-}H_9C_4\text{-}O\text{-}CF_2CHFCl$
264

▶685 VS
B (0.0104) 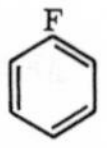
F
226

▶685 S
B (0.0104) 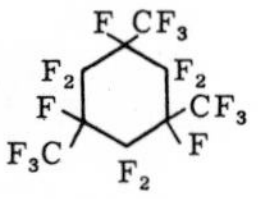
F CF₃ / F₂ / F₂ / F / CF₃ / F₃C / F / F₂
228

▶685 VS
E 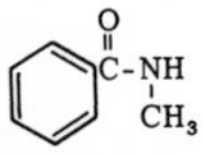
O / C-NH / CH₃
[C_{ar}-H]
2

▶685 VS
N 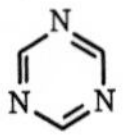
N / N / N
14

▶685 M
B (0.10) 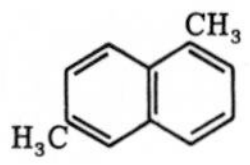
CH_3 / H_3C
184

▶685 VS
B (film) 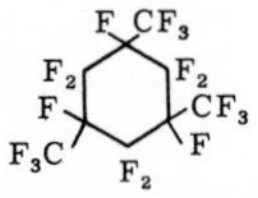
F CF₃ / F₂ / F₂ / F / CF₃ / F₃C / F / F₂
417

▶684 MB
A (200 mm Hg, 10 cm)
$CF_3CF=CF_2$
409

▶682.2 W
CS_2
$(C_6H_5)_3SiCl$
86

▶682 S
CS_2 [40%] (0.025) 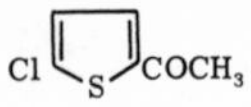
Cl / S / $COCH_3$
372

▶682 MB
CS_2 [25%]
$CH_3(CH_2)_4OH$
393

▶680 S
B 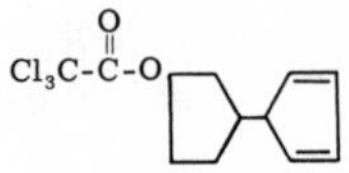
O / $Cl_3C\text{-}C\text{-}O$
379

▶680 S
CS_2 [5%] 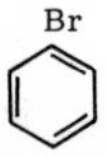
Br
414

▶680 M
E
Grating
$[(CH_3)_2CNO_2]^- Na^+$
[$C=NO_2$ wag]
328

▶679 VS
B (0.028) 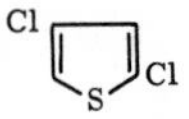
Cl / S / Cl
211

▶679 M
B (0.10)
$CH_3CH_2SCH_3$
209

▶679 
N / N
415

▶678 VS
A (100 mm Hg, 10 cm)
$CF_2Cl\text{-}CF_2Cl$
257

▶678 MW
B (0.169)
$(CH_3)_3CC(CH_3)_2\text{-}CH_2CH_2CH_3$
304

▶677.4 W

$C_6H_5HSiCl_2$

86

▶675

(structure: 1,3,5-triazine)

14

▶677 VS
B

(structure: 1,2,4,5-tetrafluorobenzene)

222

▶674 VS
B

(structure: perfluoro(ethylcyclopentane))

247

▶677 VS
CS_2 [sat.] (0.170)

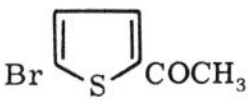

(structure: 2-acetyl-5-bromothiophene)

348

▶674 S
CS_2

(structure: benzene)

416

▶677 S
D, E
Grating

$[H_2C \cdot NO_2]^- Na^+$

[NO_2 deformation]

328

▶674 MW
B (0.0576)

H_3C-O-CF_2CHFCl

323

▶677 S
E

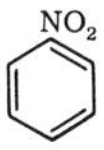

(structure: nitrobenzene)

[NO_2 symm. def.]

20

▶673 S
B (0.0104)

(structure: perfluoro-1,2-dimethylcyclohexane)

231

▶676 VS
A (100 mm Hg, 10 cm)

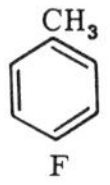

CF_2Cl-CF_2Cl

257, 258

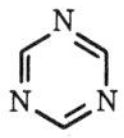

▶673 S
E

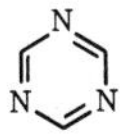

(structure: N,N-dimethylbenzamide, C_6H_5-C(=O)-$N(CH_3)_2$)

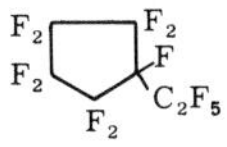

3

▶676 SSp
E

$Na_2H_2P_2O_6$

[PO_2 (OH) def.]

28

▶673 VVS
B (0.025)

(structure: cyclooctatetraene)

200

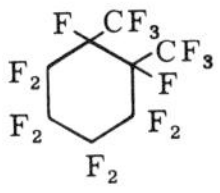

▶675 VS
B (0.064)

H_2C=CHCH(CH_3)CH_2CH_3

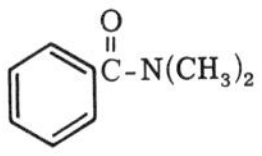

167

▶672 S
CCl_4

CBr_4

418

▶675 VW
B (0.2398)

(structure: 4-fluorotoluene)

220

▶672 M
A

$(CH_3)_2SO$

13

▶675 S
E

(structure: 1,3,5-triazine)

14

▶672 M
E
Grating

$[(CH_3)_2CNO_2]^- Na^+$

[C=NO_2 wag]

328

▶ 671 VS (Q branch)
A (10 mm Hg, 40 cm)

309

▶ 669 S
A

$HC{\equiv}C-C(=O)-H$

[HC≡C bend] 18

▶ 669 M

CBr_4

418

▶ 668 VS
A, B

$(CH_3)_2CHCH_2CH_3$

333

▶ 668 S
CS_2

CBr_4

418

▶ 667 M
B

$VOBr_3$

12

▶ 665

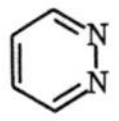

420

▶ 665 MB
B

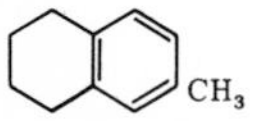

269

▶ 664 VS
B (0.0576)

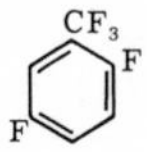

250

▶ 661 VS
B (0.056)

$CH_2{=}CHCH{=}CHCH{=}CH_2$

286

▶ 661 W
B (0.0104)

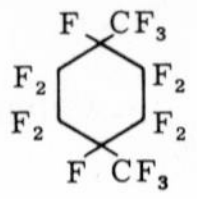

229

▶ 660 VS
B

$H_2C{=}CHCH_2C(CH_3)_3$

273

▶ 660

C_2H_5SH

[CS] 419

▶ 658 S
CS_2

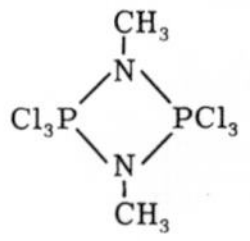

384

▶ 653 MW
B (0.0576)

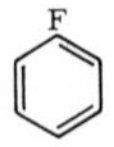

249

▶ 653 VS
E

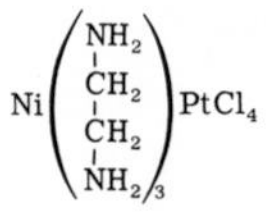

$[NH_2]$ 8

▶ 652 M
B (0.0104)

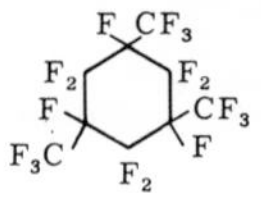

227

▶ 649 M
E

$[Coen_2SO_3]Cl$

5

▶ 649 M
E

$[Coen_2SO_3]Br$

5

▶ 649 M
E

$[Coen_2SO_3]I$

5

▶645 MB
E
$[H\text{-}S\text{-}C(NH_2)_2]NO_3$
22

▶640 VS
E 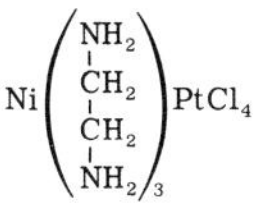
$Ni\left(\begin{matrix} NH_2 \\ CH_2 \\ CH_2 \\ NH_2 \end{matrix}\right)_3 PtCl_4$
$[NH_2]$
8

▶640 Sh
E
$[H\text{-}S\text{-}C(NH_2)_2]_2^+ SnCl_6^=$
22

▶638 W
E 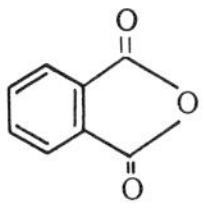
O
O
O
39

▶637 S
E
$[Coen_2SO_3 \cdot NH_3]Cl$
5

▶637 S
E 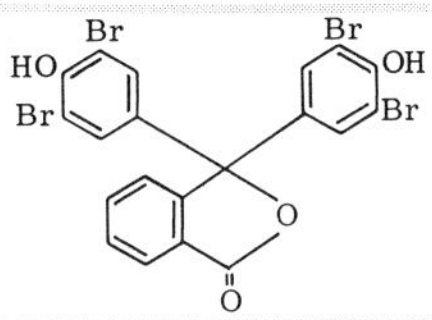
Br
HO
Br
Br
OH
Br
O
O
39

▶635 M
E
$S{=}C(NH_2)_2$
22

▶634 MW
E 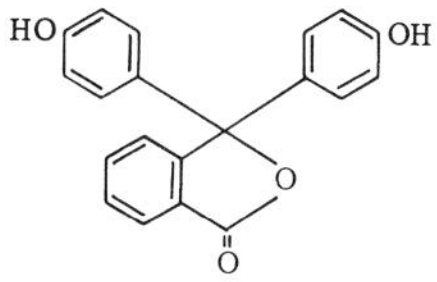
HO
OH
O
O
39

▶633 M
E 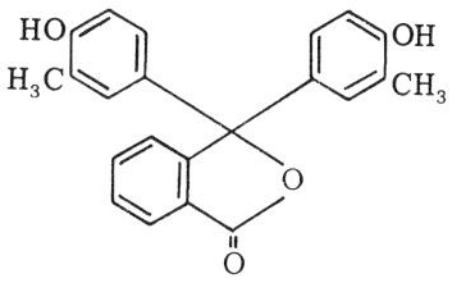
HO
H_3C
OH
CH_3
O
O
39

▶632 S
E 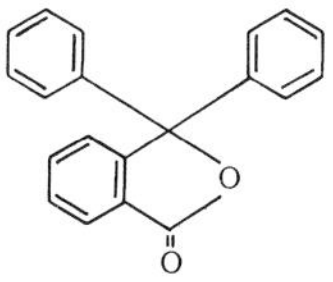
O
O
39

▶630 S
E
$[Coen_2(SO_3)_2]Na$
(trans)
5

▶630 S
P
$[(C_2H_5)_2P \cdot BBr_2]_3$
[B-Br]
30

▶629 S
E
$[Coen_2SO_3NCS]^0$
5

▶626 W
E 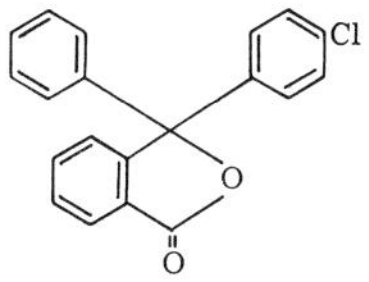
Cl
O
O
39

▶625 S
E
$[Coen_2SO_3]Cl$
5

▶625 S
E
$[Coen_2SO_3OH]^0$
5

▶625 S
E
$[Coen_2SO_3Cl]^0$
5

▶625 S
E
$[Coen_2(SO_3)_2]Na$
(cis)
5

▶624 M
E
Grating
$[(CH_3)_2CNO_2]^- Na^+$
$[NO_2$ deformation]
328

▶621 M
E
$[Coen_2SO_3]I$
5

▶621 M
E

$[Coen_2SO_3]SCN$

5

▶620 S
P

$[(C_2H_5)_2P \cdot BBr_2]_3$

[B-Br] 30

▶617 VSSh
E

$Ni\left(\begin{matrix} NH_2 \\ | \\ CH_2 \\ | \\ CH_2 \\ | \\ NH_2 \end{matrix}\right)_3 PtCl_4$

$[NH_2]$ 8

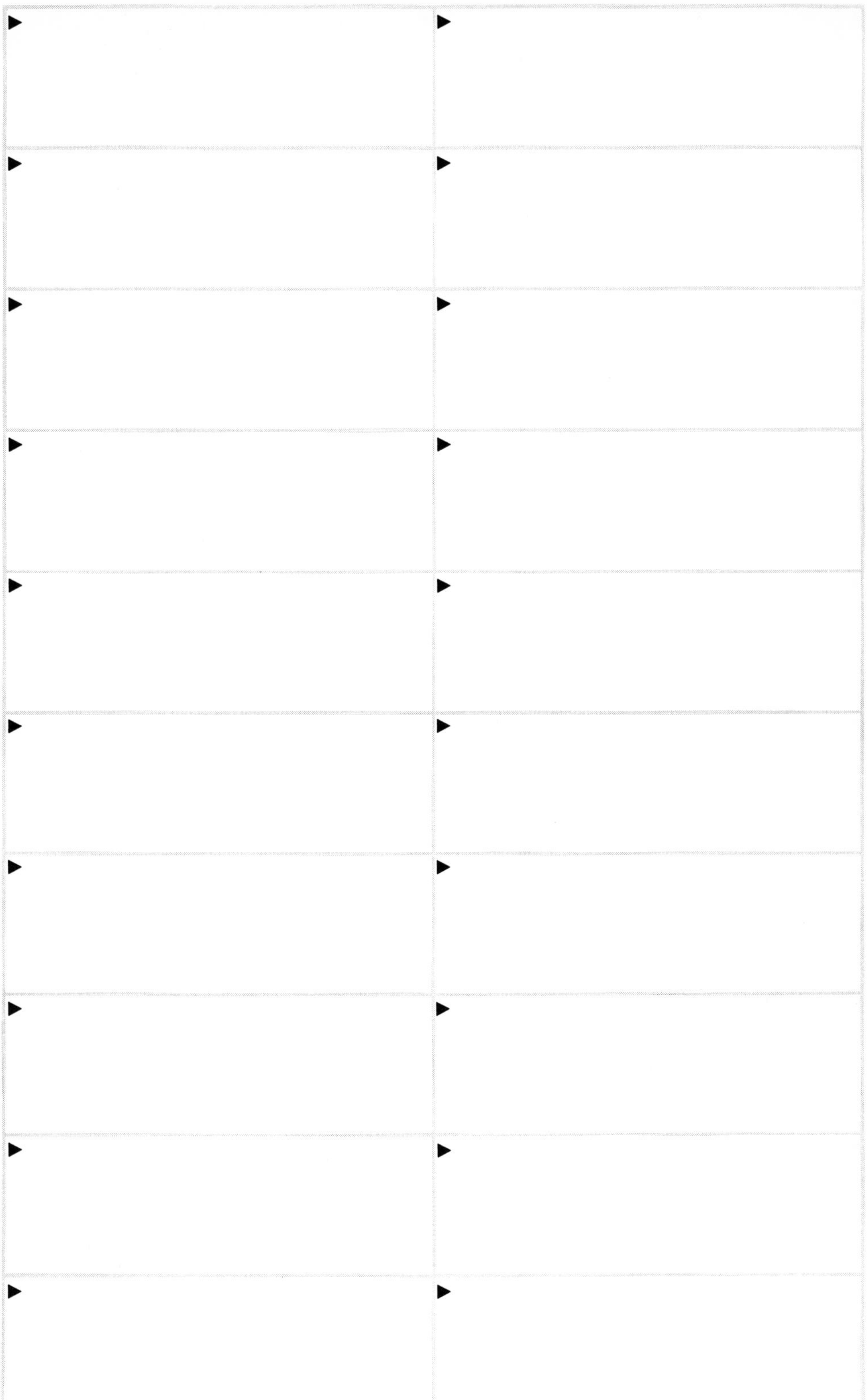

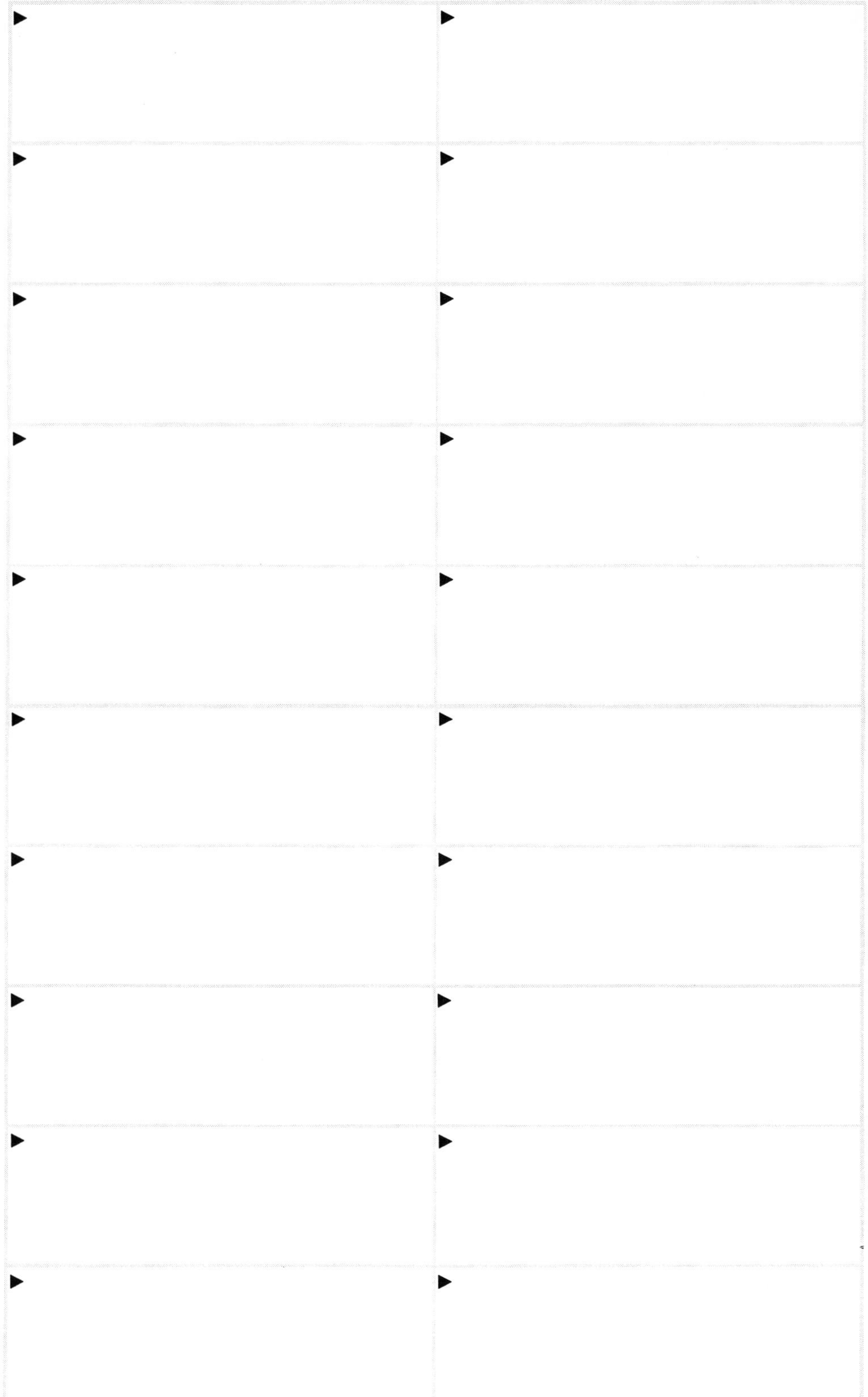

REFERENCES

The abbreviations used in this reference list are those suggested by Hershenson in his "Infrared Absorption Spectra Index for 1945-1957" (Academic Press, New York, 1959), supplemented by some similar abbreviations for several textbooks and other sources utilized in the compilation of this Handbook. One of the latter is the Canisius File. This file contains data collected over a period of several years from lectures presented at various meetings, spectra determined in Canisius laboratories, etc., that is, sources that either are not available in print or else are not generally accessible. The list contains numerous supplementary references. Many Handbook entries refer the reader, in addition to the actual source, to corroborative or analogous listings, and the Index lists still other background material for many compounds.

A list of the abbreviations used and their explanation follows:

A	The Analyst
AC	Analytical Chemistry
API	American Petroleum Research Institute, Project 44
AS	Applied Spectroscopy
B	The Journal of Biological Chemistry
BE	Chemische Berichte
BEL	Bellamy, L. J., "Infrared Spectra of Complex Molecules" John Wiley, New York, 1958
C	Journal of the Chemical Society (London)
CAN	Canisius File
CN	Canadian Journal of Chemistry
F	Bulletin de la société chimique de France
FA	Transactions of the Faraday Society
G	Angewandte Chemie
HE	Herzberg, G., "Infrared and Raman Spectra of Polyatomic Molecules," Van Nostrand, Princeton, 1945.
I	Annali di chimica (Rome)
JA	The Journal of the American Chemical Society
M	Journal of Molecular Spectroscopy
NB	Journal of Research of the National Bureau of Standards
OP	Optics and Spectroscopy
OS	Journal of the Optical Society of America
P	The Journal of Chemical Physics
PC	The Journal of Physical Chemistry
SA	Spectrochimica Acta
SC	Acta Chimica Scandinavica
WE	West, W., "Techniques of Organic Chemistry," Volume IX, Interscience, New York, 1956.

The numbers have the following significance: If the reference is a book, only the page number is given. In the case of the API file, only the card number is given. For journal references, if only two numbers are given, the first is the page number and the second, in parentheses, is the year. If more than two numbers are given, the first number is the volume number (followed in some cases by the issue number in parentheses); the next number is the page number, and the last number, always in parentheses, is the year.

1. C 2604 (1961)
2. C 2666 (1961)
3. C 3064 (1961)
4. C 3103 (1961)
5. C 3126 (1961)
6. SA 17, 40 (1961)
7. SA 17, 64 (1961)
8. SA 17, 68 (1961)
9. SA 17, 77 (1961)
10. SA 17, 82 (1961)
11. SA 17, 101 (1961)
12. SA 17, 112 (1961)
13. SA 17, 134 (1961)
14. SA 17, 155 (1961)
15. SA 17, 238 (1961)
16. SA 17, 248 (1961)
17. SA 17, 93 (1961)
18. SA 17, 286 (1961)
19. SA 17, 291 (1961)
20. SA 17, 486 (1961)
21. SA 17, 523 (1961)
22. SA 17, 530 (1961)
23. SA 17, 568 (1961)
24. C 18 (1961)
25. C 106 (1961); C 661 (1961); SA 17, 486 (1961)
26. C 489 (1961)
27. C 662 (1961)
28. C 1552 (1961)
29. C 1688 (1961)
30. C 1823 (1961)
31. C 1919 (1961)
32. C 1935 (1961)
33. C 2078 (1961)
34. C 2141 (1961)
35. C 2153 (1961)
36. C 2236 (1961)
37. C 2344 (1961)
38. C 2386 (1961)
39. AS 14(3), 61 (1960)
40. AS 15(1), 19 (1961)
41. AS 15(2), 29 (1961)
42. AS 15(6), 176 (1961)
43. C 4369 (1960)
44. C 5100 (1960)
45. AS 13(3), 61 (1959)
46. P 29(3), 611 (1958); SA 15, 77 (1959); P 29(5), 1097 (1958)
47. P 26(1), 122 (1956)
48. P 34(5), 1554 (1961)
49. P 19(7), 942 (1961)

50. SA 16, 1216 (1960)
51. P 26, 426 (1957)
52. M 2, 575 (1958)
53. CN 34, 1037 (1956)
54. SA 15, 95 (1959)
55. SA 15, 530 (1961)
56. P 26, 552 (1957)
57. P 21, 2024 (1953)
58. P 8, 369 (1940)
59. P 35, 1491 (1961)
60. P 8, 60 (1940)
61. P 35, 183 (1961)
62. P 27, 445 (1957)
63. P 29(4), 84 (1958)
64. P 24(4), 656 (1956)
65. API 129
66. API 127; API 228
67. API 126; API 227
68. API 125; API 226
69. API 124; API 267
70. API 123
71. API 120; API 239
72. API 118; API 234
73. P 27, 1168 (1957)
74. API 79
75. API 77
76. API 74
77. API 73
78. API 132
79. API 130
80. API 639; API 69; API 552; API 640
81. API 641; API 70; API 602; API 642
82. API 644
83. API 664
84. API 638
85. API 637
86. AS 14(4), 88 (1960)
87. API 45
88. API 46; CN 32, 1561 (1954)
89. API 147
90. API 121
91. API 225
92. API 196
93. API 70
94. API 67
95. API 65
96. API 40
97. API 39
98. API 37
99. API 33
100. API 31; API 635 (spectra differ slightly)
101. API 30
102. API 27; API 1063
103. API 1565; API 17; API 257; API 368; API 1137
104. API 15; API 1556; API 255; API 344; API 510; API 616
105. API 14
106. API 149; API 230
107. API 148
108. API 133
109. P 25, 303 (1956)
110. API 239; API 120
111. API 238; API 155
112. API 237; API 154
113. API 236; API 153
114. API 235; API 119
115. API 234; API 118
116. API 233; API 152
117. API 232; API 151
118. API 231; API 150
119. API 230; API 149
120. API 229; API 148
121. API 228; API 127
122. API 227; API 126
123. API 226; API 125
124. API 222
125. API 215
126. API 214
127. API 213
128. API 201; API 697
129. API 199; API 586
130. API 198
131. API 197; API 360
132. API 191; API 1587; API 1642
133. API 190; API 414; API 415; API 471
134. API 668
135. API 666; API 93
136. API 665; API 183
137. API 663; API 81; API 180; API 82
138. API 662; API 79; API 179; API 80
139. API 661; API 77, API 178; API 78
140. API 660; API 176
141. API 659; API 73, API 649; API 650
142. API 658; API 72; API 648; API 647
143. API 657; API 643
144. API 656; API 246; API 810; API 811
145. API 655
146. API 654; API 249; API 443; API 573
147. API 653; API 249; API 443; API 573; API 654
148. API 652; API 74; API 248; API 572; API 651
149. API 651; API 74; API 248; API 572
150. API 650; API 73; API 649; API 659
151. API 649; API 73; API 650; API 659
152. API 648; API 72; API 647; API 658
153. API 647; API 72; API 658; API 648
154. API 646; API 645; API 571; API 247
155. API 645; API 571; API 247; API 646
156. API 643; API 657
157. API 642; API 70, API 602, API 641
158. API 640; API 69, API 552; API 639
159. API 636
160. API 635; API 31 (spectra differ)
161. API 628
162. API 627
163. API 626; API 711
164. API 1061 (revised)
165. API 624 (revised)
166. API 623; API 708
167. API 622; API 707
168. API 621
169. API 620
170. API 617
171. API 616
172. API 790
173. API 781
174. API 780; API 1981 (spectra differ)
175. API 779
176. API 778; API 1980 (spectra differ)
177. API 777
178. API 776; API 1978 (spectra differ)
179. API 775
180. API 774; API 1596 (spectra differ)
181. API 773
182. API 772
183. API 771
184. API 770; API 1210
185. API 769; API 1595
186. API 768
187. API 767; API 1209 (spectra differ)
188. API 766
189. API 765
190. API 764; API 1977
191. API 763
192. API 762; API 1594
193. API 761
194. API 756; API 437
195. API 755; API 435
196. API 753
197. API 744
198. API 743
199. API 740; API 364
200. API 739; API 595
201. API 830
202. API 828
203. API 826
204. API 824; API 277; API 249

205. API 822
206. API 820; API 1892 (spectra differ)
207. API 818
208. API 815
209. API 813
210. API 811; API 810; API 656; API 246
211. API 801
212. API 800
213. API 799
214. API 798
215. API 795
216. API 794
217. API 1030
218. API 1028
219. API 1026
220. API 1024
221. API 1022
222. API 1020
223. API 1018
224. API 1016
225. API 1014
226. API 1012
227. API 1003
228. API 1002
229. API 1001
230. API 1000
231. API 998
232. API 996
233. API 977
234. API 980
235. API 981
236. API 982
237. API 983
238. API 984
239. API 986
240. API 987
241. API 988
242. API 989
243. API 990
244. API 991
245. API 992
246. API 993
247. API 994
248. API 995
249. API 1013
250. API 1031
251. API 1032; CN 33, 1746 (1955)
252. API 1033
253. API 1034
254. API 1035
255. API 1036
256. API 1037
257. API 1038
258. API 1039
259. API 1041
260. API 1043
261. API 1046
262. API 1047
263. API 1048
264. API 1050
265. API 1051
266. API 1052
267. API 1053
268. API 1055
269. API 1056
270. API 1069
271. API 1067
272. API 1065
273. API 1063; API 27
274. API 1061 (revised)
275. API 1059
276. API 1057
277. API 1765
278. API 1766
279. API 1767
280. API 1768
281. API 1784
282. API 1785
283. API 1786
284. API 93; API 666
285. API 667
286. API 827
287. API 816
288. API 794
289. API 28
290. API 29; API 474
291. API 35
292. API 131; API 579
293. API 133; API 583
294. API 148; API 229
295. OS 42, 570 (1952)
296. API 678
297. API 679
298. API 680; API 1840
299. API 681
300. API 683
301. API 684; API 1802
302. API 686; API 1804
303. API 687
304. API 689
305. API 690
306. API 692
307. API 693
308. API 695 (corrected)
309. API 122; API 192; CN 35, 91 (1957); JA 65, 803 (1943)
310. API 1755
311. CAN; API 1751; CN 34, 170 (1956)
312. API 796
313. OS 37, 216 (1947)
314. API 747; API 427; API 1604
315. API 746; API 426; API 1601
316. API 745; API 425
317. API 758
318. API 752; API 436
319. API 750; API 431
320. API 757
321. API 754; API 434
322. API 1049
323. API 1045
324. API 1756
325. API 1758
326. API 1757
327. M 7(2), 91 (1961); C 2202 (1958)
328. M 7(2), 105 (1961)
329. M 566 (1958)
330. API 70; API 602; API 641; API 642
331. API 69; API 552; API 639; API 640
332. API 65
333. API 64; API 441
334. API 79; API 179; API 662; API 80
335. API 73; API 649; API 650; API 659
336. API 74; API 248; API 572; API 651; API 652
337. API 75; API 177; API 524; API 76
338. API 77; API 178; API 661; API 78
339. API 81; API 180; API 663; API 82
340. API 87; API 185; API 250; API 576; API 88
341. API 91; API 186; API 252; API 578; API 92
342. API 831
343. API 1763
344. API 1761; API 1116
345. API 810
346. API 1054
347. API 1654
348. API 974
349. M(7) (1961)
350. API 433
351. CAN
352. API 1071
353. API 749
354. CAN; JA 70, 195 (1948)
355. API 386
356. API 1044
357. API 975
358. SA 17, 226 (1961)
359. CAN
360. API 574; API 83; API 181
361. API 1605

362. API 387
363. M 6, 277 (1961); P 27, 446 (1957)
364. JA 82, 557 (1960)
365. API 579; API 131
366. API 1754
367. M(7) 243 (1961)
368. CAN; SA 162 (1958); P 26, 426 (1957); SA 563 (1960)
369. CAN; C 90 (1961)
370. CAN; SA 162 (1958)
371. API 440
372. API 973
373. API 374; CN 34, 1037 (1956)
374. API 1654; SA 17, 909 (1961)
375. CAN
376. M(6) 277 (1961); CN 35, 937 (1957); M(8) 222 (1962)
377. API 732
378. CAN; C 2942 (1960)
379. C 1939 (1961)
380. C 957 (1961)
381. CAN; P 24(4), 656 (1956)
382. JA 76, 2451 (1954)
383. CAN; F 12, 741 (1945); OS 40, 96 (1950)
384. C 1825 (1961)
385. CAN; SA 12, 305 (1958); CN 39, 2452 (1961)
386. WE 558; P 21, 2024 (1953)
387. CAN; C 3372 (1961)
388. CAN
389. CAN
390. M 6, 238 (1961)
391. AC 15, 696 (1943)
392. API 1753
393. API 751; API 433
394. CAN
395. SA 17, 125 (1961)
396. CAN; JA 73, 246 (1951)
397. API 529
398. CAN; I 49, 245 (1959)
399. WE 558
400. CAN
401. API 1133
402. WE 558; P 18, 595 (1950)
403. CAN; P 34(4), 1087 (1961); SA 17, 112 (1961); C 241 (1961)
404. AC 24, 1268 (1952)
405. API 2018
406. CAN; AC 23, 1598 (1951); BE 89, 2895 (1956)
407. CAN
408. M 6, 277 (1961); OS 43, 979 (1953)
409. API 1010
410. API 676
411. CAN; API 310
412. CAN; P 26, 426 (1957); FA 36, 812 (1940)
413. CN 35, 91 (1957); JA 65, 803 (1943)
414. CAN; JA 69, 823 (1947); NB 58, 256 (1957)
415. CAN; SA 9, 113 (1957)
416. CAN
417. API 1003
418. SA 17, 125 (1961)
419. CAN; SA 18, 39 (1962)
420. CAN; SA 9, 113 (1957)
421. API 32
422. API 938
423. API 441; API 64
424. API 550
425. API 552; API 69; API 639; API 640
426. API 576; API 87; API 185; API 250; API 88
427. API 704
428. API 720 (revised)
429. API 119; API 235
430. API 976
431. CAN
432. BEL 189
433. C 3165 (1958)
434. CN 39(9), 1783 (1961); G 39(4), 745 (1961); G 39(8), 1633 (1961)
435. SA 12, 305 (1958)
436. CAN
437. CAN; SA 9, 265 (1957); P 34, 1554 (1961)
438. C 2202 (1958)
439. CAN; P 16, 1158 (1948); P 24, 1188 (1956)
440. API 41
441. CAN; CN 34, 1139 (1956)
442. CAN
443. CAN
444. P 23, 2206 (1955)
445. CAN; JA 71, 3927 (1949)
446. CAN
447. BEL 330
448. CN 39(8), 1625 (1961)
449. CAN; AC 24, 1277 (1952); AC 29, 1431 (1957)
450. JA 82, 76 (1960)
451. CAN; JA 73, 2436 (1951); C 4149 (1954)
452. SA 15, 1118 (1959)
453. API 2204
454. M 6, 277 (1961); SA 17, 91 (1961); SA 17, 102 (1961)
455. CAN; JA 81, 2568 (1959)
456. SA 17, 286 (1961)
457. CAN
458. BEL 189
459. API 1684
460. API 428; API 748
461. C 774 (1961)
462. API 979
463. CAN; SA 16, 1108 (1960)
464. API 617
465. CAN; P 29(5), 1097 (1958)
466. CN 35, 937 (1957)
467. P 27, 158 (1957)
468. SA 17, 188 (1961)
469. SA 17, 233 (1961)
470. SA 17, 365 (1961)
471. SA 17, 486 (1961); B 106 (1961)
472. API 1802; API 684
473. API 1601
474. SA 15, 195 (1959); SA 16, 1216 (1960)
475. CAN
476. CN 30, 505 (1952)
477. CAN; C 3278 (1958)
478. CAN
479. CAN
480. P 7, 563 (1939); P 8, 229 (1940); P 26, 690 (1957); SA 8, 27 (1956)
481. C 2198 (1958)
482. C 486 (1960)
483. CAN; C 2236 (1961); SA 18, 39 (1962)
484. SA 16, 1279 (1960)
485. CAN; BEL 251; BEL 255; AC 28, 1230 (1956); P 20, 138 (1952)
486. CAN; BEL 264; P 25, 203 (1956); P 20, 651 (1952)
487. CAN
488. C 3939 (1961)
489. CAN
490. CAN; P 17(6), 556 (1949)
491. CAN; CN 35, 937 (1957)
492. CAN; BEL 37; BEL 388; BE 90, 415 (1957); F 13, 33 (1946)
493. CAN; P 29(3), 484 (1958); P 24, 989 (1956); P 23, 2463 (1955)
494. CAN; P 19, 942 (1951); JA 75, 5626 (1953)
495. CAN; SA 16, 964 (1960); SA 16, 954 (1960)
496. API 472; API 719
497. CAN
498. JA 82, 98 (1960)
499. JA 82, 555 (1960)
500. C 753 (1962)
501. C 2780 (1958); AS 15(4), 116 (1961)
502. CAN
503. C 2780 (1958)
504. C 2780 (1958); C 1453 (1958)
505. C 2780 (1958); SA 16, 407 (1960); SA 16, 1314 (1960)
506. SA 17, 600 (1961)

507. CAN
508. C 1631 (1961)
509. JA 82, 1080 (1960)
510. P 27, 325 (1957)
511. BEL 131
512. AC 24, 316 (1952)
513. BEL 126
514. BEL 127
515. BEL 130
516. BEL 135
517. BEL 141
518. BEL 142
519. BEL 155
520. BEL 156
521. BEL 165
522. BEL 166
523. BEL 167
524. BEL 168
525. BEL 175
526. BEL 180
527. BEL 181
528. BEL 189
529. BEL 209
530. BEL 210
531. BEL 211
532. BEL 213
533. BEL 219
534. BEL 255
535. BEL 259
536. BEL 22; CN 27, 332 (1949)
537. BEL 23
538. BEL 24
539. BEL 25
540. BEL 31
541. SC 9, 1313 (1955)
542. BEL 37
543. BEL 38; P 17, 556 (1949)
544. BEL 40
545. BEL 41
546. BEL 42
547. BEL 44
548. BEL 48
549. BEL 50
550. P 19, 297 (1951)
551. BEL 59
552. BEL 60
553. CAN
554. P 20(11), 1720 (1952)
555. BEL 382; BEL 304; SA 17, 206 (1961)
556. C 2693 (1958)
557. CAN
558. JA 70, 2816 (1948)
559. BEL 149
560. BEL 183
561. BEL 402
562. C 1631 (1961); C 3708 (1959)
563. PC 58, 210 (1954)
564. CAN
565. CAN
566. CAN
567. BEL 270; JA 70, 194 (1948)
568. C 90 (1961)
569. BEL 150
570. BEL 268
571. BEL 185
572. BEL 37
573. CAN
574. JA 76, 2781 (1954)
575. CAN
576. CAN; SA 16, 956 (1960)
577. CAN
578. P 24(4), 656 (1956)
579. BEL 184
580. CAN
581. PC 61, 839 (1957)
582. CAN
583. CAN
584. BEL 304
585. CAN
586. CAN
587. CAN
588. C 602 (1961)
589. CAN
590. CAN
591. C 3010 (1959)
592. BEL 212
593. C 1317 (1959)
594. CAN
595. CAN
596. C 3619 (1958); SA 16, 956 (1960)
597. JA 70, 194 (1948)
598. CAN
599. CAN
600. BEL 299; SA 16, 1088 (1960)
601. SA 17, 486 (1961)
602. C 617 (1962)
603. BEL 255
604. CAN
605. CAN
606. CAN
607. C 3500 (1959)
608. C 676 (1960)
609. CAN
610. CAN
611. C 2383 (1961)
612. C 2383 (1961)
613. CAN
614. CAN; CN 39, 1214 (1961)
615. CAN
616. BEL 279
617. CAN
618. C 378 (1961)
619. CAN
620. BEL 300
621. CN 35, 1184 (1957)
622. SA 16, 1108 (1960)
623. C 13 (1959); C 1740 (1961)
624. CAN
625. SA 17, 523 (1961)
626. C 3674 (1959)
627. API 696
628. CAN; SA 16, 1165 (1960)
629. CAN
630. C 3153 (1961)
631. CAN
632. CAN; AC 28, 1259 (1956); SA 7, 101 (1955)
633. API 1556
634. BEL 305
635. C 1740 (1961)
636. CAN; CN 38, 1901 (1960)
637. M 1, 107 (1957)
638. CAN
639. CAN; SA 16, 279 (1960)
640. CAN; M 8, 126 (1962); P 35, 183 (1961); P 27, 445 (1957); P 29, 484 (1958)
641. SA 12, 305 (1958)
642. C 90 (1961); CN 35, 1199 (1957)
643. SA 17, 155 (1961)
644. CAN; CN 34, 1382 (1956)
645. C 1501 (1959)
646. CAN; CN 35, 937 (1957)
647. BEL 301
648. SA 17, 530 (1961); C 955 (1958); C 2218 (1960)
649. OP 8(1), 20 (1960)
650. BEL 274
651. CAN; SA 15, 95 (1959)
652. SA 17, 530 (1961)
653. C 3224 (1961)
654. CAN
655. CAN
656. PC 58, 1079 (1954)
657. M 1, 32 (1957)
658. AC 23, 1614 (1951)
659. OS 37, 216 (1947)
660. P 27, 403 (1957)
661. PC 56, 247 (1952)

662. P 24, 563 (1956)
663. JA 77, 5251 (1955)
664. P 23, 377 (1955)
665. AC 20, 816 (1948)
666. CAN
667. CAN
668. JA 71, 515 (1949)
669. C 1453 (1958)
670. CAN
671. C 2067 (1959)
672. BEL 49
673. C 667 (1959)
674. WE 444
675. C 661 (1961)
676. PC 61, 460 (1957)
677. HE 364
678. HE 335
679. HE 358
680. BEL 263
681. BEL 265
682. SA 16, 279 (1960)
683. SA 17, 486 (1961)
684. SA 17, 679 (1961)
685. SA 17, 634 (1961)
686. CAN
687. SA 16, 428 (1960)
688. SA 16, 279 (1960)
689. SA 17, 64 (1961)
690. CAN
691. CAN
692. CAN

INDEX

Compounds are indexed by empirical formula. The structural formula is shown only when the same empirical formula applies to two or more compounds. The compounds are arranged in alphabetical order of the symbols for the elements they contain, except that for compounds containing carbon or carbon and hydrogen the C's and H's are listed before all other elements. Within the alphabetical arrangement the compounds are listed in order of increasing frequency of the elements. Thus, all C_5... compounds come before any C_6... compound.

As has been pointed out in the Introduction, certain bands common to many hydrocarbons have been omitted to save space in the Handbook. Such bands, which are not actually listed in the pages of the Handbook, have however been included in the index and are there designated by asterisks.

The references for each compound are listed in square brackets immediately after the empirical formula. In many cases, additional references that were not the actual source of Handbook entries but which pertain to the compound and furnish valuable information are included in the brackets. Such supplementary references are also designated by asterisks.

CH_2NNaO_2 [328]

2920 MW 1263 VS 1018 VS 984 M
2847 MW 1262 VS 1017 VS 736 VS
1582 M 1261 VS 1016 VS 734 VS
1580 M 1253 M 1012 M 690 S
1445 M 1033 VS 986 M 689 S
1278 VS 1031 VS 985 M 677 S
1277 VS

CH_3Br [61, 62, 63, 363, 376, 539, 547*]

3058 2972 1305 952
3056

CH_3Cl [376, 539, 547,* 640]

1455 1355 1015

CH_3F [52,* 493, 536, 539]

2982 2965 1475 1471

CH_3I [536, 539, 547*]

1441 1255

CH_3NO [529] H-C(=O)-NH_2

1740 1709

CH_3NO [614] $CH_3N{=}O$

1582 S

CH_3NO_2 [50, 311, 601,* 647*]

3049 S 2037 MB 1420 SSh 1211 M
2976 S 1832 W 1401 VS 1103 VS
2793 S 1580 M 1379 VS 960 VW
2475 M 1558 VVS 1314 MW 920 VS
2288 MW

CH_3N_3 [58, 59, 650]

1351

CH_4NCl [487]

3075 2972

CH_4N_2O [589]

1655

CH_4N_2S [22, 541,* 652*]

1086 S 730 S 635 M

CH_4O [48, 316, 437, 510,* 539, 678]

3345 VS* 1460 VS (Raman) 1034
2950 VS 1456 1116 S 1031 VS
2833 VS 1451 VSVB 1109

CH_4S [412]

705

CH_5ClN [259]

1617

$CH_5ClN_2O_4S$ [22]

726 S 702 MWB

CH_5ClN_2S [22]

709 Sh

CH_5ISi [669]

1263 S

CH_5N [480, 539]

3470 3360 2820 S 1418

$CH_5N_3O_3S$ [22]

718 M 703 M 645 MB

CNR [681]

2180-2145

C_2ClF_3 [260]

711 MW

$C_2Cl_2F_2$ [546]

1730

$C_2Cl_2F_4$ [257, 258]

2458 VS 1736 S 1195 VS 883 VS
2407 M 1732 MS 1186 VS 864 VVW
2363 VS 1680 M 1111 SSh 850 VS
2319 VS 1679 VS 1053 VS 849 VS
2263 S 1555 S 1003 MW 768 VW
2234 S 1506 MS 981 VW 740 S
2204 VS 1477 VW 948 S 735 VS
2122 MW 1362 MB 928 S 731 S
1895 S 1274 S 923 VS 678 VS
1856 M 1233 VS 888 VS 676 VS
1846 M 1229 S

$C_2Cl_3F_3$ [253, 254*] CF_3-CCl_3

2450 S 1255 VS 1032 MS 794 VS
1515 S 1227 VS 992 M 719 VS
1471 S 1124 M 909 VS 714 VS
1431 M 1090 MSh 859 VS 709 VS
1277 WSh 1086 M

$C_2Cl_3F_3$ [255, 256*] CF_2Cl-$CFCl_2$

2380 S 1820 S 1248 VSSh 886 S
2320 S 1766 S 1209 VS 873 S
2215 VS 1720 MS 1109 VS 852 VW
2160 SSh 1637 M 1045 M 746 VW
2120 SSh 1527 M 1033 VS 734 MB
2020 M 1490 S 924 SSh 712 MW
1856 S 1458 S 902 VVS

C_2Cl_4O [553]

1820

$C_2F_3NaO_2$ [525]

1457

C_2F_4O [513]

1901

C_2F_6 [253]

2365 S

C_2HCl_3O [519]

1762

C_2HF_3 [564]

1780

$C_2HF_3O_2$ [477, 522]

3504 1820 1810

C_2H_2BrClO [557]

1799

$C_2H_2Cl_2$ [543]

1590

$C_2H_2Cl_2O$ [554]

1807 VS 1780 WSh

$C_2H_2Cl_3NO$ [531, 674]

1733 1732

$C_2H_2F_2$ [546]

1730

$C_2H_2F_3NO_2$ [584]

1736 1695

$C_2H_2N_4$ [406]

1520 725 695

$C_2H_2O_4$ [670]

1224

C_2H_3BrO [513]

1812

$C_2H_3BrO_2$ [674]

1731

C_2H_3ClO [513]

1802

$C_2H_3ClO_2$ [674]

1736

$C_2H_3Cl_3$ [51, 251, 252*]

2962 VS 1387 VS 1010 S 854 M
2953 VS 1379 VS 945 S 798 VW
2455 MS 1286 MB 936 M 767 VW
1468 M 1095 S 871 VVW 759 WSp
1458 S 1088 S 867 VVW 726 VS
1431 M 1015 S 864 VVW 689 SSh
1397 VS

C_2H_3F [546, 549*]

1650-1645

C_2H_3FO [53]

1871 1840

$C_2H_3F_3$ [234,* 447]

1290 1278 1266 1230

C_2H_3N [538, 550,* 637*]

1396

$C_2H_3NO_3$ [359]

965 940

C_2H_3NS [682]

2210 VS

$C_2H_3NaO_2$ [525]

1560

C_2H_3R [691]

1456 S

C_2H_4Br [536]

1435

$C_2H_4F_2$ [233, 462]

3001 VS 2692 VS 1812 M 1171 MSp
2963 S 2494 S 1808 M 1142 VS
2902 M 2279 VS 1659 M 1135 W
2887 MW 2264 VS 1460 M 944 VS
2877 MW 2233 MSh 1425 VS 939 VS
2822 S 2004 VS 1414 VS 883 SB
2807 S 1895 MS 1403 VS 875 MVB
2757 S 1890 MS 1264 S 873 MVB
2744 S 1881 MS 1255 S 868 VS
2711 S 1819 MS

C_2H_4O [476,* 519] CH_3-C(=O)H

1752

C_2H_4O [64, 381, 463,* 537] H_2C-CH_2 (O)

1500 1255 865

$C_2H_4O_2$ [458] H-C(=O)-OCH_3

1173

$C_2H_4O_2$ [50, 51, 368, 521, 537, 538] H_3C-C(=O)-OH

1785 1717 1381 935
1735 1418 1290

C_2H_4OS [523]

1695

C_2H_4S [537]

1471

$C_2H_5BO_2$ [4]

1605 M

C_2H_5Cl [88]

2990 VS 1399 M 984 VS 963 VS
1449 MB 1281 VS 971 VS 686 VS

C_2H_5NO [329,* 474, 500,* 530, 563]

3538 3420 1694

C_2H_5NS [54, 55, 648, 651]

1610 S 1377 S 1366 S 1314 S

C_2H_6 [56, 538, 581*]

1380 1374

$C_2H_6BCl_2N$ [503]

2800 S

$C_2H_6Cl_6N_2P_2$ [384]

2996 M 1249 M 1162 S 700 M
2941 M 1210 S 847 VS 658 S
1461 M 1184 S

$C_2H_6NNaO_2$ [328]

1437 W 1381 W

C_2H_6O [315, 473*] CH_3CH_2OH

3350 VS* 2899 VS 1335 VSB 1092 VS
3333 VS* 2875 VS* 1335 VSVB 1050 VS
2985 VS 1453 VSVB 1318 VSVB 880 VS
2967 VS* 1381 VS 1271 VSVB 802 VSB
2933 VS

C_2H_6O [47, 313,* 539] H_3C-O-CH_3

1466

C_2H_6OS [13, 653*]

2973 M 1419 M 1102 VS 1006 M
2908 M 1405 M 1094 SSh 689 M
1455 M 1304 M 1016 M 672 M
1440 MS 1111 SSh

C_2H_6S [60, 539] H_3C-S-CH_3

1323

C_2H_6S [419] C_2H_5SH

660

$C_2H_7IN_2S$ [22]

725 M 690 MW

$C_2H_7N_2S$ [22]

1080 W

C_2H_8BP [30]

1294 M

$C_2H_8Cl_2N_2Pd$ [8]

1568 VS 1284 M 1105 S 729 VS
1369 M 1165 S 1055 S

$C_2H_8Cl_2N_2Pt$ [8]

1561 VS 1192 VS 1053 S 873 M
1366 M 1131 S 991 M 756 VS
1290 S

$C_2H_8Cl_4CuN_2Pt$ [8]

1321 M

$C_2H_8Cl_4N_2NiPt$ [8]

1463 M

$C_2H_8Cl_4N_2PdPt$ [8]

1383 M

$C_2H_{10}Cl_6N_4S_2Sn$ [22]

696 MS 640 Sh

$C_3Cl_6O_3$ [513]

1832

C_3F_6 [409, 558]

1798 711 MB 706 MB 684 MB

C_3F_8 [235, 236*]

2732 S 2039 VS 1669 S 1300 W
2631 S 1949 MS 1618 M 1262 VVS
2608 S 1941 MS 1582 VS 1209 S
2571 S 1934 MS 1553 W 1155 VS
2518 VS 1872 MS 1550 S 1117 MSh
2416 S 1819 S 1437 S 1034 MS
2358 S 1774 S 1414 VSSh 1007 VS
2263 M 1708 M 1351 VVS 732 VS
2211 M

$C_3H_2N_2$ [367, 466]

1422 1322 1220 936

C_3H_2O [18, 456*]

3380 MSh 2780 M 1398 M 950 VS
3335 VS 2125 VS 1340 M 691 SSh
2869 S 1692 VS 1275 M 669 S

$C_3H_3Cl_3O_2$ [560, 561]

1776 1770

$C_3H_3F_3O$ [517]

1780

C_3H_3NO [15]

1525 W 1218 M 1028 MW 845 M
1431 VS 1129 VS 916 S 774 S
1367 MW 1088 MW

$C_3H_3N_3$ [14, 506,* 643,* 656*]

3070 M 1775 M 1251 W 921 S
3055 S 1668 M 1174 M 830 M
2285 M 1667 M 1172 S 737 S
2270 M 1617 S 1167 M 685 VS
1980 M 1556 VS 1132 VW 675 S
1957 M 1550 VS 1033 M 675 S
1780 M 1410 VS 925 VW

C_3H_4 [51, 87, 538, 657*] $H_3C\text{-}C{\equiv}CH$

2995 S 1491 M 1420 M 1387 M
2160 MS 1477 M 1407 M 1379 W
1508 M 1434 M

C_3H_4 [441, 536]

1390 1070 $H_2C{=}C{=}CH_2$

$C_3H_4ClF_3O$ [323, 356]

3019 VS 1725 M 1252 MS 838 M
2976 VS 1695 M 1232 MS 803 VS
2875 VS 1610 MSSh 1160 M 788 M
2620 M 1538 VSSh 1091 SVB 751 WB
2530 MSh 1460 VS 1025 VS 720 VS
2465 MS 1315 VSB 998 VS 701 VS
2235 MSSh 1303 VSB 985 S 701 S
2170 MS 1294 VSB 858 VVS 674 MW

$C_3H_4Cl_2O_2$ [560]

1775 1750

$C_3H_4N_2O$ [6]

1620 S 1500 SB 1423 S

$C_3H_4O_2$ [351]

980 974

$C_3H_4O_3$ [514]

1805

$C_3H_5ClO_2$ [523]

1730

C_3H_5NS [639]

1460

$C_3H_5N_5O$ [587]

1715 S

C_3H_6ClNO [533]

1565 1550 1516

$C_3H_6NNaO_2$ [328]

2907 M 1299 M 1163 VS 772 M
2855 W 1277 M 1143 VS 680 M
1608 S 1176 VS 944 S 672 M
1550 M 1166 VS 777 M 624 M

C_3H_6O [520, 658*] $CH_3CH_2\text{-}\overset{O}{C}\text{-}H$

1735

C_3H_6O [49, 516, 537, 659*] $H_3C\text{-}\overset{O}{C}\text{-}CH_3$

1742 1718 1431

$C_3H_6O_2$ [674] $H_5C_2\text{-}\overset{O}{C}\text{-}OH$

1734

$C_3H_6O_2$ [560, 644] $H_3C\text{-}\overset{O}{C}\text{-}OCH_3$

1750 1218 1204

$C_3H_6O_2$ [432] $H\text{-}\overset{O}{C}\text{-}O\text{-}C_2H_5$

1195

$C_3H_6O_3$ [201, 342, 660*]

3030 SSp 1805 MSh 1168 VS 970 S
3019 W 1412 VS 1080 VS 957 S
2869 VS 1309 S 1068 VS 944 VS
2849 VSSp 1274 M 1068 VS 935 S
2790 VSSp 1182 VS 989 S 934 M
1980 S 1172 VS 978 VS 749 M
1860 VS

$C_3H_7BO_2$ [4]

1615 M

C_3H_7NO [570] $(CH_3)_2C{=}N\text{-}OH$

1675

C_3H_7NO [465] $H_5C_2\text{-}\overset{O}{C}\text{-}NH_2$

3500 3200 1410 1140

C_3H_7NO [46, 533] $H_3C\text{-}\overset{O}{C}\text{-}NH\text{-}CH_3$

1567 1534 1413 1159 W
1565 1490

$C_3H_7NO_2$ [366] $(CH_3)_2CHNO_2$

2994 VS 2451 M 1443 SSp 1138 S
2933 VS 2398 S 1397 VSSp 1103 S
2899 VS 2268 M 1374 S 944 S
2755 VS 2079 MWB 1357 VSSp 902 S
2695 S 1996 MW 1305 S 850 VSB
2534 M 1543 VVS 1179 S 722 VW
2500 M 1462 SSp

$C_3H_7NO_2$ [377, 392] $C_3H_7NO_2$

2976 VS 1949 M 1346 MS 1045 MW
2882 VS 1546 VVS 1295 MS 922 SB
2445 M 1453 MS 1272 MS 897 MS
2141 M 1435 S 1230 S 871 S
2062 M 1385 VS 1133 S 797 MVB

$C_3H_7NO_2$ [685] $CH_3\underset{NH_3^+}{CH}COO^-$

1410 S

$C_3H_7NO_2$ [604] $NH_2\text{-}\overset{O}{C}\text{-}OC_2H_5$

1618

C_3H_8 [397, 661*]

781 VS 779 VS 736 VS 731 VS

$C_3H_8ClNO_2$ [674]

1728

$C_3H_8N_2O$ [609]

1610

C_3H_8O [314, 662*] $CH_3CH_2CH_2OH$

3356 VS 1381 VSVB 1068 VS 905 M
2967 VS 1340 VSVB 1056 VS 887 M
2933 VS 1232 VSB 1017 VS 858 S
2874 VS 1099 VS 969 VS 756 MVB
1449 VVSB

C_3H_8O [361, 460] $(CH_3)_2CHOH$

3367 VS 1464 VS 1302 VS 952 VS
2933 VS 1377 VS 1163 VS 819 VS
2882 VS 1340 S 1130 VS

C_3H_8S [209]

2975 VS 1437 VS 970 M 758 VS
2934 VS 1378 S 956 M 726 S
2738 SSh 1268 VSSp 784 S 679 M
1451 VS 1063 S

C_3H_9N [505, 663*]

2810 S

$C_3H_{13}NiSi_2$ [504]

2802 S

C_3NR [680]

2235-2215 S

$C_4CdK_2N_4$ [468]

2145 VS

C_4F_8 [243, 244*]

2766 M 2247 M 1649 S 1292 S
2718 S 2180 VS 1626 M 1269 W
2625 S 2141 MSh 1621 VS 1239 VS
2557 S 2037 M 1572 S 1156 VS
2508 M 1971 M 1531 VS 1093 MS
2437 MS 1940 VS 1502 S 1042 VS
2392 S 1866 MW 1473 VS 1037 S
2347 VS 1825 VS 1443 VS 865 SVB
2300 MW 1795 S 1403 VS

$C_4H_2Cl_2S$ [161]

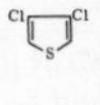

1490 VS 1348 VVS 1135 M 957 VS
1418 S 1326 SSh 1120 S 853 VS
1403 S

$C_4H_2Cl_2S$ [162]

3115 MSSp   1517 VVSSp   1168 S   884 VS
1712 SSp   1353 VS   1037 VS   697 VSB

$C_4H_2Cl_2S$ [211, 602*]

3125 SSp*   1414 VVS   1089 VS   819 VS
1515 VS   1340 VVS   876 VS   725 VS
1416 VS   1173 VS   836 VS   679 VS

C_4H_4 [544, 551*]

1600

$C_4H_4N_2$ [420]

665

$C_4H_4N_2$ [415]

1650   1610   1570   679

$C_4H_4O_3$ [556]

1872   1790

$C_4H_4O_4$ [524]

1750 (cis)   1680 (trans)

C_4H_4S [199, 607*]

3096 VS   1805 M   1408 VVS   904 M
2278 MW   1770 MS   1285 MSSh   871 S
2165 M   1590 VS   1253 VVS   835 VS
2114 M   1558 S   1082 VS   713 VS

C_4H_5ClO [672]

690

C_4H_5NO [15]

1576 S   1411 VS   1059 M   873 M
1449 M   1121 S   1036 VW

C_4H_5NO [15]

1599 VS   1445 W   916 M   788 MS
1474 VS   1160 M   870 MW   778 S
1473 VVS   1001 W

C_4H_5NS [688]

1652 M

C_4H_6 [440]

1064 VSVB

C_4H_6 [542]

3060   1570   1566

C_4H_6 [679]

1380 VS (Raman only)

$C_4H_6Br_2O$ [566]

1760

$C_4H_6ClF_3O$ [261, 262*]

3008 S   1740 M   1300 VSB   918 S
2927 MWSp   1488 S   1248 VVS   899 S
2610 MW   1452 VS   1218 SB   848 VS
2445 S   1381 VS   1086 VSB   805 VS
2280 W   1364 VS   1027 VS   734 S
2030 W   1312 VS   940 M

$C_4H_6Cl_2$ [396]

787

$C_4H_6N_2O$ [15]

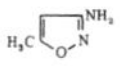

1515 S   1445 WSh   1116 W   1013 MW
1473 VVS   1423 VS   1035 VW   897 MW

$C_4H_6N_2O$ [15]

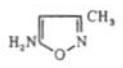

1636 VVS   1460 VS   1020 W   869 MW
1507 VS   1435 VS   1005 M

$C_4H_6N_2O$ [6]

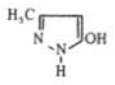

1620 S   1555   1505

C_4H_6O [388, 578,* 622*]

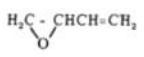

818

C_4H_6O [519, 545]

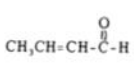

1685   1648   1638

C_4H_6O [540,* 559,* 565]

1775   1772

C_4H_6O [666]

1275   1200

$C_4H_6O_3$ [514]

1824   1748

$C_4H_6O_3$ [674]

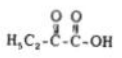

1744

$C_4H_7BrO_2$ [583]

1736

C_4H_7Cl [400]

769

$C_4H_7F_2NO$ [674]

1718

$C_4H_7NO_2$ [600]

1515

$C_4H_7NO_2$ [600]

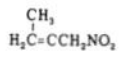

1555   1366

$C_4H_8N_2$ [567]

1664 S

$C_4H_8N_2O_2S_2$ [31]

3260 MS   1534 MS   1470 M   1049 S
3165 S   1515 S   1386 M   916 MS

C_4H_8O [445, 540*]

1090

C_4H_8O [394]

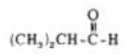

795

$C_4H_8O_2$ [451]

1111

$C_4H_8O_2$ [375]

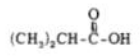

905

C_4H_9NO [529]

1720-1715   1650

$C_4H_9NO_2$ [310]

2899 VS   1548 VS   1218 S   913 M
2747 MSh   1460 S   1134 S   858 S
2132 MW   1431 VS   1064 M   833 VW
2049 M   1381 VVS   1007 S   795 VW
1980 MW   1305 M   983 W   760 S
1832 M   1290 M   954 M   740 W
1550 VS   1248 M

$C_4H_9NO_2$ [324]

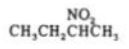

2976 VS   1546 VS   1264 S   1006 S
2933 VS   1458 S   1170 MS   971 S
2882 VS   1391 S   1144 MW   879 S
2584 WVB   1362 S   1121 SSh   845 S
2421 MW   1353 S   1110 VS   816 MW
2070 WB   1318 S   1093 S   796 VS
1996 W   1289 VS   1022 MSh   724 MB

$C_4H_9NO_2$ [325]

2976 VS   2037 W   1453 VS   1127 MWB
2933 VS   1961 MW   1404 VS   1038 MS
2857 SSh   1739 M   1374 VS   939 M
2695 MW   1639 SSh   1348 VS   858 VS
2520 MWSh   1534 VVS   1253 S   801 S
2387 MW   1473 VS   1186 VS   730 M
2164 MW

$C_4H_9NO_2$ [326]

3185 MSh   1875 M   1344 VS   933 MW
2950 VS   1818 M   1295 S   912 MSh
2882 VS   1736 S   1233 S   900 S
2558 M   1546 VS   1174 S   836 S
2445 SSp   1458 MS   1142 S   799 VW
2299 MS   1431 M   1103 MWSh   758 M
2119 W   1395 S   960 MW   734 VS
1953 MW   1379 S

C_4H_{10} [373, 649*]

2959 VS   1481 VS   913 VS   784 M
2273 W   1006 VW   799 S

$C_4H_{10}BI_2P$ [30]

1054 VS

$C_4H_{10}O$ [319] $CH_3CHCH_2CH_3$ (OH)

3400 VS*	1403 VS	1116 VSB	820 MS
2959 VS	1376 VVS	1031 VS	801 MB
2924 VS	1314 VS	992 VS	796 M
2865 VS	1290 VSB	969 S	777 S
1456 VVS	1147 VS	911 VS	

$C_4H_{10}O$ [353] $CH_3(CH_2)_3OH$

1458 VVS	1112 S	1031 S	952 S
1377 VVS	1072 VS	1010 S	846 S
1339 S	1042 S	990 S	738 S
1290 MS			

$C_4H_{10}S$ [172]

2967 VS*	2227 M	1259 VVS	781 VS
2941 VS	1621 MVB	1074 MS	764 MSh
2874 VS*	1451 VVS	1047 MS	738 M
2725 VSSh	1377 VS	973 VS	693 VS
2415 M			

$C_4H_{12}BP$ [30]

1118 M

$C_4H_{12}N_2$ [501]

2850 S

$C_4H_{12}Pb$ [402]

1169 767

$C_4H_{12}Si$ [386]

841

$C_4H_{12}Sn$ [399]

776

$C_4H_{16}B_2CoN_4O_3S$ [5]

649 M

$C_4H_{16}ClCoN_4O_3S$ [5] $[Coen_2SO_3Cl]^0$

1117 S 1075 S 984 VS 625 S

$C_4H_{16}ClCoN_4O_3S$ [5] $[Coen_2SO_3]Cl$

1119 S	1036 VS	649 M	625 S
1093 S	989 VS		

$C_4H_{16}Cl_2N_4Pd$ [8]

1609 S	1324 M	1062 S	900 M
1458 M	1280 M	1002 M	804 M
1372 M			

$C_4H_{16}Cl_2N_4Pt$ [8]

1610 S	1326 M	1138 S	999 M
1467 M	1311 M	1130 S	897 M
1454 M	1154 S	1050 S	831 M
1373 M			

$C_4H_{16}Cl_4CuN_4Pt$ [8]

1571 S	1321 M	1166 M	705 VSSh
1376 M	1282 M	1044 S	694 VS

$C_4H_{16}Cl_4N_4PdPt$ [8, 43]

1574 VS	1189 M	1065 S	750 S
1287 M	1109 S	766 S	

$C_4H_{16}Cl_4N_4Pt_2$ [8, 43]

1575 VS	1296 M	1143 S	801 S
1391 M	1219 M	1057 S	787 S

$C_4H_{16}CoIN_4O_3S$ [5]

1119 S	1042 VS	649 M	621 M
1070 S	983 VS		

$C_4H_{16}CoN_4NaO_6S_2$ [5] $[Coen_2(SO_3)_2]Na$ (trans)

1068 S 939 VS 630 S

$C_4H_{16}CoN_4NaO_6S_2$ [5] $[Coen_2(SO_3)_2]Na$ (cis)

1095 S 943 VS 625 S

$C_4H_{17}CoN_4O_4S$ [5]

1117 S 1062 S 972 VS 625 S

$C_4H_{19}ClCoN_5O_3S$ [5]

1112 S 1079 S 637 S

$C_4HgK_2N_4$ [468]

2146 VS

$C_4K_2N_4Zn$ [468]

2151.5 VS

C_5F_5N [635]

1497 S

C_5F_{10} [245, 246*]

2596 M	2366 MS	1804 VS	1295 S
2522 MS	2314 MS	1733 MW	1230 VS
2481 MW	2239 W	1618 VSB	1049 S
2434 M	2216 MW	1548 S	891 SVB
2421 M	1873 VS	1488 VS	829 MB
2383 MS			

C_5F_{12} [237, 238*] $CF_3CF_2CF_2CF_2CF_3$

2636 MSSh	2205 M	1731 S	1152 VS
2594 S	2119 VSSp	1692 MS	1136 VS
2577 S	2092 VSSp	1663 MS	1068 VW
2475 VS	2049 S	1567 VS	1046 MSh
2442 VSSh	2017 VSSp	1468 VSVB	1022 VS
2371 VSSp	1976 S	1418 VVS	882 VS
2331 S	1963 S	1339 S	834 VS
2280 MS	1920 M	1289 SSh	741 VS
2261 MW	1856 S	1259 VVS	726 VS
2237 M	1836 S	1220 VVS	

C_5F_{12} [240, 241*] $(CF_3)_2CFCF_2CF_3$

2140 M	1453 MS	1157 VS	988 VS
1885 M	1435 MSSh	1101 S	833 MSp
1585 MSB	1355 S	1095 SSh	832 VS
1553 SB	1229 VVS	1066 MW	736 VS
1493 MSh	1175 SSp		

C_5H_4BrN [433, 488,* 593,* 619*]

1589 MW	1440 MW	1096 S	1025 WSh
1576 MW	1417 VS	1086 M	1009 VS
1465 M	1117 MWSh		

C_5H_4ClN [433, 488,* 593,* 619*]

1581 MW	1419 S	1120 MWSh	1093 MW
1573 MW	1407 MW	1108 VS	1018 VS
1472 S			

$C_5H_4N_2O_2$ [433, 488, 593,* 619*]

1608 VS	1472 MW	1192 S	1021 S
1580 S	1428 S	1114 W	

$C_5H_4N_2O_3$ [24, 488,* 593,* 608,* 619,* 671*]

867 S 749 M

$C_5H_4O_2$ [7]

1684 S	1660 VS	1634 S	1414 S
1680 S	1660 S	1632 S	1317 VS
1678 SSh	1658 S	1628 S	1195 M
1675 S	1642 S	1621 S	1035 M
1675 S	1638 SSh	1614 S	1026 W
1674 S	1637 SSh	1612 S	919 VS
1672 S	1635 S	1464 W	849 VS
1662 S			

C_5H_4OS [7, 382*]

1661 SSh	1504 M	1136 VS	918 VS
1623 VS	1431 VS	1111 W	846 S
1574 S	1411 M	1021 M	814 VS
1545 M	1306 S		

C_5H_4OS [7]

1643 S 1271 M 1162 S 1124 S

C_5H_5N [405, 488,* 593,* 619,* 621,* 676*]

1580 1570 1485 750 VS

C_5H_5NO [24, 488,* 593,* 619*]

836 S 771

C_5H_6ClNO [15]

1628 M	1420 S	1125 VVS	880 MW
1456 M			

$C_5H_6N_2$ [327, 438, 488,* 593,* 596,* 619*]

3510	3095 M	1485 S	1149 M
3450 M	1634 VS	1444 VS	1143 M
3420	1621 M	1335 M	1046 M
3334	1610	1326 M	990 M
3330 M	1600 S	1318 M	775 VS
3180 M	1572 M	1278 M	769 S
3175 M	1560 M	1270 M	740 M
3100 W	1497 VS	1250 M	

$C_5H_6N_2$ [327, 433, 488,* 593,* 596,* 619*]

3400 1621 M 1445 VS 1047 M
3334 1610 1442 VS 1021 MW
3210 1590 S 1299 MB 1019 MW
3205 M 1587 S 1259 M 913 M
3095 M 1488 MS 1196 MWSh 896 M
2960 1486 S 1131 M 790 MB
1631 S 1481 M 1130 MW

$C_5H_6N_2$ [327, 488,* 593,* 596,* 619*]

3510* 3085 M 1572 M 1264 M
3435 M* 3040 S 1506 M 1220 S
3420 M 2940 1499 M 1209 S
3320 1645 S 1438 M 994 VS
3185 M 1623 M 1335 MS 843 S
3175 1604 S 1314 M 825 VS
3090 M 1597 VS 1271 MS

$C_5H_6N_2O_3$ [15]

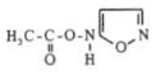

1660 VVS

$C_5H_6N_2O_3$ [15]

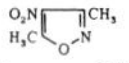

1610 VVS 1420 VVS 1167 VVS 1040 W
1442 M

$C_5H_6O_3$ [557]

1815 1766

$C_5H_7AgO_2$ [16]

1612 S

$C_5H_7CsO_2$ [16]

1613 S

$C_5H_7KO_2$ [16]

1626 S

$C_5H_7LiO_2$ [16]

1616 S

C_5H_7NO [15]

1613 VS 1414 VS 1009 M 792 MW
1459 M 1023 W 885 M

$C_5H_7NO_2$ [526, 528,* 552*]   NC-CH$_2$-C(O)-O-C$_2$H$_5$

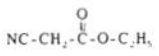

1751

$C_5H_7NO_2$ [15]

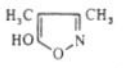

1665 VVS 1440 MW 1132 M 932 W
1537 VVS

$C_5H_7NaO_2$ [16]

1618 S

$C_5H_7O_2Tl$ [16]

1610 S

C_5H_8 [540,* 572]

1678

C_5H_8 [492, 627]

3017 1945 1611

$C_5H_8ClF_3O$ [263, 322*]

2982 VS 1393 MSh 1244 MS 979 S
2950 S* 1385 MSh 1212 VSB 958 S
2917 MSp 1368 VS 1089 VSB 915 WB
2890 S 1311 MS 1065 VSSh 906 MB
1480 M 1295 VS 1007 MW 760 W
1471 M

$C_5H_8N_2O$ [6]

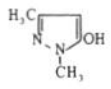

1540 1440 1385

$C_5H_8N_2O$ [6]

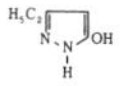

1545 1510

$C_5H_8N_2O$ [15]

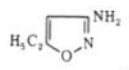

1627 VVS 1474 VVS 1120 W 906 MW
1513 S 1442 WSh 1010 W

$C_5H_8N_2O$ [15]

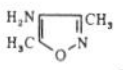

1660 M 1477 MS 1431 M 897 W
1496 MWSh

$C_5H_8N_2O$ [15]

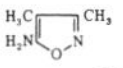

1658 VVS 1474 VVS 1135 W 1007 MW
1504 VVS 1437 S 1045 W 879 W

C_5H_8O [674]

1744

$C_5H_8O_2$ [654]

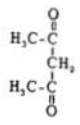

1284

$C_5H_8O_2$ [193, 527]   H$_2$C=C(CH$_3$)CO$_2$CH$_3$

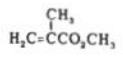

2958 SSp 1721 W 1381 S 1022 VS
2932 VSSp 1718 W 1358 S 1014 VS
2907 S 1636 VS 1324 VVS 1002 MSh
2849 SSP* 1567 VVS 1197 VS 831 SSh
1730 VVS 1440 VVS 1162 VS 812 VS

$C_5H_8O_2$ [674]

1741 1726

C_5H_{10} [92]   H$_2$C=C(CH$_3$)CH$_2$CH$_3$

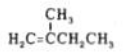

2950 VS 1453 VS 903 VS 880 VS
1650 VS 1383 S 890 VS 800 SB

C_5H_{10} [131]   H$_2$C=CHCH(CH$_3$)$_2$

2965 VS 1477 SSh 1309 MS 999 VS
2915 VS 1464 S 1111 M 990 VS
1828 MW 1422 MS 1095 M 919 VVS
1647 MS 1379 S 1010 VS

C_5H_{10} [207]   CH$_3$CH=CHCH$_2$CH$_3$ (trans)

2969 VSSp 1960 W 1670 S 1017 S
2939 VS 1840 M 1458 S 964 VS
2924 VS 1785 SSH 1379 MS 943 MSh
2730 SSh 1760 S 1290 S 876 M
2620 S 1710 S 1063 S 798 M
2400 S

C_5H_{10} [208, 548*]   CH$_3$CH=CHCH$_2$CH$_3$ (cis)

2969 VSSp 2235 S 1406 M 933 VS
2939 VS 2040 SSh 1374 M 860 M
2879 VSSp 2000 VS 1307 VS 791 S
2725 MSSh 1658 VS 1071 VS 697 VS
2400 S 1466 S 1024 S 692 M
2310 S 1458 S

$C_5H_{10}O$ [494]

2977 2936 2902

$C_5H_{10}O_2$ [495]

2970 2952

$C_5H_{10}S_3$ [434]

1074

C_5H_{12} [333, 401, 423, 424]   (CH$_3$)$_2$CHCH$_2$CH$_3$

2975 VS* 1297 S 1036 VS 766 VS
1563 S 1176 S 1012 VS 766 S
1471 VS 1163 VS 986 VS 668 VS
1387 VS 1146 S 916 S

C_5H_{12} [371]   CH$_3$CH$_2$CH$_2$CH$_2$CH$_3$

929 VS 899 VSSh 738 VSShB 733 VS

C_5H_{12} [57,* 655]   (CH$_3$)$_4$C

1370 1280

$C_5H_{12}BCl_2N$ [4]

1613 M

$C_5H_{12}O$ [194]   CH$_3$CH$_2$C(OH)(CH$_3$)$_2$

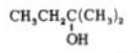

3391 MS 1376 VS 1167 M 937 VS
2967 VS* 1328 S 1059 VS 882 VS
2933 VS* 1274 S 1021 MSSh 881 VS
2882 MSh 1190 VS 1003 S 786 M
1464 VS

$C_5H_{12}O$ [195]   (CH$_3$)$_2$CHCH$_2$CH$_2$OH

3367 S 1464 VVS 1171 M 968 VS
2959 VS 1385 VS 1124 S 952 MSh
2924 VS 1368 VS 1060 VS 940 W
2865 VS 1214 S 1010 VS

$C_5H_{12}O$ [196]   CH$_3$CH$_2$CH(OH)CH$_2$CH$_3$

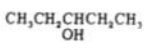

3367 VS 1376 VS 1124 VSB 967 VS
2967 VS* 1305 SB 1109 VSB 924 M
2933 VS 1247 S 1041 VS 854 MS
2874 VS 1145 VS 1009 MW 766 MVB
1460 VVS

$C_5H_{12}O$ [318]

3401 MS 1339 MS 1062 S 903 S
2967* 1297 VSB 1028 VS 891 S
2933 VS 1255 VSB 1027 VS 864 W
2874 VS 1227 MS 998 S 832 S
1456 VVS 1147 VSVB 949 S 743 S
1374 VVS 1114 VS

$C_5H_{12}O$ [321]

3367 MS 1379 VS 1111 S 932 M
2967 VS* 1255 SSh 1042 S 900 M
2924 VS 1224 B 1014 S 796 M
2874 VS* 1166 M 956 M 766 S
1456 VS

$C_5H_{12}O$ [350, 393]

3345 VS* 1462 VVS 1114 S 890 M
2959 VS 1379 VSB 1078 VS 783 MB
2933 VS 1340 SB 1056 VS 742 MB
2865 VS 1232 MS 1006 VS 729 MB
2857 VS 1200 MS 980 MS 682 MB

$C_5H_{16}CoN_5O_3S_2$ [5]

1099 S 1089 S 980 VS 629 S

$C_5H_{16}CoN_5O_3S_2$ [5]

1117 S 1036 VS 988 VS 621 M
1093 S

C_6F_6 [623]

1536 S

$C_6H_2F_4$ [222, 223]

3176 MSh 2046 VW 1513 VS 1082 VW
3088 VS 2020 MS 1451 VS 1049 M
2981 W 1969 S 1445 VS 1016 M
2918 VS 1912 VS 1439 VS 990 VW
2869 MW 1857 VS 1439 VS 982 M
2828 S 1830 VS 1377 M 963 VS
2804 S 1819 VS 1340 VS 873 VS
2771 MW 1707 MS 1305 VSSh 869 SB
2649 VS 1704 VS 1277 VSSp 867 VS
2534 VS 1701 M 1277 VS 854 VS
2490 MS 1674 S 1271 VS 822 WSh
2407 VW 1673 M 1233 VS 810 MW
2353 VS 1650 S 1225 VS 781 S
2340 VS 1630 S 1222 VS 728 MS
2279 VS 1629 S 1170 VS 720 M
2254 MS 1601 S 1167 VS 704 VS
2226 VS 1543 VS 1162 VS 700 VS
2182 W 1534 VS 1121 VW 697 VS
2105 W 1517 VS 1100 M 677 VS
2069 S

$C_6H_2F_9NaO_6$ [525]

1625

$C_6H_3F_3$ [224]

3098 S 2356 VS 1868 VS 1376 W
2914 S 2345 VS 1826 S 1287 W
2882 VS 2300 VSSp 1788 M 1250 VS
2822 MS 2246 MW 1743 S 1203 VS
2690 W 2235 M 1727 S 1144 VS
2658 M 2196 M 1709 VS 1099 VS
2615 M 2132 VW 1685 W 1059 MW
2581 S 2102 MS 1629 S 1000 SSh
2551 VS 2086 S 1613 S 932 W
2537 VS 2061 M 1585 VW 854 VS
2506 MW 2028 VS 1575 VW 809 VS
2497 MW 1961 VW 1522 VS 781 VS
2447 VS 1945 MS 1460 M 728 VS
2404 VS 1926 M 1443 VS 717 MSh
2390 VS 1893 M 1406 S 688 VS

$C_6H_3FeNO_4$ [509, 515*]

2218 S

$C_6H_4F_2$ [225]

3088 VS 2424 VS 1720 S 1085 VS
3064 VS 2296 VSSp 1634 VS 1012 VS
3030 VS 2257 VW 1511 VS 943 MB
2898 VS 2228 VS 1437 VS 928 SB
2813 W 2151 MS 1414 SSh 884 S
2786 S 2048 VS 1212 VSB 833 VS
2687 M 2021 MSh 1202 VSB 806 S
2646 M 1978 VS 1183 VS 737 VS
2527 MW 1737 M 1117 VW

$C_6H_4N_2$ [433]

1567 MW 1418 S 1122 W 1025 MS
1471 MW

$C_6H_4N_2O_4$ [620]

1560

$C_6H_4N_4O_6$ [19]

3457 MS 3456 S 3395 MW 3339 M
3457 M 3450 Sh 3344 MS 3321 M

$C_6H_4O_2$ [690]

1669

C_6H_5Br [414, 630*]

680 S

C_6H_5BrOS [348]

1672 VSSp 1319 SSp 1033 S 806 VSSh
1520 MW 1271 VVS 1017 S 796 M
1414 VVSSp 1236 MS 971 SSh 741 M
1361 SSp 1214 M 962 MSh 677 VS
1361 MS 1072 S 923 VS

$C_6H_5ClN_2O_2$ [19]

3511 M 3487 MW 3388 M 3343 MW
3504 MW 3487 Sh 3383 MW 3292 MW
3501 MW 3479 MW 3381 M 1621 VS
3497 MW 3455 MW 3373 MW 1618 S
3495 MW 3392 M

C_6H_5ClOS [372]

3095 VS 1522 M 1239 VSSh 927 M
3010 SSp 1425 VVS 1215 SSh 797 VS
2835 M 1325 VS 1041 M 682 S
1669 VS 1272 VVS 1008 VS

$C_6H_5Cl_3Si$ [86]

2702 W 1592 M 1160 W 988.4 W
1980 W 1490 W 1120 VS 846.7 W
1961 W 1466 W 1095 W 739.2 VS
1838 W 1337 M 1063 W 715.6 VS
1658 W 1304 M 1029 W 691.1 VS
1613 W 1267 W 998.2 S

C_6H_5F [226, 249, 390,* 435,* 630*]

3193 MSh 2579 Ms 2129 VW 1326 VS
3087 VS 2546 VW 2021 S 1290 W
3067 VS 2513 VW 1962 VS 1220 VS
3053 VS 2503 VW 1939 VS 1156 VS
2955 W 2484 W 1778 VS 1105 W
2915 MW 2443 S 1714 S 1066 VS
2892 MW 2375 MW 1624 SSh 1020 VS
2878 M 2313 VSSp 1600 VS 875 W
2781 MW 2302 S 1597 VS 831 W
2713 W 2261 W 1499 VS 685 VS
2649 M 2222 MS 1460 SSp 653 MW
2615 MS 2175 MW 1397 M

C_6H_5NO [433, 481,* 488,* 593,* 619*]

1595 VS 1428 S 1109 W 1025 MS
1581 MS 1120 WSh

$C_6H_5NO_2$ [20, 435,* 484,* 625,* 630*]

3096 M 1412 M 1107 S 935 S
3068 M 1351 VS 1094 S 852 S
1603 S 1316 S 1069 S 794 S
1585 S 1242 M 1020 S 704 S
1527 VS 1161 S 1002 M 677 S
1475 S

$C_6H_5N_3$ [37]

1618 MW

$C_6H_5N_3O_4$ [19, 484,* 625*]

3517 3474 MW 3381 3300
3511 3393 3368 S 1632 VS
3500 3391 3348 MS 1627
3484 M

$C_6H_5N_3O_4$ [19, 484,* 625*]

3504 Sh 3476 S 3364 M 3342 Sh
3488 W 3475 S 3362 M 3314 M
3478 S 3475 M 3361 M 1635 VVS
3478 M 3474 M 3358 M 1633 VVS
3477 M 3473 M 3358 M

C_6H_6 [309, 416, 630,* 677]

3060 VS 1500 VS 1038 S 690 VS
1970 MS 1486 S 1024 M 674 S
1820 S 1050 M 779 W 671 VS
1585 VS (Raman only)

C_6H_6ClN [534]

1613 S

C_6H_6ClN [534]

1613	1597		

$C_6H_6Cl_2Si$ [86]

2217 S	1479 W	1121 VS	810.4 VS
1908 W	1464 W	1089 W	794.2 VS
1821 W	1379 W	1068 W	730.4 VS
1773 W	1339 M	1037 W	703.2 W
1543 W	1307 W	999.1 M	692 VS
1493 W	1269 W	970.9 W	677.4 W

$C_6H_6N_2O_2$ [19, 21, 435*]

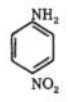

1641 MS	1592 S	1178 S	858 S
1629 S	1589 S	1112 S	842 S
1624 VVS	1503 S	1050 S	839 S
1616 S	1181 S	998 S	750 S
1599 S			

$C_6H_6N_2O_2$ [19]

1627	1626	1624 VVS	

C_6H_6O [36, 44,* 482,* 483,* 630,* 632]

1620	1350	1000 M	753 S
1515	1310	760	687 S
1495	1235		

C_6H_6S [212]

3115 SSp	1404 SSp	1082 MW	853 VS
1620 VS	1344 MW	1049 S	829 VS
1517 M	1240 S	980 VS	695 VSB
1438 VSSp	1201 VS	900 SSh	

C_6H_7N [327, 433, 453]

1600 MW	1414 MW	1102 W	1029 M
1582 M	1188 W	1043 VS	934 MVB
1479 M	1124 W	1040 W	

C_6H_7N [198, 438]

3067 VS	1433 VVS	1149 VS	980 S
3021 VS	1376 VS	1100 S	800 M
1592 VVS	1292 VVS	1052 VS	755 VVS
1575 VVS	1238 VS	1001 VS	732 VS
1473 VVS			

C_6H_7N [391,* 470, 630,* 687]

3480	3394	3376	1620
3454			

C_6H_7NO [433]

1596 MW	1478 MW	1190 MWSh	1099 W
1580 M	1424 M	1122 WSh	1042 MSSh

$C_6H_7NO_2$ [15]

1600 VS	1422 M	1003 M	916 M
1456 VS	1019 W	920 M	

$C_6H_7NO_3$ [15]

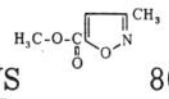

1596 M	1448 VS	1005 VS	802 M
1484 MS	1416 MW	908 M	

$C_6H_7NO_3$ [15]

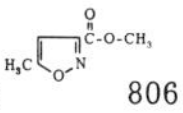

1601 VS	1414 M	1003 VS	806 S
1463 VS	1016 SSh		

$C_6H_7NO_3$ [15]

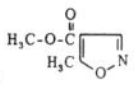

1618 VVS	1408 S	927 M	801 S
1488 S	1155 MS	896 MW	774 S
1450 VS	980 MW		

C_6H_8 [203, 286]

3095 SSp*	2335 VSSp	1374 MS	1130 S
3090 SSp*	1875 MSSh	1362 MS	1011 VS
3040 SSp	1810 VS	1311 SB	899 VS
3012 SSp	1720 M	1300 SB	748 S
2877 SSh	1623 VVS	1255 S	687 VS
2836 MSSh	1429 VVS	1166 MB	661 VS

C_6H_8BrClO [674]

1742			

$C_6H_8Br_2O$ [674]

1737			

$C_6H_8N_2$ [433]

1595 VS	1484 S	1191 MSh	1008 MW
1583 VS	1430 MWSh	1123 W	

$C_6H_8N_2O_4S_2$ [31]

3150 S	1715 VS	1440 M	1249 VS
3050 M	1529 S	1400 M	901 VS
2925 MS	1517 S	1367 MS	883 MS

C_6H_8S [213]

2995 VS	1381 S	1081 MS	851 VS
1464 VS	1318 S	1029 MS	825 VS
1445 VS	1230 S		

C_6H_9BrO [674]

1734			

C_6H_9N [344]

3385 VVS	1514 S	1291 VS	986 S
3087 SSh	1463 VS	1250 S	957 S
2925 VS	1448 VS	1149 VS	790 VVS
2870 VVS	1419 VS	1113 VS	732 S
2529	1392 VS	1039 MSB	713 WSh
1588			

C_6H_9NO [15]

1644 MS	1464 MS	1428 S	892 W
1490 WSh	1445 MS	1196 M	

$C_6H_9NO_2$ [15]

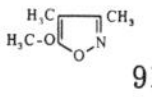

1666 VS	1480 VVS	1188 M	914 MS
1528 VVS	1451 S	1134 VS	

$C_6H_9NO_2$ [600, 625*]

1538			

C_6H_{10} [128]

3030 VSSh	1437 VS	1037 VS	876 VS
2935 VS	1323 M	1018 M	810 M
1650 VS	1266 MS	917 VS	719 VS
1608 MS	1138 VS	904 S	

C_6H_{10} [216,* 288]

3085 MSp	1385 S	1104 M	963 VS
3005 S	1293 MS	1087 M	876 VVSB
1916 M	1239 S	1047 MS	818 S
1761 S	1190 W	1002 VS	712 M
1637 S	1175 W	996 W	691 M
1456 VS			

C_6H_{10} [374] $CH_2{=}CHCH_2CH_2CH{=}CH_2$

1630 S	1419 S	995 VS	912 VS

C_6H_{10} [354] $CH_3CH{=}CHCH{=}CHCH_3$

1445 S	1054 M	945 S	813 M
1234 M	986 S	923 S	703 M
1150 M			

$C_6H_{10}ClF_3O$ [264, 265*]

2973 VS	1408 MS	1092 VS	860 S
2945 VSSh	1370 VS	1067 VS	845 MB
2882 VSSp	1308 VSB	1022 W	805 VS
2425 MB	1297 VSB	1004 WB	740 SVB
1471 M	1245 VS	969 S	686 S
1464 VS	1202 VSB	952 S	

$C_6H_{10}N_2O$ [6]

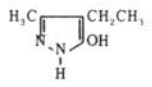

1620 S	1585	1560	

$C_6H_{10}N_2O$ [6]

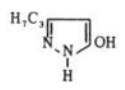

1625 S	1545		

$C_6H_{10}N_2O$ [6]

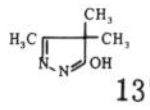

1707 S	1610	1460	1375

$C_6H_{10}N_2O_2S$ [31]

3325 S	1601 M	1240 MS	1062 M
3180 MS	1513 S	1233 MS	1028 MS
1657 S	1435 M	1105 M	

$C_6H_{10}O_3$ [674]

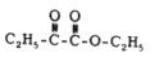

1747			

$C_6H_{10}O_3$ [436, 638]

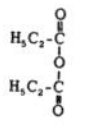

2990	1470	1020	

$C_6H_{11}NO$ [461,* 621]

1558			

$C_6H_{11}NO_2$ [600]

1357			

C_6H_{12} [103]

2922 VS 1452 VS 1038 S 905 VS
2695 MWSh 1259 MS 1014 MS 862 VS

C_6H_{12} [104, 105, 171, 633]

CH₃

3185 SSh 2627 MB 1377 S 977 VS
2995 VS 2593 M 1350 M 907 MB
2970 VS 1470 VS 1144 M 900 M
2952 VS 1461 VS 1139 S 896 MB
2866 VS 1455 VS 988 M 889 MSB
2730 M 1389 S 978 M 886 MB
2698 MSh

C_6H_{12} [65, 459]

$(CH_3)_2C{=}C(CH_3)_2$

2975 VS 2865 VS 1370 VS 1156 S
2905 VS 1447 VS 1167 S

C_6H_{12} [129]

$H_2C{=}CHC(CH_3)_3$

2975 VS 1477 VS 1362 SSh 1068 MW
1990 W 1466 VS 1267 M 1000 VS
1825 MS 1416 M 1209 S 911 VS
1645 VS 1383 S

C_6H_{12} [130]

$CH_3CH{=}CHCH(CH_3)_2$

2975 VS 1361 SSh 1256 M 967 VS
1462 VS 1335 M 1179 MW 838 W
1379 VS 1302 M 988 VSSh 720 M

C_6H_{12} [163]

$H_2C{=}C(CH_2CH_3)_2$

3085 SSh 1650 VVS 1087 S 890 VVS
2900 VVS 1456 VVS 1054 M 803 MS
1790 MSSp 1377 VVS

C_6H_{12} [275]

H_3CH_2C CH_2CH_3 C=C H H (cis)

3165 SSh 1653 VS 1299 VS 905 S
2960 VS 1462 S 1068 VS 805 SB
2875 VS 1407 S 1027 M 788 MB
2740 MSh 1377 S 1005 VW 714 VS

C_6H_{12} [168]

H_3CH_2C H C=C H CH_2CH_3 (trans)

2900 VVS 1379 VS 1312 M 1030 MSSp
1669 W 1348 S 1285 S 778 MS
1464 VVS 1326 S 1068 VS

C_6H_{12} [164, 274]

H_3C CH_3 C=C H CH_2CH_3 (cis)

2965 VS 1618 M 1206 MS 1005 S
2935 VSSh 1456 S 1114 S 918 MS
2900 VS 1379 MS 1079 S 917 S
2875 VS 1376 VVS 1078 S 812 VVS
2730 M 1355 MSh 1057 MS 788 S
1842 M 1305 M 1021 VS 739 W
1670 M 1272 M

C_6H_{12} [165, 428]

H CH_2CH_3 C=C H_3C CH_3 (trans)

3027 Sh* 1675 S 1326 SSh 999 S
2971 VS 1675 MS 1211 M 979 S
2939 VS 1642 MSh 1198 M 923 M
2924 VS* 1451 VS 1116 S 823 VVS
2900 VVS 1385 S 1072 S 785 M
2880 VSSh 1346 SSh 1030 MSh 749 M

C_6H_{12} [169, 427]

H_3C H C=C H $CH_2CH_2CH_3$ (trans)

2900 VVS 1456 VVS 1267 M 1044 MS
1730 MW 1381 VS 1087 MS 1020 M
1669 MW 1340 M 1075 MS 916 S

C_6H_{12} [276]

H_3C C_3H_7-n C=C H H (cis)

3010 SSh 1445 VS 1075 M 991 M
2960 VS* 1408 W 1057 M 965 M
2875 VSSh 1381 VS 1047 M 911 M
1656 VS 1272 VS 1035 MSh 694 VS
1460 S 1222 MW

C_6H_{12} [206]

CH_3 $H_2C{=}CCH_2CH_2CH_3$

3340 VS* 2878 VS 1377 S 819 M
3077 MSSp 1785 VS 1097 S 742 VS
2967 VVS 1653 VS 888 VS 739 VS
2934 VS 1458 S 826 M 703 W

C_6H_{12} [167]

$H_2C{=}CHCHCH_2CH_3$ CH_3

3095 SSp 1456 VVS 1292 M 909 VVS
2935 VS 1425 VS 1092 M 762 M
1828 M 1381 VVS 999 VS 675 VS
1642 VS 1319 M 960 MS

C_6H_{12} [166]

$(CH_3)_2C{=}CHCH_2CH_3$

2900 VVS 1381 VS 1122 S 908 M
2730 MSpSh 1304 S 1062 VS 833 VVS
1672 MS 1211 S 986 M 745 VS
1456 VVS

C_6H_{12} [496]

$H_2C{=}CH(CH_2)_3CH_3$

2965 VS

$C_6H_{12}ClNO$ [674]

1659

$C_6H_{12}N_2O_3$ [26]

2936 M 1462 S 1260 M 1113 S
2872 M 1449 S 1249 M 1080 S
2842 1445 S 1222 M 1042 M
1544 VS 1400 M 1150 M 897 S
1542 VS 1282 M 1128 M 851 M

$C_6H_{12}N_2O_3$ [26]

2927 S 1357 S 1064 S 849 S
1540 VS 1341 S 1012 S 796 S
1459 S 1324 S 898 S 764 M
1391 M 1241 M

$C_6H_{13}N$ [479]

3481

$C_6H_{13}NO$ [532]

1647 1615

C_6H_{14} [94, 145]

$(CH_3)_3CCH_2CH_3$

2960 VS 1730 M 1307 M 1000 S
2745 SSh 1705 MSB 1307 W 997 VS
2715 MSSh 1670 MSB 1252 S 996 VS
2665 MSh 1475 VS 1250 S 930 M
2585 MWSh 1468 S 1217 VS 929 S
2550 MWSh 1395 MSSh 1074 VS 870 MW
2415 S* 1381 MSSh 1073 S 711 S
2295 S 1376 VS 1018 VS 710 M
2235 M 1366 S 1018 S 700 M
1810 M

C_6H_{14} [144, 210, 345]

$(CH_3)_2CHCH(CH_3)_2$

3360 MWSh* 2675 VSSh 1382 VS 1036 VVS
3180 VS 2595 VSSh 1371 S 999 VS
2962 VVS 2435 S 1280 S 989 VS
2960 VS 2380 S 1196 MSh 958 W
2880 VS 2240 M 1153 VS 921 S
2877 VS 1661 SB 1129 VVS 869 VS
2725 VSSh 1464 VS 1066 S 728 VS

C_6H_{14} [95,* 332]

CH_3 $CH_3CH_2CHCH_2CH_3$

2920 VS 1385 VS 996 VS 785 S
1534 W 1308 SS 971 VS 770 S
1475 VS 1264 M 815 M 749 M
1445 VS 1160 S

C_6H_{14} [355]

$CH_3(CH_2)_4CH_3$

726 VS

$C_6H_{14}O$ [320]

3356 S* 1460 VS 1119 S 918 S
2960 VS* 1359 VS 1058 VSB 892 MW
2933 VS 1218 S 1018 VSB 725 VS
2857 VS

$C_6H_{18}B_3Cl_6P_3$ [30]

2979 S 1314 M 1298 M 738 S
2918 S 1303 M 940 VS

$C_6H_{18}B_3N_3O_3$ [32]

1575 VS 1384 VS 1180 S 1017 S
1534 VS 1342 VS 1149 S 714 S
1456 VS 1224 S 1074 S 703 S
1412 VS

$C_6H_{24}B_3P_3$ [30]

2965 S 2332 S 1268 M 1000 M
2910 S 1310 M 1111 M 938 VS
2370 S

$C_6H_{24}Cl_4N_6NiPt$ [8]

1596 S 1282 M 972 S 617 VSSh
1581 S 1101 M 653 VS
1332 S 1023 VS 640 VSSh

C_7F_{14} [247, 248*]

2554 SB 2042 M 1595 MB 1018 VS
2485 SB 1962 M 1449 S 908 VVW
2452 VSB 1874 W 1269 MB 847 MW
2356 VS 1812 M 1230 VS 814 VS
2315 VSB 1771 W 1176 MW 733 VS
2271 S 1650 M 1138 VS 674 VS
2157 MB

C_7F_{14} [232]

2380 S	1715 M	1264 VS	1170 M
2310 S	1603 MW	1247 W	1140 VS
2150 MW	1567 W	1229 MS	1027 VS
2010 VS	1524 W	1214 M	933 W
1955 W	1502 W	1205 M	732 VS
1860 M	1414 W	1188 S	688 S
1790 MS	1279 S		

C_7F_{16} [241, 242*]

2605 VSSh	2010 SB	1241 VVSB	1058 VSSp
2490 VS	1845 SB	1225 VVSB	1030 VS
2445 VS	1577 SB	1185 MSh	844 MS
2380 VS	1548 SB	1155 VS	733 VS
2210 MB	1484 SB	1139 MSh	719 VS
2180 MB	1437 S	1119 S	707 VS
2095 MW	1344 S	1096 M	

$C_7H_3F_5$ [217, 250]

3069 VS	2351 W	1795 S	1255 M
3069 VS	2313 S	1768 VS	1199 VW
2958 M	2285 MS	1716 S	1149 VSB
2918 M	2232 M	1656 VSSh	1049 VS
2900 M	2187 MW	1637 VSSp	944 M
2686 VW	2165 SSp	1617 MS	914 VS
2645 MW	2103 S	1577 W	885 VS
2612 SSp	2066 MS	1506 VS	829 VS
2548 M	2040 VS	1443 VS	779 VS
2521 M	2031 VS	1387 W	745 VS
2445 VSB*	2008 VW	1287 S	730 MSh
2415 VSB	1860 VS	1264 M	664 VS
2369 MW			

$C_7H_4ClNO_3$ [385]

3100	1720	1350	840
1750	1610		

$C_7H_4F_4$ [218]

3085 VS	2330 S	1673 W	1058 VS
2982 S	2255 M	1605 VS	1006 S
2953 M	2203 MW	1497 VS	1006 VS
2925 S	2103 S	1458 VS	994 VW
2725 VW	2076 S	1333 VS	977 MS
2621 M	2058 M	1282 VS	950 VVW
2621 MB	2025 M	1214 VS	894 VS
2592 S	1980 MSpSh	1206 VS	882 W
2543 MS	1952 SSp	1174 MS	835 W
2422 M	1880 VS	1138 VS	793 VS
2389 MW	1808 S	1135 VS	765 VVW
2367 M	1761 VS	1087 VS	746 VS
2351 MS	1698 VS		

$C_7H_5BrO_2$ [25]

1757 S 1738 S 1711 S

$C_7H_5BrO_2$ [25]

1748 S 1702 S

$C_7H_5BrO_2$ [25]

1746 S

C_7H_5ClO [38, 513]

1773 1736

$C_7H_5ClO_2$ [25, 448]

3000	1947	1565	1180 M
2910	1756 S	1480 M	1165 M
2775	1755 S	1442	1144 M
2670	1738 S	1410 M	1131 M
2655	1717 S	1332 M	1100 M
2555	1706 S	1297 M	1050 M
2530	1700 M	1270 M	1042 M
1976	1690 M		

$C_7H_5ClO_2$ [25, 585]

1748 S 1723 1703 S

$C_7H_5ClO_2$ [25]

1745 S

$C_7H_5FO_2$ [25]

1755 S 1739 S 1707 S

$C_7H_5FO_2$ [25]

1748 S

$C_7H_5FO_2$ [25]

1745 S 1699 S

$C_7H_5F_3$ [219]

3179 MSh	2349 S	1896 VS	1277 VW
3117 VS*	2328 MW	1838 M	1244 W
3100 VS	2263 MS	1838 SSh	1178 M
3077 VS	2248 MS	1815 VS	1157 SB
2957 S	2193 MS	1767 VS	1127 VSB
2884 W	2158 W	1743 M	1070 VS
2644 S	2137 W	1727 M	1027 VS
2612 M	2093 VS	1697 S	993 VW
2593 MSh	2052 VW	1663 VS	972 W
2529 MSh	2016 VW	1612 VS	925 VS
2501 MS	1982 S	1460 VS	770 VS
2395 MW	1963 VS	1391 MSh	694 VS
2381 M	1914 VS		

$C_7H_5IO_2$ [25]

1753 S 17[illegible]8 S

$C_7H_5LiO_3$ [683]

1634

C_7H_5NO [507]

2274 VS 2263

$C_7H_5NO_3$ [387]

1540	1360	1280	820
1490			

$C_7H_5NO_4$ [25]

1715 S

$C_7H_5NO_4$ [25]

1752 S 1709 S

$C_7H_5NO_4$ [25, 586]

1752 S 1720 1707 S

$C_7H_5N_3O_2$ [37]

1518 S 1344 S 1320 S

$C_7H_5ORS_2$ [684]

1650

$C_7H_5O_2$ [683]

705 – 720

$C_7H_6N_2$ [37]

1620 W	1603 M	1592 M	1591 M
1615 VVS			

$C_7H_6N_4O_6$ [19]

3334 M 3329 M 3281 Sh

C_7H_6O [519]

1704

$C_7H_6O_2$ [25, 38, 370, 448]

3005	2080	1650	1290
3000	1970	1640 M	1287
2875	1927	1620 M	1272 M
2780	1912	1600	1250
2665	1902	1590	1177
2658	1815	1455 M	1125 M
2630	1780 S	1420 M	1090 M
2590	1744 S	1355 M	1065 M
2540	1737 S	1352 M	1025 M
2345	1725 S	1320 M	935 M
2300	1697 S	1290 M	930
2197	1690		

$C_7H_6O_2$ [605]

1615

$C_7H_6O_3$ [27, 38, 448, 471, 611, 675*]

3000	1787 S	1462 M	1180 M
2918	1690 M	1442 M	1162 M
2917	1687 M	1417 M	1152 M
2870	1660 M	1382 M	1135 M
2820*	1655	1342	1071 M
2635	1642 M	1330 M	1033 M
2590	1612 M	1290 M	925 M
2570	1600 S	1264 S	903 M
2530	1482 M		

$C_7H_6O_3$ [674]

1733

$C_7H_7ClO_2S$ [398]

3090	1500	1190	812
1620	1385	1170	

C_7H_7F [220]

3069 VS	2256 MS	1709 W	1224 VS
3043 VS	2173 MW	1615 VS	1157 VVS
3001 S	2109 MW	1603 S	1099 VS
2931 VS	2054 S	1592 S	1045 M
2806 MSh	2034 MSh	1513 VS	1045 MS
2614 MW	1998 MW	1458 W	1017 S
2525 MS	1950 S	1435 MW	929 S
2510 M	1882 VVS	1383 S	842 MS
2442 VS	1766 S	1321 M	729 VS
2395 W	1746 S	1300 S	695 S
2354 MW	1736 S	1280 W	675 VW
2316 S			

C_7H_7F [221]

3067 VS	1866 MS	1460 S	1038 VS
3045 VS	1821 S	1383 S	987 MS
2964 VS	1813 S	1299 W	935 VS
2934 VS	1784 VS	1282 W	887 W
2867 M	1691 S	1267 MW	839 VS
2511 W	1678 S	1236 VS	807 MW
1979 S	1592 VS	1188 VS	754 VS
1945 S	1499 VS	1172 MSh	704 VS
1904 VS	1471 M		

C_7H_7N [41]

722

C_7H_7NO [433, 481*]

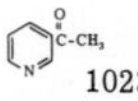

1589 VS	1193 MSh	1118 MW	1023 S
1419 VS			

C_7H_7NO [29]

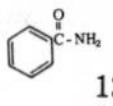

3540 MW	3080 MW	1672 VS	1358 S
3525 MW	1690 VS	1606 S	1357 VS
3520 MW	1689 VS	1584 VS	800 S
3420 MW	1675 VS	1584 S	724 S
3410 M			

$C_7H_7NO_2$ [433]

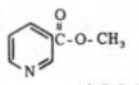

1597 S	1421 S	1122 VSSh	1025 VS
1580 MW	1188 M	1036 MW	

$C_7H_7NO_2$ [573]

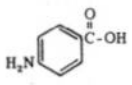

1690

$C_7H_7NO_2$ [620]

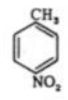

1527	1350

$C_7H_7N_3O_4$ [19]

3384 S	3380 S	3365 M	3311 MW
3383 S	3378 M	3360 MW	3285 MW
3382 M			

C_7H_8 [388, 413]

1057 S	728 S	690 S

$C_7H_8Cl_2Si$ [86]

3067 M	1770 W	1261 S	880.5 W
2899 W	1653 M	1120 VS	799.4 VS
2257 W	1548 W	1096 W	789.2 S
2169 W	1488 M	1073 W	757.6 S
2053 W	1404 M	987.4 W	734.2 VS
2015 W	1339 W		

$C_7H_8N_2O$ [433, 481*]

1599 S	1420 VS	1103 W	1024 MW
1480 VS	1123 W	1047 W	

C_7H_8OS [7, 689]

1661-1646	1447 S	1193 VS	941 MSh
1661 S	1389 S	1164 MS	936 VS
1646 VVS	1374 M	1032 MW	860 VVS
1573 S	1282 VS	1012 VS	

$C_7H_8O_2$ [7]

1682 S	1667 S	1612 S	1196 M
1679 S	1644 S	1447 M	1162 S
1675 S	1635 S	1437 MW	955 W
1674 S	1621 S	1402 VS	936 S
1673 S	1620 SSh	1376 M	867 VS
1670 VS	1615 S	1335 M	

$C_7H_8O_5$ [475]

3516 M	1787 VS	1748 S	1697 MW

$C_7H_8S_2$ [7]

1590 VS	1382 W	1175 S	991 M
1526 W	1335 MW	1030 W	867 M
1445 M	1272 W		

C_7H_9N [383, 603, 687]

3400	1510	740	700
1620	850		

C_7H_9NO [433, 481*]

1589 S	1482 SSh	1183 MWSh	1100 MWSh
1578 S	1425 S	1130 MWSh	1013 MWSh

$C_7H_9NO_3$ [15]

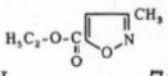

1595 MS	1446 S	910 W	768 S
1480 MS	1415 MW		

$C_7H_9NO_3$ [15]

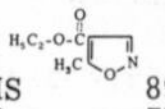

1619 VVS	1427 VS	1155 MS	897 MW
1489 VS	1405 S	930 M	776 S

$C_7H_9NO_3$ [15]

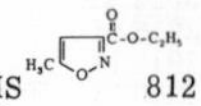

1601 VS	1464 VS	1004 MS	812 W
1473 VSSh	1426 M	920 M	

$C_7H_{10}N_2O_3$ [15]

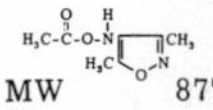

1650 SSh	1470 VS	1035 MW	877 W
1480 VS	1429 VS	1001 W	

$C_7H_{10}N_2O_3$ [15]

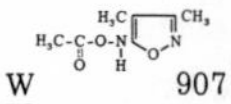

1465 VVS	1165 M	1035 W	907 W
1430 VS	1126 W	1007 M	

$C_7H_{10}O_3$ [545]

1633

$C_7H_{11}NO_2$ [15]

1666 VVS	1475 VVS	1138 VS	892 M
1521 VVS	1450 VS	1038 MSh	

$C_7H_{11}NO_2S$ [31]

1693 S	1263 MS	1112 S	1026 M
1513 VS	1233 MS	1062 M	1003 MS
1439 MS			

C_7H_{12} [542]

1651

C_7H_{12} [572]

1651

$C_7H_{12}N_2O$ [6]

1610 S	1542

$C_7H_{12}O_3$ [579, 595]

1650	1632

$C_7H_{13}N$ [596]

1644 S

C_7H_{14} [170] (cis)

2957 VS	2736 S	2685 SSp	2622 SB
2868 VS			

C_7H_{14} [170] (trans)

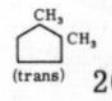

2952 VS	2729 VSSp	2700 SSh	2628 VSSp
2866 VS			

C_7H_{14} [171]

2956 VS	2736 MW	2621 MB	2476 MB
2866 VS	2647 MS		

C_7H_{14} [308, 464] (trans)

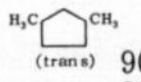

3200 M	2438 W	1308 VS	901 M
2950 VS	1456 VVS	1141 M	852 S
2832 S	1373 VVS	984 S	775 MB
2606 MSh	1349 VVS	927 MW	

C_7H_{14} [272]

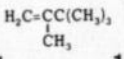

3095 MWSh*	1481 MSh	1292 MS	1017 VS
2965 VS*	1464 S	1203 VS	996 VS
2880 SSh*	1387 VSB	1183 VS	930 SSh
2730 MSh	1374 VS	1034 VS	717 VS
1639 VS			

C_7H_{14} [102, 273] — $H_2C{=}CHCH_2C(CH_3)_3$

3195 WSh	1649 VS	1401 S	1209 M
3085 VS	1637 VS	1397 MW	1205 VS
2960 VS	1534 MWSh	1374 VS	1034 MS
2950 VS	1477 VS	1368 VS	996 VS
2905 VSSh	1471 S	1290 MW	916 VS
2745 MW	1443 VS	1286 MS	913 VS
1840 S	1437 M	1245 MS	660 VS
1828 VS	1426 M	1241 VS	

$C_7H_{14}N_2O_3$ [26]

2924 M	1381 M	1105 S	946 M
2860 M	1219 M	1077 S	892 M
2810 M	1135 M	1046 M	843 M
1455 M			

$C_7H_{14}N_2O_3$ [26]

2962 S	1462 S	1212 M	1025 M
2929 M	1390 M	1085 S	947 S
2873 M	1361 S	1057 M	896 M
2843 M	1343 M	1045 M	881 M
2785 M	1260 S		

$C_7H_{15}NO$ [26]

2994 VS	1449 S	1260 VS	1097 S
2948 VS	1385 S	1212 S	1073 S
2928 S	1369 S	1188 S	1043 S
2893 S	1343 M	1146 S	1002 VS
1459 S	1292 M	1115 S	925 M

C_7H_{16} [80, 158, 331, 425] — $(CH_3)_2CH(CH_2)_3CH_3$

2957 VS	1469 VS	1349 W	919 S
2952 VS	1468 VS	1296 M	909 S
2926 VS*	1462 S	1222 MS	906 M
2874 S	1384 S	1172 MS	895 M
2740 M	1376 S	1172 S	779 MW
2699 VW	1369 S	1075 VS	728 S
2630 M	1368 S	1075 M	

C_7H_{16} [81, 93,* 157, 330] — $CH_3CH_2CH(CH_3)(CH_2)_2CH_3$

2957 VS	1381 VS	1146 M	931 M
2954 S	1379 VS	1072 MW	929 MS
2878 VS	1372 S	1021 MB	880 M
2875 VS	1355 WSh	1012 MB	879 MW
2744 MW	1308 MW	1012 M	821 M
1529 W	1295 MW	984 S	772 S
1467 S	1232 M	983 M	770 MS
1466 VS	1156 S	963 S	738 S
1465 VS	1149 S	963 MS	737 MS
1458 VS			

C_7H_{16} [82, 143, 156] — $(CH_3CH_2)_3CH$

2966 VS	2285 MS	1319 S	899 VVS
2960 VS	2215 MS	1317 M	846 S
2948 VS	2155 M	1276 S	832 S
2935 MVB	2065 M	1274 M	832 VS
2929 VS	1775 M	1261 M	791 MB
2922 MB	1481 VS	1167 S	789 M
2884 VS	1464 VS	1153 S	787 MB
2863 MB	1463 VS	1129 S	772 VS
2862 VS	1462 VS	1127 M	765 VS
2730 SSh	1381 S	1042 S	765 S
2605 SSh	1380 VS	1005 MS	752 S
2500 MW	1337 S	1004 VS	733 M
2470 MW	1334 M		

C_7H_{16} [84, 85, 362] — $CH_3(CH_2)_5CH_3$

2957 VS	1468 VS	1280 MW	774 M
2927 VS	1459 VSSh	1140 M	743 VSVB
2738 MW	1458 VS	1076 M	739 MSh
2672 MW	1380 M	987 VW	723 VS
1469 VS	1379 VS	774 S	

C_7H_{16} [77,* 141, 150, 151, 335] — $(CH_3)_2CHCH_2CH(CH_3)_2$

2966 VS	2652 MS	1369 VS	920 VS
2960 VS	2620 MS	1368 S	919 VS
2958 S	1471 VS	1328 VS	870 VS
2905 M	1470 VS	1328 M	868 W
2871 M*	1387 SSp	1279 MWSh	865 W
2731 S	1387 S	1172 S	811 M
2725 S	1386 VS	986 M	809 VS
2655 MS	1374 S	985 VS	

C_7H_{16} [76,* 148, 149, 336] — $CH_3CH_2C(CH_3)_2CH_2CH_3$

2966 VS	1381 VS	1198 S	912 M
2953 VS	1379 S	1195 S	855 M
2882 MSB	1378 MS	1082 S	785 S
2859 S	1365 S	1081 S	784 S
1577 M	1294 M	1004 VS	767 M
1477 VS	1289 M	1000 VS	766 M
1464 VS	1217 SSh	915 M	

C_7H_{16} [142, 152, 153] — $(CH_3)_2CHCH(CH_3)CH_2CH_3$

3180 VSSh*	2405 S	1330 S	1121 S
2961 VS	2290 MBSh	1329 M	1062 MS
2960 VS	1664 SB	1309 S	1025 S
2931 VS	1466 VS	1289 M	1014 VS
2880 VS	1462 S	1284 M	995 S
2876 VS	1461 VS	1266 M	965 VS
2874 VSSh	1387 S	1253 M	918 S
2741 SSh	1386 S	1188 MS	801 MS
2730 VSSh	1381 S	1166 MS	786 S
2610 VSSh	1379 S	1144 MSSh	785 M
2609 SSh	1370 S	1132 SSh	751 M
2515 MB	1368 M		

C_7H_{16} [146, 147] — $(CH_3)_3CCH(CH_3)_2$

2962 VS	2619 MS	1381 VS	1106 VS
2952 VS	1465 VS	1370 S	1084 VS
2914 MB	1463 VS	1320 M	999 MS
2871 MS	1399 MSh	1211 S	923 MSB
2796 MWVB	1398 MSSh	1161 VS	834 MS
2720 MSSh	1382 S		

C_7H_{16} [154, 155] — $(CH_3)_3CCH_2CH_2CH_3$

3202 SSh*	2727 MSSh	1393 S	1270 MSh
2955 VS	1478 VS	1381 VS	1249 S
2947 VS	1476 VS	1378 S	948 S
2909 M	1469 VS	1365 VS	928 S
2880 VS*	1468 VS	1364 VS	741 S
2873 S*			

C_8F_{16} [229, 230]

2320 VS	1587 M	1121 M	866 VS
2255 W	1531 W	1109 MW	851 VS
2240 W	1462 M	1079 S	824 M
2210 M	1299 VS	1067 M	806 W
2035 M	1264 S	1035 MW	747 W
1875 W	1244 S	1003 WSh	734 VS
1845 M	1200 MS	914 M	711 M
1810 MS	1171 MS	873 W	687 VS
1661 W	1157 MS	869 VS	661 W
1626 W			

C_8F_{16} [231]

2490 S	1422 W	1110 M	838 W
2400 VS	1271 VSB	1068 M	737 M
2035 VW	1245 SB	1032 S	732 M
2005 W	1224 MB	933 W	725 M
1955 W	1206 M	883 S	711 M
1720 VW	1182 M	864 S	687 S
1495 W	1159 S	849 VS	673 S
1445 W	1136 M		

$C_8F_{18}O$ [266, 267]

2610 SSh	1381 MSh	1060 VS	819 S
2475 VSVB	1350 SSh	991 VS	803 MW
2390 VS	1321 MSh	956 VS	769 VVW
2195 MW	1304 VSB	937 MW	747 VS
2115 S	1232 VSVB	917 S	735 VS
1950 SB	1151 VS	843 W	712 VS
1548 VWB			

$C_8HF_7N_2$ [37]

1604 S	1566 VVS	1548 VVS	

$C_8H_4ClF_3N_2$ [37]

1558 M	1325 S	1195 S	1170 S

$C_8H_4F_3N_3O_2$ [37]

1558 M	1492 S	1322 S	1163
1533 S	1348 S	1192 S	

$C_8H_4O_3$ [39]

1773 M	1719 S	1006 W	638 W

$C_8H_5F_3N_2$ [37]

1492 M	1188 S	1122 S	1110 S
1338 S	1178 S		

$C_8H_5F_3N_2$ [37]

1335 S	1175 S	1125 M	1100 S

$C_8H_5F_3N_2$ [37]

1577 M	1538 MW	1192 S	1143 MS
1555 M	1329 MS	1172 S	1132 MS

$C_8H_6N_2$ [598]

1628-1618 S
1581-1566 S

$C_8H_6N_2O_2$ [37] 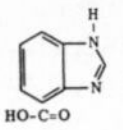
1625 S 1600 S 1540 M

$C_8H_6N_2O_2$ [37] 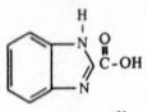
1650 S 1520 M

$C_8H_6N_2O_2$ [37] 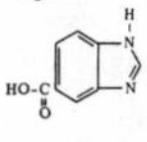
1625 M

$C_8H_6N_4O_7$ [19]
3380 M 3349 MW 3305 MW 3220 M
3359 MW

$C_8H_6O_2$ [39]
1749 S 1284 MS 1109 W 1017 M

$C_8H_6O_5$ [25]
1747 S 1702 S

$C_8H_7BrO_2$ [25] 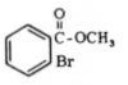
1744 S

$C_8H_7BrO_2$ [25] 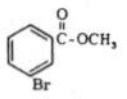
1734 S

$C_8H_7BrO_2$ [25] 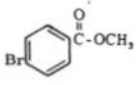
1734 S

$C_8H_7ClO_2$ [25] 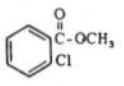
1744 S 1727 S

$C_8H_7ClO_2$ [25] 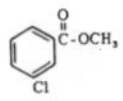
1735 S

$C_8H_7ClO_2$ [25] 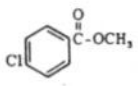
1731 S

$C_8H_7FO_2$ [25] 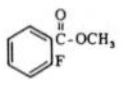
1749 S 1741 S 1726 S

$C_8H_7FO_2$ [25] 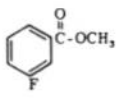
1733 S

$C_8H_7FO_2$ [25] 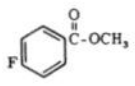
1732 S

$C_8H_7IO_2$ [25]
1740 S

C_8H_7N [312] 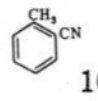
3074 S 1603 SSp 1292 VS 1044 VS
2940 S 1488 VVSSp 1215 VS 758 S
2890 MSh 1460 VSSp 1166 S 711 VS
2225 VSSp 1385 SSp 1112 VS

C_8H_7N [214, 641]
3040 VS 1456 S 1272 S 951 M
2940 VS 1450 S 1179 VS 815 MS
2230 VS 1410 S 1120 S 815 VS
1919 M 1385 S 1041 S 704 S
1608 VS 1292 S 1022 S 703 VS
1506 VS

$C_8H_7NO_4$ [25] 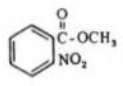
1747 S

$C_8H_7NO_4$ [25] 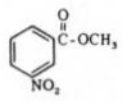
1738 S

$C_8H_7NO_4$ [25] 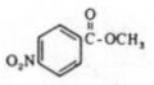
1737 S

C_8H_8 [200]
2995 VS 1751 S 1399 S 969 VS
2950 VSSh 1732 S 1224 VS 943 VS
1919 MS 1639 VVS 1205 VS 799 VS
1845 MS 1613 SSh 1038 MS 673 VVS
1779 MS 1582 MSh 992 M

C_8H_8ClN [41]
3090 1575 1180 842
1594 1468 1088 720

$C_8H_8N_2$ [37] 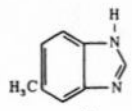
1543 MS

$C_8H_8N_2$ [37] 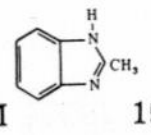
1628 M 1620 VVS 1562 M 1542 VS
1622 MW

$C_8H_8N_2O$ [37]
1349 M

$C_8H_8O_2$ [25] 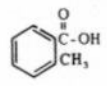
1742 S 1696 S

$C_8H_8O_2$ [25] 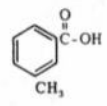
1698 S

$C_8H_8O_2$ [25] 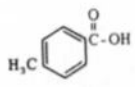
1740 S 1697 S

$C_8H_8O_2$ [610] 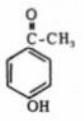
1600 S

$C_8H_8O_2$ [25, 38, 667] 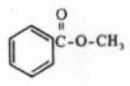
1730 S 1270

$C_8H_8O_3$ [25] 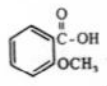
1760 Sh 1751 S 1741 S 1702 W

$C_8H_8O_3$ [25] 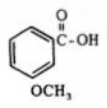
1698 S

$C_8H_8O_3$ [25] 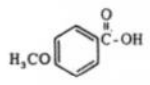
1737 S 1691 S

$C_8H_8O_3$ [579] 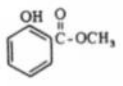
1684

$C_8H_9FeGeNO_4$ [509]
2135 S

$C_8H_9FeNO_4Sn$ [509]
2142 S

C_8H_9N [41]
1610 1463 1088 786
1587

C_8H_9NO [2, 3, 433] 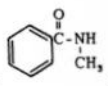
3480 W 1657 VS 1421 MW 935 S
3460 W 1653 VS 1415 MW 830 S
3350 W 1579 MW 1415 MW 802 S
3080 W 1578 MW 1413 MW 797 S
3050 W 1575 MW 1282 M 698 VS
2940 W 1524 M 1276 M 696 VS
1675 VS 1524 M 1215 MW 685 VS
1673 VS 1484 M

C_8H_9NO [41] 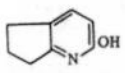
1630 1482 725

$C_8H_9NO_2$ [433]
1598 VS 1193 MSh 1039 MS 1027 VS
1422 S

$C_8H_9N_3S$ [380]
3400 S 1540 S 1375 S 1299 S

C_8H_{10} [68, 123] 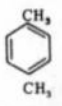
3052 S 2940 S 2880 M 1520 S
3030 S 2936 VS 1525 S 1515 S
3025 S 2890 M 1520 VS 895 VS
3003 S

C_8H_{10} [69]
3040 S 1615 SSp 1387 MW 690 S
2940 S 1500 S 769 VS

C_8H_{10} [70]
3050 M 2890 VS 1460 M 752 M
3049 S 1500 S 773 S 697 S
2990 VS

C_8H_{10} [202] — $CH_2=CHCH=CHCH=CHCH_3$

3093 SSp	2971 MSSp	1406 VSVSp	958 M
3032 VS	1805 S	1140 M	898 VS
3014 VS	1636 VS	1008 VS	

$C_8H_{10}N_2O$ [433]

1590 S	1480 VS	1190 MWSh	1040 MWSh
1581 S	1420 VS	1104 MW	1023 M

$C_8H_{10}N_2O_2$ [21, 620]

2815 S	1506 S	1112 S	824 S
1597 S	1484 S	995 S	820 S
1595 S	1332 S	947 S	751 S
1580 S	1200 S	940 S	696 S
1527 S	1115 S		

$C_8H_{11}ClSi$ [86]

3012 W	2045 W	1883 W	1109 M

$C_8H_{11}N$ [431, 455] — 2-C_3H_7-n pyridine

2933 S	1581 S	1433 S	1374 S
2841 S	1471 S		

$C_8H_{11}N$ [431] — 3-C_3H_7-n pyridine

2907 S	1462 S	1186 S	1025 S
2833 S	1450 S		

$C_8H_{11}N$ [431] — 4-C_3H_7-n pyridine

2857 S	1605 S	1466 S	1412 S
1670 W	1497 MS		

$C_8H_{11}NO_3$ [15]

1618 VVS	1408 S	978 M	897 M
1489 S	1155 MS	931 M	774 S
1427 VS			

$C_8H_{12}Cl_2N_2O_2$ [617]

1570

$C_8H_{12}N_2$ [597]

1592 S	1439 S

$C_8H_{12}O_2$ [518]

1700	1605

$C_8H_{12}O_2S_2$ [33]

1092 M	1010 S

$C_8H_{13}N$ [343]

3379 VVS	1524 MS	1304 S	981 M
2908 VS	1450 VVS	1260 W	957 S
2539 W	1399 W	1237 VS	786 M
2492 W	1384 VS	1189 S	724 VS
2470 W	1372 VS	1096 VS	701 S
1591 S	1327 W	1063 VS	

C_8H_{14} [125] — ring with H_3C, CH_3, CH_3

2970*	1276 S	940 M	844 M
1450 VS	1176 M	906 M	800 VS
1334 S	1015 VS	897 M	763 VS
1319 S	969 S	877 M	

C_8H_{14} [126] — ring with CH_3, CH_3, CH_3

2940 S	1220 S	1000 S	823 VS
1475 VS	1117 VS	976 M	794 VS
1385 VS	1025 VS	921 M	702 M
1309 M			

C_8H_{14} [127] — ring with CH_3, H_3C, CH_3

2740 VS	1188 S	996 M	840 M
1433 VS	1156 S	920 S	761 VS
1342 VS	1106 S	901 M	745 VS
1266 M	1075 S	876 M	

C_8H_{14} [542] — ring with CH_3

1673

$C_8H_{14}N_2O$ [6]

1705 S	1605	1380

$C_8H_{14}N_2O_2$ [31]

3290 S	2846 M	1535 S	1449 MS
2914 MS	1654 VS		

$C_8H_{14}N_2OS$ [31]

3372 MS	2846 MS	1512 VS	1385 M
3228 S	1677 S	1393 MS	1025 MS
2920 MS			

$C_8H_{14}O_4$ [478, 571]

3480	1733	1470

$C_8H_{15}ClO$ [513]

1790

$C_8H_{15}N$ [596]

1644 S

$C_8H_{15}NO$ [592]

1649

C_8H_{16} [96] — $H_2C=C(CH_3)CH_2C(CH_3)_3$

2960 VS	1441 VSSh	1325 M	979 S
2920 VS*	1391 SSh	1263 M	896 VS
1645 VS	1372 VSSh	1238 S	826 MW
1478 VS	1364 VS	1202 M	765 S
1464 VSSh			

C_8H_{16} [97] — $H_2C=CH(CH_2)_3CH(CH_3)_2$

2975 VS	1468 VS	1372 S	911 VVS
1835 MS	1440 VSSh	1174 M	810 MW
1649 VS	1387 S	995 VS	

C_8H_{16} [98] — $H_2C=C(CH_3)(CH_2)_4CH_3$

3005 VS	1315 MVB	1122 MW	846 W
1727 SSh	1263 WVB	1113 MW	830 W
1656 VS	1222 MW	971 M	770 MB
1460 VS	1193 W	887 VS	727 S
1380 VS	1153 W		

C_8H_{16} [124] — ring with H_3C, CH_3, CH_3

2940 S	1370 VS	1190 M	996 M
1460 VS	1315 S		

C_8H_{16} [204] — $(CH_3)_2C=CHC(CH_3)_3$

2963 VS*	1479 S	1362 SSp	1029 VS
2933 VS	1466 S	1256 S	984 S
2730 VSSh	1453 S	1229 VS	940 MW
1664 MS	1395 MS	1202 VS	923 MS
1645 SSh	1385 MS	1159 VS	824 VS
1575 W	1374 MS	1075 VS	

C_8H_{16} [291, 422] — $CH_3(CH_2)_2CH=CH(CH_2)_2CH_3$ (cis)

1650 VS	1268 SSp	972 VS	821 VW
1456 S	1145 MS	895 MB	783 MB
1381 SSp	1098 M	886 MB	760 M
1343 VS	1059 VS	861 MW	741 S
1308 MSSp			

C_8H_{16} [205] — $CH_3(CH_2)_2CH=CH(CH_2)_2CH_3$ (trans)

2965 VVS	1670 MS	1342 VS	1059 VS
2933 VVS	1464 VS	1266 S	968 S
2877 VS	1462 S	1208 S	760 M
2843 VS	1444 S	1147 MW	759 VSB
1860 VS	1379 S	1098 MS	740 VS
1745 VS			

C_8H_{16} [289, 290] — $H_2C=CH(CH_2)_5CH_3$

2950 VS	1464 VS	1381 S	993 VS
1832 M	1457 S	1299 W	912 VS
1649 S	1442 VSSh	1114 W	910 VS
1643 VS	1390 M	994 S	725 S

C_8H_{16} [99, 421] — $CH_3CH_2CH=CH(CH_2)_3CH_3$

2995 VS	1298 MS	1070 MW	891 W
1466 VS	1244 MW	1047 M	776 MB
1379 S	1103 MW	968 VS	737 Sb
1344 MS	1076 W	931 MS	

C_8H_{16} [100, 101,† 160] — $CH_3CH=CH(CH_2)_4CH_3$ (cis)

†Mixture of cis and trans compounds

2975 VS	1456 VS	1378 S	925 M
2970 VS	1448 VSSh	1307 MB	730 S
2900 VVS	1406 S	969 VS	727 M
1656 S	1389 M	966 VS	701 MB
1466 S	1381 VS	965 VS	692 VS
1460 VS			

$C_8H_{16}BrN$ [596]

1646 S

$C_8H_{16}Cl_2O_2PdS_2$ [33]

1127 S	1068 S

$C_8H_{16}Cl_2O_2PtS_2$ [33]

1155 S	1140 M	1087 M	1077 M

$C_8H_{16}N_2O_3$ [26] — 5-methyl-5-nitro-3-(n-propyl)tetrahydro-1,3-oxazine (H_3C, O_2N, N-C_3H_7-n)

2936 S 1381 M 1105 S 953 S
2872 S 1350 S 1079 S 897 S
2824 S 1277 M 1049 S 869 M
2787 S 1217 S 1005 M 850 S
1458 S 1135 S 990 M

$C_8H_{16}N_2O_3$ [26] — (H_3C, O_2N, N-C_3H_7-i)

2979 S 2821 M 1338 M 1031 M
2937 S 1459 S 1237 S 946 S
2878 S 1389 S 1218 S 893 S
2844 M 1365 S 1081 VS 855 M

$C_8H_{16}N_2O_3$ [26] — (H_5C_2, O_2N, N-C_2H_5)

2980 S 2774 M 1360 M 937 M
2943 M 1460 S 1222 S 911 M
2880 M 1442 S 1082 S 896 M
2826 M 1384 M 1051 S 842 M

C_8H_{18} [83] — $CH_3CH_2CH(CH_3)-CH(CH_3)CH_2CH_3$

3170 MSh 1381 VS 1071 S 949 VSB
2960 VS 1325 M 1071 S 799 SSh
2880 VS 1292 MW 1063 VS 778 VS
2730 S 1272 MW 1017 S 778 S
2620 S 1181 MSh 997 SVB 763 MW
2160 MW 1160 MW 996 VSB 738 M
1466 VS 1122 VVS 954 VSB

C_8H_{18} [89, 91, 664*] — $CH_3(CH_2)_6CH_3$

2967 VS 2868 S 1380 S 1081 W
2933 VSB 1470 VS 1346 W 765 W
2920 VS 1466 VS 1305 W 723 M

C_8H_{18} [90, 341] — $(CH_3)_2CHC(CH_3)_2-CH_2CH_3$

2980 S 1464 S 1096 MS 1008 VVS
2890 Sh 1188 VS 1088 VS 931 M
2890 VS 1157 M 1087 MS 777 M
2604 MSh 1109 VS 1035 MS 776 M
1470 VSB

C_8H_{18} [106, 119] — $CH_3CH_2CH(CH_3)(CH_2)_3CH_3$

2970 VS 2750 WSh 1304 MW 1000 MB
2934 VS 1470 VS 1218 MW 966 M
2930 VS 1469 M 1151 M 780 MB
2882 VS 1380 VS 1080 W 730 MB
2880 SSh 1350 MSh

C_8H_{18} [71, 110, 340, 436] — $(CH_3)_3CCH(CH_3)CH_2CH_3$

2973 S 1470 VS 1205 S 1080 VS
2970 S 1370 VS 1159 MS 1027 MS
2885 S 1242 S 1156 S 1002 S
2880 VS* 1221 S 1117 W 780 S
1477 VS 1208 M 1083 MS 717 W
1472 S

C_8H_{18} [112, 360] — $CH_3CH_2C(CH_3)_2-CH_2CH_2CH_3$

3191 W 2887 S 1389 S 856 M
3032 M 2725 SSh 1192 VS 784 S
2968 VS* 1466 VS 1015 S 741 S
2910 S 1471 VS 953 S

C_8H_{18} [111] — $CH_3CH_2CH(CH_3)CH(CH_3)CH_2CH_3$

3167 MS 2973 VS 2885 VS* 1467 VS

C_8H_{18} [113, 137, 339] — $(CH_3)_2CHCH_2CH_2CH(CH_3)_2$

3340 MSh 2385 S 1224 VS 935 WSh
3180 Sh 1690 SVB 1171 VS 920 VS
2966 VS 1473 VS 1092 S 919 S
2960 VS 1472 S 1091 S 871 W
2927 S 1385 S 1044 MS 839 W
2882 M 1368 S 1040 S 813 M
2880 S 1339 VS 1037 MS 776 WSh
2725 SSh 1294 MS 954 M 755 M
2505 MSh 1261 MS 949 VS 753 VS

C_8H_{18} [116, 337] — $(CH_3)_3C(CH_2)_3CH_3$

2965 VS 1471 VS 1206 S 894 S
2879 M 1379 VS 1107 M 784 W
2710 SSh 1250 S 912 MS 730 S
1476 VS

C_8H_{18} [117, 140] — $(CH_3CH_2)_2CHCH_2CH_2CH_3$

2969 VS 1464 VS 1227 MW 913 VS
2960 VS 1381 S 1155 SB 888 VS
2934 VS 1344 MSh 1131 SB 865 MWSh
2920 VS* 1323 M 1074 MS 821 VS
2884 VS 1311 M 1043 SB 776 VS
2880 VS 1294 M 1028 SB 755 W
2730 VSSh 1277 MW 1011 SB 749 S
2595 SSh 1269 MW 921 VS 734 VS
1468 S 1250 MW

C_8H_{18} [118] — $CH_3(CH_2)_2CH(CH_3)-(CH_2)_2CH_3$

2969 VS 2937 S 2883 S 1467 VS

C_8H_{18} [120, 294] — $(CH_3)_2CH(CH_2)_4CH_3$

2967 VS 2881 S 1470 VS 1170 MS
2934 VS 2750 W 1380 S 938 MB
2930 VS 1480 VS 1342 M 727 S

C_8H_{18} [135, 284] — $(CH_3)_2CHCH(CH_3)CH(CH_3)_2$

3170 VSSh* 1661 SB 1295 S 1041 VS
3058 S 1639 W 1295 M 1040 VS
2960 VS 1475 VS 1272 M 999 M
2880 VS 1471 VS 1271 S 997 VS
2730 VSSh 1449 VS 1186 M 973 VS
2703 MSh 1387 S 1163 S 973 S
2611 M 1377 VS 1163 M 957 WSh
2326 W 1376 S 1124 VS 916 VS
2325 VS 1368 S 1124 VS 891 W
2190 MWB 1321 S 1100 S 814 MS
1667 W 1318 S 1075 S 754 VS

C_8H_{18} [136] — $(CH_3)_2CHCH(CH_2CH_3)_2$

3180 VSSh 1468 VS 1129 VS 909 VS
2960 VS 1385 S 1041 SVB 905 VS
2880 VS 1368 S 1025 SVB 904 VSSh
2730 VSSh 1332 MS 1014 SVB 868 VS
2620 VSSh 1312 M 1006 SVB 822 VS
2500 MWB 1295 M 946 VS 777 SSh
2410 VS 1276 S 919 S 769 S
2290 VS 1248 M 916 S 733 M
2240 MSSh 1184 S 914 VS 718 M
1658 SB 1161 S

C_8H_{18} [74,* 114, 138, 334, 429] — $(CH_3)_2CHCH_2CH(CH_3)CH_2CH_3$

3180 SSh 2210 MSVB 1171 VS 926 M
3067 VS 1675 SVB 1153 VS 922 VS
2966 VS 1629 SB 1085 M 880 M
2950 VS 1470 VS 1050 M 864 W
2915 S 1468 VS 1047 MS 861 S
2884 VS 1468 VS 1013 S 826 S
2725 SSh 1385 S 1000 S 813 S
2645 SSh 1374 VS 996 VS 805 S
2640 SSh 1368 S 973 S 769 M
2410 SB 1299 MS 970 VS 767 VS
2240 M 1284 M

C_8H_{18} [72, 75,* 115, 139, 338] — $(CH_3)_2CHCH(CH_3)CH_2CH_2CH_3$

3180 SSh 1667 SVB 1255 MSh 944 M
2980 VS 1471 VS 1235 M 937 M
2968 VS 1470 VS 1188 M 918 M
2960 VS 1464 VS 1148 VSSh 907 M
2939 S 1389 S 1130 S 904 S
2884 S 1381 VS 1127 VVS 899 S
2880 VS 1381 S 1053 S 870 S
2730 VSSh 1380 VS 1034 S 853 M
2680 VSSh 1370 S 1009 S 758 M
2673 MSh 1311 S 996 S 740 VS
2605 VSSh 1271 M 973 S 718 MW
2597 MSh 1267 MS 955 M

$C_8H_{18}O$ [317, 665*]

3356 VS* 2857 M 1119 S 954 M
2959 VS 1462 VS 1056 VSB 724 S
2924 VS 1376 VS 988 M

$C_8H_{19}N$ [388]

1116 S

$C_8H_{22}N_4O_2$ [498, 535*]

2905 SB

$C_8H_{24}B_2P_2$ [30]

1043 M

C_9F_{18} [227, 228, 417]

2050 M 1252 W 1010 S 827 VS
1850 W 1225 MW 950 M 749 M
1830 W 1203 M 936 M 735 M
1800 MW 1176 S 925 MW 711 VS
1725 W 1125 MW 912 MSh 685 S
1480 W 1068 VW 882 M 685 VS
1425 W 1055 W 841 VS 652 M
1282 S 1037 M

$C_9H_4F_6N_2$ [37] — (F_3C, CF_3 benzimidazole)

1336 S 1190 S 1150 S

$C_9H_4F_6N_2$ [37] — (F_3C, F_3C benzimidazole)

1660 M 1350 S 1170 S 1138 S
1610 M 1193 S 1144 S 1120 S

$C_9H_4F_6N_2$ [37]
1610 M 1526 M 1168 S 1137 S
1554 S 1330 S

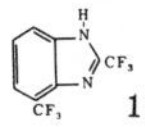

$C_9H_4F_6N_2$ [37]
1560 M 1329 MS 1166 S 1120 S
1550 S 1199 S 1136 S

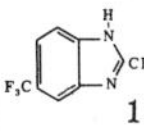

$C_9H_5BrCrO_3$ [17, 588,* 645*]
1450 S

$C_9H_5ClCrO_3$ [17, 588,* 645*]
1450 S 1410 S 805 S

$C_9H_5F_3N_2O_2$ [37]
1543 M

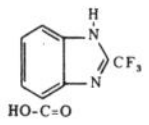

$C_9H_5F_3N_2O_2$ [37]
1648 S 1520 M

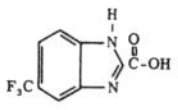

$C_9H_5F_3N_2O_2$ [37]
1623 S

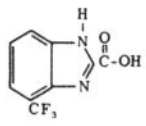

$C_9H_5F_5N_2$ [37]
1593 W 1538 M 1345 MS 1154 S

$C_9H_6CrO_3$ [17, 588,* 645*]
1450 S 795 S

$C_9H_6F_2N_2O_2$ [37]
1690 S 1625 M

$C_9H_6N_2O_2$ [507]
2265 VS

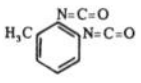

$C_9H_6N_2O_2$ [507]
2265 VS

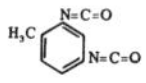

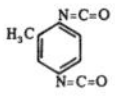

$C_9H_6N_2O_2$ [507]
2265 VS

$C_9H_7ClN_2$ [41]
1594 1540 1157 938

$C_9H_7F_3N_2$ [37]
1627 M 1545 VS 1195 S 1119 S
1610 MW 1510 M 1127 S

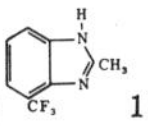

$C_9H_7F_3N_2$ [37]
1540 M 1170 S 1130 S 1110 S
1337 S 1165 S

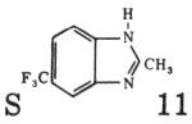

$C_9H_7F_3N_2$ [37]
1619 W 1550 M 1188 S 1170 S
1601 MW 1340 MS 1182 S 1142 S
1551 M 1325 S

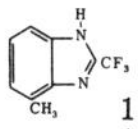

$C_9H_7F_3N_2$ [37]
1552 MS 1325 S 1175 S 1144 S
1515 M

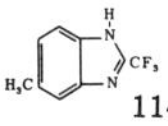

$C_9H_7F_3N_2O$ [37]
1642 M 1574 VVS 1175 S 1130 S
1630 VS 1550 MS 1163 S 1118 S
1619 S 1344 M

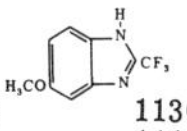

$C_9H_7F_3N_2O$ [37]
1635 M 1538 M 1198 S 1122 S
1610 M 1334 S 1170 S

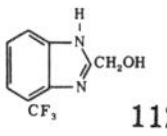

$C_9H_7F_3N_2O$ [37]
1635 M 1332 S 1168 S 1125 S
1545 M

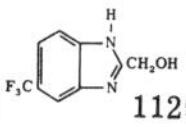

C_9H_7N [197, 591,* 668*]
3030 VS 1372 VVS 1140 S 864 VS
1618 VS 1271 VVS 1038 S 829 VS
1575 VS 1248 VS 1017 S 803 S
1486 VS 1230 S 975 S 781 VS
1445 MS 1211 VS 944 VS 745 VS
1427 MS 1176 M

C_9H_7NO [398, 591,* 616]
1577 780

$C_9H_7N_3$ [37]
1547 M 1325 M

C_9H_8 [9]
3297.0 M 1943.2 S 1457.8 VS 1067.9 S
3110 S 1915.0 S 1393.2 VS 1018.6 VS
3068.5 VS 1884.5 S 1361.3 S 947.2 VS
3025.6 S 1856.7 S 1332.5 M 942.3 VS
3015 S 1825.5 S 1312.5 S 914.8 VS
2943.8 S 1686 S 1287.8 MS 861.3 S
2887 VS 1609.6 VS 1226.2 S 830.5 M
2771.0 S 1587.7 M 1205.2 S 765.4 S
2598.6 S 1574.3 M 1166.2 S 730.1
2304.8 S 1553.3 S 1122.7 S 718.2
2090 M 1483.2 VS 1106.4 M 692.8 M
2049 S

C_9H_8ClN [41]
1080

$C_9H_8ClNO_2$ [41]
1608 1160 923 762
1571 1130

$C_9H_8N_2$ [41]
1558 1240 1032 913

$C_9H_8N_2O$ [41]
1616 1465 958 773
1589

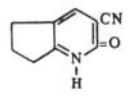

$C_9H_8N_2O$ [15]
1635 VVS 1452 VS 1004 W 886 M
1488 VS 1433 VS 949 M

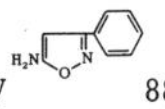

$C_9H_8N_2O$ [15]
1629 VVS 1518 VS 1469 VVS 914 MW

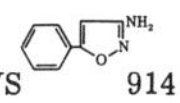

$C_9H_8N_2O$ [6]
1610 1575 1515 S 1435

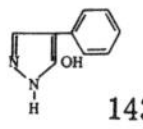

$C_9H_8N_2O$ [6]
1625 S 1543 1505

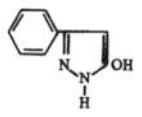

$C_9H_8N_2O_2$ [37]
1670 S

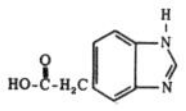

$C_9H_8N_2O_2$ [37]
1645 S 1612 S 1565 M

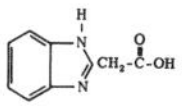

$C_9H_8O_2$ [545]
1626

$C_9H_8O_3$ [514]
1808 1745

$C_9H_8O_5$ [25]
1733 S

$C_9H_9FeNO_4$ [509]
2186

$C_9H_9NO_2$ [41]
1462 1050 774

$C_9H_9NO_3$ [41]
1620 1562 1486 1181

$C_9H_{10}ClN$ [41]
3080 1572 S 1195 716
3070 1468 858 710
1590 1467 848

$C_9H_{10}N_2$ [37]
1622 M 1553 MS

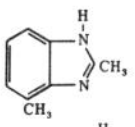

$C_9H_{10}N_2$ [37]
1620 MW 1548 S

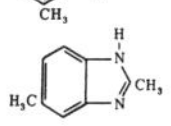

$C_9H_{10}N_2$ [37]
1626 M 1614 VVS 1547 M 1510 M
1620 MW 1595 M 1540 VVS 1330 M

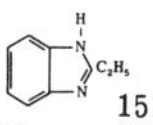

$C_9H_{10}O_2$ [25]
1728 S

$C_9H_{10}O_3$ [25]
1745 Sh 1736 S 1718 VS

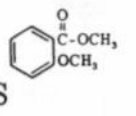

$C_9H_{10}O_3$ [25]
1728 S

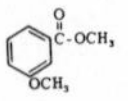

$C_9H_{10}O_3$ [25]
1723 S

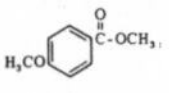

$C_9H_{11}N$ [41]
1588 1580 1572 1570 S 1558 1184 1102 782 728

$C_9H_{11}NO$ [634]
1495

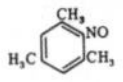

$C_9H_{11}NO$ [3]
2940 MW 1640 VS 1622 VS 1578 M 1504 M 1445 M 1388 S 1267 M 1214 MW 1075 M 791 S 733 S 696 S 673 S

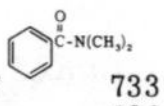

$C_9H_{11}NO$ [618]
1570 S

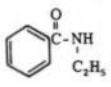

$C_9H_{11}NO$ [41]
1648 1620 1036 890 720

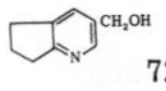

$C_9H_{11}NO$ [41]
1640 1562 S 1460 S 1198 854 722

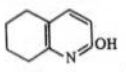

$C_9H_{11}NO_2$ [25]
1740 S 1693 S

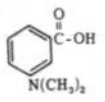

$C_9H_{11}NO_2$ [433]
1600 VS 1589 MW 1424 S 1193 MSh 1037 MW 1027 VS

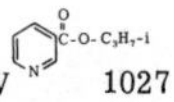

$C_9H_{11}NO_2$ [433, 481*]
1595 VS 1578 MW 1419 S 1191 MSh 1037 MW 1024 VS

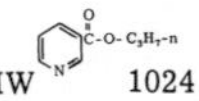

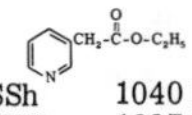

$C_9H_{11}NO_2$ [433]
1598 MW 1580 M 1479 M 1426 MS 1178 VSSh 1117 MWSh 1040 MWSh 1027 VS

$C_9H_{11}NS$ [606]
1622

C_9H_{12} [67, 122]
3090 MW 3075 M 3040 M 3037 M 2971 S 2950 S 2939 S 2881 M 2880 M 1500 M 1458 M 742 MS 698 MS

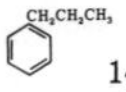

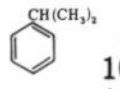

C_9H_{12} [66, 121]
3090 S 3076 M 3038 S 2980 VS 2970 VS 2941 S 2904 M 2884 M 1500 M 1499 S 1497 W 1460 M 1459 M 1080 MW 1056 W 1029 M 761 S 698 VS

C_9H_{12} [411]
705 S

$C_9H_{12}ClN$ [41]
3080 1553 802 795 790

$C_9H_{12}ClNO$ [41]
1048

$C_9H_{12}N_2$ [41]
1612 1478 1040 897 720

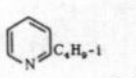

$C_9H_{13}N$ [431, 438, 455]
2941 S 2865 S 2857 S 1587 S 1565 S 1471 S 1387 S 1364 S 1335 M 1299 MS 1170 S 1047 S

$C_9H_{13}N$ [431]
2857 S 1604 S 1466 S 1412 S

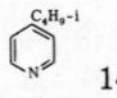

$C_9H_{13}N$ [455]
2900 S

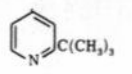

$C_9H_{13}N_3S$ [380]
3400 S 3200 S 1600 S 1440 S 868 S

C_9H_{14} [599]
1626

$C_9H_{14}Cl_2N_2$ [41]
896

$C_9H_{15}NO_3$ [613]
1597

$C_9H_{17}N$ [574]
1665 S

C_9H_{18} [271]
3085 MWSh 2925 VS 2855 S 2730 SSh 2680 VSSh 2045 VWVB 1821 S 1642 VS 1439 MSh 1379 S 1355 MWSh 1261 W 1215 WB 1055 MSVB 851 MWVB 829 WB 724 S 723 VS

$C_9H_{18}ClNO_4$ [574]
1686 S

$C_9H_{18}N_2O_3$ [26]
2938 S 2765 M 1541 VS 1453 S 1399 M 1376 M 1277 M 1213 S 1080 S 1045 S 947 S 893 S

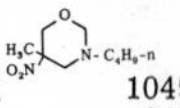

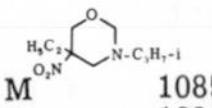

$C_9H_{18}N_2O_3$ [26]
2963 S 1543 VS 1460 S 1443 S 1386 M 1362 S 1324 M 1231 S 1210 M 1085 S 1031 M 938 M

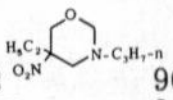

$C_9H_{18}N_2O_3$ [26]
2954 VS 2920 S 2812 S 2761 S 1461 S 1440 S 1390 S 1378 S 1359 M 1273 M 1244 M 1211 S 1080 VS 1053 S 1030 M 1015 M 967 M 948 S 884 S 849 S 789 M

C_9H_{20} [292, 365]
2950 VS 2645 MSSh 1469 VS 1393 MSh 1386 S 1368 VS 1342 VS 1330 S 1315 S 1283 M 1277 M 1250 VS 1206 VS 1169 S 1121 S 1099 M 950 M 920 VS 912 MSh 760 M

C_9H_{20} [78, 300]
3180 SSh 2950 VS 2880 VSSh 2740 MSh 2730 SSh 2650 WSh 2605 MSSh 2500 MWB 2405 S 2295 S 2230 MW 2110 MW 1705 S 1479 VS 1474 VS 1397 S 1395 S 1380 VS 1379 VS 1366 VS 1307 S 1304 S 1236 VS 1218 VS 1195 VS 1148 MS 1126 S 1086 M 1022 S 1020 VS 931 MSh 919 VS 905 SSh 905 MSh 850 M 833 VW 785 MS 772 VS 729 S 728 M

C_9H_{20} [79, 285, 410]
3180 S 2940 VS 2725 MSh 2725 MSh 2295 M 2110 W 1667 WB 1477 VS 1468 VS 1466 VS 1395 M 1394 MS 1380 VS 1379 M 1366 VS 1355 WSh 1247 VS 1247 VS 1205 VS 1205 VS 1156 VS 1115 MB 1096 MSh 1027 M 1014 M 996 S 969 VS 968 VS 949 M 926 MS 780 VS 780 S 773 VS 773 S 744 VS 711 M

C_9H_{20} [108,* 293]
3185 S* 2950 VS 2880 VS* 2720 S 2675 S 2620 S 1472 VS 1388 VS 1376 SSh 1368 VS 1310 M 1296 M 1220 S 1201 VS 1179 MSh 1141 VS 1114 VS 1085 VS 1028 MWSh 1022 MW 990 SSh 983 S 927 M 913 M 786 M

C_9H_{20} [299]

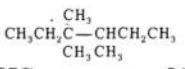

3180 VSSh	1464 VS	1182 VS	975 M
2970 VS	1381 VS	1152 VS	964 VS
2890 VS	1368 VSSh	1119 M	932 S
2740 VSSh	1339 MSh	1082 VS	917 S
2660 SBSh	1302 S	1050 MS	878 W
2620 SBSh	1269 MW	1043 MSh	828 M
2405 VS*	1224 MSh	1006 VS	775 VS
2295 MS	1205 MS	983 VWSh	704 M
1925 M			

C_9H_{20} [134]

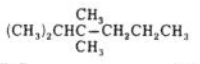

2730 VSSh	1370 S	1070 M	928 SB
2620 VS	1314 M	1049 M	922 SB
2415 S*	1299 MS	1041 S	903 MSh
2335 MSh	1261 MS	1029 MS	893 S
2295 M	1225 MS	995 M	871 MW
2185 M	1206 MS	961 M	837 MW
2055 M	1186 VS	947 M	818 M
1675 S	1186 M	935 SB	715 VW
1471 VS	1112 MS	933 SB	711 W
1393 SSh	1094 S	932 SB	689 MW
1375 S			

C_9H_{20} [301, 472]

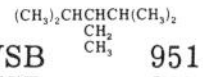

3380 MWSh	2250 MB	1175 VSB	951 S
3180 VSSh	1661 SB	1161 VSB	937 VS
3100 SSh*	1475 VS	1144 SShB	919 S
2890 VS	1385 S	1126 VS	880 S
2870 VS*	1368 S	1080 MS	867 VW
2730 VSSh	1321 VS	1062 MS	842 S
2625 VSSh	1274 S	1034 S	806 M
2405 VS*	1256 S	1018 MSh	782 W
2295 MB	1183 VSB	965 VS	758 S

$C_{10}H_3F_9N_2$ [37]

1570 M	1345 S	1162 S	1148 S
1553 MW	1190 S		

$C_{10}H_5F_7N_2$ [37]

1179 MS

$C_{10}H_6F_4N_2O_2$ [37]

1676 S

$C_{10}H_6F_6N_2$ [37]

1642 W	1545 VS	1176 S	1135 S
1555 M	1333 S	1148 S	

$C_{10}H_7ClCrO_4$ [17]

815 S

$C_{10}H_7F_3N_2O$ [37]

1700 S

$C_{10}H_7NO_4$ [673]

1740 S

$C_{10}H_8CrO_3$ [17]

1470 S	805 S

$C_{10}H_8CrO_4$ [17]

1475 S

$C_{10}H_8N_2O$ [590]

1590 S	1560 S

$C_{10}H_8N_2O_2$ [19]

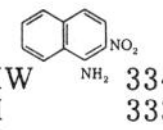

3520 MW	3501 MW	3363 MW	3348 M
3519 MW	3490 MW	3362 M	3336 M
3516 Sh	3464 MW	3360 M	

$C_{10}H_8N_2O_2$ [597]

1635 S	1579 S	1552 S	1473 S

$C_{10}H_8N_2O_2$ [590]

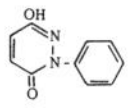

1650 S

$C_{10}H_8N_2O_3$ [673]

1728 S

$C_{10}H_8N_4O_6$ [577]

1750 S	1730 S	1620

$C_{10}H_9ClN_2$ [41]

1593 S	1543	1040	699
1593	1171	949	694

$C_{10}H_9F_3N_2$ [37]

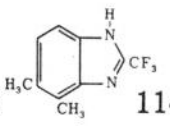

1545 VS	1188 S	1170 S	1146 S
1328 S			

$C_{10}H_9F_3N_2$ [37]

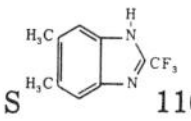

1555 MS	1191 S	1172 S	1107 S
1332 MS			

$C_{10}H_9N$ [378]

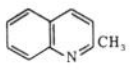

817 S	742 S

$C_{10}H_9N$ [378]

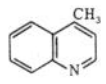

836 S	815

$C_{10}H_9N$ [378]

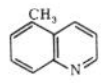

813 S

$C_{10}H_9N$ [378]

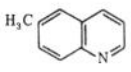

874 S	830 S	828 S

$C_{10}H_9N$ [378]

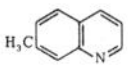

885 S	828 S	799 S

$C_{10}H_9NO_2$ [15]

1613 VVS	1452 VS	1022 S	946 MW
1489 VVS	1042 M	963 M	897 M

$C_{10}H_{10}BrNO$ [26]

2928 M	1471 M	1263 S	1008 S
2846 M	1390 M	1170 M	835 S
1644 VS	1343 S	1098 S	835 S
1585 S	1276 S	1067 S	728 M

$C_{10}H_{10}ClNO$ [26]

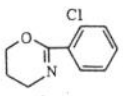

2926 S	1431 S	1034 S	805 M
2819 M	1377 M	1023 Sh	764 S
1652 VS	1290 VS	927 M	733 S
1589 M	1105 VS	852 M	719 S
1465 S	1071 S		

$C_{10}H_{10}ClNO$ [26]

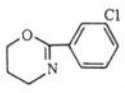

1590 M	1255 VS	941 Sh	816 M
1469 S	1131 VS	923 M	795 S
1422 S	1109 S	896 M	740 S
1345 VS	1090 S	882 M	711 S
1284 S	1039 M	862 S	

$C_{10}H_{10}ClNO$ [26]

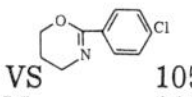

2921 S	1470 Sh	1264 VS	1058 M
2873 M	1433 M	1168 M	1014 S
1644 VS	1394 S	1100 VS	843 S
1591 S	1345 VS	1090 VS	843 S
1482 S	1280 S		

$C_{10}H_{10}ClNO_2$ [41]

1602	1178	939	742
1563	1148		

$C_{10}H_{10}FNO$ [26]

2930 M	1348 S	1128 VS	844 S
2878 M	1280 S	1103 S	844 S
1599 S	1270 S	1060 M	823 S
1468 M	1150 S	1014 M	737 M

$C_{10}H_{10}Fe$ [450]

3083 S

$C_{10}H_{10}N_2$ [41]

1604 S	1560 S	932	706
1604	1006		

$C_{10}H_{10}N_2O$ [6]

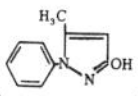

1590 S	1575 S	1510 S	1325

$C_{10}H_{10}N_2O$ [6]

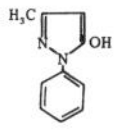

1600	1520 S	1450	1383

$C_{10}H_{10}N_2O$ [6]

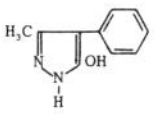

1595 S	1445

$C_{10}H_{10}N_2O$ [6]

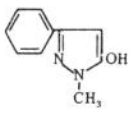

1565 S	1543 S	1445	1375

$C_{10}H_{10}N_2O$ [41]

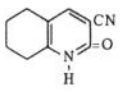

1627	1465	964	770

$C_{10}H_{10}N_2O$ [41]			
1607	1007	933	762
1217			

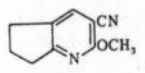

$C_{10}H_{10}N_2O_2$ [37]			
1635 M	1600 M	1550 MS	

$C_{10}H_{10}N_2O_3$ [26]			
2884 M	1466 M	1127 VS	860 M
2851 M	1434 M	1107 S	841 M
1659 VS	1346 VS	1086 S	783 S
1604 M	1286 S	1065 M	757 M
1572 M	1269 S	927 M	721 S
1522 VS	1252 S		

$C_{10}H_{10}O_3$ [25]
1703 S

$C_{10}H_{10}O_4$ [25]
1733 S

$C_{10}H_{10}O_5$ [450]
3095 S

$C_{10}H_{10}Ru$ [450]
3078 S

$C_{10}H_{11}NO$ [502]
2835 S

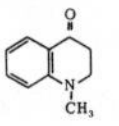

$C_{10}H_{11}NO$ [26]			
2928 M	1448 M	1132 S	1024 M
2878 M	1349 S	1105 S	933 M
1580 M	1274 S	1073 M	694 S
1492 M			

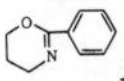

$C_{10}H_{11}NO_2$ [433]			
1592 MW	1418 S	1122 MW	1044 S
1575 M	1186 VSSh	1111 W	1025 VS
1477 M			

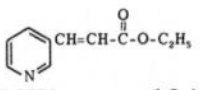

$C_{10}H_{11}NO_2$ [41]			
1610	1174	945	764
1588	1065		

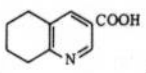

$C_{10}H_{11}NO_3$ [41]			
1620	1480	1033	724
1562	1198		

$C_{10}H_{12}ClN$ [41]
1078

$C_{10}H_{12}N_2$ [37]			
1622 W	1540 M	1535 VS	1324 M
1615 VVS			

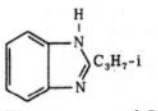

$C_{10}H_{12}N_2$ [37]	
1622 MW	1540 M

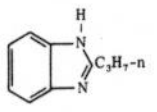

$C_{10}H_{12}N_2$ [37]
1346 S

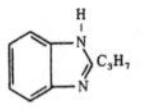

$C_{10}H_{12}N_4S$ [380]			
3200 S	1371 S	1253 S	878 S
1444 S	1297 S		

$C_{10}H_{13}N$ [41]			
3085	1576	1150	793
1588 S	1457	1085	767
1585	1193	803	

$C_{10}H_{13}NO$ [41]		
1610	1006	916

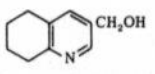

$C_{10}H_{13}NO$ [41]			
3020	1560	866	731
1613			

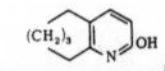

$C_{10}H_{13}NO_2$ [433]			
1599 VS	1420 S	1037 MW	1026 VS
1588 MW	1192 MSSh		

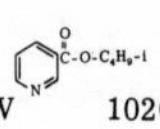

$C_{10}H_{13}NO_2$ [433]			
1594 VS	1420 S	1035 MW	1025 VS
1578 MW	1190 MSh		

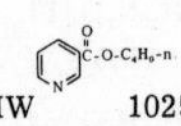

$C_{10}H_{13}NO_2$ [433]			
1597 S	1420 S	1035 MWSh	1025 VS
1580 MW	1191 MSSh		

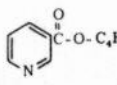

$C_{10}H_{13}NO_2$ [433]			
1600 MW	1479 M	1124 MWSh	1028 MS
1582 M	1426 MS	1104 MWSh	

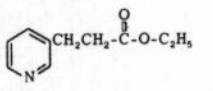

$C_{10}H_{13}NO_2$ [25]
1727 S

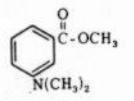

$C_{10}H_{14}$ [132]			
2950	1462 VS	1279 MSSp	1021 VS
1893 MS	1385 VSSp	1209 M	820 VVSB
1518 VS	1365 VSSp	1111 S	722 VS
1513 VS	1302 SSp	1058 VS	720 VS

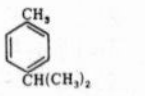

$C_{10}H_{14}$ [133]			
3050 VS	1471 VS	1242 MSh	1032 VS
2950 VS	1446 VS	1204 S	845 V
1601 VS	1394 MS	1115 S	763 VVSB
1495 VS	1368 VS	1081 S	762 VSB
1471 VS	1269 VS		

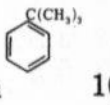

$C_{10}H_{14}BaO_4$ [16]
1630 S

$C_{10}H_{14}CaO_4$ [16]
1618 S

$C_{10}H_{14}ClNO$ [41]	
1630	899

$C_{10}H_{14}CoO_4$ [16]
1613 S

$C_{10}H_{14}CuO_4$ [16]	
1585 VS	1560 S

$C_{10}H_{14}MgO_4$ [16]
1629 S

$C_{10}H_{14}N_2$ [41]		
1202	922	712

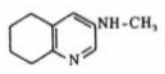

$C_{10}H_{14}N_2$ [41]			
3020	1576	922	712
1611	1053		

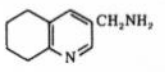

$C_{10}H_{14}N_2O_2$ [21]			
1600 S	1515 S	1192 S	950 S
1597 S	1476 S	1113 S	752 S
1580 S	1470 S	998 S	751 S
1516 S	1198 S	991 S	695 S

$C_{10}H_{14}N_3S$ [380]
1378 S

$C_{10}H_{14}NiO_4$ [16]
1616 S

$C_{10}H_{14}O$ [379]			
3279 S	2817 S	1355 S	725 S
2985 S	1445 S	1005 S	699 S
2907 S			

$C_{10}H_{14}O_2$ [430]			
3436*	1437 VVS	1269 MB	1060 SSh
2925 S	1383 VVS	1242 MB	1047 S
1698 VS	1339 S	1190 MS	1011 S
1650 VS	1314 S	1089 M	

$C_{10}H_{14}O_4Pd$ [16]
1561 S

$C_{10}H_{14}O_4Sr$ [16]
1613 S

$C_{10}H_{14}O_4Zn$ [16]
1600 S

$C_{10}H_{14}O_5V$ [16]
1567 S

$C_{10}H_{15}N$ [455]

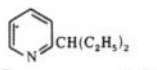

2857 S 1475 S 1295 MS 1053 S

$C_{10}H_{15}N$ [431]

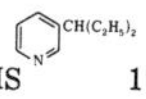

2874 S 1471 S 1189 MS 1186 M

$C_{10}H_{15}N$ [431]

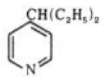

2857 S 1429 MS 1377 S 782 M
1462 S 1408 S 1042 S

$C_{10}H_{15}N_3S$ [380]

3400 S 1440 S 1299 S 876 S
3200 S

$C_{10}H_{16}Cl_2N_2$ [41]

928

$C_{10}H_{18}O_2S_4$ [434]

1030 S 1026 S 1023 S 1020 S
1028 S 1024 S 1022 S 1020 S
1027 S 1023 S 1021 S

$C_{10}H_{18}O_2S_4Zn$ [434]

1062 S 1046 S 1039 S 1037 S
1060 S 1044 S 1038 S 1030 S
1050 S 1040 S 1037 S

$C_{10}H_{18}O_4$ [579]

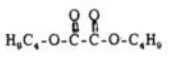

1746

$C_{10}H_{18}O_4$ [571]

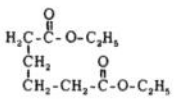

1740

$C_{10}H_{20}$ [215]

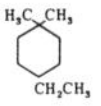

2940 VS 1346 SSh 1152 S 929 S
2725 SSh 1285 M 1107 M 917 S
1468 VS 1272 SSp 1056 M 908 S
1458 VS 1247 SSp 979 VS 817 S
1389 SSp 1187 VS 937 MS 767 M
1368 SSp 1180 VS

$C_{10}H_{20}$ [270]

3085 VS 1825 M 1235 W 1075 M
2935 VS 1642 VS 1181 MWB 1050 S
2875 VS 1464 S 1136 MWB 722 W
2025 VWB 1379 S 1116 MW

$C_{10}H_{20}N_2O_3$ [26]

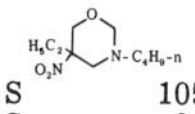

2942 S 1441 S 1326 S 1050 M
2919 S 1388 M 1203 S 943 M
1543 VS 1375 M 1086 S 894 M
1460 S 1341 M

$C_{10}H_{20}N_2O_3$ [26]

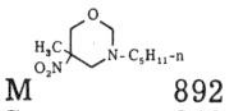

2941 VS 1400 M 1210 M 892 S
2907 S 1376 S 1076 S 848 S
1453 S 1348 S 1046 S

$C_{10}H_{21}BO_2$ [4]

1610 M

$C_{10}H_{22}$ [305]

3220 SSh* 2470 M 1700 SB 1185 VS
2950 VS 2410 S* 1479 VS 1135 S
2875 VS* 2295 S 1471 VSSh 1070 W
2740 VSSh 2105 S 1395 S 1013 MS
2715 SSh 1990 M 1368 VVS 930 VS
2655 MSh 1960 M 1300 VS 910 VS
2565 MB 1890 M 1285 S 755 MB
2505 MB 1800 M 1247 VVS

$C_{10}H_{22}$ [303]

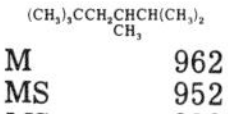

2960 VS 1471 VS 1176 M 962 SB
2610 MSh 1395 SSh 1143 MS 952 MSB
2610 SSh 1389 SSh 1114 MS 928 S
2295 S 1379 SSh 1094 W 920 MSh
2240 M 1330 S 1082 MW 909 S
2170 M 1312 MSSh 1041 VS 877 M
2140 M 1284 S 1030 SSh 856 M
2115 M 1245 VS 1004 S 772 M
1481 VSSh 1203 S 975 S 743 MW

$C_{10}H_{22}$ [307]

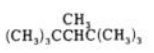

2970 VS 2390 S 1368 VS 1015 M
2915 VS 2295 S 1238 VS 978 VS
2880 VSSh 2250 S 1214 VS 930 MSB
2725 SSh 1477 VS 1163 S 889 M
2605 MSSh 1460 S 1134 MW 748 M
2500 MW 1395 VS 1085 VS 696 M
2465 MW 1377 S

$C_{10}H_{22}$ [302]

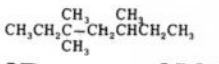

3180 MSh 1466 VS 1115 MB 956 VSSh
2960 VS 1379 S 1096 MB 933 SBSh
2730 Sh 1311 MS 1057 S 922 SBSh
2665 MSSh 1295 MS 1040 S 885 MWB
2390 SB 1274 S 1011 VVSB 859 MB
2295 M 1186 VS 1000 VVSB 721 M
2235 MB 1155 VS 967 VS 719 M
1667 SVB

$C_{10}H_{22}$ [304]

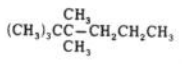

3220 SSh* 1850 M 1297 M 994 MW
2960 VS 1690 S 1267 M 954 S
2730 VSSh 1613 S 1230 S 927 S
2640 SSh 1477 VS 1174 VVS 916 VS
2550 MSh 1458 SSh 1159 VVS 863 S
2385 S 1403 MS 1059 M 789 S
2295 M 1381 VS 1021 MS 743 VS
2060 MS 1372 VS 1008 S 678 MW
2025 MS 1319 M

$C_{10}H_{22}$ [306]

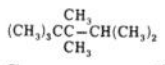

3230 SSh* 2255 MS 1328 S 1019 S
2970 VS 1685 SVB 1230 S 993 M
2725 VSSh 1484 SB 1168 VS 930 VS
2645 MSSh 1406 MSh 1133 VS 913 S
2610 MSh 1389 VS 1086 S 894 MS
2385 S 1381 VSSh 1064 VS 841 M
2295 S 1374 VSSh

$C_{11}H_5FeNO_4$ [509]

2174 S

$C_{11}H_7F_7N_2$ [37]

1325 MS 1195 S 1153 S 1150 S

$C_{11}H_7NO$ [507]

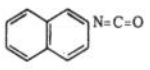

2267 VS

$C_{11}H_7NO$ [507]

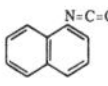

2275 2270

$C_{11}H_8CrO_5$ [17]

1460 S 815 S

$C_{11}H_8N_2$ [37]

1632 VVS 1596 VVS

$C_{11}H_8N_2O_2$ [433]

1572 MW 1428 M 1128 WSh 1025 MW
1471 MW 1188 MW 1109 M 1001 M

$C_{11}H_9N$ [433]

1584 MW 1408 S 1023 MW 1006 M
1470 MW

$C_{11}H_9NO$ [24]

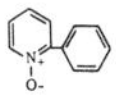

844 S

$C_{11}H_9NO$ [24]

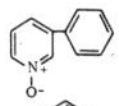

759 M

$C_{11}H_9NO$ [24]

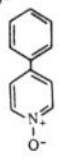

770 M

$C_{11}H_{10}$ [42, 191, 452]

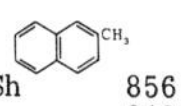

1922 M 1374 MSh 1155 MSh 856 S
1807 M 1351 S 1145 M 818 S
1595 S 1272 S 1041 S 787 M
1455 M 1255 M 964 S 768 S
1437 SSh 1240 M 953 S 745 S
1423 S 1200 M 892 S 732 VS
1405 MSh 1174 MS

$C_{11}H_{10}$ [42, 192]

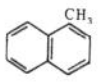

1922 S 1599 VS 1216 S 972 MS
1880 M 1512 S 1168 VS 953 MS
1838 M 1460 VVS 1146 MSh 926 MW
1815 M 1436 VVS 1082 S 856 S
1759 M 1397 VVS 1037 S 795 VS
1696 M 1381 VVS 1026 VS 777 VS
1654 M 1344 MS 980 MS 733 S
1630 M 1270 VS

$C_{11}H_{10}CrO_3$ [17]

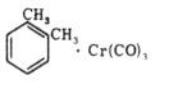

1480 S

$C_{11}H_{10}CrO_3$ [17]
1460 S

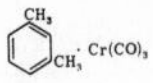

$C_{11}H_{10}CrO_3$ [17]
1490 S 835 S

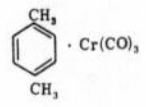

$C_{11}H_{10}CrO_4$ [17]
1495 S

$C_{11}H_{10}FeO_2$ [450]
2624 S

$C_{11}H_{10}N_4$ [508]
2227 M 2216 M 2212 Sh

$C_{11}H_{10}N_2O_3$ [15] 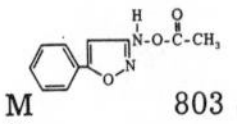
1624 VVS 1131 W 1000 M 803 S
1432 VVS 1039 W 919 M

$C_{11}H_{10}N_2O_3$ [15] 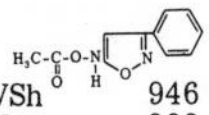
1617 VVS 1446 M 1034 WSh 946 M
1481 M 1407 M 989 W 890 W

$C_{11}H_{10}N_4O_7$ [575]
1695 S 1580 S

$C_{11}H_{10}O_2Ru$ [450]
2611 S 2538 S

$C_{11}H_{11}ClN_2$ [41]
1595 1036 1036 935

$C_{11}H_{11}Cl_3O_2$ [379]
1238 S

$C_{11}H_{11}N$ [378]
898 S 840 S 755

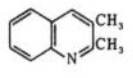

$C_{11}H_{11}N$ [378]
856 S 743 S

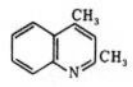

$C_{11}H_{11}N$ [378]
807 S

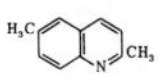

$C_{11}H_{11}N$ [378]
831

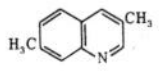

$C_{11}H_{11}NO_2$ [15]
1647 VVS 1462 MS 1036 MW 999 VS
1487 VVS 1451 S

$C_{11}H_{11}NO_4$ [349, 576, 652]
2980 1610 1350 980
1710 1447 S 1310 845
1650

$C_{11}H_{11}N_3O$ [508]
2216 M 2210 M

$C_{11}H_{11}N_3O_2$ [562]
1760 S 1733 S 1713 S 1687 S

$C_{11}H_{12}ClNO_2$ [41]
1460 S 932 932 742

$C_{11}H_{12}N_2$ [41]
1562 S 1074 919 708

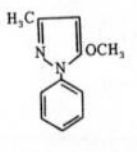

$C_{11}H_{12}N_2O$ [6]
1600 1573 1450

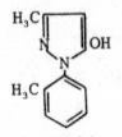

$C_{11}H_{12}N_2O$ [6]
1545 S 1445 1365

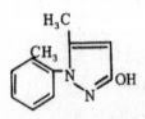

$C_{11}H_{12}N_2O$ [6]
1558 S 1515 S 1310

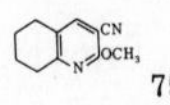

$C_{11}H_{12}N_2O$ [41]
1605 S 1210 934 757
1605 1007

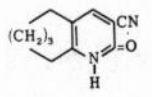

$C_{11}H_{12}N_2O$ [41]
1156 774

$C_{11}H_{12}N_2O_5$ [37]
1686 S

$C_{11}H_{12}O_3$ [25]
1748 Sh 1718 VS

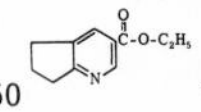

$C_{11}H_{13}NO_2$ [41]
3090 1575 950 778
1610 1026

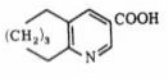

$C_{11}H_{13}NO_2$ [41]
3050

$C_{11}H_{13}NO_3$ [41]
1615 1564 1472 960

$C_{11}H_{14}$ [269]
3003 VSSh 1451 VSB 1152 M 939 VS
2933 VS 1441 VSB 1079 M 909 VS
2865 VS 1379 S 1066 M 901 VS
2732 MSh 1355 VS 1038 S 823 VS
2667 M 1297 VS 1005 MW 698 VS
1618 S 1285 VS 987 S 665 MB
1504 VS

$C_{11}H_{14}ClNO_2$ [41]
1018

$C_{11}H_{14}N_2$ [37]
1532 M 1126 S

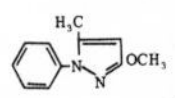

$C_{11}H_{14}N_2O$ [6]
1605 1570 1505 1455

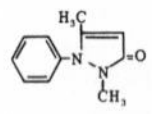

$C_{11}H_{14}N_2O$ [6]
1655 S 1570 1385

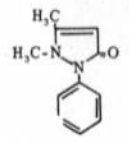

$C_{11}H_{14}N_2O$ [6]
1675 S 1590 1390

$C_{11}H_{15}NO$ [41]
1612 1478 1025 715
1582 1227 903

$C_{11}H_{16}$ [277]
3030 SSh 1460 VSVB 1182 VS 946 S
2970 VS 1381 S 1106 MW 928 VS
2930 SSh 1362 S 1080 VS 846 VS
2870 SSh 1334 M 1037 VS 808 W
1770 S 1257 S 1000 S 703 VS
1610 VS

$C_{11}H_{16}ClNO$ [41]
1628 1018 893

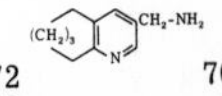

$C_{11}H_{16}N_2$ [41]
3040 1575 1072 709
1610 1473 912

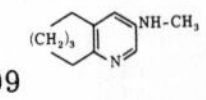

$C_{11}H_{16}N_2$ [41]
1192 910 709

$C_{11}H_{16}O_3$ [475]
3507 M 1702 MW

$C_{11}H_{18}Cl_2N_2$ [41]
928

$C_{11}H_{19}NO$ [497]
2915 VS

$C_{11}H_{19}NO_3$ [594]
1650

$C_{11}H_{22}$ [352]
3075 WSh 1418 SSp 1116 MW 870 MSh
2740 MSSh 1379 SSp 1075 MSB 837 VWB
1821 M 1305 M 992 S 750 SShB
1464 M 1236 M 908 VS 722 VS
1439 M 1178 MW

$C_{11}H_{22}N_2O_3$ [26]

2920 S	1457 S	1349 S	1049 M
2845 S	1400 M	1215 M	949 M
2769 M	1379 M	1081 S	895 M
1545 VS			

$C_{11}H_{22}N_2O_3$ [26]

2953 S	1378 M	1207 S	947 M
2927 S	1359 M	1082 S	914 M
1542 VS	1338 M	1052 S	894 M
1460 S	1325 S	1028 M	848 M
1443 S			

$C_{11}H_{23}BO_2$ [4]

1625 M

$C_{12}F_{27}N$ [268, 346]

2585 VSB	1577 M	1098 S	797 VS
2500 VSB	1451 SVB	1048 S	787 VSSh
2450 VSB	1357 MS	994 SSh	751 VSSh
2340 VSB	1308 VSB	985 SB	732 VVS
2200 MWVB	1284 VSVB	973 S	721 VS
1940 SB	1241 VSVB	948 VS	710 WB
1850 M	1217 VSVB	833 VW	701 VS
1705 S	1157 VS		

$C_{12}Fe_3O_{12}$ [511]

1833

$C_{12}H_6N_2O_2$ [507]

2270	2265

$C_{12}H_8N_4O_6$ [19]

3304 MW	3300 MW	3288 W	3236 W
3303 MW	3295 M		

$C_{12}H_{10}Cl_2Si$ [86]

2688 W	1335 M	1067 W	850.2 W
1658 W	1304 M	1029 W	830.7 W
1590 S	1263 W	997.3 S	739.6 VS
1567 W	1188 M	985.6 W	718.3 VS
1431 VS	1121 VS	973.1 W	693.4 VS
1377 W	1107 S	950.6 W	

$C_{12}H_{10}FeO_4$ [450]

2625 S	2544 S

$C_{12}H_{10}N_2$ [37]

1635 VVS	1595 VVS	1524 M

$C_{12}H_{10}N_2O$ [433]

1591 S	1420 VS	1126 W	1026 M
1481 VS	1190 MWSh	1098 MW	

$C_{12}H_{10}O_2$ [452,* 629]

1625

$C_{12}H_{10}O_4Ru$ [450]

2637

$C_{12}H_{11}NO_4$ [673]

1745 S	1738 S	1723 S

$C_{12}H_{12}$ [42, 179, 452*]

1927 M	1655 M	1455 S	1048 S
1883 M	1609 S	1437 S	974 S
1828 M	1560 M	1388 M	946 M
1812 S	1541 S	1366 M	922 S
1750 M	1512 S	1339 S	840 VVS
1708 M	1495 M	1281 S	771 M
1682 M	1474 M	1182 S	

$C_{12}H_{12}$ [180, 452*]

1910 M	1505 M	1338 M	960 S
1776 S	1450 S	1269 M	886 VS
1623 S	1442 S	1164 MS	816 VS
1607 S	1378 M	1036 S	

$C_{12}H_{12}$ [42, 181, 452*]

1914 S	1605 S	1433 VS	1086 M
1845 S	1563 M	1375 S	1023 VS
1796 M	1527 S	1330 M	948 S
1769 S	1510 S	1274 S	872 VS
1731 S	1460 VS	1145 S	748
1646 S			

$C_{12}H_{12}$ [42, 182, 452*]

1927 M	1432 VS	1208 M	941 S
1682 M	1381 VS	1167 S	906 M
1593 VS	1373 VVS	1112 S	814 VS
1547 M	1335 S	1040 VS	802 VSB
1523 M	1282 M	991 W	789 S
1466 VVS	1262 VS	971 M	773 VS
1444 VS	1235 M		

$C_{12}H_{12}$ [42, 183, 452*]

1925 S	1447 VS	1142 S	897 S
1923 S	1365 S	1072 VS	870 VS
1734 M	1334 S	1036 VS	823 VVS
1643 VS	1267 S	1005 M	776 S
1611 VS	1207 M	975 M	752 VS
1517 VVS	1196 MS	965 S	705 MW
1474 MS	1163 VS		

$C_{12}H_{12}$ [42, 184, 452*]

1916 S	1344 VSSh	1038 VS	868 VVS
1637 VS	1265 S	1004 M	836 S
1610 VVS	1213 S	978 S	816 VS
1523 VVS	1163 VS	963 S	750 VS
1455 VVS	1078 S	924 M	720 M
1433 VVS	1054 S	897 S	685 M
1376 VVS			

$C_{12}H_{12}$ [42, 185, 452*]

1929 M	1524 S	1379 M	1035 S
1726 M	1485 M	1335 MS	1014 MS
1709 M	1464 S	1258 S	968 M
1662 M	1452 S	1205 M	891 M
1608 S	1435 M	1171 VS	871 MW
1573 M	1420 M	1066 S	790 VS
1555 M			

$C_{12}H_{12}$ [42, 186, 452*]

1856 M	1428 VS	1139 S	948 S
1808 M	1394 VS	1111 M	873 MW
1711 M	1335 MS	1089 M	854 S
1604 VVS	1270 VS	1034 VS	824 VS
1468 VS	1217 VS	1023 VS	797 W
1448 VS	1162 VS	995 VS	755 VS

$C_{12}H_{12}$ [187, 452*]

1611 VVS	1377 S	1131 S	889 S
1585 SSh	1278 S	1028 VS	859 VS
1516 VVS	1217 VS	983 S	845 VS
1472 VS	1178 S	948 S	774 VS
1448 VS	1159 MS	922 MW	747 VS
1416 VS			

$C_{12}H_{12}$ [42, 188, 452*]

1871 M	1433 VS	1170 SSh	966 MS
1848 M	1371 VS	1153 VS	865 S
1718 M	1335 M	1129 MSh	808 VS
1638 S	1269 VS	1083 S	786 VS
1611 VVS	1236 VS	1038 VS	743 VS
1586 VS	1213 S	998 MS	724 W
1550 M	1185 VS		

$C_{12}H_{12}$ [42, 189, 452*]

1601 S	1264 VS	1057 S	854 VS
1528 VS	1166 SB	1018 S	818 VS
1493 VS	1153 SB	974 MSh	782 VS
1435 VVS	1140 SB	949 S	764 VSSh
1361 VS	1122 SSp	890 S	750 VVS
1308 S			

$C_{12}H_{12}$ [190, 452*]

1916 M	1450 VVS	1255 VS	967 MS
1801 M	1382 VVS	1164 S	953 MS
1590 VS	1372 VSSh	1059 S	790 VS
1528 W	1346 SSh	1028 M	732 S
1508 VS	1312 S	1009 S	696 M
1460 VVS			

$C_{12}H_{12}CrO_3$ [17]

1465 S

$C_{12}H_{12}FeO$ [450]

1658 S	1115 M

$C_{12}H_{12}N_4$ [508]

2221 M	2214 M	2210 Sh

$C_{12}H_{12}N_4$ [508]

2224 M	2212 M	2211 Sh	2202 Sh

$C_{12}H_{12}Os$ [450, 452]
1670 S 1116 M

$C_{12}H_{12}ORu$ [452]
1658 S 1116 M

$C_{12}H_{13}Cl_3O_2$ [379]
1754 S 833 S 826 S 680 S
877 S

$C_{12}H_{13}N_3O$ [508] 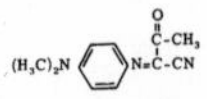
2220 M 2212 M

$C_{12}H_{13}N_3O$ [508] 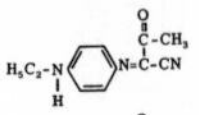
2214 M 2210 M

$C_{12}H_{13}N_3O$ [508] 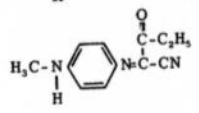
2216 M

$C_{12}H_{13}N_3O_2$ [562] 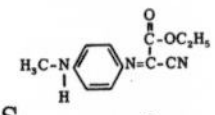
1756 S 1741 S 1727 S

$C_{12}H_{13}N_3O_2$ [562] 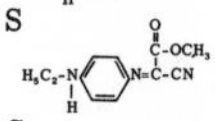
1760 S 1733 S 1731 S

$C_{12}H_{13}N_3O_2$ [562] 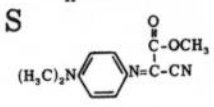
1757 S 1731 S 1708 S

$C_{12}H_{13}N_5O_6S$ [570, 580]
1665 S 1650 S 1580 S

$C_{12}H_{14}ClNO_2$ [41]
1460

$C_{12}H_{14}N_2O$ [41]
1602 1002 930 776
1195

$C_{12}H_{15}NO_2$ [41]
3000 1027 945 770
1613

$C_{12}H_{15}NO_3$ [26]
2835 Sh 1415 S 1084 VS 807 M
1655 VS 1306 S 1033 S 793 M
1575 S 1261 VS 1004 S 762 M
1469 VS 1181 S 867 M 750 S
1439 S 1119 S

$C_{12}H_{16}INOS$ [615]
1580

$C_{12}H_{16}N_2OS$ [31]
3267 MS 1650 S 1511 S 1442 S
3080 M

$C_{12}H_{18}$ [296] 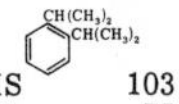
2959 VS 1383 VS 1224 MS 1031 VS
1488 VVS 1366 VS 1074 VS 754 VS
1462 VVS 1266 MS 1047 M

$C_{12}H_{18}$ [297] 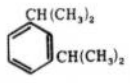
2941 VS* 1379 VS 1188 S 893 S
1597 VS 1362 VS 1050 VS 793 VS
1486 VVS 1314 VS 924 VS 703 VS
1453 VVS

$C_{12}H_{18}$ [298] 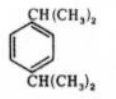
2941 VS 1404 VS 1299 S 1070 S
2855 VS* 1379 S 1277 M 1050 S
1511 VS 1361 S 1189 S 1016 S
1460 VVS 1330 MS 1101 S 830 VS
1418 VVS

$C_{12}H_{19}N$ [455]
1565 S 1429 S 1299 MS 1152 S
1466 S

$C_{12}H_{22}O_3$ [514]
1825 1760

$C_{12}H_{24}N_2O_3$ [26]
2912 VS 1390 M 1325 S 943 S
2846 S 1373 S 1237 M 914 S
1542 VS 1359 S 1208 S 889 S
1460 S 1337 S 1053 S 788 M

$C_{12}H_{30}B_3Br_6P_3$ [30]
2940 S 1410 S 1239 M 630 S
2905 S 1382 S 1050 VS 620 S
2852 S 1258 M

$C_{12}H_{30}B_3Cl_6P_3$ [30]
2940 S 2859 S 1388 M 1050 S
2911 S 1412 M 1244 W 691 VS

$C_{12}H_{30}B_3I_6P_3$ [30]
2950 S 2855 S 1386 S 1250 MW
2910 S 1410 S 1268 MW

$C_{12}H_{33}B_3N_6$ [32]
2932 VS 1452 VS 1342 S 1104 M
2857 S 1417 VS 1269 VS 717 MSh
1493 VS 1376 VS 1180 S 710 S

$C_{12}H_{36}B_3P_3$ [30]
2955 S 2380 S 1388 M 1003 M
2922 S 2345 S 1260 M 982 MW
2870 S 1422 M 1242 M

$C_{13}H_8O_6$ [624]
1595

$C_{13}H_{12}N_2$ [37]
1635 VVS 1596 VVS 1525 M

$C_{13}H_{12}N_2O$ [433]
1580 S 1423 VS 1102 MSSh 1023 M
1488 VS 1175 MWSh 1044 MW

$C_{13}H_{13}ClSi$ [86]
2959 W 1653 W 1408 W 1256 S
1976 W 1613 W 1337 W 1119 VS
1958 W 1567 W

$C_{13}H_{13}NO_4$ [673]
1748 S 1735 S

$C_{13}H_{14}$ [42, 177] 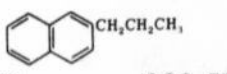
1707 M 1353 SSh 1122 MS 893 VS
1643 S 1324 MS 1089 M 848 VS
1607 S 1262 S 1017 MS 793 VS
1442 VS 1164 MS 958 M 764 S
1429 VS 1148 MS 943 S 744 VS
1362 S 1139 MS

$C_{13}H_{14}$ [42, 178] 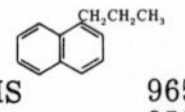
1905 MS 1454 VS 1195 MS 965 M
1807 MS 1441 VS 1160 S 952 M
1749 M 1379 VS 1090 M 782 VVS
1717 M 1364 VS 1079 SSh 732 S
1589 VS 1341 S 1026 MS 711 MW
1505 M 1327 SSh 1006 MS

$C_{13}H_{14}N_4$ [508]
2223 M 2213 M 2209 Sh

$C_{13}H_{15}N$ [378] 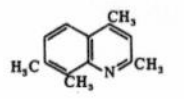
860 815

$C_{13}H_{15}N$ [378] 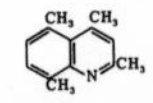
857 818

$C_{13}H_{15}N_3O$ [508] 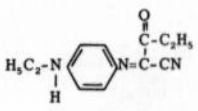
2220 M 2211 M 2202 M

$C_{13}H_{15}N_3O$ [508] 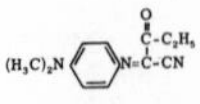
2213 M 2200 M

$C_{13}H_{15}N_3O_2$ [562] 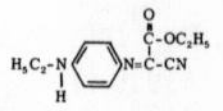
1756 S 1750 S 1726 S

$C_{13}H_{15}N_3O_2$ [562] 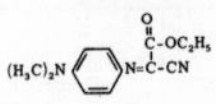
1754 S 1740 S 1724 S

$C_{13}H_{16}N_2O$ [6]
1605 S 1530 S 1505 S

$C_{13}H_{17}NO_2$ [41]
1577 1202 1032 777
1472 1192 932

$C_{14}H_{12}N_2OS$ [31]
3244 M 1517 S 1447 S 1058 M
1672 S 1453 S 1374 MS

$C_{14}H_{12}N_2O_2$ [597] 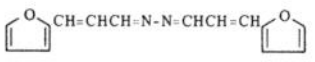
1628 S 1575 S 1561 S

$C_{14}H_{12}N_2O_2$ [31] 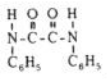
3300 MS 1664 S 1526 VS 1440 S

$C_{14}H_{13}NS$ [606]
1611

$C_{14}H_{14}FeO_2$ [452]
1115 M

$C_{14}H_{14}N_2$ [37]
1633 1597 VVS 1524 M

$C_{14}H_{16}$ [175] 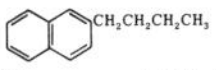
1712 MS 1272 S 1147 MS 956 MS
1646 VS 1242 M 1130 S 905 S
1608 S 1214 M 1117 MS 888 S
1513 VS 1203 M 1108 MS 853 S
1458 VVS 1173 MS 1022 S 752 VSB
1377 VVS 1159 MS 961 S

$C_{14}H_{16}$ [176] 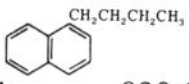
1604 VS 1336 MSh 1043 MSh 820 SSh
1518 S 1277 S 1028 S 778 VS
1464 VVS 1260 MS 976 M 749 S
1404 VS 1220 M 958 M 732 S
1384 VS 1180 S 860 MS 711 MW
1370 SSh 1093 MS

$C_{14}H_{16}N_4$ [508]
2223 M 2214 M 2210 Sh

$C_{14}H_{18}N_2OS$ [31]
3200 MS 1527 VS 1447 S 1038 MS
2920 MS 1516 S 1390 M 1030 M
1674 S

$C_{14}H_{18}N_2O_2$ [31]
3280 S 2845 M 1512 VS 1511 S
2918 MS 1657 S

$C_{14}H_{23}N$ [431, 455]
2857 S 1471 S 1387 S 1168 S
1705 W 1433 S 1368 S 1152 S

$C_{14}H_{24}N_2OS$ [31]
3236 MS 1665 S 1449 M 1054 MS
3200 MS 1510 VS 1396 M 1041 M
2920 MS 1453 M

$C_{14}H_{24}N_2O_2$ [31]
3280 S 2842 M 1511 S 1445 S
2912 MS 1645 S 1448 M

$C_{14}H_{24}N_2S_2$ [31]
3140 MS 1509 S 1447 M 1362 M
2908 MS 1503 S 1388 MS 973 MS
2840 M

$C_{14}H_{25}NO$ [357, 497]
2865 VS 1550 MW 1441 SSh 1272 M
1678 VS 1511 MW 1215 M 1215 M
1642 VS 1466 M 1372 M 968 VS

$C_{15}H_{10}N_2O_2$ [507]
2272 VS

$C_{15}H_{11}NO$ [15]
1616 S 1468 VS 1044 W 915 M
1496 S 1406 VS 948 MS

$C_{15}H_{12}N_2O$ [6]
1600 S 1565 S 1520 1320

$C_{15}H_{18}$ [42, 173] 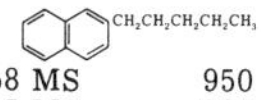
1630 S 1339 SSh 1158 MS 950 S
1598 S 1298 MS 1145 MS 890 VS
1501 VS 1268 VS 1130 S 857 VS
1452 VVS 1238 M 1113 MS 805 VVS
1434 VVS 1210 M 1021 S 783 VSSh
1370 VS 1171 MS 968 MS 757 VVS

$C_{15}H_{18}$ [174] 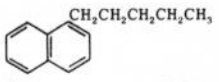
1818 M 1406 VS 1217 MS 865 M
1700 M 1386 VS 1168 S 776 VS
1606 VS 1354 SSh 1083 MS 732 S
1512 VS 1264 S 1018 S 711 M
1462 VVS

$C_{15}H_{21}AlO_6$ [16]
1621 S 1555 S

$C_{15}H_{21}CoO_6$ [16]
1592 S

$C_{15}H_{21}CrO_6$ [16]
1587 S

$C_{15}H_{21}FeO_6$ [16]
1582 S

$C_{15}H_{21}LaO_6$ [16]
1610 S

$C_{15}H_{21}MnO_6$ [16]
1592 S

$C_{15}H_{21}O_6Sc$ [16]
1565 S

$C_{15}H_{21}O_6Y$ [16]
1610 S

$C_{15}H_{24}$ [278]
2995 VS 1531 MS 1248 VS 1057 S
2890 M 1470 VSB 1190 VS 1002 W
2730 MW 1385 VSSp 1146 MB 920 VS
1782 MS 1365 VSSp 1108 VS 870 VS
1764 S 1316 S 1098 VVS 830 VS
1695 W 1270 MS 1073 VS 817 MSh
1608 VS

$C_{16}H_{11}ClN_2O$ [35] 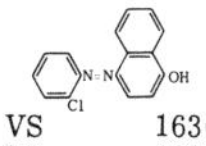
3370 M 1653 VS 1650 VS 1636 VS
3350 M 1652 VS 1644 VS 1625 VS
3340 M

$C_{16}H_{11}ClN_2O$ [35] 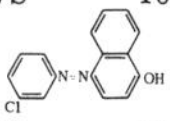
3580 M 3340 M 1643 VS 1635 VS
3570 VW 3280 M 1640 VS 1630 VS
3350 M

$C_{16}H_{11}ClN_2O$ [35] 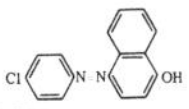
3220 M 3170 M 1627 VS

$C_{16}H_{11}ClN_2O$ [35] 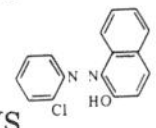
1622 VS 1620 VS 1617 VS

$C_{16}H_{11}ClN_2O$ [35] 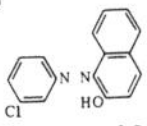
1625 VS 1622 VS 1621 VS 1620 VS
1623 VS

$C_{16}H_{11}ClN_2O$ [35] 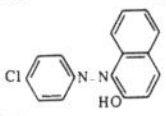
1625 VS 1623 VS 1622 VS

$C_{16}H_{11}N_3O_3$ [35] 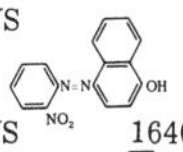
3330 MS 3300 M 1651 VS 1640 VS
3320 MS 1655 VS 1643 VS

$C_{16}H_{11}N_3O_3$ [35] 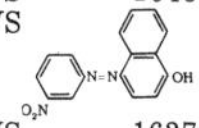
3330 M 1648 VS 1638 VS 1627 VS
3280 M 1643 VS

$C_{16}H_{11}N_3O_3$ [35] 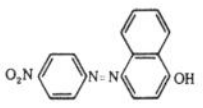
3280 M 1628 VS

$C_{16}H_{11}N_3O_3$ [35] 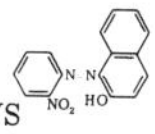
1628 VS 1624 VS 1622 VS

$C_{16}H_{11}N_3O_3$ [35] 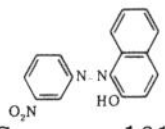
1625 VS 1624 VS 1621 VS 1615 VS

$C_{16}H_{11}N_3O_3$ [35] 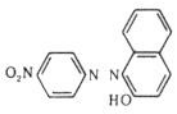
1624 VS

$C_{16}H_{12}N_2O$ [35]

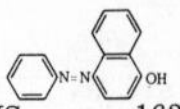

3580 S   3260 M   1641 VS   1629 VS
3570 MS   3140 M   1634 VS   1625 VS
3360 W

$C_{16}H_{12}N_2O$ [35]

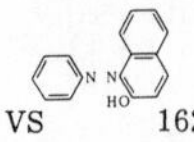

1624 VS   1622 VS   1622 VS   1621 VS
1623 VS

$C_{16}H_{14}N_2O$ [6]

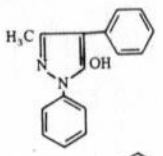

1607 S   1558   1435

$C_{16}H_{14}N_2O$ [6]

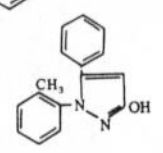

1565 S   1505   1310

$C_{16}H_{14}N_2O$ [6]

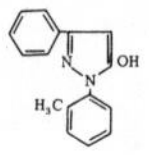

1440   1370

$C_{16}H_{14}O_2$ [489, 582]

3050   1740   1670

$C_{16}H_{15}NO_2$ [597]

1646

$C_{16}H_{16}O_5$ [556]

1735 SB

$C_{16}H_{20}N_2O_3$ [673]

1745 S

$C_{16}H_{21}N_3O$ [508]

2207 M   2200 M

$C_{16}H_{21}N_3O_2$ [562]

1750 S   1722 S   1694 S

$C_{16}H_{32}Br_2NiO_4S_4$ [33]

970 S

$C_{16}H_{34}$ [281]

2955 VS   1467 VS   1140 MS   969 M
2925 VS   1459 VS   1099 MVB   895 S
2865 VS   1379 SSp   1075 MVB   845 MVB
2730 S   1341 WSh   1029 MWVB   783 S
2165 W   1301 W   1005 WVB   725 VS
2035 W   1191 MSh

$C_{17}H_{14}FeO$ [450]

1631 S   1626 S   1105 S

$C_{17}H_{14}N_2O$ [35]

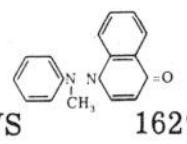

1653 VS   1633 VS   1632 VS   1629 VS
1636 VS

$C_{17}H_{14}N_2O$ [35]

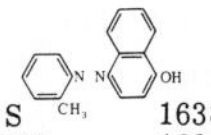

3610 S   3360 W   1648 S   1638 VS
3600 S   3300 M   1647 VS   1626 VS
3570 MS   3240 M   1639 VS   1625 VS
3380 MW

$C_{17}H_{14}N_2O$ [35]

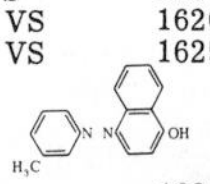

3575 MS   3250 M   1641   1628 VS
3340 M   3140 M   1639 VS

$C_{17}H_{14}N_2O$ [35]

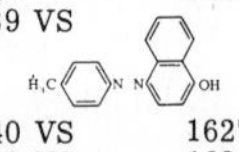

3580 S   3270 M   1640 VS   1627 VS
3570 MS   3140 M   1633 VS   1625 VS
3350 M

$C_{17}H_{14}N_2O$ [35]

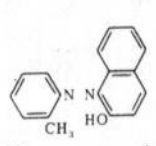

1624 VS   1621 VS   1621 VS   1618 VS
1622 VS

$C_{17}H_{14}N_2O$ [35]

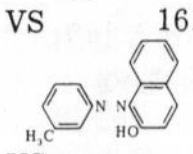

1624 VS   1622 VS   1620 VS
1623 VS

$C_{17}H_{14}N_2O$ [35]

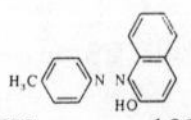

1624 VS   1621 VS   1620 VS   1617 VS
1623 VS

$C_{17}H_{14}N_2O_2$ [35]

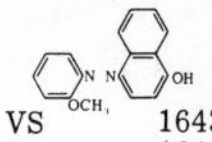

3380 M   1650 VS   1649 VS   1643 VS
3370 M   1650 VS   1648 VS   1641 VS
3320 M

$C_{17}H_{14}N_2O_2$ [35]

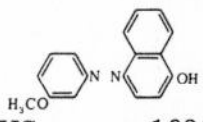

3560 M   3250 M   1642 VS   1630 VS
3350 M   3130 M   1630 VS

$C_{17}H_{14}N_2O_2$ [35]

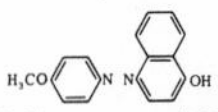

3575 S   3250 M   1639 S   1626 VS
3570 S   3130 M   1630 S   1625 VS
3340 W

$C_{17}H_{14}N_2O_2$ [35]

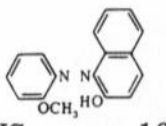

1625 VS   1624 VS   1622 VS   1615 VS

$C_{17}H_{14}N_2O_2$ [35]

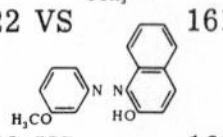

1625 VS   1623 VS   1622 VS   1621 VS

$C_{17}H_{14}N_2O_2$ [35]

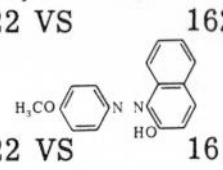

1624 VS   1623 VS   1622 VS   1612 VS
1624 VS

$C_{17}H_{14}OOs$ [450]

1627 S

$C_{17}H_{16}O_4$ [556]

1813 S   1776   1766 S   1726

$C_{17}H_{18}O_4$ [556]

1751 S   1734 S

$C_{17}H_{18}O_5$ [556]

1730   1715 Sh

$C_{17}H_{34}$ [159]

3085 SSpSh*   1642 SSp   1381 S   909 VS
2900 VVS   1464 VVS   993 VS   722 S

$C_{18}H_{15}ClSi$ [86]

1972 W   1592 W   1261 W   915.6 W
1905 W   1488 W   1115 VS   737.4 S
1890 W   1433 S   1065 M   712.4 VS
1770 W   1376 W   1030 W   682.2 W
1656 W   1332 W

$C_{18}H_{16}Ge$ [499]

2040 S

$C_{18}H_{16}N_2O_2$ [597]

1608 S

$C_{18}H_{16}Si$ [499]

2135 S

$C_{18}H_{20}N_4O$ [41]

3060   1248   933   735
1581   1130

$C_{18}H_{20}O_4$ [556]

1752 S   1750 S   1749 S   1736 S

$C_{18}H_{20}O_5$ [556]

1739 S   1712 S

$C_{18}H_{21}N_3$ [41]

1112   910   721

$C_{18}H_{22}ClN_3$ [41]

902

$C_{18}H_{30}$ [279]

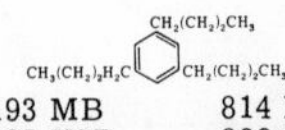

3170 WSh   1462 VSB   1193 MB   814 MB
3020 SSh   1377 SSp   1105 VSB   800 SB
2940 VS*   1360 W   1010 M   780 M
2860 VS   1339 SSp   936 SB   745 VS
2730 MW   1297 SB   900 Sh   728 M
1767 MB   1245 M   853 SB   708 S
1605 VS

$C_{18}H_{30}$ [280]		$CH_3CHCH_2CH_3$ / $CH_3CH_2HC(CH_3)$–C_6H_3–$CHCH_2CH_3(CH_3)$	
3030 SSh	1531 MW	1180 VS	909 VS
2970 S	1461 VS	1091 VS	872 VS
2920 MS	1454 VS	1080 S	842 M
2870 MS	1377 VS	1068 MB	815 M
1783 M	1328 MB	1017 VSSh	778 VS
1764 M	1287 S	1004 VS	714 VS
1602 VS	1233 M	962 VS	

$C_{19}H_{15}OP$ [34]			
3040 M	1330 M	1140 M	830 M
1600 M	1208 M	1120 M	780 S
1480 M	1185 S	1060 S	740 S
1440 S			

$C_{19}H_{15}O_2P$ [34]			
3040 M	1350 M	1120 M	770 M
1600 M	1205 M	1065 S	750 S
1482 M	1175 S	830 M	735 S
1440 S	1140 M		

$C_{19}H_{16}$ [499]			
2890 S			

$C_{20}H_{10}Br_4O_4$ [39]			
1742 S	1289 S	1018 W	637 S

$C_{20}H_{13}ClO_2$ [39]			
1775 S	1112-1100 S	1015 MS	626 W
1280 S			

$C_{20}H_{14}O_2$ [39]			
1770 S	1106 M	1014 W	632 S
1286 M			

$C_{20}H_{14}O_4$ [39]			
1740 S	1287 MS	1013 W	634 MW
1725 S	1106 MS		

$C_{20}H_{18}$ [457]			
1185			

$C_{20}H_{18}Sn$ [364]			
950 S			

$C_{20}H_{18}Pb$ [364]			
942 S			

$C_{20}H_{21}N_5O_4S$ [570, 580]			
1745 S	1740	1680 S	1590 S

$C_{20}H_{24}N_4O$ [41]			
1573	1245	1125	724
1474			

$C_{20}H_{25}N_3$ [41]			
3030	1115	917	713
1610	922		

$C_{20}H_{26}ClN_3$ [41]			
924			

$C_{20}H_{28}O_8Zr$ [16]			
1613 S			

$C_{20}H_{49}BrP_4Ru$ [1]			
1945 S			

$C_{20}H_{49}ClP_4Ru$ [1]			
1938 S			

$C_{20}H_{49}IP_4Ru$ [1]			
1948 S			

$C_{21}H_{15}NO$ [15]			
1625 M	1463 MSh	1407 VVS	916 MW
1525 W			

$C_{21}H_{18}FeORu$ [452]			
1623 S			

$C_{21}H_{18}Fe_2O$ [450]			
1612 S			

$C_{22}H_{18}O_4$ [39]			
1710 S	1117 S	633 M	

$C_{22}H_{20}N_2O_2$ [597]			
1583 S	1477 S		

$C_{22}H_{28}N_4O$ [41]			
3010	1265	1129	935
1578			

$C_{22}H_{28}O_2$ [40]			
1607 S	1604 M	1593 M	1591 M

$C_{22}H_{29}N_3$ [41]			
1616	1097	905	713
1206			

$C_{22}H_{30}ClN_3$ [41]			
903			

$C_{24}H_{18}FeO_2$ [450]			
1631 S			

$C_{24}H_{18}O_2Ru$ [450]			
1634 S			

$C_{24}H_{23}BrN_2O_2$ [39]			
1011 M			

$C_{24}H_{24}N_2O_2$ [39]			
1755 S	1282 M		

$C_{24}H_{32}O_2$ [40]			
1599	1592		

$C_{24}H_{48}Br_2CoO_6S_6$ [33]			
967 S			

$C_{24}H_{48}Br_4Co_2O_6S_6$ [33]			
966 S			

$C_{24}H_{48}Cl_4Ni_2O_6S_6$ [33]			
971 S			

$C_{24}H_{48}CoI_2O_6S_6$ [33]			
963 S			

$C_{24}H_{48}Co_2I_4O_6S_6$ [33]			
965 S			

$C_{26}H_{28}N_2O_2$ [39]			
1757 S			

$C_{26}H_{36}O_2$ [40, 569*]			
1601 S	1597 S	1596 S	

$C_{26}H_{54}$ [282]		$(C_2H_5)_2CH(CH_2)_{20}CH_3$	
2915 VS	1465 VS	1039 MB	900 S
2850 VS	1379 VSSp	1008 MB	773 SSh
2680 VW	1147 M	958 MB	721 VVS
2325 M	1078 MB		

$C_{26}H_{54}$ [283]		$C_{26}H_{54}$ (5, 14-di-n-butyloctadecane)	
2925 VS	1459 VS	1198 MSh	969 MW
2850 VS	1379 VS	1139 MS	894 S
2730 W	1341 VSB	1115 VW	873 MSh
2155 W	1299 VSSh	1083 SB	777 MSh
2030 MW	1257 SSh	1016 MVB	727 VVS
1467 VS	1220 SSh		

$C_{28}H_{30}O_4$ [39]			
1287 S	1107 S		

$C_{28}H_{40}O_2$ [40]			
1603 S			

$C_{37}H_{74}OS_2$ [434]			
1061 S	1060 S		

$C_{38}H_{34}GeSn$ [364]			
1070 S			

$C_{38}H_{34}SiSn$ [364]			
1070 S			

$C_{38}H_{34}Sn_2$ [364]			
1068 S			

$C_{52}H_{48}GeSn_2$ [364]

1070 S

$C_{52}H_{48}SiSn_2$ [364]

1070 S

$C_{n+2}H_{2n+5}NO$ [530]

1679

$C_{3n}H_{3n}N_n$ [23]

2927 S   1252 S   1075 M

$C_{4n}H_{6n}$ [628]

1630

CaH_4O_6S [449]

1100

CaN_2O_6 [45]

1770   1365   1040   818
1640   1330   912   748
1630   1050   824   738
1380

$Ca_2H_4P_2O_8$ [28]

922 S

CdN_2O_6 [45]

1760   1620   1425   815

$CeH_2O_7P_2$ [28]

1080 VSB   915 S

$CeH_8N_8O_{18}$ [45]

2320   1385   815   746
1615   1038   804

CeN_3O_9 [45]

2320   1630   1040   812
1770   1325   1038   746
1640   1315   813   723

CeN_6O_{18} [631]

1530   1420

$ClCrKO_3$ [358]

963 VS   950 VS   948 S   905 S

$ClNO$ [555]

1815   1800

Cl_3PS [403]

752 S

$CoH_{26}N_6NaO_{10}P_2$ [28]

1058 VSB   911 S

CoN_2O_6 [45]

1780   1357   833   717
1620   1058   827

$CrH_{15}N_8O_9$ [45]

1770   1055   832   731
1625   1052   828   715
1505   1020   781   712
1275   1016   772

$CrH_{15}N_{11}D_{18}$ [45]

1620

CrN_3O_9 [45]

2330   1615   1015   826
1630   1395   845

$CsNO_3$ [45]

2370   1620   1370   825
1760   1380

$CuH_{10}N_2O_{11}$ [45]

2400   1390   1340   835
2370   1360   1053   822
1785

FeN_3O_9 [45]

1785   1390   835   825
1620

GdN_3O_9 [45]

2240   1330   1038   812
1640   1295   1025   812
1632   1042   826   748
1480

GeH_2O_4Sr [690]

2700 M   1835 MB   1278 M

HNO_3 [642]

1668 S   1401 M   1299 VS   929 S

H_2KO_4P [28]

2750 M   1580 M   1300 M

$H_2Li_4P_2O_7$ [28]

935 SSp   870 W

$H_2Mg_2O_7P_2$ [28]

1094 VSB   940 S

$H_2Na_2O_6P_2$ [28]

2252 MB   1186 VSSp   1026 SSp   867 SSp
1695 MB   1050 VSSp   943 SSp   676 SSp
1339 SSp

H_4N_2 [485]

3325   3314

$H_4Na_2O_8S_2$ [28]

2198 M   2083 S   1240 VSSp   760 S

$H_{12}LaN_9O_{18}$ [45]

2340   1760   1042   816
2050

$H_{14}Li_4P_2O_{13}$ [28]

1064 VSSp   929 S   877 S

$H_{20}K_2Na_2O_{16}P_2$ [28]

911 S

$H_{20}Na_4O_{16}P_2$ [28]

916 S

HgN_2O_6 [45]

1780   1376   1035   825
1752   1090   835   806
1605

$Hg_2N_2O_6$ [45]

2390   1382   996   808
2360   1282   962   797
1760   1262   835   791
1630   1032   825   741
1475 S

KNO_3 [45]

2370   1380   826   825
1760   1370

$K_2O_6S_2$ [28]

2208 S   1240 VSSp   996 SSp   710 M
2083 M   1212 MSp

$K_4O_6P_2$ [28]

1075 VSSp   1037 SSh   923 SVSp   879 Sh

LaN_2O_6 [45]

1045   920   816

LaN_3O_9 [45]

1780   1475   1140   910
1765   1392   1058   752
1650   1346   1035   742
1630   1328   1032   740
1620   1305   914

$LiNO_3$ [45]

2400   1630   1365   826
1790   1370   838   736
1640

MgN_2O_6 [45]

2350   1625   1365   825
1650   1372   1045

$MnN_2O_6 \cdot xH_2O$ [45] 01077

1320	1035	904	753
1117			

$NNaO_3$ [45]

2400	1775	1355	835
1785	1375	836	

NO_3Rb [45]

2360	1760	1375	1052
2350	1620	1370	836
1770			

N_2NiO_6 [45]

2320	1620	1033	758
1785	1615	816	744
1635	1390		

N_2O_6Pb [45]

1345	1013	805	722
1065	1008	800	

N_2O_6Sr [45]

1785	1352	1012	815
1780	1045	823	738

N_2O_6Zn [45]

2390	1630	1375	838
1790	1620	1052	825
1760	1395	994	725

N_3NdO_9 [45]

1392	1040	815	

N_3O_9Pb [45]

1640

N_3O_9Pd [45]

742

N_3O_9Pr [45]

2230	1450	920	785
2220	1306	809	752
1785	1300	805	737
1620	1024	803	735-743
1455	1022		

N_3O_9Sm [45]

1650	1320	832	745
1630	1041	815	738
1390	1035	812	721
1375			

N_3O_9Y [45]

1780	1355	835	780
1640	1320	832	752
1635	1300	815	750
1485	1038	814	

$N_4O_{12}Th$ [45]

2280	1520	1030	745
1620	1390	808	

$Na_2O_6S_2$ [28]

2203 S	1235 VSSp	1000 VSSp	707 S
2088 M			

$Na_4O_6P_2$ [28]

2150 M	2010 S	1420 S	942 S
2114 M	2000 S	1085 VSB	770 Sh
2105 M	1450 S	1083 VSSp	710 M

$Na_4O_7P_2$ [28]

1152 VSSp	1030 SSp	917 M	735 VSSp
1121 VSSp	987 SSp		

$O_6P_2Pb_2$ [28]

1081 VSSh	1047 VSSp	896 S

$O_6P_2Tl_4$ [28]

1053 SB	873 S

APPENDIX I

Correlation Tables for
Methyl Deformation Frequencies
(Compiled by Lowell Karre)

TABLE I. Methyl—Carbon Frequencies

Compound	Asymmetric Bend	Symmetric Bend	Reference
	A. Four Methyls on One Carbon Atom		
$(CH_3)_4C$ (V)	1475 S	1372	[32]
$(CH_3)_4C$ (V)	1430	1254	[33]
	B. Three Methyls on One Carbon Atom		
$(CH_3)_3CH$ (V)	1477	1394	[34, 35]
$(CH_3)_3CH$ (L)	1450	1373	[35]
$(CH_3)_3C$-OH	1447	–	[34]
$(CH_3)_3C$-F	1464	1374	[34]
$(CH_3)_3C$-SH	1449	1369	[34]
$(CH_3)_3C$-Cl	1445	1361	[34]
$(CH_3)_3C$-Br	1454	1358	[34]
$(CH_3)_3C$-I	1448	1366	[34]
$(CH_3)_3CD$ (L)	1450	1380, 1358	[35]
$(CH_3)_3CD$ (V)	1477	1390	[35]
$(CH_3)_3C$-$C(CH_3)_3$ (soln.)	1477 VS, 1468 VS	1383 VS, 1373 VS	[36]
	C. Two Methyls on One Carbon Atom		
$(CH_3)_2CCl$-CH_2Cl (V)	1450 S	1387 VS	[37]
$(CH_3)_2CCl$-CH_2Cl (L)	1456 VS	1373 VS	[37]
$(CH_3)_2CCl$-CH_2Cl (S)	1460 VS	1372 VS	[37]
C_6H_5-$CH_2(CO)OCH(CH_3)_2$	1468	1375	[38]
C_6H_5-$(CO)OCH(CH_3)_2$	1467	1372	[38]
* 3-P-$(CO)OCH(CH_3)_2$	1470, 1456	1389, 1375	[38]
* 3-P-CH=$CH(CO)OCH_2CH(CH_3)_2$	1470	1396, 1377	[38]
* 3-P-$(CO)OCH_2CH(CH_3)_2$	1471	1393, 1380	[38]
$H(CO)OCH_2CH(CH_3)_2$	1469	1392, 1370	[38]
$(CH_3)_2CH$-SNO	1460	1375	[38]
$(CH_3)_2CH(CH_2)_3CH$=CH_2 (V)	1475 S	1387 S	[40]
$(CH_3)_2CH(CH_2)_3CH$=CH_2 (L)	1468 S, 1440 S	1387 S	[40]
$(CH_3)_2CH$-$CH(CH_3)_2$ (L)	1459	1381	[41]
$(CH_3)_2CH$-$CH(CH_3)_2$ (S)	1459	1381	[41]
spiro-$(CH_2)_2C(CH_3)_2$ (V)	1470 S, 1460 VS	1380 m	[42]
$(CH_3)_2CBr$-$CBr(CH_3)_2$ (soln.)	1442 VS	1374 VS	[36]
	D. One Methyl on One Carbon Atom		
CH_3CN (V)	1454 VS	1389	[43, 44]
CH_3CN (S)	1410	1373	[44]
CH_3CN (L)	1456	1414	[45]
$CH_3(CO)OCH_3$	1427	1375	[17]
$CH_3(CO)NH_2$ (soln.)	–	1337 S	[46]
$CH_3(CO)NH_2$ (S)	1461 M	1355 MS	[46, 47]
$CH_3(CO)NH_2 \cdot HCl$ (S)	1474 S	1377 S	[46]
$CH_3(CS)NH_2$ (S)	1481 M (1393)	1363 S (1364)	[46, 48]
$CH_3(CS)NH_2$ (soln.)	(1377)	1366 S	[46, 48]
NH_3-$CH(CH_3)$-$(CO)O^-$	1451 M	1356 S	[49]
ND_3-$CH(CH_3)$-$(CO)O^-$	1457 M	1344 S	[49]
NH_3-$CH(CH_3)$-$(CO)OH$	1462 Sh	1359 S	[49]
CH_3CH_2-Hg-CH_2CH_3 (L)	1468 S	1375 S	[50]
CH_3CH_2-Zn-CH_2CH_3 (L)	1465 M	1373 M	[50]
CH_3CH_2-Cd-CH_2CH_3 (L)	1465 S	1373 S	[50]
$(CH_3CH_2)_4Sn$ (L)	1472 VS	1380 S	[50]

*P = Pyridine; number designates position at which it is substituted.

TABLE I (continued)

Compound	Asymmetric Bend	Symmetric Bend	Reference
D. One Methyl on One Carbon Atom			
$(CH_3CH_2)_3P$ (L)	1468 S	1380 S	[50]
$CH_3CH_2CH_2CH_3$	1461	1382	[51]
$CH_3CH_2O(CO)CH_2OH$ (aq.)	1450	1387	[52]
$CH_3CH_2O(CO)CH_2OH$ (soln.)	1460	1389	[52]
$CH_3CH_2O(CO)CH(CH_3)OH$ (aq.)	1464	1383	[52]
$CH_3CH_2O(CO)CH(CH_3)OH$ (soln.)	1471	1387	[52]
$Zn^{+}\,{}^{-}O(CO)CH(CH_3)OH$ (aq.)	1477	--	[52]
$[CH_3\text{-}C(NH_2)_2]Cl^{-}$	1425	1378	[47]
$[CH_3\text{-}C(NH_2)_2]SbCl_6^{--}$	1416	1378	[47]
$[CH_3\text{-}C(NH_2)_2]SnCl_6^{--}$	1415	1382	[47]
$[CH_3\text{-}C(NH_2)_2]PtCl_6^{--}$	1411	1379	[47]
$[CH_3\text{-}C(ND_2)_2]PtCl_6^{--}$	1412	1381	[47]
$C_6H_5\text{-}CH_2CH_3$ (L)	1456	1374	[53]
$C_6H_5\text{-}OCH_2CH_3$ (L)	1460	1390	[53]
$C_6H_5\text{-}SCH_2CH_3$ (L)	1437	1378	[53]
$C_6H_5\text{-}NH(C_2H_5)$	1477	1429	[54]
$C_6H_5\text{-}NH(C_4H_9)$	1475	1429	[54]
$CH_3(CH_2)_3CH_3$ (L)	1464	1379	[41]
$CH_3(CH_2)_3CH_3$ (S)	1474, 1452	1379	[41]
$CH_3(CH_2)_4CH_3$ (L)	1467	1379	[41]
$CH_3(CH_2)_4CH_3$ (S)	1478, 1456	1379	[41]
$CH_3(CH_2)_5CH_3$ (L)	1470	1377	[41]
$CH_3(CH_2)_5CH_3$ (S)	1473, 1467	1377	[41]
$CH_3(CH_2)_6CH_3$ (S)	1460	–	[55]
$(CH_3CH_2\text{-}O)_3B$ (V)	1500 S (CH_2)	1383 VS	[21]
$(CH_3CH_2\text{-}O)_2BH$ (V)	1500 S (CH_2)	1383 VS	[21]
$(CH_3CH_2\text{-}O)_2BD$ (V)	1500 S (CH_2)	1381 VS	[21]
$CH_3(CO)NH(CH_3)$	1445 M	1373 S	[56]
$(CH_3CH_2)_3B$ (V)	1471 S	1391 MW	[22]
$CH_3\text{-}CH_3$ (V)	1472	1379	[57]
$CH_3\text{-}CF_3$	1443	1408	[58]
CH_3CCl_3	–	1381	[18]
CH_3CHCl_2	–	1382	[18]
$CH_3C{\equiv}CH$	–	1382	[18]
$CH_3CH{\equiv}CH$	–	1370	[18]
$CH_3C{\equiv}CCl$	–	1378	[18]
$CH_3(CO)OH$	–	1340	[18]
$CH_3(CO)NH\text{-}NH(CO)CH_3$ (S)	1437	1368	[59]
$CH_3(CO)ND\text{-}ND(CO)CH_3$ (S)	–	1366	[59]
$CH_3(CO)NH\text{-}CH_3$ (V)	1413	1373	[59]
$CH_3(CO)ND\text{-}CH_3$ (V)	1401	1375	[59]
$CH_3(CO)\text{-}ONa$ (S)	1488	1425	[60]
$CH_3\text{-}CH_2\text{-}CN$ (V)	1462 S	1383	[61]
$CH_3\text{-}CH_2\text{-}CN$ (L)	1461 VS	1386	[61]
$CH_3(C{\equiv}C)_2CH_3$ (soln.)	1437 (1428 S)	1370 (1378 VS)	[62, 63]
$CH_3(C{\equiv}C)_3CH_3$ (soln.)	1432	1380	[62]
$CH_3(C{\equiv}C)_4CH_3$ (soln.)	1419	1374	[62]
$CH_3(C{\equiv}C)_5CH_3$ (soln.)	1466	1379	[62]

TABLE II. Methyl—Silicon Frequencies

Compound	Asymmetric Bend	Symmetric Bend	Reference
A. Four Methyls on One Silicon Atom			
$(CH_3)_4Si$ (V)	1430 S	1253 S	[32]
$(CH_3)_4Si$ (V)	–	1259 VS	[64]
B. Three Methyls on One Silicon Atom			
$(CH_3)_3Si$-$Si(CH_3)_3$ (L)	1412 S	1245 S	[65]
$(CH_3)_3Si$-NH-$Si(CH_3)_3$ (L)	1398 S	1250 S	[65]
$(CH_3)_3Si$-O-$Si(CH_3)_3$ (L)	1410 S	1255 S	[65]
$(CH_3)_3Si$-CH_2-$Si(CH_3)_3$ (L)	1422 S	1251 S	[65]
$(CH_3)_3Si$-$Si(CH_3)_3$	–	1245	[66]
$(CH_3)_3Si$-Cl (V)	1415 M (1414)	1260 S (1248)	[67, 68]
$(CH_3)_3SiH$ (V)	1467 or (1425)	1267	[69]
$(CH_3)_3SiD$ (V)	1437 or 1419	1270	[69]
$(CH_3)_3SiCl$ (V)	–	1261 VS	[64]
C. Two Methyls on One Silicon Atom			
$(CH_3)_2SiCl_2$	1405	1258	[70]
$(CH_3)_2SiCl_2$ (V)	1412 M	1261 S (1264)	[64, 67]
$(CH_3)_2SiD_2$ (V)	1434 or 1423	1270	[69]
$(CH_3)_2SiH_2$ (V)	1440, 1380	1260	[71]
$(CH_3)_2SiH_2$ (V)	1440	1260	[69]
Cl-$Si(CH_3)_2C_6H_5$ (soln.)	–	1253 VS	[72]
D. One Methyl on One Silicon Atom			
CH_3SiH_2F (V)	1440 M, 1418 M	1265 S	[71]
CH_3SiH_2Cl (V)	1442 M, 1410 M	1265 S	[71]
CH_3SiH_2Br (V)	1425 M	1265 S	[71]
CH_3SiH_2I (V)	1418 M	1264 S	[71]
CH_3SiH_2NC (V)	1422 M	1266 S	[71]
$(CH_3SiH_2)_2O$ (V)	1418, 1375	1264 S	[71]
$(CH_3SiH_2)_3N$ (V)	1453, 1418, 1380	1260 S	[71]
$(CH_3SiH_2)_2S$ (V)	1375	1260 S	[71]
CH_3SiCl_3 (V)	1417 M (1416)	1271 M (1271)	[67, 73]
$C_6H_5Si(Cl)_2CH_3$ (soln.)	–	1261 S	[72]
$(C_6H_5)_2Si(Cl)CH_3$ (soln.)	–	1256 S	[72]
CH_3SiH_3	1430 (1412)	1260	[69, 74]
CH_3SiD_3	1411	1264	[69]
CH_3SiCl_3 (V)	–	1266 S	[64]

TABLE III. Methyl—Nitrogen Frequencies

Compound	Asymmetric Bend	Symmetric Bend	Reference
A. Four Methyls on One Nitrogen Atom			
$(CH_3)_4N^+$	1455	–	[75]
B. Three Methyls on One Nitrogen Atom			
$(CH_3)_3N$	1466	1402	[76]
$(CH_3)_3NH^+$ (S)	1468	1389	[77]
C. Two Methyls on One Nitrogen Atom			
$(CH_3)_2NH_2^+$ (S)	1471	1405	[77]
$(CH_3)_2NH$ (V)	1466	1404	[78]
$(CH_3)_2NH_2I$ (S)	1472 S, 1464 VS	1407 S	[79]
$(CH_3)_2ND_2I$ (S)	1474 M, 1463 M	1409 M	[79]
$(CH_3)_2NB_2H_5$ (V)	1458 VS	1387 M	[80]
NH_2-$N(CH_3)_2$ (V)	1457 M	1405	[81]
CH_3NH-$N(CH_3)_2$ (V)	1477 S	1398	[81]
D. One Methyl on One Nitrogen Atom			
CH_3NO_2 (V)	1488, 1449	1413	[82, 83]
$H(CO)NH(CH_3)$ (V)	–	1411	[84]
$H(CO)NH(CH_3)$ (soln.)	–	1416	[84]
$H(CO)NH(CH_3)$ (L)	1450	1415	[84, 85]
$H(CO)ND(CH_3)$ (L)	1467	1405	[85]
CH_3NC (V)	1459 (1456)	(1414)	[44, 86]
CH_3NC (V)	1467	1429	[87]
$CH_3(CO)NH(CH_3)$ (L)	1445 M	1413 S	[88]
$Cd[CH_3(CO)NH(CH_3)]Cl_2$	1445 M	1418 S	[88]
CH_3NCS (V)	1470	1426	[45]
CH_3NCO (L)	1453	1377	[45]
CH_3NCO (V)	1453	1377	[89]
$CH_3NH_3^+$ (S)	1467	1408	[77]
C_6H_5-$NH(CH_3)$	1473	1418	[54]
$CH_3N{=}NCH_3$ (V)	1430	–	[90]
CH_3NH_2 (V)	1460 (1470)	1426 (1385)	[91, 92]
CH_3NH_2 (V)	1435, 1473	1430	[93]
CH_3ND_2 (V)	1485, 1468	1430	[93]
CH_3N_3 (V)	1482, 1434 M (1452)	1351 M (1417)	[89, 94]
$(CH_3)NH$-NH_2 (V)	1464 M	–	[95]
$(CH_3)NH$-NH_2 (L)	1474 S	–	[95]
CH_3NH-$NH(CH_3)$	1476 M	–	[95]
CH_3NH-$NH(CH_3)$	1477 S	–	[95]
$CH_3NH_3^+Cl^-$ (S)	1463	1428	[96]
$CH_3NH(CO)(CO)NH(CH_3)$ (S)	1462	1404	[59]
$CH_3ND(CO)(CO)ND(CH_3)$ (S)	1465	1402	[59]
$H(CO)NH(CH_3)$ (V)	1453	1414	[59]
$H(CO)ND(CH_3)$ (V)	–	1406	[59]
$CH_3(CO)NH(CH_3)$ (V)	1445	1413	[59]
$CH_3(CO)ND(CH_3)$ (V)	1442	1401	[59]
CH_3NSO (L)	1468 Sh	1438 S	[97]

TABLE IV. Methyl—Phosphorus Frequencies

Compound	Asymmetric Bend	Symmetric Bend	Reference
	A. Three Methyls on One Phosphorus Atom		
$(CH_3)_3PO$ (S)	1420, 1437	1340, 1305, 1292	[98]
$(CH_3)_3PO$ (L)	1460 SB	–	[99]
$(CH_3)_3PO$	1420	1340	[76]
$(CH_3)_3P$	1417 M	1310 M	[76]
$(CH_3)_3P$	1430 M, 1417 M	1310 M	[100]
$(CH_3)_3P \cdot BF_3$	1425 W	1300 W	[100]
	B. Two Methyls on One Phosphorus Atom		
$(CH_3)_2PH$ (V)	1440 S, 1415 W	1304 W, 1284 M	[101]
$(CH_3)_2PH$	1440	–	[76]
$(CH_3)_2PCF_3$	1440 M, 1425 W	1310 W	[100]
$(CH_3)_2PCF_3 \cdot BF_3$	1430 W	1320 M	[100]
	C. One Methyl on One Phosphorus Atom		
CH_3-PH_2	1450 M?	1346 W	[102]
CH_3PH_2	1450	1346	[76]
$CH_3P(CF_3)_2$	1465 M	1307 M	[100]

TABLE V. Methyl—Boron Frequencies

Compound	Asymmetric Bend	Symmetric Bend	Reference
	A. Three Methyls on One Boron Atom		
$(CH_3)_3B$ (V)	1470 M	1306 VS	[103]
$(CH_3)_3B$ (V)	~1480 WB	1305 VS	[23]
$(CH_3)_3B$ (V)	~1460, 1440	1306, 1295	[24]
$(CH_3)_3B^{10}$ (V)	~1460 WB	1310 VS	[23]
$(CH_3)_3B$-NH_2NH_2 (S)	1440 WB	1282 VS	[104]
$(CH_3)_3B$-NH_3 (S)	–	1283 VS	[104]
	B. Two Methyls on One Boron Atom		
$(CH_3)_2B^{10}H$-BH_3	1437 M	1330 S, 1319 Sh	[28]
$(CH_3)_2BH$-BH_3	1441 M	1326 S, 1316 Sh	[28]
$(CH_3)_2BD$-BD_3	1443 M	1328 S, 1321 Sh	[28]
$(CH_3)_4B_2^{10}H_2$	1433 M	1318 S, 1266 W	[26]
$(CH_3)_4B_2H_2$	1437 M	1312 VS, 1260 W	[26]
$(CH_3)_4B_2D_2$	~1430	1321 VS	[26]
	C. One Methyl on One Boron Atom		
$(CH_3BH_2)_2$	1437 M	1328 M, 1315 M	[25]
$(CH_3B^{10}H_2)_2$	1433 M	1328 M, 1316 M	[25]
$(CH_3BD_2)_2$	1428 M	1328 S, 1316 S	[25]
$CH_3B_2H_5$	1424 M	1319 M	[27]
$CH_3B_2^{10}H_5$	1428 M	1321 M	[27]
$CH_3B_2D_5$	1433 or 1404	1319	[27]
$[-N(CH_3)-B(CH_3)-]_3$ (soln.)	~1450	1326 S	[105]

TABLE VI. Methyl—Oxygen Frequencies

Compound	Asymmetric Bend	Symmetric Bend	Reference
A. Two Methyls on One Oxygen			
$(CH_3)_2O$ (L)	1466 VS?	1466 VS?	[106]
B. One Methyl on One Oxygen			
CH_3OH (V)	1477 M	1455 M	[107]
CH_3OH (L)	1480 Sh	1455 M	[107]
CH_3OH (S)	1468 Sh	~1445 MSh	[107]
CH_3OD (V)	1427, 1500	1458 M	[107]
CH_3OD (L)	1472 M	1452 M	[107]
CH_3OD (S)	1475 M, 1443	1465 M	[107]
$(CH_3O)_3B$ (L)	1490 VS	1178 S	[108]
$(CH_3O)_3B$ (V)	1493 VS	1198 S	[108]
$CD_3(CH_2)_9CH_2COOCH_3$ (soln.)	1458	1436	[109]
$CD_3(CH_2)_9CD_2COOCH_3$ (soln.)	1458	1435	[109]
$CCl_3(CH_2)_9CH_2COOCH_3$ (soln.)	1457	1435	[109]
$HC{\equiv}C(CH_2)_7CH_2COOCH_3$ (soln.)	1460	1436	[109]
$CH_3OCH_2OCH_3$ (soln.)	1473, 1465	1448	[16]
$CH_3OCH_2OCH_3$ (V)	1487, 1464	1458	[16]
$H(CO)OCH_3$ (soln.)	1465, 1454	1445	[17]
$CH_3(CO)OCH_3$ (soln.)	1469, 1450	1440	[17]
$H(CO)NH$-OCH_3 (S)	1464 S	1441 S	[110]
$H(CO)NH$-OCH_3 (L)	1468 M	1433 M	[110]
$H(CO)NH$-OCH_3 (V)	1476 M	1439 M	[110]
$CH_2{=}CH(CO)OCH_3$	1462	1443	[38]
o-NO_2-$C_6H_4CH{=}CH(CO)OCH_3$	1462	1441	[38]
o-HO-$C_6H_4(CO)OCH_3$	1465	1446	[38]
o-Cl-$C_6H_4(CO)OCH_3$	–	1439	[38]
o-NO_2-$C_6H_4(CO)OCH_3$	–	1438	[38]
$H(CO)OCH_3$	1455	1436	[38]
$Cl(CO)OCH_3$	1457	1437	[38]
$CH_3(CO)OCH_3$ (soln.)	1456	1435	[111]
NH_2-OCH_3 (L)	1464 S	1438 S	[112]
NH_2-OCH_3 (V)	1475 M	1439 M	[112]
$C_6H_5(CO)OCH_3$ (soln.)	–	1437 M	[31]
* 4-P-$(CO)OCH_3$ (soln.)	–	1440 M	[31]
* 3-P-$(CO)OCH_3$ (soln.)	–	1438 M	[31]
* 2-P-$(CO)OCH_3$ (soln.)	–	1446 M	[31]
** 4-PO-$(CO)OCH_3$ (soln.)	–	1437 M	[31]
** 3-PO-$(CO)OCH_3$ (soln.)	–	1442 M	[31]
** 2-PO-$(CO)OCH_3$ (soln.)	–	1442 S	[31]
* 4-P-BCl_3-$(CO)OCH_3$ (soln.)	–	1439 Sh	[31]
* 3-P-BCl_3-$(CO)OCH_3$ (soln.)	–	1441 M	[31]
$C_6H_5CH_2(CO)OCH_3$ (soln.)	–	1439 M	[31]
* 4-P-$CH_2(CO)OCH_3$ (soln.)	–	1440 M	[31]
$C_6H_5CH_2CH_2(CO)OCH_3$ (soln.)	–	1440 M	[31]
* 4-P-$CH_2CH_2(CO)OCH_3$ (soln.)	–	1438 M	[31]
$C_6H_5CH{=}CH(CO)OCH_3$ (soln.)	–	1436 M	[31]
* 4-P-$CH{=}CH(CO)OCH_3$ (soln.)	–	1435	[31]
m-NH_2-C_6H_4-OCH_3 (soln.)	1460 M	1440 M	[30]
m-CH_3O-C_6H_4-OCH_3 (soln.)	1459 M	1437 M	[30]
m-HO-C_6H_4-OCH_3 (soln.)	1464 M	1442 M	[30]
m-CH_3-C_6H_4-OCH_3 (soln.)	1467 M	1436 M	[30]
m-$CH_3O(CO)CH{=}CHC_6H_4$-OCH_3 (soln.)	1460 M	1438 M	[30]
m-$CH_3CH_2O(CO)CH{=}CHC_6H_4$-$OCH_3$ (soln.)	1465 M	1432 (overlapped)	[30]

TABLE VI (continued)

Compound	Asymmetric Bend	Symmetric Bend	Reference
m-NO_2-C_6H_4-OCH_3 (soln.)	1458 M	1440 ShM	[30]
p-NH_2-C_6H_4-OCH_3 (soln.)	1468 M	1443 M	[30]
p-CH_3O-C_6H_4-OCH_3 (soln.)	1464 M	1442 M	[30]
p-HO-C_6H_4-OCH_3 (soln.)	1467 M	1436 M	[30]
p-CH_3-C_6H_4-OCH_3 (soln.)	1465 M	1442 ShM	[30]
*p-(CH=CH_4-P)-C_6H_4-OCH_3 (soln.)	1465 M	1445 M	[30]
**p-(CH=CH_4-PO)-C_6H_4-OCH_3 (soln.)	1465 ShM	1450 ShM	[30]
*p-(C≡C_4-P)-C_6H_4-OCH_3 (soln.)	1463 M	1442 M	[30]
p-NO_2-C_6H_4-OCH_3 (soln.)	1459 M	1442 M	[30]
*4-P-OCH_3 (soln.)	1466 M	1445 M	[30]
*2-P-OCH_3 (soln.)	1470 ShM	1445 M	[30]
**4-PO_3(-CH_3)-OCH_3 (soln.)	1460 ShM	1442 ShM	[30]
**4-PO-OCH_3 (soln.)	1464 M	1440 M	[30]
**3-PO-OCH_3 (soln.)	1460 M	1438 M	[30]
**2-PO-OCH_3 (soln.)	1459 ShM	1440 ShM	[30]
**4-(PN-BCl_3)-OCH_3 (soln.)	1466 M	1440 M	[30]
$CH_3(CH_2)_9CH_2COOCH_3$ (soln.)	1458	1436	[29]
$CD_3(CH_2)_9CH_2COOCH_3$ (soln.)	1458	1436	[29]
$CH_3(CH_2)_9CD_2COOCH_3$ (soln.)	1458	1436	[29]
$CD_3(CH_2)_9CD_2COOCH_3$ (soln.)	1458	1435	[29]
$CCl_3(CH_2)_9CH_2COOCH_3$ (soln.)	1457	1435	[29]
$CH_3(CH_2)_9CCl_2COOCH_3$ (soln.)	1458	1436	[29]
HC≡$C(CH_2)_7CH_2COOCH_3$ (soln.)	1460	1436	[29]

*P = Pyridine; number designates position at which it is substituted.
**PO = Pyridine N-oxide; number designates position at which it is substituted.

TABLE VII. Methyl—Sulfur Frequencies

Compound	Asymmetric Bend	Symmetric Bend	Reference
	A. Two Methyls on One Sulfur Atom		
CH_3-S-CH_3 (V)	1445 VS	1323 VS	[113]
$CH_3(SO)CH_3$	1455 or 1440	1319 or 1304	[114]
	B. One Methyl on One Sulfur Atom		
CH_3-CH_2-S-CH_3 (V)	1456 S or 1445 S	1328 W	[115]
CH_3-CH_2-S-CH_3 (L)	1453 VS, 1446 VS, 1435 VS	1323 W	[115]
CH_3-CH_2-S-CH_3 (S)	1449 S	1324 W	[115]
CH_3-SH	1475	–	[116]
CH_3-SH	–	1335	[18]
CH_3SCN (L)	1428	1316	[45]
CH_3SNO (V)	1430	1300	[39, 117]
$[CH_3\text{-S-C}(NH_2)_2]^+I^-$	1409	1328	[118]
C_6H_5-S-CH_3 (L)	1441	1316	[53]
CH_3SO_3Li (aq.)	1428 M	1333 S	[119]
CH_3SO_3Na (aq.)	1428 M	–	[119]
CH_3SO_3K (aq.)	1429 M	1328 S	[119]
$CH_3SO_3CH_3$	1420	1333	[120]
$CH_3SO_3C_2H_5$	1420	–	[120]

TABLE VIII. Methyl—Halogen Frequencies

Compound	Asymmetric Bend	Symmetric Bend	Reference
	A. Methyl – Fluorine Frequencies		
CH_3F (V)	1467	1198 ?	[121]
CH_3F (V)	1475	–	[23, 19]
CH_3F (V)	1476	1460	[122]
CH_3F (V)	1476	1460	[86]
CH_3F (V)	1476	1200 ?	[123]
	B. Methyl – Chlorine Frequencies		
CH_3Cl (V)	1455	1355	[124, 125]
CH_3Cl (V)	1455	1355	[126]
CH_3Cl (S)	1445, 1442, 1438	1345, 1336	[124]
CH_3Cl (S)	1445, 1441, 1437	1346, 1336	[126]
CH_3Cl (S)	1441	–	[44]
	C. Methyl – Bromine Frequencies		
CH_3Br (V)	1444	1305	[124]
CH_3Br (V)	1445	1305	[126, 125]
CH_3Br (S)	1435, 1421	1294, 1291	[124]
CH_3Br (S)	1435, 1421	1296, 1293	[126]
	D. Methyl – Iodine Frequencies		
CH_3I (V)	1440	1251	[124, 125]
CH_3I (V)	1440	1251	[125, 126]
CH_3I (S)	1425, 1419, 1401, 1396	1240, 1235	[124]
CH_3I (S)	1426, 1420, 1401, 1396	1241, 1236	[126]
CH_3I (L)	1427	1239	[127]
CH_3I (L)	1429	1242	[128]

TABLE IX. Methyl—Metal Frequencies

Compound	Asymmetric Bend	Symmetric Bend	Reference
A. Methyl – Lead Compounds			
$(CH_3)_4Pb$ (V)	1453 M	1392 M	[129]
$(CH_3)_4Pb$ (L)	1440 S	1390 S	[129]
B. Methyl – Zinc Compounds			
CH_3-Zn-CH_3 (V)	1444 W	1185 S	[130]
C. Methyl – Mercury Compounds			
CH_3-Hg-CH_3 (V)	1475 M	1205 W	[130]
D. Methyl – Germanium Compounds			
$[(CH_3)_2GeO]_3$ (soln.) (cyclic)	–	1237	[131]
$[(CH_3)_2GeO]_4$ (soln.) (cyclic)	1408	1238	[131]
$[(CH_3)_3Ge_2]O$ (L)	1408	1236	[131]
$[(CH_3)_2GeS]_2$ (soln.)	1406	1228	[131]
$(CH_3)_4Ge$ (V)	1420	1235	[131]
$(CH)_6Ge_2$ (L)	1407	1236	[131]
$(CH_3)_2Ge(CF{=}CF_2)_2$ (L)	1414 W	1254 W	[132]
E. Methyl – Tin Compounds			
$(CH_3)_2SnO$ (S)	1410	1206	[131]
$(CH_3)_2SnS_3$ (soln.)	1409	1193	[131]
$[Cl(CH_3)_2Sn]_2O$ (S)	1408	1198	[131]
$(CH_3)_2Sn(CF{=}CF_2)_2$ (L)	1408 W	1194 W, 1203 W	[132]
F. Methyl – Copper Compounds			
Cu-CH_3 (S)	1405	1328	[133]

SUMMARY

The frequency ranges within which the asymmetric and symmetric methyl bending frequencies absorb, based on the data presented in the preceding tables, are tabulated below. The summary is presented in the same order and manner as the tables.

Configuration	Asymmetric Bend	Symmetric Bend
I. Methyl – Carbon Frequencies		
A. Four methyls on carbon	1475-1430	1372-1254
B. Three methyls on carbon	1477-1445	1394-1358
C. Two methyls on carbon	1475-1440	1396-1370
D. One methyl on carbon	1488-1377	1429-1337
II. Methyl – Silicon Frequencies		
A. Four methyls on silicon	1430	1253
B. Three methyls on silicon	1467-1398	1270-1245
C. Two methyls on silicon	1440-1380	1270-1253
D. One methyl on silicon	1453-1375	1271-1256
III. Methyl – Nitrogen Frequencies		
A. Four methyls on nitrogen	1455	–
B. Three methyls on nitrogen	1468-1466	1402-1389
C. Two methyls on nitrogen	1477-1458	1409-1387
D. One methyl on nitrogen	1488-1430	1430-1377
IV. Methyl – Phosphorus Frequencies		
A. Three methyls on phosphorus	1460-1417	1340-1292
B. Two methyls on phosphorus	1440-1415	1320-1284
C. One methyl on phosphorus	1465-1450	1346-1307
V. Methyl – Boron Frequencies		
A. Three methyls on boron	1480-1440	1310-1282
B. Two methyls on boron	1443-1430	1330-1260
C. One methyl on boron	1450-1404	1328-1315
VI. Methyl – Oxygen Frequencies		
A. Two methyls on oxygen	1466?	1466?
B. One methyl on oxygen	1493-1450	1458-1178
VII. Methyl – Sulfur Frequencies		
A. Two methyls on sulfur	1455-1440	1323-1304
B. One methyl on sulfur	1475-1409	1335-1300
VIII. Methyl – Halogen Frequencies		
A. Methyl–fluorine	1476-1467	1460-1198
B. Methyl–chlorine	1455-1437	1355-1336
C. Methyl–bromine	1445-1421	1305-1291
D. Methyl–iodine	1440-1396	1251-1235
IX. Methyl – Metal Frequencies		
A. Methyl–lead compounds	1453-1440	1392-1390
B. Methyl–zinc compounds	1444	1185
C. Methyl–mercury compounds	1475	1205
D. Methyl–germanium compounds	1420-1406	1254-1228
E. Methyl–tin compounds	1410-1408	1206-1194
F. Methyl–copper compounds	1405	1328

CONCLUSIONS

The following conclusions are presented as a table of approximate group frequency assignments for the methyl bending modes, based on the data in Tables I through IX. It must be realized that these adjustments are broad generalizations and that the tables should be consulted for more specific information on individual compounds.

Configuration	Asymmetric Bend	Symmetric Bend
	I. Methyl – Carbon Frequencies	
$(CH_3)_4C$	1475	1372
$(CH_3)_3C$-R	1455 ± 10	1365 ± 10
$(CH_3)_2C$-R_2	1460 ± 10	1380 ± 10*
CH_3C-R_3	1460 ± 20	1375 ± 15*
	II. Methyl – Silicon Frequencies	
$(CH_3)_4Si$	1430	1255 ± 5
$(CH_3)_3Si$-R	1410 ± 15	1250 ± 10
$(CH_3)_2Si$-R_2	1410 ± 30	1260 ± 10
CH_3Si-R_3	1410 ± 30	1260 ± 10
	III. Methyl – Nitrogen Frequencies	
$(CH_3)_4N$	1455	–
$(CH_3)_3N$	1465 ± 5	1395 ± 10
$(CH_3)_2N$-R	1465 ± 10	1395 ± 10
CH_3N-R_2	1450 ± 20	1410 ± 30
	IV. Methyl – Phosphorus Frequencies	
$(CH_3)_3P$	1420 ± 10	1320 ± 20
$(CH_3)_2P$-R	1430 ± 10	1300 ± 20
CH_3P-R_2	1460 ± 10	1325 ± 20
	V. Methyl – Boron Frequencies	
$(CH_3)_3B$	1460 ± 20	1295 ± 15
$(CH_3)_2B$-R	1435 ± 10	1320 ± 10
CH_3B-R_2	1430 ± 20	1320 ± 10
	VI. Methyl – Oxygen Frequencies	
$(CH_3)_2O$	1466?	1466?
CH_3O-R	1465 ± 10	1440 ± 10
	VII. Methyl – Sulfur Frequencies	
$(CH_3)_2S$	1435 ± 20	1315 ± 10
CH_3S-R	1430 ± 20	1315 ± 15
	VIII. Methyl – Halogen Frequencies	
CH_3F	1475 ± 5	1460
CH_3Cl	1450 ± 10	1345 ± 10
CH_3BR	1435 ± 10	1300 ± 10
CH_3I	1420 ± 20	1245 ± 10
	IX. Methyl – Metal Frequencies	
CH_3-Ge	1410 ± 10	1245 ± 10
CH_3-Sn	1410 ± 10	1200 ± 10

*A doublet is often observed in connection with the $(CH_3)_3C$- and $(CH_3)_2C$- symmetric bending modes.

BIBLIOGRAPHY

[1] Nielsen, A.H., Recent Advances in Infrared Spectroscopy, Office of Ordnance Research Tech. Memo. 53-2, December, 1953.
[2] Julius, W.H., Verhandl. Akad. Wetenschappen Amsterdam 1, 1 (1892).
[3] Coblentz, W.W., Phys. Rev. 20, 273 (1905).
[4] Randall, H.M., Fowler, R.C., Fuson, N., and Dangl, J.R., Infrared Determination of Organic Structures, Van Nostrand, New York (1949).
[5] Bjerrum, N., Verhankl. deut. physik. Ges. 16, 737 (1916).
[6] Wigner, E., Nachr. Ges. Wiss. Gottinger, 133 (1930).
[7] Wilson, E.B., Phys. Rev. 45, 706 (1934).
[8] Delahay, P., Instrumental Analysis, MacMillan, New York (1957).
[9] Barnes, R.B., Gore, R.C., Liddel, U., and Williams, V.Z., Infrared Spectroscopy, Reinhold (1944).
[10] Rasmussen, R.S., J. Chem. Phys. 16, 712 (1948).
[11] Colthup, N.B., J. Opt. Soc. Amer. 40, 397 (1950).
[12] Bellamy, L.J., Infrared Spectra of Complex Molecules, Wiley, New York (1953).
[13] Cross, A.D., Practical Infrared Spectroscopy, Butterworths, London (1960).
[14] King, W.T., and Crawford, B., J. Mol. Spectroscopy 5, 421 (1960).
[15] King, W.T., and Crawford, B., J. Mol. Spectroscopy 8, 58 (1962).
[16] Wilmshurst, J.K., Can. J. Chem. 36, 285 (1958).
[17] Wilmshurst, J.K., J. Mol. Spectroscopy 1, 201 (1957).
[18] Wilmshurst, J.K., J. Chem. Phys. 26, 426 (1957).
[19] Wilmshurst, J.K., J. Chem. Phys. 23, 2463 (1955).
[20] Wright, N., and Hunter, M.J., J. Amer. Chem. Soc. 69, 803 (1947).
[21] Lehmann, W.J., Weiss, H.G., and Shapiro, I., J. Chem. Phys. 30, 1222 (1959).
[22] Lehmann, W.J., Wilson, C.O., Jr., and Shapiro, I., J. Chem. Phys. 28, 781 (1958).
[23] Lehmann, W.J., Wilson, C.O., Jr., and Shapiro, I., J. Chem. Phys. 28, 777 (1958).
[24] Lehmann, W.J., Wilson, C.O., Jr., and Shapiro, I., J. Chem. Phys. 31, 1071 (1959).
[25] Lehmann, W.J., Wilson, C.O., Jr., and Shapiro, I., J. Chem. Phys. 33, 590 (1960).
[26] Lehmann, W.J., Wilson, C.O., Jr., and Shapiro, I., J. Chem. Phys. 34, 783 (1961).
[27] Lehmann, W.J., Wilson, C.O., Jr., and Shapiro, I., J. Chem. Phys. 32, 1088 (1960).
[28] Lehmann, W.J., Wilson, C.O., Jr., and Shapiro, I., J. Chem. Phys. 34, 476 (1961).
[29] Jones, R.N., Can. J. Chem. 40, 301 (1962).
[30] Katritzky, A.R., and Coats, N.A., J. Chem. Soc. 2062 (1959).
[31] Katritzky, A.R., Monro, A.M., Beard, J.A.T., Dearnley, D.P., and Earl, N.J., J. Chem. Soc. 2182 (1958).
[32] Shull, E.R., Oakwood, T.S., and Rank, D.H., J. Chem. Phys. 21, 2024 (1953).
[33] Shimizu, K., and Murata, H., J. Mol. Spectroscopy 5, 40 (1960).
[34] Mann, D.E., Acquista, N., and Lide, D.R., Jr., J. Mol. Spectroscopy 2, 575 (1958).
[35] Evans, J.C., and Bernstein, H.J., Can. J. Chem. 34, 1037 (1956).
[36] Cleveland, F.F., Lamport, J.E., and Mitchell, R.W., J. Chem. Phys. 18, 1320 (1950).
[37] Hayashi, M., Ichishima, I., Shimanouchi, T., and Mizushima, S., Spectrochim. Acta 10, 1 (1957).
[38] Katritzky, A.R., Lagowski, J.M., and Beard, J.A.T., Spectrochim. Acta 16, 954 (1960).
[39] Philippe, R.J., and Moore, H., Spectrochim. Acta 17, 1004 (1961).
[40] Rasmussen, R.S., Brattain, R.R., and Zucco, R.S., J. Chem. Phys. 15, 135 (1947).
[41] Axford, D.W.E., and Rank, D.H., J. Chem. Phys. 18, 51 (1950).
[42] Cleveland, F.F., Murray, M.J., and Gallaway, W.S., J. Chem. Phys. 15, 742 (1947).
[43] Parker, F.W., and Nielsen, A.H., J. Mol. Spectroscopy 1, 107 (1957).
[44] Milligan, D.E., and Jacox, M.E., J. Mol. Spectroscopy 8, 126 (1962).
[45] Ham, N.S., and Willis, J.B., Spectrochim. Acta 15, 360 (1959).
[46] Spinner, E., Spectrochim. Acta 15, 95 (1959).
[47] Mecke, R., and Kutzelnigg, W., Spectrochim. Acta 16, 1216 (1960).
[48] Kutzelnigg, W., and Mecke, R., Spectrochim. Acta 17, 530 (1961).
[49] Fukushima, K., Onishi, T., Shimanouchi, T., and Mizushima, S., Spectrochim. Acta 15, 236 (1959).
[50] Kaesz, H.D., and Stone, F.G.A., Spectrochim. Acta 15, 360 (1959).
[51] Powell, D.B., Spectrochim. Acta 16, 241 (1960).
[52] Goulden, J.D.S., Spectrochim. Acta 16, 715 (1960).
[53] Green, J.H.S., Spectrochim. Acta 18, 39 (1962).
[54] Gerrard, W., Mooney, E.F., and Willis, H.A., Spectrochim. Acta 18, 155 (1962).
[55] Pimentel, G.E., and Klemperer, W.A., J. Chem. Phys. 23, 376 (1955).
[56] Miyazawa, T., Shimanouchi, T., and Mizushima, S., J. Chem. Phys. 29, 611 (1958).
[57] Nyquist, I.M., Mills, I.M., Person, W.B., and Crawford, B., Jr., J. Chem. Phys. 26, 552 (1957).
[58] Pan, C.Y., and Nielsen, J.R., J. Chem. Phys. 21, 1426 (1953).
[59] Miyazawa, T., Shimanouchi, T., and Mizushima, S., J. Chem. Phys. 24, 408 (1956).
[60] Jones, L.H., J. Chem. Phys. 23, 2105 (1955).
[61] Duncan, N.E., and Janz, G.J., J. Chem. Phys. 23, 434 (1955).
[62] Jones, E.R.H., J. Chem. Soc., 754 (1950).
[63] Weber, A., Ferigle, S.M., and Cleveland, F.F., J. Chem. Phys. 21, 1613 (1953).
[64] Ignat'eva, L.A., Bazhulin, P.A., and Balva, I.K., Vestnik Moskov. Univ., Ser. Mat., Mekh., Astron., Fiz. i Khim. (1959), No. 6, p. 127.
[65] Cerato, C.C., Lauer, J.L., and Beachell, H.C., J. Chem. Phys. 22, 1 (1954).
[66] Murata, H., and Shimizu, K., J. Chem. Phys. 23, 1968 (1955).
[67] Smith, A.L., J. Chem. Phys. 21, 1997 (1953).
[68] Shimizu, K., and Murata, H., J. Mol. Spectroscopy 4, 201 (1960).
[69] Ball, D.F., Goggin, P.L., McKean, D.C., and Woodward, L.A., Spectrochim. Acta 16, 1358 (1960).
[70] Shimizu, K., and Murata, H., J. Mol. Spectroscopy 4, 214 (1960).
[71] Ebsworth, E.A.V., Onyszchuk, M., and Sheppard, N., J. Chem. Soc. 1453 (1958).
[72] Grenoble, M.E., and Launer, P.J., Applied Spectroscopy 14, 85 (1960).

[73] Burnelle, L., and Duchesne, J., J. Chem. Phys. 20, 1324 (1952).
[74] Randic, M., Spectrochim. Acta 18, 115 (1962).
[75] Silver, S., J. Chem. Phys. 8, 919 (1940).
[76] Halmann, M., Spectrochim. Acta 16, 407 (1960).
[77] Bellanato, J., Spectrochim. Acta 16, 1344 (1960).
[78] Barcelo, J.R., and Bellanato, J., Spectrochim. Acta 8, 27 (1956).
[79] Ebsworth, E.A.V., and Sheppard, N., Spectrochim. Acta 13, 261 (1959).
[80] Mann, D.E., J. Chem. Phys. 22, 70 (1954).
[81] Shull, E.R., Wood, J.L., Aston, J.G., and Rank, D.H., J. Chem. Phys. 23, 1191 (1954).
[82] Ito, K., and Bernstein, H.J., Can. J. Chem. 34, 170 (1956).
[83] Wells, A.J., and Wilson, E.B., Jr., J. Chem. Phys. 9, 314 (1941).
[84] Jones, R.L., J. Mol. Spectroscopy 2, 581 (1958).
[85] DeGraaf, D.E., and Sutherland, G.B.B.M., J. Chem. Phys. 26, 716 (1957).
[86] Linnett, J.W., J. Chem. Phys. 8, 91 (1940).
[87] Williams, R.L., J. Chem. Phys. 25, 656 (1956).
[88] Martinette, Sr.M., Mizushima, S., and Quagliano, J.V., Spectrochim. Acta 15, 77 (1959).
[89] Eyster, E.H., and Gillette, R.H., J. Chem. Phys. 8, 369 (1940).
[90] West, W., and Killingsworth, R.B., J. Chem. Phys. 6, 1 (1938).
[91] Cleaves, A.P., and Plyler, E.K., J. Chem. Phys. 7, 563 (1939).
[92] Owens, R.G., and Barker, E.F., J. Chem. Phys. 8, 229 (1940).
[93] Gray, A.P., and Lord, R.C., J. Chem. Phys. 26, 690 (1957).
[94] Milligan, D.E., J. Chem. Phys. 35, 1491 (1961).
[95] Axford, D.W.E., Janz, G.J., and Russell, D.E., J. Chem. Phys. 19, 704 (1951).
[96] Waldron, R.D., J. Chem. Phys. 21, 734 (1953).
[97] Glass, W.K., and Pullin, A.D.E., Trans. Faraday Soc. 57, 546 (1961).
[98] Daasch, L.W., and Smith, D.C., J. Chem. Phys. 19, 22 (1951).
[99] Mortimer, F.S., Spectrochim. Acta 9, 270 (1957).
[100] Beg, M.A.A., and Clark, H.C., J. Chem. Phys. 27, 182 (1957).
[101] Beachell, H.C., and Katlafsky, B., J. Chem. Phys. 27, 182 (1957).
[102] Linton, H.R., and Nixon, E.R., Spectrochim. Acta 15, 146 (1959).
[103] Woodward, L.A., Hall, J.R., Dixon, R.N., and Sheppard, N., Spectrochim. Acta 15, 249 (1959).
[104] Paterson, W.G., and Onyszchuk, M., Can. J. Chem. 39, 2324 (1961).
[105] Watanabe, H., Narisada, M., Nakagawa, T., and Kubo, M., Spectrochim. Acta 16, 78 (1960).
[106] Crawford, B.L., Jr., and Joyce, L., J. Chem. Phys. 7, 307 (1939).
[107] Falk, M., and Whalley, E., J. Chem. Phys. 34, 1554 (1961).
[108] Servoss, R.R., and Clark, H.M., J. Chem. Phys. 26, 1179 (1957).
[109] Jones, R.N., Can. J. Chem. 40, 301 (1962).
[110] Parsons, A.E., J. Mol. Spectroscopy 2, 566 (1958).
[111] Nolin, B., and Jones, R.N., Can. J. Chem. 34, 1382 (1956).
[112] Davies, M., and Spiers, N.A., J. Chem. Soc. 3971 (1959).
[113] Fonteyne, R., J. Chem. Phys. 8, 60 (1940).
[114] Horrocks, W.D., Jr., and Cotton, F.A., Spectrochim. Acta 17, 134 (1961).
[115] Hayashi, M., Shimanouchi, T., and Mizushima, S., J. Chem. Phys. 26, 608 (1957).
[116] Thompson, H.W., and Skerrett, N.P., Trans. Faraday Soc. 36, 812 (1940).
[117] Philippe, R.J., J. Mol. Spectroscopy 6, 492 (1961).
[118] Kutzelnigg, W., and Mecke, R., Spectrochim. Acta 17, 530 (1961).
[119] Simon, A., Kriegsmann, H., and Dutz, H., Chem. Ber. 89, 2378 (1956).
[120] Simon, A., Kriegsmann, H., and Dutz, H., Chem. Ber. 89, 1883 (1956).
[121] Andersen, F.A., Bak, B., and Brodersen, S., J. Chem. Phys. 24, 989 (1956).
[122] Crawford, B.L., Jr., and Brinkley, S.R., J. Chem. Phys. 9, 69 (1941).
[123] Adel, A., and Barker, E.F., J. Chem. Phys. 2, 627 (1934).
[124] Jacox, M.E., and Hexter, R.M., J. Chem. Phys. 35, 183 (1961).
[125] Dickson, A.D., Mills, I.M., and Crawford, B., Jr., J. Chem. Phys. 27, 445 (1957).
[126] Dows, D.A., J. Chem. Phys. 29, 484 (1958).
[127] Mador, I.L., and Quinn, R.S., J. Chem. Phys. 20, 1837 (1952).
[128] Fenlon, P.F., Cleveland, F.F., and Meister, A.G., J. Chem. Phys. 19, 1561 (1951).
[129] Sheline, R.K., and Pitzer, K.S., J. Chem. Phys. 18, 595 (1950).
[130] Gutowsky, H.S., J. Chem. Phys. 17, 128 (1949).
[131] Brown, M.P., Okawara, R., and Rochow, E.G., Spectrochim. Acta 16, 595 (1960).
[132] Stafford, S.L., and Stone, F.G.A., Spectrochim. Acta 17, 412 (1961).
[133] Costa, G., and DeAlti, G., Gazz. chim. ital. 87, 1273 (1957).

APPENDIX II

Correlation Tables for
C–N Stretching Frequencies
(Compiled by Eugene Tyma)

TABLE I. Primary Aliphatic Amines

Compound	Liquid	Gas	Solution	Reference
Amine Group Substituted on the Primary Alpha Carbon				
Methylamine		1040, 1044		[7, 11, 19]
Ethylamine		1085		[11]
n-Propylamine	1077	1090	1075	[11]
n-Butylamine	1085			[11]
iso-Butylamine	1068	1069	1068	[11]
n-Amylamine	1072	1075	1072	[11]
iso-Amylamine	1075			[11]
n-Hexylamine	1075			[11]
n-Heptylamine	1072		1073	[11]
Ethylenediamine	1090			[11]
Hexamethylenediamine	1083*			[11]
Amine Group Substituted on the Secondary Alpha Carbon				
iso-Propylamine	1039	1041		[11]
sec-Butylamine	1043			[11]
Cyclohexylamine	1038			[11]
Amine Group Substituted on the Tertiary Alpha Carbon				
t-Butylamine	1038	1033		[11]
t-Octylamine	1031			[11]
Menthanediamine	1023			[11]
t-"Nonylamine"	1033			[11]
Primene J M-T	1037			[11]

*Solid phase.

TABLE II. Secondary Aliphatic Amines

Compound	Liquid	Gas	Solution	Reference
Amine Group Substituted on the Primary Alpha Carbon				
Dimethylamine		1160		[11]
Diethylamine	1146	1140		[11]
Di-n-propylamine	1133			[11]
Di-n-butylamine	1138			[11]
Di-iso-butylamine	1133			[11]
Di-n-amylamine	1132		1132	[11]
Di-iso-amylamine	1133		1133	[11]
Di-n-hexylamine	1133		1133	[11]
Di-n-heptylamine	1133		1133	[11]
Amine Group Substituted on the Secondary Alpha Carbon				
Di-iso-propylamine	1190			[11]
Di-sec-butylamine	1172			[11]
Allyl-iso-propylamine	1173			[11]
N,N'-Di-iso-propylhexa-methylenediamine	1171			[11]

TABLE III. Nitro–Alkanes

Compound	cm^{-1}	Intensity	Reference
Bromopicrin	845, 840, 834	S	[20]
Fluoropicrin	855, 863, 871	M	[20]
Chloropicrin	853, 848, 840	M	[20]
Trifluoronitrosomethane	810.7	S	[21]
Nitromethane	918, 917	M	[15, 16]
Nitroethane	876	M	[15]
1-Nitropropane	804	M	[15]
2-Nitropropane	851	M	[15]
Tetranitromethane	802	VS	[22]
n-Nitropropane	870	M	[16]
i-Nitropropane	851	S	[16]
n-Nitrobutane	857	S	[16]
n-Nitropentane	876	S	[16]
n-Nitrohexane	836	S	[16]
Methyl-2-nitroethylether	848	S	[16]
1-Chloro-1-nitropropane	848	S	[16]
2-Chloro-2-nitropropane	850	S	[16]
1-Chloro-1,1,2,2-tetrafluoro-2-nitroethane	909	S	[16]
Ethylesternitroacetiacid	859	S	[16]

TABLE IV. Compounds Containing CH_3–N Group

Compound	cm^{-1}	Intensity	Reference
Trimethylamine-borane	915	M	[27]
Ammonium methylnitramide	1089	S	[8]
O,N-Dimethylnitramide	1042, 1050	S	[8]
Silver methylnitramide	1062, 1060	S	[8]
N-Methyltoluene-p-sulphonamide	1060		[18]
N-Methylacetamide	1160		[29]
N-Methylformamide	1160, 1040		[23]
•N-Methylhydroxylamine	1034, 994	VS	[26]
•O,N-Dimethylhydroxylamine	1057	VS	[26]

TABLE V. $C\text{–}N\begin{smallmatrix}O\\O\end{smallmatrix}$ in Ring Compounds

Type of compound	Number of compounds	Position cm^{-1}
Nitrobenzene derivatives	(17)	868-827
Nitrodiphenyl derivatives	(6)	857-850
Nitroparaffins	(9)	876-836 *
Nitro-halogen paraffins	(7)	863-838 * *

*918 for nitromethane (CH_3NO_2); 804 for 1-nitropropane ($C_2H_5CH_2NO_2$).
**909 for $CClF_2CF_2NO_2$; 802 for tetranitromethane.

TABLE VI. HN=C–N Group in Acetamidines

Compound	cm^{-1}	Intensity	Reference
N-o-Tolyltrichloroacetamidine	1239	M	[24]
N-m-Tolyltrichloroacetamidine	1265	M	[24]
N-p-Tolyltrichloroacetamidine	1237	M	[24]
N-Phenyltrichloroacetamidine	1237, 1240	M	[24]
N-p-Ethoxyphenyltrichloro-acetamidine	1232, 1236	VS	[24]
N,N-Dimethyltrichloroacetamidine	1267	VS	[24]
N-Methyltrichloroacetamidine	1313	VS	[24]
N-n-Butyltrichloroacetamidine	1313	S	[24]
N-n-Amyltrichloroacetamidine	1312	S	[24]
N-Ethyltrichloroacetamidine	1314	VS	[24]
N-Benzyltrichloroacetamidine	1338	S	[24]
N-Morpholyltrichloroacetamidine	1275	VS	[24]

TABLE VII. Some Ethanamides (Amido Group)

Compound	cm^{-1}	Intensity	Reference
N-Butylethanamide	1555	VS	[14]
N-Butyl-2-chloroethanamide	1552	S	[14]
N-Butyl-2-chloro-2-fluoroethanamide	1553	S	[14]
N-Butyl-2,2-dichloroethanamide	1552	S	[14]
N-Butyl-2,2-difluoroethanamide	1554	VS	[14]
N-Butyl-2,2,2-trifluoroethanamide	1557	VS	[14]

TABLE VIII. C_{alk}–N Stretching Frequencies of Acetamidines

Compound	cm^{-1}	Intensity	Reference
N,N-Dimethyltrichloroacetamidine	1090	VS	[24]
N-Methyltrichloroacetamidine	1152	VS	[24]
N-n-Butyltrichloroacetamidine	1155	VS	[24]
N-n-Amyltrichloroacetamidine	1155	S	[24]
N-Ethyltrichloroacetamidine	1168	VS	[24]
N-Benzyltrichloroacetamidine	1168	VS	[24]
N-Morpholyltrichloroacetamidine	1117	VS	[24]

TABLE IX. General Ranges of C–N Vibration in Various Compounds

Type of Compound	ν (C_{alk} – N), cm^{-1}
Primary aliphatic amines	1090-1023
Secondary aliphatic amines	1190-1095
N-Butylethanamides	1065-1095
N,N-Dibutylethanamides	1098-1122
Nitramides	1089-1042
N-Substituted trichloroacetamidines	1168-1117
N,N-Disubstituted trichloroacetamidines	1090

TABLE X. Nitro Group Attached to Monophenyl Ring

Compound	cm^{-1}	Intensity	Reference
Nitrobenzene	854, 852	S	[17, 28]
o-Nitroaniline	848	W	[17]
m-Nitroaniline	868	S	[17]
p-Nitroaniline	857	S	[17]
o-Trifluoromethylnitrobenzene	854	S	[17]
p-Trifluoromethylnitrobenzene	837	VS	[17]
2-Nitro-3-trifluoromethylaniline	851	M	[17]
4-Nitro-2-trifluoromethylaniline	832	M	[17]
5-Nitro-3-trifluoromethylaniline	837	M	[17]
5-Nitro-o-toluidine	827	S	[17]
o-Nitroacetanilide	856	M	[17]
m-Nitroacetanilide	840	M	[17]
p-Nitroacetanilide	848	VS	[17]
4-Nitro-2-trifluoromethylacetanilide	850	M	[17]
2-Nitro-3-trifluoromethylacetanilide	850	M	[17]
4-Nitro-3-trifluoromethylacetanilide	840	S	[17]
5-Nitro-o-acetotoluidide	837	M	[17]

TABLE XI. Nitro Group Attached to Diphenyl Ring

Compound	cm^{-1}	Intensity	Reference
4-Nitrodiphenyl	854	VS	[17]
4-Nitro-3-trifluoromethyldiphenyl	850	S	[17]
2-Nitro-3'-trifluoromethyldiphenyl	855	M	[17]
4-Nitro-3'-trifluoromethyldiphenyl	857	S	[17]
4:4'-Dinitro-3-trifluoromethyldiphenyl	852	S	[17]
4:4'-Dinitro-3:3'-bistrifluoromethyldiphenyl	854	S	[17]

TABLE XII. Secondary Aromatic Amines

Compound	cm^{-1}	Reference
N-Methylaniline	1262	[18]
N-Ethylaniline	1256	[18]
Diphenylaniline	1241	[1]
N-Ethyl-o-toluidine	1261	[18]
N-Ethyl-m-toluidine	1259	[18]
N-Methyl-p-toluidine	1261	[18]
p-Chloro-N-methylaniline	1261	[18]
Diphenylamine	1241	[18]

TABLE XIII. Amide II Band

Compound	cm^{-1}	Deuterated	Reference
Acetylglycine-N-methylamide	1563	1479	[31]
Diacetylhydrazine	1506	1413	[12]
N-Methylformamide	1490, 1545, 1575	1435, 1372	[12]
N-Ethylacetamide	1565	1473	[12]
N-Methylacetamide	1487, 1500, 1567	1475, 1385	[12]
	1566, 1583		
Polyglycine	1555, 1535	1447	[12]
N,N'-Dimethyloxamide	1532	1445	[12]
Diformylhydrazine	1480	1338	[12]

BIBLIOGRAPHY

[1] McMenamy, C. A., "A Listing of the Carbon—Nitrogen Single Bond Frequencies in the Infrared," Unpublished papers of Canisius College, Buffalo, New York.
[2] Bellamy, L. J., Infrared Spectra of Complex Molecules, London (1958).
[3] Colthup, J. Opt. Soc. Amer. 40, 397 (1950).
[4] Parsons, A. E., J. Mol. Spec. 6, 201 (1961).
[5] Tatsuo, Miyazawa, J. Mol. Spec. 4, 155 (1960).
[6] Neville, Jonathan, J. Mol. Spec. 6, 205 (1961).
[7] Techniques of Infrared Spectroscopy, p. 112, Mass. Institute of Technology, Cambridge (1955).
[8] Neville, Jonathan, J. Mol. Spec. 5, 101 (1960).
[9] Stewart, James E., J. Chem. Phys. 26, 248 (1957).
[10] Miyazawa, Shimanouchi, and Mizushima, J. Chem. Phys. 29, 611 (1958).
[11] Stewart, James E., J. Chem. Phys. 30, 1259 (1959).
[12] Miyazawa, Takehiko, Shimanouchi, and Mizushima, J. Chem. Phys. 24, 408 (1956).
[13] Evans, J. C., J. Chem. Phys. 22, 1228 (1954).
[14] Letaw, Jr., and Gropp, J. Chem. Phys. 21, 1621 (1953).
[15] Smith, Pan, and Nielsen, J. Chem. Phys. 18, 706 (1950).
[16] Haszeldine, J. Chem. Soc. (London) 2525 (1953).
[17] Randle and Whiffen, J. Chem. Soc. (London) 4153 (1952).
[18] Hadži and Skrbljak, J. Chem. Soc. (London) 843 (1957).
[19] Gray and Lord, J. Chem. Phys. 26, 690 (1957).
[20] Mason, Banus, and Dunderdale, J. Chem. Soc. (London) 759 (1956).
[21] Mason, Banus, and Dunderdale, J. Chem. Soc. (London) 754 (1956).
[22] Lindenmeyer and Harris, J. Chem. Phys. 21, 408 (1953).
[23] DeGraaf and Sutherland, J. Chem. Phys. 26, 716 (1957).
[24] Grivas and Taurins, Can. J. of Chem. 37, 795 (1959).
[25] Orville-Thomas and Parsons, Trans. Faraday Soc. 54, 460 (1958).
[26] Mansel, Davies, and Spiers, J. Chem. Soc. (London) 3071 (1959).
[27] Rice, Galiano, and Lehmann, J. Phys. Chem. 61, 1222 (1957).
[28] Green, Kynaston, and Lindsey, Spectrochim. Acta 17, 486 (1961).
[29] Mizushima, Shimanouchi, Nagakura, Kuratani, Tsuboi, Baba, and Fujioka, J. Am. Chem. Soc. 72, 3490 (1950).
[30] Davies, Evans, and Jones, Trans. Far. Soc. 51, 761 (1955).
[31] Moriwaki, Tsuboi, Shimanouchi, and Mizushima, J. Am. Chem. Soc. 81, 5914 (1959).
[32] Brown, J. F., Jr., J. Am. Chem. Soc. 77, 6341 (1955).